Advanced Methods of Fatigue Assessment

Dieter Radaj · Michael Vormwald

Advanced Methods of Fatigue Assessment

Springer

apl. Prof. Dr.-Ing. habil. Dieter Radaj
Technische Universität Braunschweig
Braunschweig
Germany

Universitätsprofessor Dr.-Ing. Michael Vormwald
Technische Universität Darmstadt
Darmstadt
Germany

ISBN 978-3-642-30739-3 ISBN 978-3-642-30740-9 (eBook)
DOI 10.1007/978-3-642-30740-9
Springer Heidelberg New York Dordrecht London

Library of Congress Control Number: 2012947847

Printed on acid-free paper

Springer is part of Springer Science+Business Media (www.springer.com)

Foreword

When I started my Ph.D. course at the University of Padova almost 30 years ago and was faced with the topic of fatigue strength of welded joints and notched components for the first time, I was impressed by the work carried out by some German researchers. A special position was immediately occupied by the local approach based on fictitious notch rounding where Neuber's concept of a microstructural support length was combined by Radaj with the worst case concept (notch tip radius equal to zero) to assess the fatigue strength of welded joints. Starting from Neuber's idea, which is a milestone in Notch Mechanics, the first chapter of this book presents some recent developments extending the theory and application of the fictitious notch rounding concept.

In the mid-1980s, some Italian researchers guided by Bruno Atzori suggested to model the weld toe region as a pointed V-notch and to assess the fatigue limit of welded joints on the basis of the asymptotic stress distribution ahead of the point of stress singularity. Doing so, the scale effect was fully included in the stress field being understood as governed by the degree of singularity of Williams' linear-elastic solution. The corresponding intensity was later on quantified by means of the notch stress intensity factors, separating the influence of the three loading modes on the overall stress distribution ahead of the V-notch tip. Since the notch stress intensity factors represent odd singularities, which depend on the notch opening angle, a comparison between the fatigue failures of different weld geometries could be carried out by using the strain energy density averaged over a small control volume surrounding the point of stress singularity. The averaged strain energy density is always bounded, independent of the notch acuity, and can be easily determined under linear-elastic conditions by using the nodal displacements and not their derivatives, which allows coarse meshing in finite element analysis without a major loss in accuracy. The radius of the control volume was understood as a welded material property, which can be estimated on a statistical basis taking advantage of a large body of fatigue test data from welded joints of different geometry. In many cases of practical interest, not only the fatigue crack initiation life if correlated to the locally averaged strain energy density values but also the total life, at least when most of the fatigue life is assigned to initiation and microcrack propagation inside the zone governed by the V-notch singularity. A link between the locally averaged strain energy density and Rice's J-integral surely exists, provided a convenient, material-dependent, circular path around the V-notch tip on its material side is defined.

As it is well known, the stress intensity factors were used by Irwin also in the case of 'blunt cracks' for determining the maximum elastic stress at crack tips with a very small radius. It is evident that the notch stress intensity factor concept can also be used in the case of blunt notches when the local stress field is known with sufficient accuracy. Under mode I and mode III loading conditions, the notch stress intensity factor includes not only the point-related stress concentration factor K_t but also the effect of the stress gradient ahead of the notch root.

The concepts of notch stress intensity factor and averaged strain energy density, both under linear-elastic and small-scale yielding conditions, represent the focus of Chaps. 2 and 3 of this book where, together with the original formulations, a large number of results from a long-term collaboration involving one of the authors of the book, my collaborators and me are documented. I wish to express my personal gratitude to Dieter Radaj to whom I feel deeply indebted, not only for the attention paid to some papers of mine, sharing generously ideas and knowledge with me, but also for his friendship and proximity.

In the last chapter of the book, the focus is on the role played by cyclic plasticity during fatigue crack growth. Here, a powerful synthesis of the main results obtained over a period of 50 years is presented, from the Dugdale model and ΔK-based Paris propagation law dated 1960 to the most recent advances dated 2011 where the crack evolution is described in terms of blunting and V-shaped micronotch formation during the loading phase and subsequent resharpening, i.e., shearing of the activated slip bands, during the unloading phase. Plasticity-induced crack closure greatly influences the crack growth rate. Small-scale and large-scale yielding conditions are analysed separately, showing how, in the latter case, some advantages come from the usage of the ΔJ-integral with respect to other formulations. However, the matter remains really very complex, due to a number of interaction effects bound up with the loading conditions, material behaviour laws, crack length, stress gradients in the notched components and so on. In particular, multiaxial and mixed mode loading conditions are a continuing challenge for actual research. In the conclusions of Chap. 4 the authors underline that the general problem of fatigue crack growth under large-scale cyclic yielding conditions and non-proportional combined loading is far from being solved. Together with an exhaustive review of the state of the art, the reader will appreciate the assessment of the trends governing the new research activities.

Finally, I am glad to express my thanks to the authors for giving me the opportunity to introduce their new book which will surely become a reference work for researchers and engineers engaged on the design of structural components with regard to static and fatigue strength as well as on the analysis of fatigue crack propagation under complex service loading conditions.

Vicenza, April 2012

Paolo Lazzarin
University of Padova

Preface

The book in hand presents advanced methods of brittle fracture and fatigue assessment. It continues the expositions of two well-known reference works for designers of machines and constructions, structural analysts, testing engineers and code-responsible experts. The one book in English language with the title 'Fatigue assessment of welded joints by local approaches' is authored by Radaj et al. (^{2}2006). The other book in German language with the title 'Ermüdungsfestigkeit—Grundlagen für Ingenieure' ('Fatigue strength—fundamentals for engineers') is authored by Radaj and Vormwald (^{3}2007).

A further separate new book was found necessary after the two works mentioned above had reached a certain limit size. The second edition of the application-related English book has 639 pages based on 997 references. The third edition of the fundamentals-related German book consists of 688 pages with 1761 references.

This new book is conceived to review and explain advanced methods of brittle fracture and fatigue assessment which are not sufficiently well represented in the two books mentioned above, especially in respect of the major steps in methods development and application which have taken place in the recent few years.

The authors do not claim that the addressed scientific fields are the only ones in which major advance has been achieved. The personal involvement in actual research efforts was also decisive for the selection of the fields. Nonetheless, the authors believe that these fields will remain most influential on the methods development in the future. The following comments may elucidate the significance of the selected fields.

In Chap. 1 headed 'Generalised Neuber Concept of Fictitious Notch Rounding' and authored by Radaj, this analytically demanding approach is brought to a completion by extending it to three new areas, to in-plane shear loading with out-of-bisector crack propagation, to varying notch opening angles and to V-notches with root hole as an alternative to blunt V-notches. The reference notch concept for the fatigue assessment of welded joints is based on the above approach.

In Chap. 2 headed 'Extended Stress Intensity Factor Concepts' and authored by Radaj, the stress intensity factor approach which is well known from fracture mechanics is extended in two directions. One extension refers to pointed V-notches whose singular notch stress field is described by mode-related and angle-dependent notch stress intensity factors. The other extension is related to rounded V-notches (blunt shape or root hole shape) whose non-singular notch stress field is governed by generalised notch stress intensity factors. Additionally, the extension into the elastic-plastic range is performed. Rigid inclusion problems are treated in a similar way as the open notch problems, defining the corresponding stress intensity factors.

In Chap. 3 headed 'Local Strain Energy Density Concept' and authored by Radaj, the stress intensity factor approaches just mentioned are transferred into a local energy concept which is well suited for the assessment of the brittle fracture or fatigue failure limit of sharply notched structural members. The local energy concept is more appropriate for strength assessments than the conventional stress concentration concept, because the local average strain energy density, based on stress intensity factors, characterises the strength-relevant field as opposed to stress concentration factors which are related to strength-irrelevant point stresses. The size effect in strength assessments is naturally included in the local strain energy density concept. Most important for applications is the fact that finite element procedures based on extremely coarse meshes can be used without a major loss in accuracy of the results.

In Chap. 4 headed 'Elastic-Plastic Fatigue Crack Growth' and authored by Vormwald, the various implications of cyclic plasticity with regard to the fatigue crack growth are explained. Conventional fatigue crack propagation approaches rely on similitude arguments and relationships between the stress intensity factor range and the crack growth rate. The application limit of this approach is specified by small-scale yielding conditions. Still within these limits, an explanation and straightforward modelling of the mean stress influence and the influence of variable amplitudes requires consideration of cyclic plasticity. Plasticity-induced crack closure is of great influence on the crack growth rate. Modelling tools and algorithms are presented. Outside the small-scale yielding limits, the stress intensity factor range must be substituted by a crack driving force parameter of elastic-plastic fracture mechanics. Various proposals are presented and discussed with a focus on the J-integral. Together with an adequate consideration of crack closure, advances in simulating fatigue crack growth in this regime more realistically are presented. Multiaxial and mixed mode loading are a continuing challenge for actual research. These topics are discussed against the background of current expertise and available computational resources.

The following formal aspects of the presentation in the book is important to know. Each of the four chapters above is a separate unit within the book with its own abstract at the beginning and its own reference list, symbol list and list of contents at the end. The reason for this is that the book is offered by the publisher not only in the printed, but also in the electronic version inclusive of a pay-per-chapter option. It is also introduced for the reader's convenience.

Our colleagues from the University of Padova in Vicenza, Paolo Lazzarin together with Filippo Berto and Michele Zappalorto, have immensely contributed to the first three chapters of the book. They are the original developers of the advanced concepts of stress intensity factors and local strain energy density leading to successful applications. They were available for discussing some intriguing scientific problems associated with their methods development. They have patiently answered innumerable questions related to their original publications. They have contributed about one-half of the large number of figures in the book, which were only slightly modified for a uniform appearance. We hope that the unequalled scientific merit of these persons will become apparent to the readers of the book. The authors acknowledge the received support with deep respect.

Additionally, the first author appreciates 10 years of challenging cooperation with the eminent personalities mentioned above. These years with nearly daily contacts have become an especially valuable final phase of his scientific career. He wants to express his thanks and his respect by presenting this book, summarising and elucidating the achievements of his colleagues in Vicenza.

Other colleagues have given support to special items of the book contents. Dietmar Gross from the Technical University of Darmstadt has provided a curve plot illustrating the Ramberg–Osgood material law. Mikkel Pedersen from the Aalberg University in Denmark has supplied a plot of fatigue test data supporting the notch stress design curve for fatigue-loaded welded joints. We appreciate the kindness of these two persons.

The manuscript of the first three chapters of this book was put into well-executed typescript by Claudia Raschke whose commitment to this task facilitated the first author's work substantially. This person has also prepared the graphical artwork, mainly based on plots provided by the original authors. It was a tremendous task to keep track with the great bulk of written or drawn material delivered by the author for further electronic processing. This support given to the book cannot be overrated. Any less qualified person would have brought the project to a breakdown. The first author pays his respects to Claudia's contribution to the project.

Both authors express their thanks to Kerstin Breidenbach for helping to bring the fourth chapter into its final form.

Finally, we express our thanks to the publishing house Springer-Verlag, who has agreed to publish this work and has supported our efforts in every respect.

Stuttgart and Darmstadt, March 2012

Dieter Radaj
Michael Vormwald

Contents

Chapter 1
Generalised Neuber Concept of Fictitious Notch Rounding

D. Radaj

Abstract Sharp or pointed notches reduce the fatigue strength or life of structural components drastically, but not to the extent of the elastic notch stress increase. Microstructural support is observed for crack initiation at the notch root. The support effect may be described by averaging the maximum notch stresses in a small material volume (length ρ^*) at the notch root (radius ρ), which can be expressed by the maximum stress of a corresponding notch of a slightly enlarged, fictitious radius, $\rho_f = \rho + s\rho^*$ (Neuber 1937, 1968). The support factor s is derived for elementary notches and V-notches in the three loading modes: in-plane tensile and shear loading as well as out-of-plane shear loading (modes 1, 2 and 3). Out-of-bisector crack initiation and propagation is basic for mode 2 loading. The dependency of s on the notch opening angle 2α is recognised, besides its correlation with multiaxiality conditions and failure criteria. The Neuber concept of fictitious notch rounding is thus generalised. Application-relevant issues such as reference notches, design *S–N* curves, non-singular stress components, seam-welded cruciform joints and spot-welded lap joints are also dealt with.

Contents

D. Radaj and M. Vormwald, *Advanced Methods of Fatigue Assessment*,
DOI: 10.1007/978-3-642-30740-9_1,

1.1 Survey of Chapter Contents

Fatigue crack initiation and propagation at the root of pointed or sharply rounded notches is a highly localised process of material damage accumulation initiated on a microstructural and submicrostructural level. This process is described by dislocation movement, dislocation arrest at grain boundaries, microcrack initiation and propagation traversing the grain boundaries, taking crack closure and crack interaction effects into account (Radaj and Vormwald 2007). The relevant microstructural cracking theories contribute substantially to an understanding of this complicated process and reveal the essential microstructural parameters which may be used for optimising structural materials and their processing.

The traditional ‘continuum theories’ of fatigue, Neuber’s microstructural support concept among them, have quite another field of application. They are directed to engineering methods of fatigue life or strength assessment of structural components. Design optimisations may be performed on this basis.

The Neuber concept of fictitious notch rounding is applied by engineers to the fatigue assessment of welded joints using a uniform reference radius $\rho_r = 1$ mm. The fatigue-critical weld notches are actually (not only fictitiously) rounded using this radius for the purpose of a notch stress analysis. The resulting maximum notch stress is considered to be fatigue-effective to the extent indicated by the notch stress *S–N* curve scatter band derived mainly from fatigue testing of standard welded joints. These endurable notch stresses do not actually occur if the reference notch radius is chosen far from Neuber’s fictitious radius based on the microstructural support effect. Thus, the more recently proposed reference radius $\rho_r = 0.05$ mm which does not include the support effect results in unrealistically high notch stresses.

Neuber proposed the hypothesis of microstructural support in 1937 and combined it with the concept of fictitious notch rounding. The microstructural support hypothesis assumes that a macrocrack initiates at the notch as soon as the averaged stress in a small material volume defined by the ‘substitute microstructural length’ ρ^* exceeds the endurance limit of the unnotched specimen (finite life is not considered). The concept of fictitious notch rounding was introduced in order to avoid the (at that time) laborious stress averaging procedure at the notch root. The notch is analysed with a fictitiously enlarged notch radius resulting directly in the averaged stress of the smaller real notch radius.

The enlargement of the notch radius is defined by the microstructural length ρ^* of the material multiplied by a support factor s which depends on the stress multiaxiality at the notch root (plane stress, plane strain, pure shear) and the failure criterion (Rankine, von Mises, Beltrami or others). Neuber got the microstructural length ρ^* from fatigue test results for reversed tensile and bending loading of notched specimens without taking the crack closure effect into account. The support effect of the material is thus overrated. Using Neuber’s data, the reference radius for sharply notched weld toes was roughly approximated with $\rho_r = 1$ mm.

It makes sense to further investigate the somewhat ‘old-fashioned’ Neuber concept in order to get an indication, how the widely used reference radius concept might be improved and where it will definitely fail (Radaj et al. 2013). This is valuable information for engineers in charge of fatigue assessment guidelines. Additionally, there is always an academic interest in completing well-founded theories.

In Sect. 1.2, the effect of the notch opening angle on the support factor in tensile loading is clarified. Also, the method is extended to pointed notches where a substantial deviation of the support factor as compared with rounded notches may occur. Elementary solutions are included. Additionally, a jump in notch stress multiaxiality may have to be taken into account when substituting the pointed notch by the fictitiously rounded notch.

In Sect. 1.3, the effect of the notch opening angle on the support factor in out-of-plane shear loading is investigated, once more inclusive of pointed notches.

In Sect. 1.4, the support factor for in-plane shear loading is determined assuming crack propagation either in the bisector or out of the bisector plane.

In Sect. 1.5, application-relevant issues of fictitious or reference notch rounding are dealt with: mixed mode loading, short cracks or slits, effect of non-singular stresses at pointed crack or slit tips, effect of cross-sectional weakening, effect of mean stress and residual stress, fictitious notch compared with reference notch, design *S–N* curves, reference notches at weld roots and the theoretical *S–N* curve of a spot-welded lap joint.

The chapter is concluded with a list of symbols followed by a list of references.

1.2 Fictitious Notch Rounding for Tensile Loading

1.2.1 Survey of Section Contents

On the basis of comprehensive and accurate stress field equations for sharply rounded V-notches under tensile loading derived by Lazzarin and Tovo and improved by Filippi et al., the microstructural support factor s of these blunt notches is determined which quantifies the fictitious notch radius in Neuber's microstructural support concept. Considering different failure criteria (Rankine, von Mises and Beltrami) the fictitious notch radius is evaluated for different notch opening angles as a function of the actual notch radius and the microstructural support length. Plane stress and, alternatively, plane strain conditions are introduced. Once the fictitious radius has been determined, the support factor s is derived from it. The derivations are preceded by elementary solutions for pointed notches in general. The investigation of blunt V-notches is supplemented by considering V-notches with root hole.

The support factor s is found to be very sensitive to the notch opening angle. Constant 'plateau values' are determined for notch radii greater than the microstructural support length. The dependence of s on the multiaxial failure criterion and on the multiaxiality condition (plane stress or plane strain) is also investigated.

Various numerical analyses using the FE method have been carried out to compare the stress concentration factor of the fictitiously rounded notch using the plateau values of s with the effective stress concentration factor obtained by integrating the relevant equivalent stress over the microstructural support length along the bisector of the pointed V-notch. The eventual 'jump' in stress multiaxiality between the rounded and the pointed notch in terms of the considered fracture criterion has to be taken into account. It is shown that the plateau values of s are well suited for engineering usage in structural strength assessments.

The expositions are primarily based on the analytical developments and their application to specimens with V-notches carried out by Lazzarin and Berto (Berto et al. 2008, 2009). A list of symbols and a list of references are available at the end of Chap. 1.

1.2.2 Neuber's Microstructural Support Concept

It is a well-known fact that the static and fatigue strength of structural components is substantially reduced by the stress concentration at notches. But the strength reduction at sharp notches is not as severe as the elastic notch stress concentration indicates. A 'microstructural support effect' is observed, named after its physical origin. This chapter deals with the microstructural support concept in the form introduced by Neuber (Neuber 1936, 1958, 1968 and 1985). Starting from the idea that the averaged stress in an 'elementary material particle' at the notch root is decisive for the strength of notched structural members, a material-dependent microstructural length parameter is defined which is named '(substitute) microstructural support length'.

The concept of microstructural support applied to high-cycle fatigue loading considers the fact that not the theoretical maximum notch stress is fatigue-effective in the case of sharp notches but the notch stress averaged over a short distance normal to the notch contour where the fatigue crack is initiated. Neuber's further idea was to determine the fatigue-effective notch stress directly (i.e. without explicit notch stress averaging) by performing the notch stress analysis with a fictitiously enlarged notch radius ρ_f corresponding to the relevant microstructural support. The hypothesis of microstructural support applied to static loading is related to brittle fracture. Here, Wieghardt was the earlier, but thereafter forgotten founder of the notch stress averaging concept (Wieghardt 1907).

The fictitious notch radius is given by Neuber in the following form:

$$\rho_f = \rho + s\rho^* \tag{1.1}$$

where ρ is the actual notch radius, s is the support factor which takes multiaxial stress effects on failure into account and ρ^* is the 'microstructural support length'. The length ρ^* exclusively depends on the material and its microstructural condition, expressed by the yield limit σ_Y in fatigue loading (Neuber 1968) or by the fracture toughness K_{Ic} and the averaged local fracture stress $\overline{\sigma}_c$ in static loading (Weiss 1971):

$$\rho^* = \frac{2}{\pi}\left(\frac{K_{Ic}}{\overline{\sigma}_c}\right)^2 \tag{1.2}$$

Another relationship for static loading refers to the yield limit σ_Y at distance ρ^* which is approximately equal to the ultimate strength in the case of brittle fracture (Neuber 1985):

$$\rho^* = \frac{1}{2\pi}\left(\frac{K_{Ic}}{\sigma_Y}\right)^2 \tag{1.3}$$

This relationship corresponds to the 'critical distance approach' in fatigue (Peterson 1950) while the aforementioned formula given by Weiss corresponds to the notch stress averaging concept. As shown by Taylor and further discussed by Radaj, the parameter ρ^* should be four times as large in the latter case which is confirmed by the two relationships above (Taylor 1999; Radaj and Vormwald 2007).

Further details referring to Neuber's notch stress averaging concept describing fatigue can be found in the authors' just quoted book, ibid. Chap. 4.7: the microstructural length ρ^* dependent on the material and its yield limit, formulae for fatigue notch factors and notch sensitivity as well as support factor s dependent on failure criterion and notch stress multiaxiality.

Following Neuber's concept, it has been proposed to predict the high-cycle fatigue strength of welded joints (toe and root failures) based on fictitious notch rounding (Radaj 1969, 1990; Radaj et al. 2006). Within a worst case assessment for mild steel, $\rho = 0$ mm, $s = 2.5$ and $\rho^* = 0.4$ mm were introduced. This procedure has been proven to be generally applicable to welded joints of structural steels and aluminium alloys. It has become a standardised procedure within the IIW design recommendations (Hobbacher 2009).

1.2.3 Objectives, Method and Scope of Analysis Related to V-Notches

The support factor s has been tabulated by Neuber considering different failure criteria both for tensile and bending loading (plane stress and plane strain conditions) as well as for torsional loading (Neuber 1968). The values of s are given without reference to the notch opening angle while an influence of this angle was originally stated by Neuber (at least for pointed notches) considering the normal stress criterion both for tensile and torsional loading (Neuber 1958). Also in the more recent edition of his book, the influence of the notch opening angle has disappeared and a single value of s seems to be proposed to describe fictitious notch rounding for any notch opening angle (Neuber 1985). The values of s given in a separate article (Neuber 1968) for tensile (and bending) loading are referring to the plane stress condition of flat specimens and to the (approximately) plane strain condition of axisymmetric specimens with a circumferential notch. In conclusion, some doubts remain with respect to the application of the approach, in particular when dealing with sharp notches with arbitrary notch opening angles.

With the objective to reanalyse this intriguing subject based on highly accurate equations for the notch stresses in tensile loading (Filippi et al. 2002) and considering different failure criteria (normal stress according to Rankine, distortional strain energy density according to von Mises, total strain energy density according to Beltrami) under plane stress and plane strain conditions, the fictitious notch radius is evaluated for different notch opening angles (Berto et al. 2008, 2009). The plateau values of the support factor s are derived, and, using these values, the fictitious notch rounding approach is applied to pointed V-notches in tensile loading.

The elementary material particle at the notch root is characterised by the material-dependent microstructural support length ρ^*. A rough first approximation from

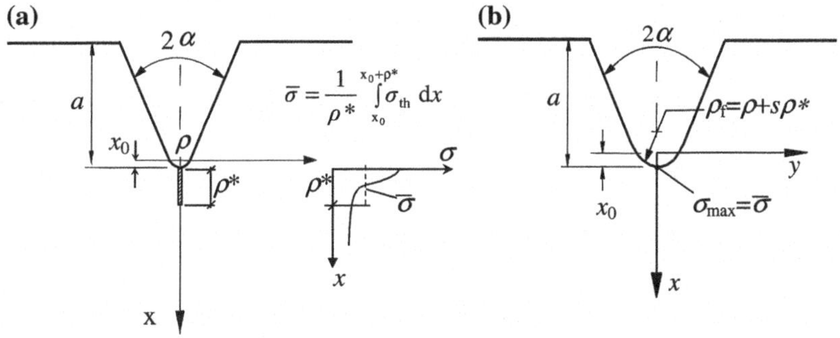

Fig. 1.1 Fictitious notch rounding concept applied to tensile loaded blunt V-notches; real notch with stress $\bar{\sigma}$ averaged over ρ^* (**a**) and substitute notch with fictitious notch radius ρ_f producing $\sigma_{max} = \bar{\sigma}$ (**b**); (Berto et al. 2009)

experimental results states $\rho^* \approx 0.1$ mm for mild steel and aluminium alloys in fatigue. Stress averaging is performed over this distance solving the notch problem and integrating the theoretical (equivalent) notch stresses σ_{th} over ρ^*, Fig. 1.1a:

$$\bar{\sigma} = \frac{1}{\rho^*} \int_{x_0}^{x_0+\rho^*} \sigma_{th}\, dx \tag{1.4}$$

Alternatively, the following equation is used where the equivalent stress is represented by a square root expression:

$$\bar{\sigma}^2 = \frac{1}{\rho^*} \int_{x_0}^{x_0+\rho^*} \sigma_{th}^2\, dx \tag{1.5}$$

In order to simplify this procedure, Neuber has introduced the concept of fictitious notch rounding (Neuber 1968). The averaged notch stress $\bar{\sigma}$ can directly be determined by analysing a substitute notch with fictitiously enlarged notch radius ρ_f, Fig. 1.1b. The factor s results from performing the averaging process for different multiaxiality conditions and failure criteria ($s = 2.0$–3.0 according to Neuber).

The analytical method given by Neuber (1958, pp. 166–174) has been used for determining the fictitious notch radius ρ_f (and therefrom the support factor s) based on closed-form expressions for the notch stresses at one-sided V-notches in tensile loading. The procedure consists of the following steps:

- to describe the relevant (equivalent) notch stress σ_{th} in the bisector plane (the assumed crack path) by the closed-form expressions for the sharply rounded V-notch (without microstructural support),
- to determine the notch stress $\bar{\sigma}$ averaged over the microstructural support length ρ^* (other closed-form expressions),
- to equate the expression of $\bar{\sigma}$ for finite ρ^* with that of $\bar{\sigma}$ for $\rho^* = 0$, in order to derive the fictitious radius ρ_f dependent on ρ, ρ^* and 2α by numerical procedures,

- to derive the factor $s = (\rho_f - \rho)/\rho^*$ which is also dependent on ρ, ρ^* and 2α in general.

Not Neuber's stress field equations are used in the following, but more accurate equations later on derived (Filippi et al. 2002). The influence of further geometrical parameters such as a finite notch depth or cross-sectional width are not taken into account.

The following conditions are considered in the numerical investigation:

- plane stress and plane strain conditions,
- normal stress criterion (Rankine), distortional strain energy criterion (von Mises) and total strain energy criterion (Beltrami),
- actual notch radius ρ varied between 0 and 10 mm,
- notch opening angle 2α varied between 0 and 150°,
- microstructural support length ρ^* varied between 0.05 and 0.5 mm.

1.2.4 Elementary Solutions for Pointed Compared with Rounded Notches under Tensile Loading

Before reviewing the general solution for V-notches, more elementary solutions for pointed compared with rounded notches under tensile loading, mainly crack-like, but also extended to a V-shape, are presented. They illustrate the fictitious notch rounding concept without major demands on mathematical understanding.

The concept is demonstrated first for an elliptical notch within a tensile loaded infinite plate. Only the pointed crack-like notch ($\rho = 0$, $2\alpha = 0$) is considered in comparison to the elliptical notch representing fictitious notch rounding. This limit case ($\rho = 0$) allows elementary closed-form solutions (Radaj and Zhang 1992, ibid. Appendix).

The tensile loaded infinite plate with a transverse internal crack and a corresponding elliptical hole is considered (plane stress or plane strain conditions). The notch stress concentration factor K_t of the elliptical hole is given by (Inglis 1913):

$$K_t = \frac{\sigma_{max}}{\sigma_n} = 1 + 2\sqrt{\frac{t}{\rho}} \tag{1.6}$$

with maximum notch stress σ_{max}, nominal stress σ_n, notch depth or semiaxis t and notch radius ρ. The stress $\overline{\sigma}$ averaged over the support length ρ^* ahead of the corresponding crack ($\rho = 0$) is given as follows (Weiss 1971):

$$\overline{\sigma} = \sqrt{\frac{2}{\pi\rho^*}}\,\sigma_n\sqrt{\pi t} \tag{1.7}$$

Fictitious notch rounding by $\rho_f = s\rho^*$ should result in $\sigma_{max} = \overline{\sigma}$:

$$\sqrt{\frac{2t}{\rho^*}} = 1 + 2\sqrt{\frac{t}{s\rho^*}} \tag{1.8}$$

thus defining the support factor s for fictitious notch rounding:

$$s = \frac{4}{\left(\sqrt{2} - \sqrt{\rho^*/t}\right)^2} \tag{1.9}$$

One gets $s = 2.0$ for $\rho^* \rightarrow 0$ (no microsupport) or $t \rightarrow \infty$ (deep notches). For $0 \leq \rho^* \leq 0.5$ mm (the range of ρ^* in fatigue given by Neuber), the range of s is $2.0 \leq s \leq 2.8$. In case the simplified form of Eq. (1.6) for sharp notches is used, neglecting the term 1, the result of Eq. (1.8) is $s = 2.0$ independent of ρ^*/t. The conclusion is that s depends on ρ^*/t if the restriction to sharp notches is deleted. Using the sharp notch formula, the constant value $s = 2.0$ is appropriate. The derivation above for internal notches can be transferred to sharp external notches. The free surface factor both for deep external notches and for external cracks is identical ($\kappa_0 = 1.1215$) with the effect that s remains unchanged.

The factor $s = 2.0$ is also confirmed by the approximation of σ_{max} for blunt cracks (Creager and Paris 1967) which may be interpreted as a parabolic notch solution. It proceeds from the stress intensity factor K_I of the pointed crack tip, equating σ_{max} and $\overline{\sigma}$:

$$\sigma_{max} = \frac{2K_I}{\sqrt{\pi\rho}} \quad (\rho = \rho_f = s\rho^*) \tag{1.10}$$

$$\overline{\sigma} = K_I\sqrt{\frac{2}{\pi\rho^*}} \tag{1.11}$$

A further comparison value of s can be obtained from the solution for the tensile loaded keyhole notch (Kullmer 1992; Radaj et al. 2001):

$$\sigma_{max} = \frac{3}{\sqrt{2}}\frac{K_I}{\sqrt{\pi\rho}} = 2.12\frac{K_I}{\sqrt{\pi\rho}} \quad (\rho = \rho_f = s\rho^*) \tag{1.12}$$

Equating σ_{max} from Eq. (1.12) for $\rho = \rho_f$ with $\overline{\sigma}$ from Eq. (1.11) for $\rho = 0$, the result is $\rho_f = 2.25\rho^*$ or $s = 2.25$.

The normal stress criterion is implicitly assumed to be valid when using the averaged stress $\overline{\sigma}$ for strength assessments. Now, the von Mises equivalent stress is alternatively considered:

$$\sigma_{vM} = \frac{1}{\sqrt{2}}\sqrt{(\sigma_1 - \sigma_2)^2 + (\sigma_2 - \sigma_3)^2 + (\sigma_1 - \sigma_3)^2} \tag{1.13}$$

with the principal stresses $\sigma_1 = \sigma_y$, $\sigma_2 = \sigma_x = \sigma_y$ and $\sigma_3 = \sigma_z = 0$ for the crack tip under plane stress conditions, or $\sigma_1 = \sigma_y$, $\sigma_2 = \sigma_x = \sigma_y$ and $\sigma_3 = \sigma_z =$

$(\sigma_x + \sigma_y) = 0.6\sigma_y$ (with $\nu = 0.3$) for the crack tip under plane strain conditions. Then the averaged von Mises stresses are $\overline{\sigma}_{\mathrm{vM}} = \overline{\sigma}$ for plane stress and $\overline{\sigma}_{\mathrm{vM}} = 0.4\overline{\sigma}$ for plane strain.

When substituting the averaged stress at the crack tip by the maximum notch stress of the fictitiously rounded notch, the multiaxiality condition of the stresses at the notch root compared with that at the crack tip changes abruptly. Under plane stress conditions, the relation of the three principal stresses is 1 : 1 : 0 (crack tip) compared with 1 : 0 : 0 (notch root), but σ_{vM} remains unchanged. Under plane strain conditions, the relation is 1 : 1 : 0.6 (crack tip) compared with 1 :0 :0.3 (notch root), with $\overline{\sigma}_{\mathrm{vM}} = 0.4\overline{\sigma}$ compared with $\overline{\sigma}_{\mathrm{vM}} = 0.89\overline{\sigma}$, where $\overline{\sigma}$ is the averaged one-dimensional stress (1 : 0 : 0). We may term the factor $\kappa = 0.89/0.4$ 'jump factor' because it characterises the jump in multiaxiality when the pointed crack tip is substituted by a very small but finite notch radius. The multiaxiality condition over the stress averaging length ρ^* remains the same within the validity range of the stress intensity factor.

Considering the von Mises stress at the crack tip under plane stress, the jump factor is $\kappa = 1.0$, i.e. the Eqs. (1.6–1.11) can be converted to the von Mises stress situation without a jump factor, so that the support factor s remains unchanged. Another situation occurs with the plane strain condition characterised by $\kappa = 0.89/0.4 = 2.225$. The simplified form of Eq. (1.8) for sharp notches in terms of von Mises stresses must then be stated as follows:

$$0.4\sqrt{\frac{2t}{\rho^*}} = 0.89\sqrt{\frac{4t}{s_{\mathrm{vM}}\rho^*}} \tag{1.14}$$

resulting in $s_{\mathrm{vM}} = 9.90$. The following general formula is valid within the sharp notch simplification:

$$s_{\mathrm{vM}} = \kappa_{\mathrm{vM}}^2 s \tag{1.15}$$

where the jump factor κ_{vM} depends on the multiaxiality condition (plane stress or plane strain).

Whereas the maximum stress at the fictitious notch is directly reduced by the jump factor, the fictitious radius ρ_{f} and the support factor s are enlarged by the squared jump factor. Without the simplification of Eq. (1.8), the factor s_{vM} has to be calculated from the modified Eq. (1.9):

$$s_{\mathrm{vM}} = \frac{4}{\left(\sqrt{2}/\kappa_{\mathrm{vM}} - \sqrt{\rho^*/t}\right)^2} \tag{1.16}$$

For example, with $\rho^* = 0.5$ mm, $t = 10$ mm and $\kappa_{\mathrm{vM}} = 2.225$ for plane strain we get $s_{\mathrm{vM}} = 23.56$ or $\rho_{\mathrm{f}} = s_{\mathrm{vM}}\rho^* = 11.78$ mm, i.e. $\rho_{\mathrm{f}} > t$. This is not an incorrect result, but it characterises not only mere notch rounding but also substitution of the crack by a fictitious elliptical hole.

Beltrami's equivalent stress is found from the following formula (with Poisson's ratio v):

$$\sigma_{\mathrm{B}} = \sqrt{\sigma_1^2 + \sigma_2^2 + \sigma_3^2 - 2v(\sigma_1\sigma_2 + \sigma_2\sigma_3 + \sigma_1\sigma_3)} \tag{1.17}$$

Considering the multiaxiality condition at the crack tip and, in comparison, at the notch root, the jump factors are $\kappa_{\mathrm{B}} = 0.85$ for plane stress and $\kappa_{\mathrm{B}} = 0.94$ for plane strain corresponding to $s_{\mathrm{B}} = 1.45$ and $s_{\mathrm{B}} = 1.77$, respectively, for $s = 2.0$. In analogy to Eq. (1.15), the following general formula is applied:

$$s_{\mathrm{B}} = \kappa_{\mathrm{B}}^2 s \tag{1.18}$$

where κ_{B} depends on the multiaxiality condition (plane stress or plain strain).

The aforementioned values of κ_{B} have been gained proceeding in analogy with the von Mises stress consideration above. The value $\kappa_{\mathrm{B}} = 0.85$ under plane stress conditions results from $\overline{\sigma}_{\mathrm{B}} = 1.18\overline{\sigma}$ (crack tip) compared with $\overline{\sigma}_{\mathrm{B}} = \overline{\sigma}$ (notch root), $1/1.18 = 0.85$, whereas $\kappa_{\mathrm{B}} = 0.94$ under plane strain conditions is based on $\overline{\sigma}_{\mathrm{B}} = 1.02\overline{\sigma}$ compared with $\overline{\sigma}_{\mathrm{B}} = 0.95\overline{\sigma}$, $0.95/1.02 = 0.94$.

The procedure above for tensile loaded pointed crack-like notches with $2\alpha = 0°$ can easily be transferred to tensile loaded pointed V-notches with $2\alpha \neq 0$ using the notch stress intensity factor K_1 instead of the (crack) stress intensity factor K_{I}. The transverse stress σ_{y} along the bisector $(r - r_0)$ reads:

$$\sigma_y = \frac{1}{\sqrt{2\pi}} K_1 r^{(\lambda_1 - 1)} \tag{1.19}$$

with the eigenvalue λ_1 according to Williams' solution. Therefore, the stress σ_y averaged over the distance ρ^* from the V-notch tip is:

$$\overline{\sigma} = \frac{1}{\sqrt{2\pi}} \frac{K_1 \rho^{*(\lambda_1 - 1)}}{\lambda_1} \tag{1.20}$$

The maximum stress $\sigma_{\max}$ at the corresponding rounded notch is given as follows (Lazzarin and Filippi 2006):

$$\sigma_{\max} = \frac{K_{1,\rho} r_0^{\lambda_1 - 1}}{\sqrt{2\pi}} (1 + \tilde{\omega}_1) \tag{1.21}$$

with the generalised notch stress intensity factor $K_{1,\rho}$ (which is considered here coincident, for the sake of simplicity, with the notch stress intensity factor K_1 of the pointed notch), with the distance r_0 between the origin of the coordinate system and the notch root which depends on the notch radius and the notch opening angle 2α, and with the auxiliary parameter $\tilde{\omega}_1$ which depends on the notch opening angle 2α. Equating Eqs. (1.20) and (1.21) with $K_{1,\rho} = K_1$ and introducing $r_0 = (1 - 1/q)s\rho^*$ results in the following relationship:

$$s = \frac{q}{q-1}\left[\lambda_1(1+\tilde{\omega}_1)\right]^{1/(1-\lambda_1)} \tag{1.22}$$

Inserting $\tilde{\omega}_1 = 0.810$ for $2\alpha = 90°$ and $\tilde{\omega}_1 = 0.432$ for $2\alpha = 135°$, the result is $s = 2.91$ and $s = 4.48$, respectively. This result is based on the normal stress criterion. Other criteria require the differentiation between plane stress and plane strain as well as the introduction of the jump factor.

1.2.5 General Solution for Sharp V-Notches under Tensile Loading

The sharp V-notch with arbitrary notch opening angle under tensile loading, inclusive of the pointed V-notch, is considered (Berto et al. 2008, 2009). The notch is designated as sharp if the notch radius is small in relation to the notch depth. For determining the support factor s, accurate stress field equations reduced to the bisector of the notch, are required. Such equations were derived by Neuber for sharp V-notches based on a suitable Airy stress function inclusive of Williams' eigenvalues for pointed V-notches (Neuber 1958; Williams 1952). The stress function had to be developed into a binomial series. A solution with higher accuracy was presented by Filippi et al. extending a solution by Lazzarin and Tovo based on the Kolosov–Muskhelishvili complex stress function method (Filippi et al. 2002; Lazzarin and Tovo 1996).

A comparison of the stress ratio $\sigma_\theta/\sigma_{ng}$ along the notch bisector according to the two solutions above (solid lines) and according to a FE analysis (point symbols) is shown in Fig. 1.2 for the notch opening angle $2\alpha = 90°$ and for the notch radii $\rho = 0.2$ and 1.0 mm. The two solutions above refer to the one-sided V-notch in an infinite plate whereas the FE analysis results are found for the two-sided V-notch specimen which was used in the parametric investigation below. The high accuracy of the solution by Filippi et al. (thick solid line), especially too at some distance from the notch root, can clearly be seen.

For determining the support factor s, Eq. (1.4) for $\overline{\sigma}$ or its squared form, Eq. (1.5), has to be evaluated based on the above mentioned accurate stress equations. The averaged (equivalent) stress $\overline{\sigma}(\rho, \rho^*)$ at the real notch has to be equated to its limit value $\overline{\sigma}(\rho_f, \rho^* \to 0)$ at the fictitiously rounded notch:

$$\overline{\sigma}(\rho, \rho^*) = \overline{\sigma}(\rho_f, \rho^* \to 0) \tag{1.23}$$

From this equation, $\rho_f(\rho,\ \rho^*)$ is gained and therefrom the support factor s:

$$s = \frac{\rho_f - \rho}{\rho^*} \tag{1.24}$$

Neuber derived the following expression for $\overline{\sigma}(\rho, \rho^*)$ valid for any notch opening angle 2α integrating the stress σ_θ along the notch bisector (this corresponds to the normal stress criterion):

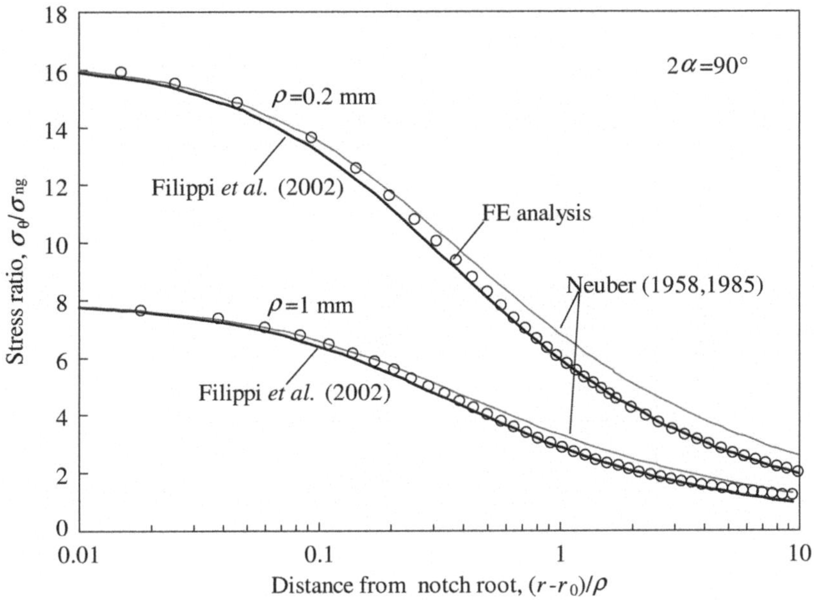

Fig. 1.2 Comparison of the stress ratio $\sigma_\theta/\sigma_{ng}$ along the notch bisector according to solutions by Neuber and by Filippi et al. (solid lines) as well as by FE analysis (point symbols); notch opening angle $2\alpha = 90°$, notch radii $\rho = 0.2$ and 1 mm; (Berto et al. 2008)

$$\begin{aligned} \overline{\sigma} &= \frac{1}{\rho^*}\int\limits_{x_0}^{x_0+\rho^*} \sigma_{\text{th}}\,\mathrm{d}x \\ &= C_1(1+C_2)\frac{r+q}{q}\frac{1}{\rho^*}\left[(x_0+\rho^*)^{\frac{r}{q}} - x_0^{\frac{r+q-1}{q}}(x_0+\rho^*)^{\frac{1-q}{q}}\right] \end{aligned} \tag{1.25}$$

with $r = f(q) \approx 1.0$, $q = f(2\alpha)$ and $x_0 = f(\rho)$.

In the limit case $\rho^* \to 0$ for $\rho = \rho_{\text{f}}$, Eq. (1.25) has the following form (Neuber 1958, p. 172):

$$(\overline{\sigma})_{\rho^*\to 0} = \left[C_1(1+C_2)\frac{r+q}{q}\,\frac{r+q-1}{q}\left[\frac{q-1}{q}\rho_{\text{f}}\right]^{\frac{r-q}{q}}\right] \tag{1.26}$$

By equating $\overline{\sigma}$ from Eq. (1.25) and (1.26), the fictitious notch radius ρ_{f} is obtained:

$$\rho_{\text{f}} = \left\{\frac{q-1}{(q+r-1)\rho^*}\left[\left(\rho+\frac{q}{q-1}\rho^*\right)^{\frac{t}{q}} - \left(\rho+\frac{q}{q-1}\rho^*\right)^{\frac{1-q}{q}}\rho^{\frac{r+q-1}{q}}\right]\right\}^{\frac{q}{r-q}} \tag{1.27}$$

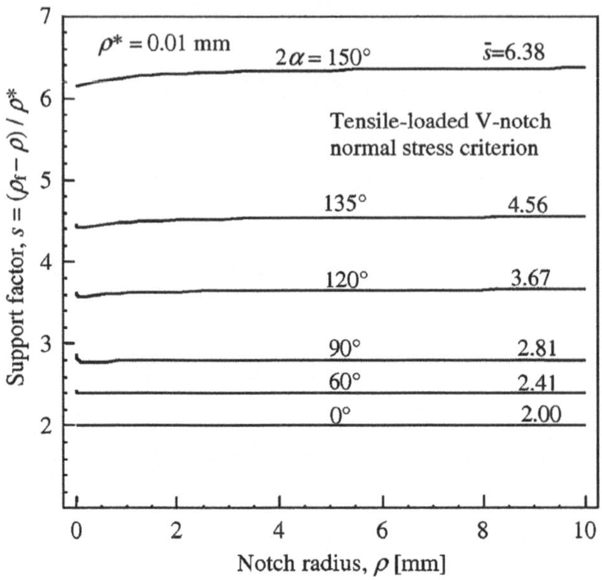

Fig. 1.3 Support factor s as function of the real notch radius ρ for different notch opening angles 2α and for $\rho^* = 0.01$ mm; tensile loaded V-notch, normal stress criterion; plateau values $\bar{s}$; (Berto et al. 2009)

As already noted, the parameters q and r depend on the notch opening angle 2α. In the case of the parabolic notch ($2\alpha = 0°$), Eq. (1.27) is greatly simplified:

$$\rho_{\mathrm{f}} = \rho + 2\rho^* \tag{1.28}$$

resulting in the support factor $s = 2.0$.

The support factor s is plotted for different notch opening angles 2α as a function of the actual notch radius ρ for the normal stress criterion in Fig. 1.3 and for the von Mises criterion under plane stress conditions in Fig. 1.4. Plateau values of s for larger values of ρ are revealed. The strong influence of the notch opening angle is evident, the plateau values of s ranging from 2.0 (for $2\alpha = 0°$) to 6.38 (for $2\alpha = 150°$) in the former case and from 2.5 (for $2\alpha = 0°$) to 7.4 (for $2\alpha = 150°$) in the latter case. It is also evident that the factor s is highly variable for small values of the actual notch radius in the case of the von Mises criterion. In Fig. 1.4, the courses of s obtained with $\rho^* = 0.01$ mm and $\rho^* = 0.4$ mm are compared. As ρ^* increases, the plateau value of s, indicated by the range $\rho \geq \rho^*$, is approximated more slowly. By applying the von Mises criterion in combination with plane strain conditions, the plot of s shown in Fig. 1.5 is generated. Finally, in Fig. 1.6, the factor s is shown as derived by applying the Beltrami criterion under plane strain conditions. Once again, the factor s converges to a plateau value in the range $\rho > \rho^*$.

The curve shape $s = s(\rho)$ for $\rho < \rho^*$ and especially for $\rho \to 0$ may deviate substantially from the plateau value of s. Two effects may be superimposed in this

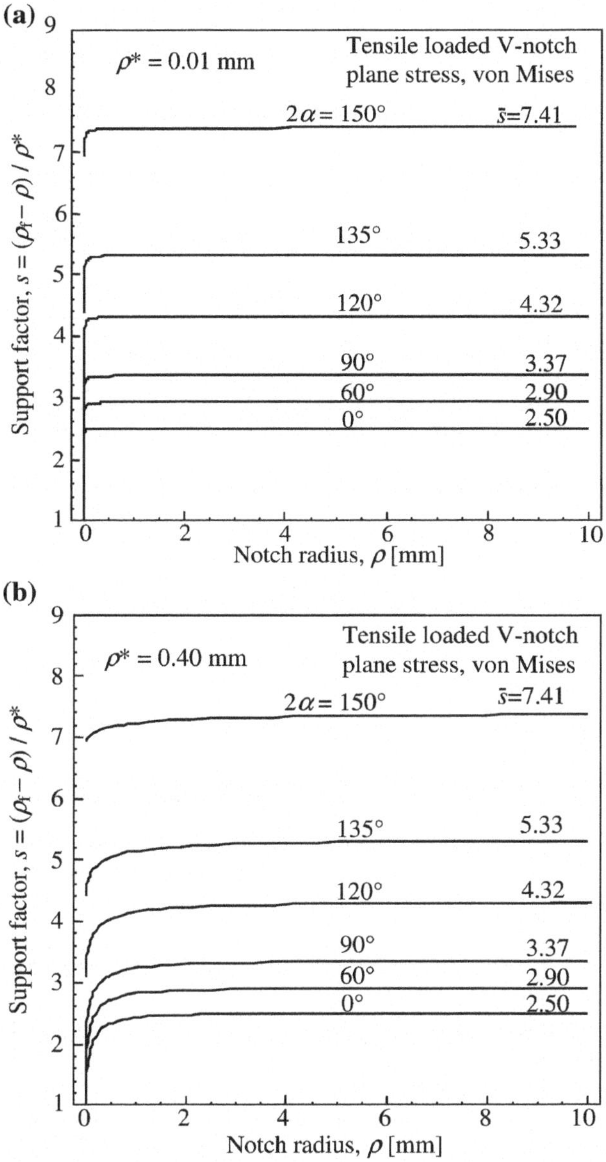

Fig. 1.4 Support factor s as function of the real notch radius ρ for different notch opening angles 2α and for two values for the microstructural support length, $\rho^* = 0.01$ mm (**a**) and $\rho^* = 0.40$ mm (**b**); tensile loaded V-notch, plane stress, von Mises equivalent stress; plateau values $\bar{s}$; (Berto et al. 2009)

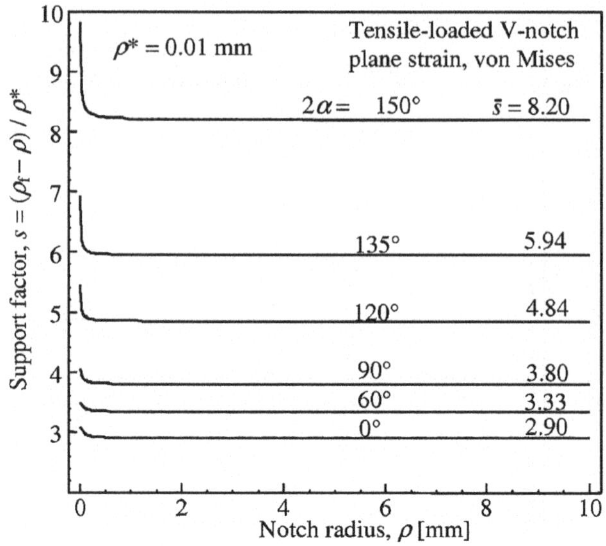

Fig. 1.5 Support factor s as function of the real notch radius ρ for different notch opening angles 2α and for $\rho^* = 0.01$ mm; tensile loaded V-notch, plane strain, von Mises equivalent stress; plateau values $\bar{s}$; (Berto et al. 2009)

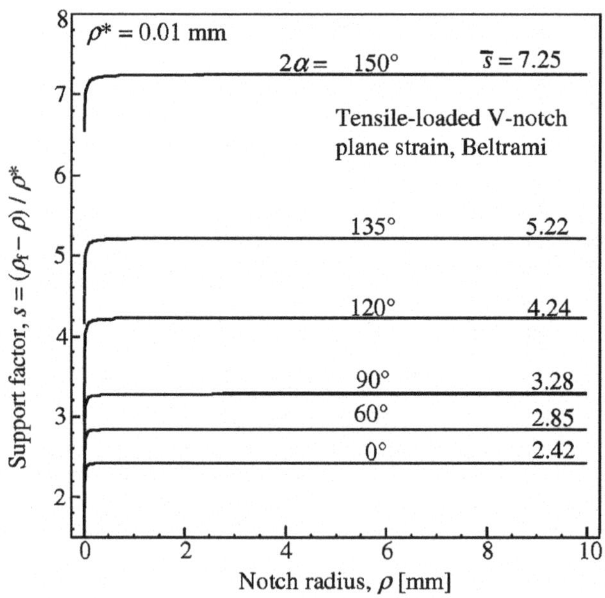

Fig. 1.6 Support factor s as function of the real notch radius ρ for different notch opening angles 2α and for $\rho^* = 0.01$ mm; tensile loaded V-notch, plane strain, Beltrami equivalent stress; plateau values $\bar{s}$; (Berto et al. 2009)

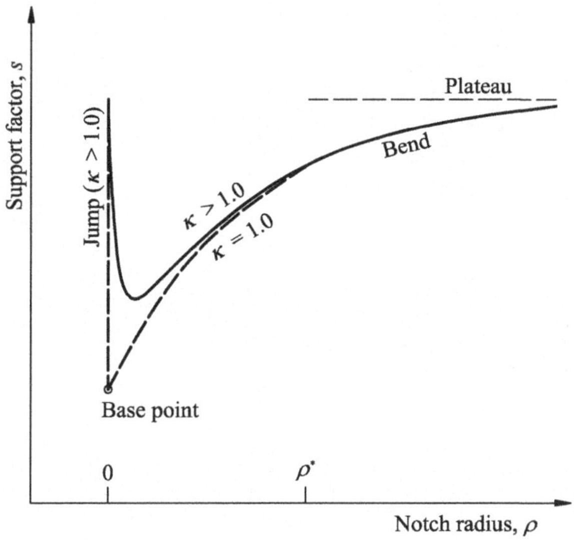

Fig. 1.7 Details of the support factor curve in the range of small notch radii (hypothetical diagram); with microstructural support length ρ^* and jump factor κ referring to notch stress multiaxiality; (Radaj, unpublished)

range, Fig. 1.7. The first effect is a curve bend resulting from the notch shape and represented in the analytical solution for small values of ρ. This curve bend has nothing to do with a change in multiaxiality. The second effect is a curve jump resulting from a jump in multiaxiality when comparing the averaged stress of the pointed notch with the maximum stress of the (fictitiously) rounded notch. It occurs only with $\kappa \neq 1.0$, i.e. only with the von Mises or Beltrami criterion, not with the normal stress criterion. An upward jump corresponds to $\kappa > 0$, a downward jump to $\kappa < 0$. But all this needs further clarification. Additionally, it has to be noted that the evaluation of Eq. (1.23) for $\rho \to 0$ is not always numerically stable.

The plateau value of s and the details of the downward or upward curve bend depend on the kind of analytical solution and the appertaining notch shape (besides the dominating influence of the notch opening angle 2α). Slightly different plateau values are gained using the stress field equations derived by Neuber and by Filippi et al. (see below).

The plateau values $\bar{s}$ as a function of the notch opening angle 2α are plotted in Fig. 1.8 for the different failure criteria assuming plane stress or plane strain conditions. These plateau values are also summarised in Table 1.1. In the case of the normal stress criterion, the factor s has additionally been determined by using Neuber's equations. The difference is minor for $2\alpha < 90°$ but increases for larger values of 2α, the maximum deviation being about 10 % for $2\alpha = 150°$. These differences are due to the fact that Williams' eigenvalues λ_1 differ slightly from Neuber's eigenvalues t/q after binomial series development.

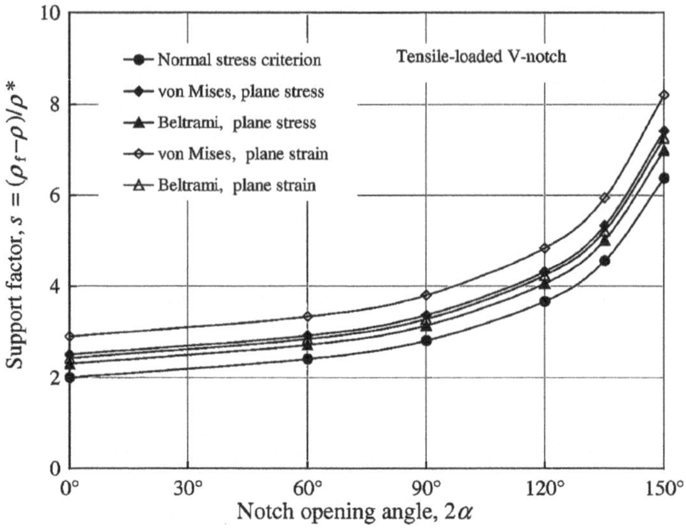

Fig. 1.8 Support factor $\bar{s}$ (plateau value) as function of notch opening angle 2α for different failure criteria under plane stress and plane strain conditions; (Berto et al. 2009)

Table 1.1 Microstructural support factor $\bar{s}$ (plateau value) of tensile loaded V-notches dependent on notch opening angle 2α for different failure criteria (ns, vM, B) under plane stress (ps) and plane strain (pn) conditions; stress field equations according to Neuber and Filippi et al.; normal stress criterion (ns), von Mises criterion (vM), Beltrami criterion (B); (Berto et al. 2009)

	Neuber	Filippi, Lazzarin and Tovo				
2α	s(ns)	$\bar{s}$(ns)	$\bar{s}$(vM, ps)	$\bar{s}$(vM, pn)	$\bar{s}$(B, ps)	$\bar{s}$(B, pn)
0°	2.00	2.00	2.50	2.90	2.30	2.42
60°	2.36	2.41	2.90	3.33	2.72	2.85
90°	2.72	2.81	3.37	3.80	3.14	3.28
120°	3.47	3.67	4.32	4.84	4.06	4.24
135°	4.21	4.56	5.33	5.94	5.02	5.22
150°	5.73	6.38	7.41	8.20	6.99	7.25

Neuber has given several expressions of s obviously derived for the parabolic notch (Neuber 1968). He considered different multiaxial failure criteria and also different types of specimen under tensile loading. The values of s obtained from the expressions suggested by Neuber match those obtained from the amended equations derived by Filippi et al. which collapse into the Creager–Paris solution in the case of $2\alpha = 0°$, interpreted as a parabolic notch solution (see again Table 1.1 for $2\alpha = 0°$). The values for flat bars according to Neuber coincide with the plateau values obtained from the parabolic notch solution under plane stress conditions. Also, the values of s obtained by Neuber for axisymmetric specimens assuming plane strain conditions coincide with the parabolic notch solution under plane strain conditions. The values of s for the pointed V-notch dependent on the notch opening angle are also available (Neuber 1958, ibid. Fig. 98b).

Summarising the numerical results, the following trends have been found for the support factor s in the case of tensile loading:

- the factor s depends on the real radius ρ, starting with lower, higher or identical values for $\rho = 0$, but converging to a plateau value for $\rho > \rho^*$;
- the factor s depends only slightly on ρ^* for $\rho \geq \rho^*$, contrary to the range $\rho < \rho^*$ where the non-converged factor s may depend heavily on ρ^*;
- the factor s determined for pointed V-notches ($\rho = 0$) is deviating from the corresponding plateau value in general;
- the factor s depends on the multiaxiality condition at the notch tip and on the failure criterion, as already stated by Neuber;
- the factor s depends on the notch opening angle 2α, as originally stated by Neuber, but later on deleted for applications.

1.2.6 Parametric Investigation of Tensile Loaded V-Notch Specimens

It should be clear from the above that the support factor s is not only dependent on the loading mode, the multiaxiality condition and the failure criterion, but also, to a varying extent, from further influence parameters such as the notch opening angle, the notch depth, the notch radius, the cross-sectional width, the microstructural length, the geometry of the fictitious notch (circular, elliptical, parabolic) and the degree of notch stress approximation. Not all of these parameters can be taken into account in the analytical solution for the fictitious notch radius. A first conclusion from this is that direct integration over ρ^* is safer to use than fictitious notch rounding. On the other hand, direct integration may be too tedious for engineers. Therefore, uniform values of s are demanded for approximate evaluations. It is proposed and proven feasible to use the plateau values of s for that purpose. When doing so, it has to be kept in mind that for actually pointed cracks or corner notches ($\rho = 0$), the plateau values of s may be substantially larger than $s(\rho = 0)$ and are then non-conservative.

The fictitious notch rounding concept is applied in the following to pointed V-notches ($\rho = 0$) with different notch opening angles and analysed by the FE method (Berto et al. 2009). The geometry and dimensions of the specimen are shown in Fig. 1.9, whereas the concept applied to sharp notches is demonstrated in Fig. 1.1. By evaluating the integral of the theoretical (equivalent) stress σ_{th} along the distance ρ^*, it is possible to define an effective stress concentration factor $\overline{K}_t$ (corresponding to the theoretical fatigue notch factor K_f) for a sharp notch as follows:

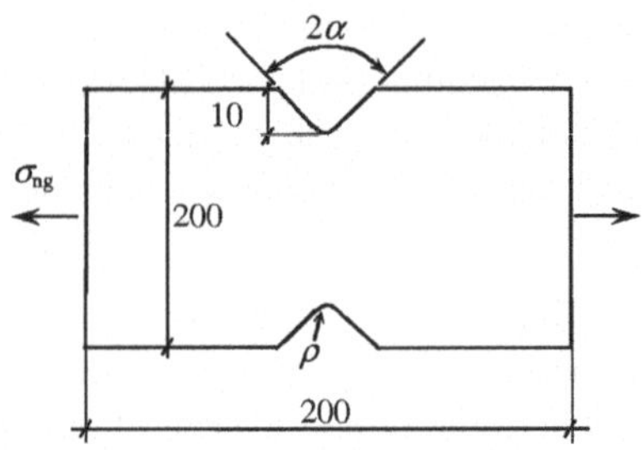

Fig. 1.9 Geometry and dimensions of the V-notch specimen considered in the FE analyses, tensile loaded flat bar with two-sided V-notch, $\rho = 0$ compared with $\rho = \rho_f$; the nominal stresses σ_{ng} relating to the gross cross-sectional area; (Berto et al. 2009)

$$\overline{K}_t = \frac{\overline{\sigma}}{\sigma_{ng}} = \frac{1}{\rho^* \sigma_{ng}} \int_{x_0}^{x_0+\rho^*} \sigma_{th}\, dx \tag{1.29}$$

where $\overline{\sigma}$ is the effective or averaged (equivalent) notch stress and σ_{ng} is the applied nominal stress acting in the gross cross-sectional area.

Considering now the fictitious notch radius $\rho_f = s\rho^*$ (for $\rho = 0$) the stress concentration factor $K_t(\rho_f)$ assumes the following form:

$$K_t(\rho_f) = \frac{\sigma_{max}(\rho^*, s)}{\sigma_{ng}} \tag{1.30}$$

where the maximum equivalent stress σ_{max} at the fictitiously rounded notch root should be equal to $\overline{\sigma}$. Thus, the stress concentration factor $K_t(\rho_f)$ depends both on the material-dependent microstructural support length ρ^* and the failure criterion.

In the case of pointed notches, an adjustment is necessary. The stress multiaxiality at the pointed notch, analysed by the FE method in the following, deviates from the stress multiaxiality at the corresponding fictitiously rounded notch, also analysed by the FE method. Therefore, the jump factor κ has to be introduced when evaluating the von Mises or Beltrami equivalent notch stresses or the corresponding stress concentration factors. The effective stress concentration factor $\overline{K}_t$ has to be compared with the modified stress concentration factor $K_t(\rho_f)/\kappa$, because κ was not taken into account in the s values directly according to Eq. (1.15) or (1.18).

Referring to the schematic diagram in Fig. 1.7, the factor κ should be applied to the base point value of s. This value may be determined based on Eq. (1.22). On the other hand, the strictly circular notch shape in the FE analysis is only approximated within the analytic stress field solution. Therefore, applying the jump factor κ on the plateau values of s is also a tenable approximation. The procedure is confirmed by the fact that the resulting deviations between the two stress concentration factors above, $\overline{K}_t$ and $K_t(\rho_f)/\kappa$, remain sufficiently small.

An attempt to neglect the jump factor κ completely gave extremely large deviations in the case of von Mises and plane strain, corresponding to the large κ values under these conditions. The deviations remain small in the case of the Beltrami criterion.

In all cases considered in this subsection ($\rho = 0$), the plateau values reported in Table 1.1 have been introduced for the support factor s. Three different values of the microstructural support length ρ^* have been considered in the FE analyses: $\rho^* = 0.05$, 0.1 and 0.5 mm. The notch opening angles are $2\alpha = 0$, 90 and 135°. By considering different failure criteria combined with plane stress and plane strain conditions, respectively, a comparison has been carried out between the effective stress concentration factor $\overline{K}_t$ given by Eq. (1.29) and the stress concentration factor $K_t(\rho_f)$ evaluated with fictitious notch rounding, Eq. (1.30), modified by κ. In both cases, the stresses in the specimens are obtained from the FE analysis. The first step is to evaluate Eq. (1.29), considering the normal or equivalent stress obtained by FE analysis along the bisector of the pointed V-notch. The second step is to introduce the fictitious notch radius for different notch opening angles and failure criteria by using the values of s from Table 1.1. The third step is to determine the value of $K_t(\rho_f)/\kappa$ from FE analysis, modelling the rounded notch with the depth of the pointed notch. All results are summarised in Table 1.2 where the relative deviation Δ has been determined using the following expression:

$$\Delta = \frac{K_t(\rho_f)/\kappa - \overline{K}_t}{\overline{K}_t} \tag{1.31}$$

The results in Table 1.2 are discussed based on the stated Δ values:

- Most Δ values remain within the ±5 % range and nearly all Δ values within the ±10 % range; this is acceptable for engineering applications.
- Positive and negative Δ values occur with approximately the same frequency, mainly positive Δ values for $2\alpha = 0°$ and $2\alpha = 135°$, mainly negative Δ values for $2\alpha = 90°$; positive Δ values indicate too high $K_t(\rho_f)/\kappa$ values, i.e. conservative results in strength assessments.
- The Δ values change continuously between small and large ρ^* values with a tendency of larger Δ values for larger ρ^* values in the case of $2\alpha = 0°$; this can be explained by an influence of t/ρ^* with the notch depth t, see Eq. (1.9).
- The introduction of the jump factor κ is essential in the case of the von Mises equivalent stresses under plane strain conditions, but also useful to a minor extent in other cases.

By using the plateau values of s, two basic diagrams also used by Neuber are redrawn. In Fig. 1.10 the ratio ρ_f/ρ^* is plotted as a function of ρ/ρ^* considering the normal stress criterion. The diagram was intended by Neuber for engineering applications and is plotted here using double-logarithmic scales. The point

Table 1.2 Fictitious notch rounding applied to pointed V-notch specimen ($\rho = 0$) subjected to tensile loading using the plateau values $\bar{s}$ and the jump factor κ, comparison of $K_t(\rho_f)/\kappa$ with $\overline{K}_t$, both obtained from FE analysis (relative deviation Δ); variation of notch opening angle 2α, support length ρ^*, failure criterion and plane stress (ps) versus plane strain (pn); normal stress (ns), von Mises (vM) and Beltrami (B) criterion; (Berto et al. 2009)

			$\rho^* = 0.05$ mm				$\rho^* = 0.1$ mm				$\rho^* = 0.5$ mm			
	$\bar{s}$	κ	ρ_f [mm]	$K_t(\rho_f)/\kappa$	$\overline{K}_t$	Δ %	ρ_f [mm]	$K_t(\rho_f)/\kappa$	$\overline{K}_t$	Δ %	ρ_f [mm]	$K_t(\rho_f)/\kappa$	$\overline{K}_t$	Δ %
$2\alpha = 0°$														
ns, ps, pn	2.00	1.00	0.10	24.54	22.63	8.44	0.20	17.52	16.02	9.36	1.00	8.19	7.17	14.23
vM, ps	2.50	1.00	0.13	21.69	22.55	−3.81	0.25	15.72	15.88	−1.01	1.25	7.38	7.04	4.83
vM, pn	2.90	2.22	0.15	9.12	9.02	1.08	0.29	6.59	6.35	3.78	1.45	3.10	2.83	9.51
B, ps	2.30	0.85	0.12	26.54	26.69	−0.56	0.23	19.18	18.79	2.06	1.15	9.02	8.33	8.33
B, pn	2.42	0.94	0.12	23.44	22.94	2.21	0.24	16.94	16.16	4.80	1.21	7.97	7.17	11.13
$2\alpha = 90°$														
ns, ps, pn	2.81	1.00	0.14	19.08	19.48	−2.05	0.28	14.00	14.20	−1.41	1.41	6.82	6.81	0.15
vM, ps	3.37	1.13	0.17	15.58	17.18	−9.34	0.34	11.42	12.52	−8.75	1.69	5.57	6.01	−7.38
vM, pn	3.80	2.04	0.19	8.18	8.59	−4.81	0.38	6.00	6.26	−4.15	1.90	2.92	3.00	−2.61
B, ps	3.14	0.97	0.16	18.73	19.91	−5.92	0.31	13.74	14.51	−5.29	1.57	6.70	6.96	−3.72
B, pn	3.28	1.07	0.16	16.64	17.36	−4.12	0.33	12.21	12.66	−3.52	1.64	5.95	6.07	−1.92
$2\alpha = 135°$														
ns, ps, pn	4.56	1.00	0.23	9.93	9.94	-0.10	0.46	7.93	7.92	0.13	2.28	4.70	4.68	0.43
vM, ps	5.33	1.06	0.27	8.90	8.79	1.21	0.53	7.10	7.01	1.34	2.67	4.22	4.14	1.86
vM, pn	5.94	1.38	0.30	6.59	6.41	2.87	0.59	5.27	5.11	3.09	2.97	3.12	3.02	3.42
B, ps	5.02	1.05	0.25	9.16	9.48	−3.36	0.50	7.31	7.56	−3.25	2.51	4.33	4.47	−3.06
B, pn	5.22	1.10	0.26	8.64	8.63	0.07	0.52	6.89	6.87	0.30	2.61	4.08	4.06	0.54

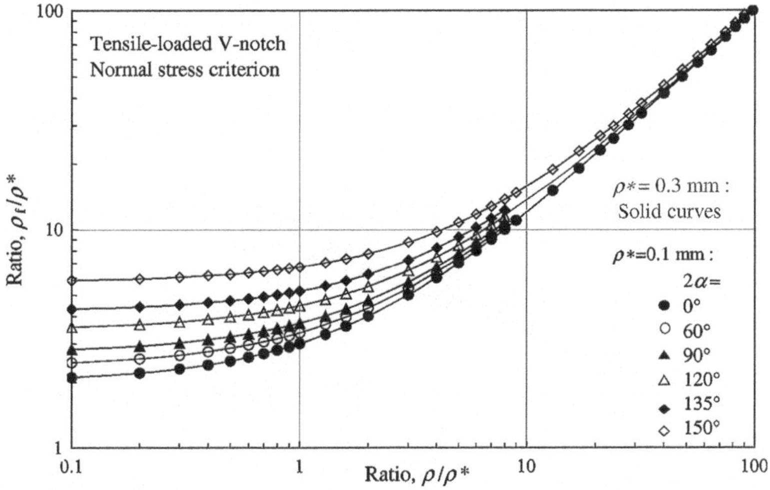

Fig. 1.10 Ratio ρ_f/ρ^* as function of ratio ρ/ρ^* for $\rho^* = 0.3$ mm (solid curves) and $\rho^* = 0.1$ mm (point symbols) as well as for different notch opening angles 2α obtained with plateau values $\bar{s}$; tensile loaded V-notch, normal stress criterion; (Berto et al. 2009)

symbols have been obtained for $\rho^* = 0.1\,\text{mm}$, but the diagram is valid also for other values of ρ^* (as documented here by the solid curves obtained for $\rho^* = 0.3\,\text{mm}$).

Considering now the dependency of ρ_f/ρ^*, one can write:

$$\frac{\rho_f}{\rho^*} = \frac{\rho + s\rho^*}{\rho^*} = s + \frac{\rho}{\rho^*} \tag{1.32a}$$

thus showing that ρ_f/ρ^* is a linear function of ρ/ρ^*. For different notch opening angles, Eq. (1.32a) is represented by a set of parallel straight lines in linear scales, and the intersection with the axis $\rho/\rho^* = 0$ corresponds to the support factor s. It is important to note that, to obtain the linear function, the fictitious notch radius has to be expressed in the form $\rho_f = \rho + s\rho^*$, independently of the notch opening angle. The idea to apply this expression for ρ_f to any notch opening angle is mentioned by Neuber, but not proven to be applicable (Neuber 1958). To consider this expression valid for any notch opening angle is consistent with the decision to use the plateau values of s in the analysis.

Another generally applicable diagram is depicted in Fig. 1.11. Once the notch opening angle 2α and the failure criterion is chosen, the ratio ρ_f/ρ can be plotted as a function of ρ^*/ρ, according to the following expression:

$$\frac{\rho_f}{\rho} = \frac{\rho + s\rho^*}{\rho} = s\frac{\rho^*}{\rho} + 1 \tag{1.32b}$$

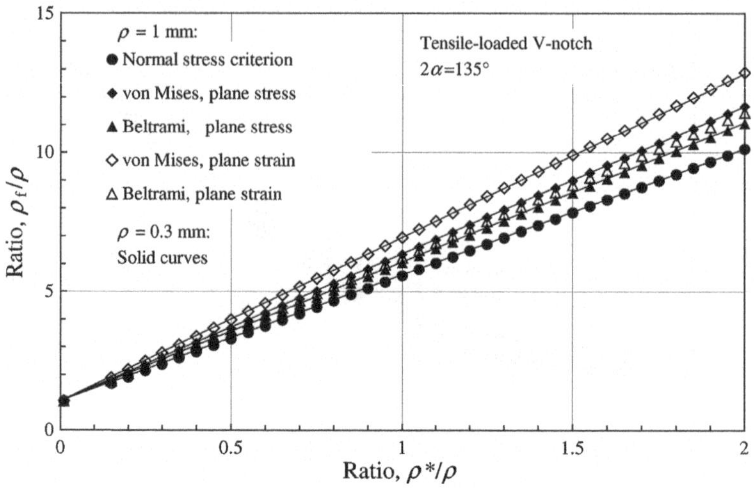

Fig. 1.11 Ratio ρ_f/ρ as function of ratio ρ^*/ρ for notch opening angle $2\alpha = 135°$; real notch radius $\rho = 0.3$ mm (solid curves) and $\rho = 1$ mm (point symbols) obtained with plateau values $\overline{s}$; tensile loaded V-notch, different failure criteria under plane stress and plane strain conditions; (Berto et al. 2009)

In this case the set of straight lines intercepts the axis $\rho^*/\rho = 0$ at $\rho_f/\rho = 1$ and the support factor s, depending on the failure criterion and stress multiaxiality considered, gives the slope of the lines. Also here, the expression $\rho_f = \rho + s\rho^*$ is used independently of the notch opening angle, but with s depending on it. The point symbols shown in Fig. 1.11 have been obtained for $2\alpha = 135°$ and $\rho = 1$ mm. Varying the notch radius ρ, the diagram remains the same (see the solid curves which are related to $\rho = 0.3$ mm).

1.2.7 Solution for Tensile Loaded V-Notches with Root Hole

A further solution for the fictitious notch rounding problem referring to tensile loading has recently been published for V-notches with root hole (Berto and Zappalorto 2011, 2012). The basic procedure is illustrated by Fig. 1.12, where the real notch with stress $\overline{\sigma}$ averaged over ρ^* is contrasted with the fictitiously rounded notch with the maximum stress $\sigma_{max} = \overline{\sigma}$.

For determining the support factor s, Eq. (1.4) for $\overline{\sigma}$ is evaluated based on the stress field equations for the V-notch with root hole subjected to tensile loading (Zappalorto and Lazzarin 2011, reviewed in Sect. 2.4.6). The averaged stress $\overline{\sigma}(\rho, \rho^*)$ at the real notch has to be equated to its limit value $\overline{\sigma}(\rho, \rho^* \to 0) = \sigma_{max}(\rho_f)$ at the fictitiously rounded notch, Eq. (1.23). The resulting expression for $\rho_f(\rho, \rho^*)$ reads:

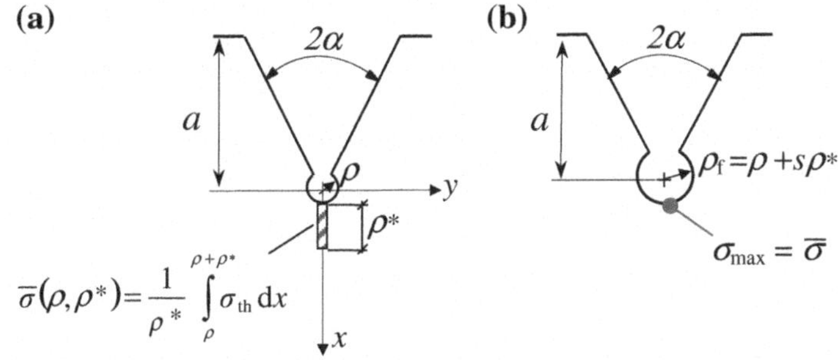

Fig. 1.12 Fictitious notch rounding concept applied to tensile loaded V-notches with root hole; real notch with stress $\overline{\sigma}$ averaged over ρ^* (**a**) and substitute notch with fictitious notch radius ρ_f producing $\sigma_{max} = \overline{\sigma}$ (**b**); (Berto and Zappalorto 2009)

$$\rho_f = \left\{ \frac{1}{4(\lambda\chi - \chi - 1)} \left[\left(\frac{2\rho^*\rho^\lambda}{(\rho+\rho^*)^2} + \frac{4\rho^{1+\lambda}}{(\rho+\rho^*)^2} + \frac{2\rho^{2+\lambda}}{\rho^*(\rho+\rho^*)^2} \right) \left(2 + \frac{1}{\lambda} - \frac{6}{1+\lambda} \right) \right.\right.$$
$$+ \frac{(\rho+\rho^*)^\lambda}{\rho^*} \left(-1 - \frac{1}{\lambda} + \chi - \frac{\chi}{\lambda} \right) + \left(\frac{\rho}{\rho+\rho^*} \right)^{2\lambda} (\rho+\rho^*)^{\lambda-1} \left[\left(8 - 2\lambda - \frac{12}{1+\lambda} \right.\right.$$
$$\left.\left.\left.\left. -2\chi(1-\lambda) \right) + \frac{\rho+\rho^*}{\rho^*} \left(-3 + \frac{12}{1+\lambda} - \chi + \frac{\chi-1}{\lambda} \right) + \frac{\chi\rho^*(1-\lambda)}{(\rho+\rho^*)} \right] \right] \right\}^{\frac{1}{\lambda-1}} \tag{1.33a}$$

For $2\alpha = 0$, $\rho \neq 0$ ($\chi = \chi_1 = 1.0$, $\lambda = \lambda_1 = 0.5$), Eq. (1.33a) results in:

$$\rho_f = \frac{144(\rho+\rho^*)^5}{(12\rho^2 + 19\rho\rho^* + 8\rho^{*2})^2} \tag{1.33b}$$

For $\rho = 0$ and arbitrary values of 2α, Eq. (1.33a) is simplified:

$$\rho_f = \left\{ \frac{\chi(\lambda-1) - (\lambda+1)}{4\lambda[\chi(\lambda-1) - 1]} \right\}^{\frac{1}{\lambda-1}} \rho^* \tag{1.33c}$$

The values of s for $\rho = 0$ and $\overline{s}$ for $\rho = 10$ mm (plateau values) are listed in Table 1.3 for different notch opening angles 2α and compared with the relevant $\overline{s}$ values for blunt V-notches from Table 1.1. The factors s and $\overline{s}$ rise with the notch opening angle. The plateau values for the root hole notch are slightly smaller than for the blunt notch, but the values of s for $\rho = 0$ for the root hole notch are larger.

The support factor s for the tensile loaded V-notch with root hole is also plotted as function of the real notch radius ρ for different notch opening angles 2α in Fig. 1.13 and for different values of ρ^* in Fig. 1.14. The dependency of s on ρ and ρ^* (besides 2α) is strong for small values of ρ, but weak for the plateau values.

A validation of the method has been performed based on FE analyses of the tensile loaded flat bar specimen with two-sided V-notch with root hole, $\rho = 0$

Table 1.3 Microstructural support factor s for different notch opening angles 2α; tensile loaded V–notch with root hole: limit value s for $\rho = 0$ and plateau value $\bar{s}$ for $\rho = 10$ mm, $\rho^* = 0.1$ mm; in comparison blunt V-notch; normal stress criterion; (Berto and Zappalorto 2011)

			Root hole V-notch		Blunt V-notch
2α	λ_1	χ_1	$s(\rho = 0)$	$\bar{s}(\rho = 10$ mm$)$	$\bar{s}(\rho = 10$ mm$)$
0°	0.500	1.000	2.25	1.84	2.00
30°	0.501	1.071	2.29	1.86	2.18
45°	0.505	1.166	2.36	1.89	2.28
60°	0.512	1.312	2.49	1.94	2.41
90°	0.545	1.841	3.13	2.15	2.81
120°	0.616	3.003	5.42	2.67	3.67
135°	0.674	4.153	9.64	3.25	4.56
150°	0.752	6.362	30.2	4.43	6.38

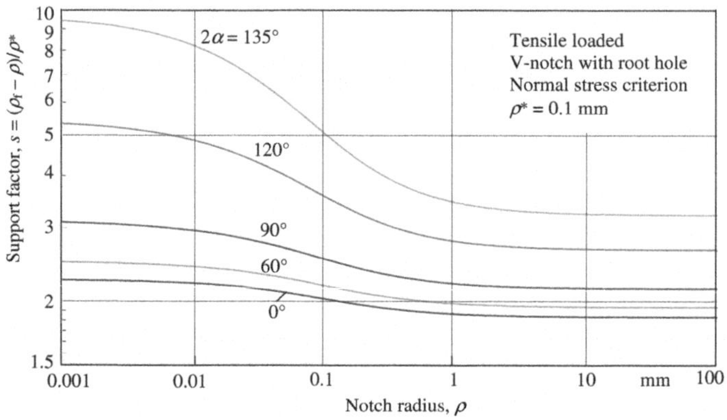

Fig. 1.13 Support factor s as function of notch radius ρ for different notch opening angles 2α; $\rho^* = 0.1$ mm; tensile loaded V-notch with root hole; (Berto and Zappalorto 2011, unpublished)

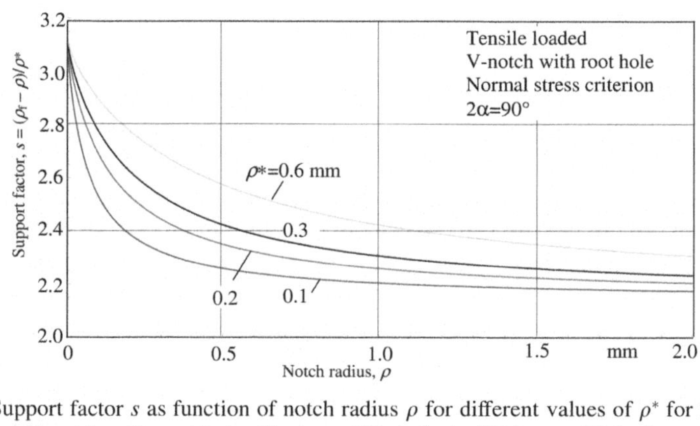

Fig. 1.14 Support factor s as function of notch radius ρ for different values of ρ^* for $2\alpha = 90°$; tensile loaded V-notch with root hole; (Berto and Zappalorto 2011, unpublished)

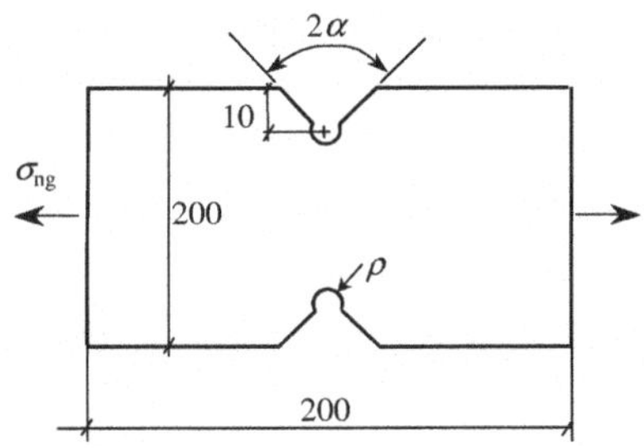

Fig. 1.15 Geometry and dimensions of the V-notch specimen considered in the FE analyses; tensile loaded flat bar with two-sided V-notch with root hole, $\rho = 0$ compared with $\rho = \rho_f$; nominal stresses σ_{ng} relating to the gross cross-sectional area; (Berto and Zappalorto 2011)

Table 1.4 Fictitious notch rounding with root hole applied to V-notch specimens ($\rho = 0$) subjected to tensile loading using the relevant values of s; comparison of $K_t(\rho_f)$ with $\overline{K}_t$, both obtained from FE analysis (relative deviation Δ); for different notch opening angles 2α; normal stress criterion; (Berto and Zappalorto 2011)

		$\rho = 0$, $\rho^* = 0.1$ mm			
2α	s	ρ_f [mm]	$K_t(\rho_f)$	$\overline{K}_t$	Δ %
0°	2.25	0.225	16.69	16.21	2.5
120°	5.42	0.543	10.37	10.25	1.12
135°	9.64	0.964	7.93	7.92	0.12
150°	30.2	3.022	5.32	5.30	0.4

compared with $\rho = \rho_f$, Fig. 1.15. The result of the validations is presented in Table 1.4 in the form already known from Table 1.2. The relative deviations Δ are sufficiently small.

1.2.8 Conclusions

Based on accurate stress field solutions for blunt V-notches and V-notches with root hole under tensile loading, the microstructural support factor s is evaluated and plotted as a function of the real notch radius ρ for different notch opening angles 2α. Plateau values of s are revealed for larger values of ρ. These plateau values of s depend on the multiaxiality of the stress state (plane stress or plane strain), the adopted multiaxial failure criterion and the considered notch opening angle. For $\rho < \rho^*$, the factor s may deviate from the plateau value and then depends on ρ.

As demonstrated by the parametric investigation of pointed V-notch specimens ($\rho = 0$) subjected to tensile loading, the plateau values of s combined with the jump factor κ are well suited for engineering usage in structural strength assessments. These s values are ready for incorporation in those codes and design recommendations for structural components under static or cyclic loading conditions (brittle or fatigue fracture) which comprise local approaches such as the notch stress or notch strain approach.

1.3 Fictitious Notch Rounding for Out-of-Plane Shear Loading

1.3.1 Survey of Section Contents

This section is closely related to the precedent section in respect of contents and representation. Actually, reference is made to one and the same publication below. The analytical frame for out-of-plane shear stress fields at notches is less complicated than for in-plane stress fields. The former stress fields are governed by potential functions, the latter by bipotential functions (real functions according to Airy or complex functions according to Kolosov and Muskhelishvili). In consequence, closed form solutions can be derived to a larger extent and approximation formulae are easier to gain.

On the basis of Neuber's stress field equations for sharply rounded V-notches in out-of-plane shear loading, the microstructural support factor s of these notches is determined which quantifies the fictitious notch radius in Neuber's microstructural support concept. The fictitious notch radius is evaluated for different notch opening angles as a function of the actual notch radius and the microstructural support length. Once the fictitious radius has been determined, the support factor s is derived from it. The derivations are preceded by elementary solutions for pointed notches in general. The investigation of blunt V-notches is supplemented by considering V-notches with root hole.

The support factor s is found to be sensitive to the notch opening angle, but constant 'plateau values' are determined for notch radii greater than the microstructural support length. There is no dependence of s on the failure criterion in out-of-plane shear loading.

Various numerical analyses using the FE method have been carried out to compare the stress concentration factor of the fictitiously rounded notch using the plateau values of s with the effective stress concentration factor obtained by integrating the relevant stress over the microstructural support length along the bisector of the pointed V-notch. There is no 'jump' in stress multiaxiality between the rounded and the pointed notch set into comparison in the case of out-of-plane shear loading. It is shown that the plateau values of s are well suited for engineering usage in structural strength assessments.

The expositions are primarily based on the analytical developments and their application to specimens with V-notches carried out by Lazzarin and Berto (Berto et al. 2008, 2009). A list of symbols and a list of references are available at the end of Chap. 1.

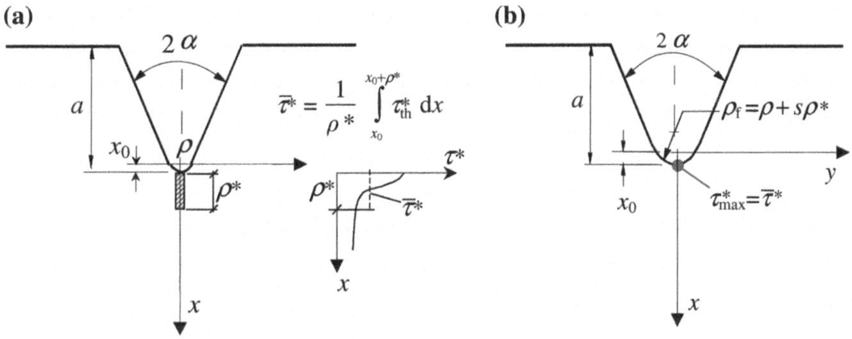

Fig. 1.16 Fictitious notch rounding concept applied to out-of-plane shear-loaded blunt V-notches; real notch with shear stress $\bar{\tau}^*$ averaged over ρ^* (**a**) and substitute notch with fictitious notch radius ρ_f producing $\tau^*_{max} = \bar{\tau}^*$ (**b**); (Berto et al. 2009)

1.3.2 Objectives, Method and Scope of Analysis Related to V-Notches

As has already been stated for tensile loaded notches, the dependency of the support factor s on the notch opening angle has to be investigated. This is true also for out-of-plane shear-loaded notches. Additionally, there is an unsettled discrepancy between the s values given in the literature for pointed compared with rounded notches. For out-of-plane shear loading of notches, Neuber (1968) gives $s = 1.0$ contrary to Neuber (1958) stating $s = 0.5$. On the other hand, $s = 0.5$ is derived by (Radaj and Zhang 1992) for out-of-plane shear loading of crack tips.

The basic procedure is illustrated by Fig. 1.16, where the real notch with shear stress $\bar{\tau}^*$ averaged over ρ^* is contrasted with the fictitiously rounded notch with maximum shear stress $\tau^*_{max} = \bar{\tau}^*$. Notch stress averaging over the support length ρ^* is performed in analogy to Eq. (1.4):

$$\bar{\tau}^* = \frac{1}{\rho^*}\int_{x_0}^{x_0+\rho^*} \tau^*_{th}\,dx \tag{1.34}$$

where $\bar{\tau}^*$ is the averaged and τ^*_{th} the theoretical shear stress in the notch bisector line. The asterisk on τ is used to distinguish out-of-plane shear from in-plane shear (without asterisk).

The values of s and ρ_f derived based on Eq. (1.34) refer to the shear stress failure criterion. They are also valid for other failure criteria because the same criterion-dependent factor is connected with $\bar{\tau}^*$ and τ^*_{th} when using the equivalent stresses in Eq. (1.4). There is no jump in stress multiaxiality occurring under out-of-plane shear loading when substituting pointed notches by (fictitiously) rounded notches.

The analytical method given by Neuber (1958, pp. 166–169) has been used for determining the fictitious notch radius ρ_f and therefrom the support factor s according to Eq. (1.24) for one-sided V-notches in out-of-plane shear loading. Closed-form expressions were derived. The procedural steps are identical with those described for tensile loaded notches. The following conditions are considered in the numerical investigation:

- the real notch radius ρ varying between 0 and 10 mm,
- the notch opening angle 2α varying between 0 and 135°,
- the microstructural support length ρ^* varying between 0.05 and 0.5 mm.

1.3.3 Elementary Solutions for Pointed Compared with Rounded Notches under Out-of-Plane Shear Loading

Before reviewing the general solution for V-notches, more elementary solutions for pointed compared with rounded notches under out-of-plane shear loading are presented. They illustrate the fictitious notch rounding concept without major demands on mathematical understanding.

The out-of-plane shear loaded crack (mode 3) and the corresponding fictitious elliptical notch in comparison can be dealt with in analogy to Eqs. (1.6) to (1.9), considering the out-of-plane shear stress τ^* as relevant (Radaj and Zhang 1992):

$$K_t = \frac{\tau^*_{max}}{\tau^*_n} = \left(1 + \sqrt{\frac{t}{\rho}}\right) \tag{1.35}$$

$$\bar{\tau}^* = \sqrt{\frac{2}{\pi \rho^*}} \tau^*_n \sqrt{\pi t} \tag{1.36}$$

$$\sqrt{\frac{2t}{\rho^*}} = 1 + \sqrt{\frac{t}{s\rho^*}} \tag{1.37}$$

$$s = \frac{1}{\left(\sqrt{2} - \sqrt{\rho^*/t}\right)^2} \tag{1.38}$$

The result for $\rho^* \to 0$ or $t \to \infty$ (pointed notches) is $s = 0.5$. Considering $0 \leq \rho^* \leq 0.5$ mm combined with $t = 10$ mm one gets $0.5 \leq s \leq 0.71$. The result is independent of the applied failure criterion, as already stated by Neuber.

The factor $s = 0.5$ is confirmed by the approximation of τ^*_{max} for blunt cracks (Creager and Paris 1967) which may be interpreted as a parabolic notch solution. It proceeds from the stress intensity factor K_{III} of the pointed crack tip, equating τ^*_{max} and $\bar{\tau}^*$:

$$\tau^*_{\max} = \frac{K_{\mathrm{III}}}{\sqrt{\pi\rho}} \qquad (\rho = \rho_{\mathrm{f}} = s\rho^*) \tag{1.39}$$

$$\bar{\tau}^* = K_{\mathrm{III}}\sqrt{\frac{2}{\pi\rho^*}} \tag{1.40}$$

Further comparison values of s can be obtained from the analytical solution for the out-of-plane shear-loaded parabolic notch (Neuber 1985, pp. 25–26):

$$\rho_{\mathrm{f}} = \frac{1}{2}\left[\rho + \rho^* + \sqrt{\rho(\rho + 2\rho^*)}\right] \tag{1.41}$$

By inserting $\rho_{\mathrm{f}} = \rho + s\rho^*$, the following relationship for the support factor s is obtained:

$$s = \frac{1}{2}\left[1 - \frac{\rho}{\rho^*} + \sqrt{\left(\frac{\rho}{\rho^*}\right)^2 + 2\left(\frac{\rho}{\rho^*}\right)}\right] \tag{1.42}$$

The numerical evaluation of Eq. (1.42) gives $s = 0.5$ for $\rho = 0$, steeply rising with $\rho > 0$ ($s = 0.87$, 0.96, 0.998 for $\rho/\rho^* = 1$, 5, 100, respectively) and converging to the plateau value $s = 1.0$ for $\rho^* \rightarrow 0$ or $\rho \rightarrow \infty$ (confirmed by the curve for $2\alpha = 0°$ in Fig. 1.17).

An ambiguity (not a contradiction) in the values of s reported by Neuber for out-of-plane shear loading should be mentioned. On the one hand, the factor $s = 1.0$ is recommended for structural design, considering structural notches with finite notch radius (Neuber 1968). On the other hand, $s = 0.5$ results from Eq. (1.42) with $\rho = 0$ and $\rho^* \ll t$. Evaluating this equation tentatively with $\rho = \rho^*$ instead of $\rho = 0$, the result is $s = 0.87$. Once more, the tendency of enlarged values of s for increasing values of ρ is found.

Another comparison value of s can be obtained from the solution for the out-of-plane shear-loaded keyhole notch (Kullmer 1992; Radaj et al. 2001):

$$\tau^*_{\max} = \frac{\sqrt{2}K_{\mathrm{III}}}{\sqrt{\pi\rho}} = 1.41\frac{K_{\mathrm{III}}}{\sqrt{\pi\rho}} \qquad (\rho = \rho_{\mathrm{f}} = s\rho^*) \tag{1.43}$$

Equating $\tau^*_{\max}$ from Eq. (1.43) for $\rho = \rho_{\mathrm{f}}$ with $\bar{\tau}^*$ from Eq. (1.40) for $\rho = 0$, the result is $\rho_{\mathrm{f}} = \rho^*$ or $s = 1.0$. This shows that the limit value of s for $\rho \rightarrow 0$ depends on the path which the limiting process takes. The limit value $s = 0.5$ is gained from making an elliptical or parabolic notch narrower and narrower whereas $s = 1.0$ results from making a keyhole smaller and smaller.

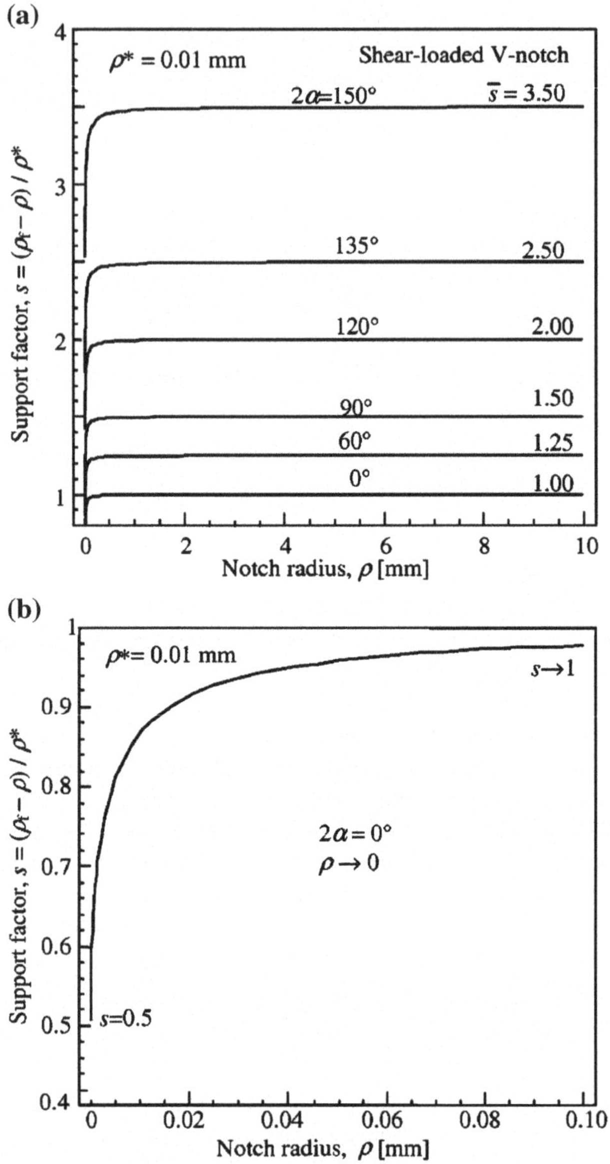

Fig. 1.17 Support factor s as function of the real notch radius ρ with plateau values $\bar{s}$ for different notch opening angles 2α and $\rho^* = 0.01$ mm (**a**); out-of-plane shear-loaded V-notch; plot (**b**) refers to the limit value process $\rho \to 0$ for the parabolic notch ($2\alpha = 0°$); (Berto et al. 2009)

1.3.4 General Solution for Sharp V-Notches under Out-of-Plane Shear Loading

The sharp V-notch with arbitrary notch opening angle under out-of-plane shear loading, inclusive of the pointed V-notch, is considered (Berto et al. 2009). A closed-form solution is given by Neuber (1958, pp. 166–169). According to this solution, the averaged shear stress $\overline{\tau}^*$ over the distance ρ^* is:

$$\begin{aligned}\overline{\tau}^* &= \frac{1}{\rho^*}\int\limits_{x_0}^{x_0+\rho^*}\frac{1}{q}x^{\frac{1-q}{q}}\mathrm{d}x = \frac{1}{\rho^*}\left[(x_0+\rho^*)^{\frac{1}{q}} - (x_0)^{\frac{1}{q}}\right] \\ &= \frac{1}{\rho^*}\left[\left(\frac{q-1}{q}\rho+\rho^*\right)^{\frac{1}{q}} - \left(\frac{q-1}{q}\rho\right)^{\frac{1}{q}}\right]\end{aligned} \tag{1.44}$$

with the parameter q linked to the notch opening angle 2α (e.g. $q = 2.0$ for $2\alpha = 0°$, parabolic notch).

In the limit case $\rho^* \to 0$ for $\rho = \rho_\mathrm{f}$, Eq. (1.44) has the following form (Neuber 1958, p. 168):

$$(\overline{\tau}^*)_{\rho^*\to 0} = \frac{1}{q}\left(\frac{q-1}{q}\rho_\mathrm{f}\right)^{\frac{1-q}{q}} \tag{1.45}$$

By equating $\overline{\tau}^*$ from Eq. (1.44) and (1.45), Neuber obtained the following relationship for ρ_f:

$$\rho_\mathrm{f}(\rho,\rho^*) = \frac{q}{q-1}\left\{\frac{q}{\rho^*}\left[\left(\frac{q-1}{q}\rho+\rho^*\right)^{\frac{1}{q}} - \left(\frac{q-1}{q}\rho\right)^{\frac{1}{q}}\right]\right\}^{\frac{q}{1-q}} \tag{1.46}$$

The support factor s is gained from Eq. (1.24) inserting ρ_f according to Eq. (1.46).

The factor s dependent on notch radius ρ for different notch opening angles 2α is plotted in Fig. 1.17a with a detail in Fig. 1.17b. The differences in the values s mentioned in the paragraph preceding Eq. (1.34) can be explained by the curve plot for $2\alpha = 0°$. The intersection with the axis $\rho = 0$ really gives $s = 0.5$, independently of ρ^*, whereas $s = 1.0$ is the plateau value, which is approximated more or less rapidly (depending on ρ^*) for $\rho > 0$. The evaluation of s for $\rho^* \to 0$ is numerically stable in contrast to the corresponding evaluation in the case of tensile loading. As shown earlier by evaluation of Eq. (1.43), the limit value of s for $\rho \to 0$ is 1.0 when considering a keyhole. This means that the limit value is dependent on the notch shape.

Summarising the numerical results, the following trends have been found for the support factor in the case of out-of-plane shear loading:

- the factor s depends on the real radius ρ, starting with lower values for $\rho = 0$, but converging to a plateau value for $\rho > \rho^*$;

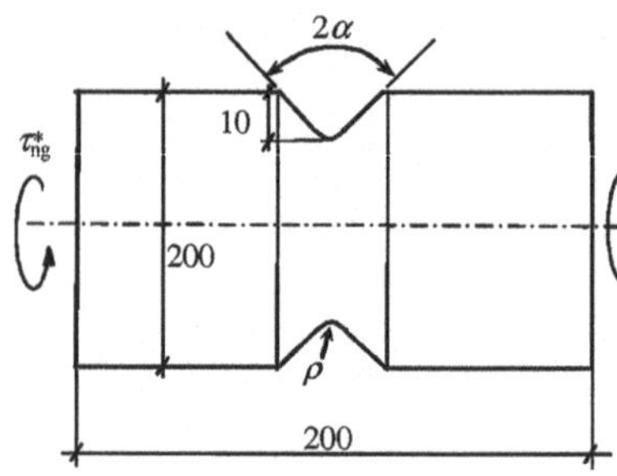

Fig. 1.18 Geometry and dimensions of the V-notch specimen considered in the FE analyses, torsional loaded round bar with circumferential V-notch, $\rho = 0$ compared with $\rho = \rho_f$; the nominal stresses τ^*_{ng} relating to the gross cross-sectional area; (Berto et al. 2009)

- the factor s depends only slightly on ρ^* for $\rho \geq \rho^*$, contrary to the range $\rho < \rho^*$ where the non-converged factor s may depend heavily on ρ^*;
- the factor s determined for pointed V-notches ($\rho = 0$) is smaller than the corresponding plateau value (Neuber 1958, ibid. Fig. 96b);
- the factor s does not depend on the failure criterion;
- the factor s depends on the notch opening angle 2α, as originally stated by Neuber, but later on deleted.

1.3.5 Parametric Investigation of Out-of-Plane Shear-Loaded V-Notch Specimens

The parametric investigation for the out-of-plane shear-loaded V-notch specimen is conducted on the same line as shown in Sect. 1.2.6 for the tensile loaded V-notch specimen (Berto et al. 2009). The specimen considered is a torsional loaded round bar with circumferential V-notch, Fig. 1.18. The averaged stress $\bar{\tau}^*$ at the pointed notch is compared with the maximum stress τ^*_{max} at the fictitiously rounded notch, using the plateau values of s. Both, the pointed and rounded notch specimens are analysed by the FE method.

The effective stress concentration factor $\overline{K}_t$ for a sharp, rounded notch subjected to out-of-plane shear loading is defined as follows:

$$\overline{K}_t = \frac{\bar{\tau}^*}{\tau^*_{ng}} = \frac{1}{\rho^* \tau_{ng}} \int_{x_0}^{x_0+\rho^*} \tau^*_{th} \, dx \qquad (1.47a)$$

where $\bar{\tau}^*$ is the effective (averaged) notch shear stress and τ^*_{ng} the applied nominal shear stress acting in the gross cross-section area.

Table 1.5 Fictitious notch rounding of round bar specimen with pointed V-notch ($\rho = 0$) subjected to torsional loading using the plateau values $\bar{s}$; comparison of $K_t(\rho_f)$ with $\overline{K}_t$, both obtained from FE analysis (relative deviation Δ); variation of notch opening angle 2α and support length ρ^*; the dimension of ρ_f is [mm]; (Berto et al. 2009)

		$\rho^* = 0.05$ mm				$\rho^* = 0.1$ mm				$\rho^* = 0.5$ mm			
2α	$\bar{s}$	ρ_f	$K_t(\rho_f)$	$\overline{K}_t$	Δ %	ρ_f	$K_t(\rho_f)$	$\overline{K}_t$	Δ %	ρ_f	$K_t(\rho_f)$	$\overline{K}_t$	Δ %
0°	1.0	0.050	21.08	22.00	−4.18	0.10	15.0	15.59	−3.78	0.50	6.96	7.01	−0.71
90°	1.5	0.075	10.27	10.14	1.28	0.15	8.18	8.04	1.74	0.75	4.83	4.69	2.99
135°	2.5	0.125	5.01	4.85	3.30	0.25	4.36	4.22	3.32	1.25	3.17	3.06	3.59

The stress concentration factor $K_t(\rho_f)$ with $\rho_f = s\rho^*$ (for $\rho = 0$) assumes the form:

$$K_t(\rho_f) = \frac{\tau^*_{max}(\rho^*, s)}{\tau^*_{ng}} \tag{1.47b}$$

where the maximum shear stress τ^*_{max} at the fictitiously rounded notch root should be equal to $\bar{\tau}^*$. Thus, the stress concentration factor $K_t(\rho_f)$ depends on the material-dependent support length ρ^* and the support factor s. In case of a failure criterion other than the shear stress criterion, $\overline{K}_t$ and $K_t(\rho_f)$ are changed by an identical factor.

The parametric FE analysis is performed for the notch opening angle $2\alpha = 0$, 90 and 135°, in combination with the microstructural support length $\rho^* = 0.05$, 0.1 and 0.5 mm. The plateau values $\bar{s}$ are adopted from the theoretical V-notch analysis described above. The results of the FE analysis are summarised in Table 1.5. They are discussed based on the stated Δ values:

- All Δ values remain within the ±5 % range; this is acceptable for engineering applications.
- On an average, the Δ values in torsional loading are lower than those in tensile loading.
- The Δ values are positive, and therefore conservative in strength assessments, for $2\alpha = 90$ and 135°; they slightly increase for increasing values of ρ^*.
- The Δ values are negative only for $2\alpha = 0°$, with Δ decreasing as ρ^* increases.
- The Δ values change continuously with ρ^* and 2α, respectively, showing a maximum absolute value of 4.18 % for $2\alpha = 0°$ and $\rho^* = 0.05$ mm.

1.3.6 Solution for Out-of-Plane Shear-Loaded V-Notches with Root Hole

A further solution for the fictitious notch rounding problem referring to out-of-plane shear loading of V-notches with root hole is available from a publication (Berto 2012). The basic procedure is illustrated by Fig. 1.19, where the real notch with shear stress $\bar{\tau}$ averaged over ρ^* is contrasted with the fictitiously rounded notch with maximum shear stress $\tau^*_{max} = \bar{\tau}^*$.

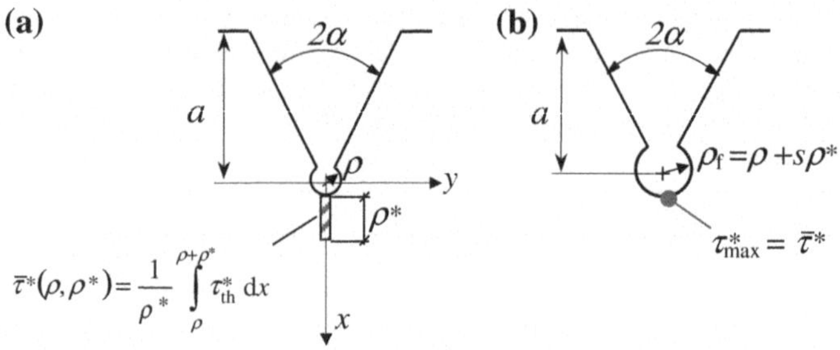

Fig. 1.19 Fictitious notch rounding concept applied to out-of-plane shear loaded V-notches with root hole; real notch with stress $\bar{\tau}^*$ averaged over ρ^* (**a**) and substitute notch with fictitious notch radius ρ_f producing $\tau_{max} = \bar{\tau}^*$ (**b**); (Berto 2012)

Table 1.6 Microstructural support factor s for different notch opening angles; out-of-plane shear loaded V-notch with root hole: limit value s for $\rho = 0$ and plateau value $\bar{s}$ for $\rho = 10$ mm, $\rho^* = 0.1$ mm; in comparison blunt V-notch; (Berto 2012)

		Root hole V-notch		Blunt V-notch
2α	λ_3	$s(\rho = 0)$	$\bar{s}(\rho = 10$ mm)	$\bar{s}(\rho = 10$ mm)
0°	0.500	1.00	1.00	1.00
60°	0.600	1.58	1.25	1.20
90°	0.667	2.37	1.50	1.50
120°	0.750	5.06	2.00	2.00
135°	0.800	10.49	2.52	2.50

For determining the support factor s, Eq. (1.34) for $\bar{\tau}^*$ is evaluated based on the stress field equations for the V-notch with root hole subjected to out-of-plane shear loading (Zappalorto and Lazzarin 2011, reviewed in Sect. 2.4.6). The averaged shear stress $\bar{\tau}^*(\rho, \rho^*)$ at the real notch has to be equated to its limit value $\bar{\tau}^*(\rho_f, \rho^* \to 0) = \tau^*_{max}(\rho_f)$ at the fictitiously rounded notch. The resulting expression for $\rho_f(\rho, \rho^*)$ reads:

$$\rho_f = 2^{\frac{1}{1-\lambda_3}} \left\{ \frac{(\rho + \rho^*)^{\lambda_3}}{\rho^* \lambda_3} \left[1 - \left(\frac{\rho}{\rho + \rho^*} \right)^{2\lambda_3} \right] \right\}^{\frac{1}{\lambda_3 - 1}} \tag{1.48a}$$

For $2\alpha = 0$, $\rho \neq 0$ ($\lambda_3 = 0.5$), Eq. (1.48a) results in:

$$\rho_f = \rho + \rho^* \tag{1.48b}$$

For $\rho = 0$ and arbitrary values of 2α, Eq. (1.48a) is simplified:

$$\rho_f = (2\lambda_3)^{\frac{1}{1-\lambda_3}} \rho^* \tag{1.48c}$$

The values of s for $\rho = 0$ and $\bar{s}$ for $\rho = 10$ mm (plateau values) are listed in Table 1.6 for different notch opening angles 2α and compared with the relevant $\bar{s}$

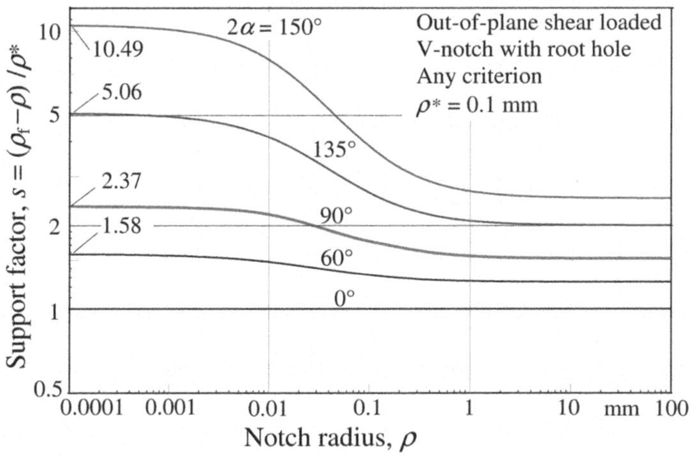

Fig. 1.20 Support factor s as function of the real notch radius ρ for different notch opening angles 2α for $\rho^* = 0.1$ mm; out-of-plane shear-loaded V-notch with root hole; (Berto 2012, unpublished)

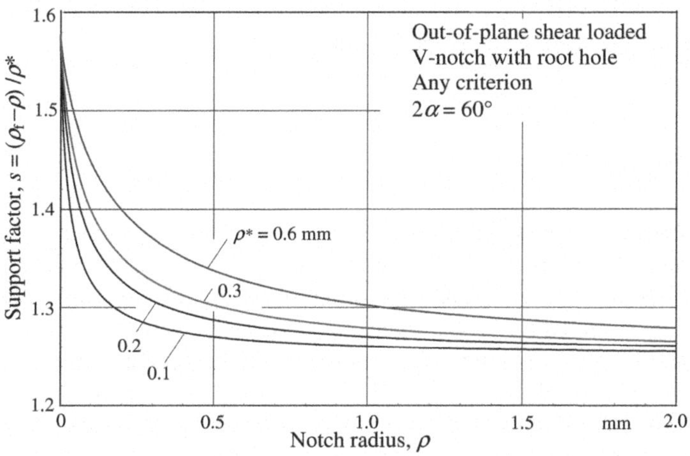

Fig. 1.21 Support factor s as function of the real notch radius ρ for different values of ρ^* for $2\alpha = 60°$; out-of-plane shear-loaded V-notch with root hole; (Berto 2012, unpublished)

values for blunt V-notches from Table 1.3. The factors s and $\bar{s}$ rise with the notch opening angle. The plateau values for root hole notch and blunt notch are identical. The limit values of s for $\rho = 0$ for the root hole notch are larger than for the blunt notch.

The support factor s for the out-of-plane shear-loaded V-notch with root hole is plotted as function of the real notch radius ρ for different notch opening angles 2α,

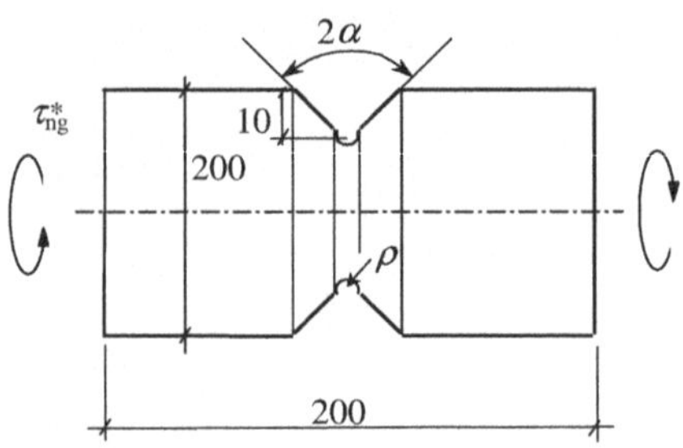

Fig. 1.22 Geometry and dimensions of the V-notch with root hole specimen considered in the FE analyses; torsional loaded round bar with circumferential V-notch, $\rho = 0$ compared with $\rho = \rho_f$; the nominal stresses τ^*_{ng} relating to the gross cross-sectional area; (Berto 2012)

Fig. 1.20, and for different values of ρ^*, Fig. 1.21. The dependency of s on ρ and ρ^* (besides of 2α) is strong for small values of ρ, but weak for the plateau values.

A validation of the method has been performed based on FE analyses of the torsional loaded round bar specimen with circumferential V-notch with root hole, $\rho = 0$ compared with $\rho = \rho_f$, Fig. 1.22. The result of the validation is presented in Table 1.7 in the form already known from Table 1.2. The relative deviations Δ are sufficiently small, at least for $\rho^* = 0.1$ mm.

1.3.7 Conclusions

Based on Neuber's stress field solution for V-notches subjected to out-of-plane shear loading, the microstructural support factor s is evaluated and plotted as a function of the real notch radius ρ for different notch opening angles 2α. Plateau values of s are revealed for larger values of ρ. These plateau values of s depend on the considered notch opening angle, but not on the applied multiaxial failure criterion. For $\rho < \rho^*$ the factor s may deviate from the plateau value and then depends on ρ. Additionally, the s value may depend on the ratio ρ^*/t and on the appropriateness of the applied stress concentration factor formula or its substitute, the FE analysis.

As demonstrated by the parametric numerical investigation of pointed V-notch round bar specimens ($\rho = 0$) subjected to torsional loading, the plateau values of s are well suited for engineering usage in structural strength assessments. These s values are ready for incorporation in those codes and design recommendations for structural components under static or cyclic loading conditions (brittle fracture or fatigue failure) which comprise local approaches such as the notch stress or notch strain approach.

Table 1.7 Fictitious notch rounding with root hole applied to pointed V-notch round bar specimen ($\rho = 0$) subjected to torsional loading using the relevant values of s; comparison of $K_t(\rho_f)$ with $\overline{K}_t$, both obtained from FE analysis (relative deviation Δ); different notch opening angles 2α, two values of ρ^*; (Berto 2012)

		$\rho^* = 0.1$ mm				$\rho^* = 0.3$ mm			
2α	$\overline{s}$	ρ_f	$K_t(\rho_f)$	$\overline{K}_t$	Δ %	ρ_f	$K_t(\rho_f)$	$\overline{K}_t$	Δ %
0°	1.00	0.100	15.71	15.59	0.76	0.300	9.29	9.03	2.80
60°	1.58	0.158	10.72	10.68	0.37	0.474	6.98	6.87	1.60
90°	2.37	0.237	8.07	8.01	0.75	0.711	5.66	5.57	1.62
120°	5.06	0.506	5.48	5.44	0.74	1.518	4.24	4.13	2.66
135°	10.49	1.049	4.28	4.21	1.66	3.147	3.54	3.39	4.42

1.4 Fictitious Notch Rounding for In-Plane Shear Loading

1.4.1 Survey of Section Contents

This section deviates in its structuring of contents from the preceding two sections, because the notch stress situation is more complex for in-plane shear loading than for in-plane tensile and out-of-plane shear loading. Elementary solutions and evaluations make up the main body of this section. More general theories are also available and included.

Sections 1.4.3–1.4.9 refer to in-bisector crack propagation because of its relevance for engineers, because simple mixed mode loading formulae are possible and because this approach is matching with Neuber's original concept. The presented relationships between maximum notch stress and stress intensity factor and also for the support factors s are restricted to slits or cracks ($2\alpha = 0°$) as limit cases of elliptical, parabolic or hyperbolic notches as well as of keyhole notches and blunt cracks. Sections 1.4.10 and 1.4.11 refer to out-of-bisector crack propagation at blunt V-notches and root hole V-notches subjected to pure in-plane shear loading based on analytical and numerical solutions. Plateau value curves for the support factor s are derived.

The expositions are primarily based on the analytical developments and their numerical evaluation and confirmation carried out by Berto, Lazzarin and Radaj (Berto and Lazzarin 2010; Berto et al. 2011; Radaj and Zhang 1993$^{(2)}$). A list of symbols and a of list of references are available at the end of Chap. 1.

1.4.2 Pecularities with In-Plane Shear Loading

This Section 1.4 on in-plane shear loading (i.e. mode 2) concludes the theoretical development of Neuber's fictitious notch rounding concept with reference to V-notches of different notch opening angles inclusive of cracks or slits.

Section 1.2 is devoted to in-plane tensile loading (i.e. mode 1) and Section 1.3 to out-of-plane shear loading (i.e. mode 3). These two loading modes prevail at structural notches, slits or cracks and must primarily be taken into account when assessing the structural strength or life. Analytical solutions for the elastic V-notch problem in the two loading cases mentioned above are available and can be used to derive accurate values of the microstructural support factor s, which evolves as mainly dependent on the loading mode (1 or 3), the notch opening angle 2α and, in case of mode 1, the multiaxial failure criterion (normal stress, von Mises or Beltrami). The values of s are more or less independent of the real notch radius ρ (plateau values of s) for $\rho > \rho^*$ (with the microstructural length ρ^*). It is shown by a parametric investigation related to pointed V-notches using the FE method that the plateau values of s give sufficiently accurate results even for pointed notches ($\rho = 0$), provided the jump in stress multiaxiality between pointed and rounded notches is taken into account (jump factor κ).

In engineering applications, the loading modes 1 and 3 are predominant in general whereas in-plane shear loading (mode 2) is an exceptional case or a secondary effect. Predominant in-plane shear loading occurs in tensile-shear loaded lap joints. The mentioned secondary effect constitutes itself under certain mixed mode loading conditions.

Dealing with in-plane shear loading is complicated by the fact that two potential modes of crack propagation have to be taken into account. The case considered in the original Neuber concept is the ligament or bisector direction, i.e. coplanar crack propagation. On the other hand, nonplanar, i.e. branching or kinking crack propagation may occur, eventually connected with lower basic stresses than coplanar crack propagation. The crack propagation stabilises into a direction where pure mode 1 loading is acting. This direction is generally approximated by the Erdogan–Sih maximum tangential stress criterion (Erdogan and Sih 1963). Both cracking directions occur in structural components with slits, cracks or sharp V-notches (pointed or rounded).

The same dual cracking behaviour is observed under out-of-plane shear loading (mode 3). Here the macro-appearance of the crack is coplanar in general whereas the micro-appearance reveals the nonplanar cracking steps. Contrary to mode 3 loading with a three-dimensional cracking topography, the nonplanar crack propagation under mode 2 loading is easier to be treated analytically.

In structural components, the coplanar or in-bisector crack propagation behaviour under mode 2 loading is prevailing under one of the following conditions:

- Circumferential notches force the crack into the bisector plane because crack coalescence and cross-sectional weakening by the cracks occur mainly in the bisector plane.
- Mixed mode loading with higher contents of mode 1 enforces coplanar crack propagation.
- Mode 2 loading with static preloads in mode 1 causes coplanar crack propagation.

- Experimental research shows that coplanar crack propagation may occur under pure mode 2 loading with propagation rates substantially larger than in mode 1 loading.

1.4.3 Preliminaries to the Elementary Theory

It is a well known fact that the stress intensity factor K_{I} of cracks under tensile loading transverse to the crack (mode 1) can be determined by considering the maximum notch stress $\sigma_{\max}$ of the relevant notch problem in which the notch radius is decreasing to zero (Irwin 1957):

$$K_{\mathrm{I}} = \lim_{\rho \to 0} \frac{1}{2} \sigma_{\max} \sqrt{\pi \rho} \quad (K_{\mathrm{II}} = K_{\mathrm{III}} = 0) \tag{1.49}$$

Similar relationships have been derived for in-plane and out-of-plane shear loading (mode 2 and mode 3):

$$K_{\mathrm{II}} = \lim_{\rho \to 0} \sigma'_{\max} \sqrt{\pi \rho} \quad (K_{\mathrm{I}} = K_{\mathrm{III}} = 0) \tag{1.50}$$

$$K_{\mathrm{III}} = \lim_{\rho \to 0} \tau^{*}_{\max} \sqrt{\pi \rho} \quad (K_{\mathrm{II}} = K_{\mathrm{I}} = 0) \tag{1.51}$$

The maximum stress $\sigma_{\max}$ is connected with the symmetric notch stress distribution in mode 1 loading and the maximum stress $\sigma'_{\max}$ with the antimetric notch stress distribution in mode 2 loading. The Eq. (1.50) for mode 2 loading has the following abnormal characteristics:

- A tensile notch stress is correlated with a shear stress intensity.
- This tensile stress does not occur in the apex point of the notch but at some (small) distance from it at the curved notch edge.
- This tensile stress, which acts in the direction of the notch edge, is inclined against the crack and ligament direction.

The abnormal characteristics result in a problem when the fictitious notch radius ρ_{f} introduced by Neuber is considered, which takes the microstructural support effect of sharp notches into account. There are no values of s available for in-plane shear loading, because Neuber has not considered this loading case within his concept.

There is further indication that the in-plane shear loading case is occasionally considered abnormal. Hasebe et al., for example, give general relationships between notch stress concentration and stress intensity for V-notches (inclusive of the crack) with free and rigidly fixed edges comprising the symmetric and antimetric stress states with the exception of the antimetrically loaded (i.e. in-plane shear loaded) free edge notch (Hasebe et al. 1990).

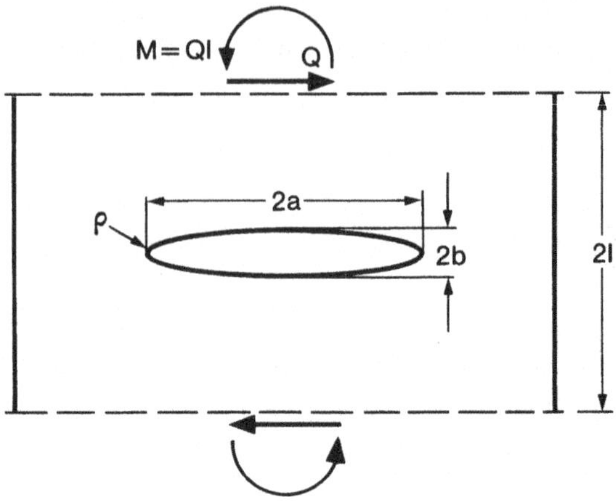

Fig. 1.23 Shear-bending loading of infinite strip with elliptical hole; (Neuber 1958)

The abnormal situation associated with the crack subjected to in-plane shear loading has been investigated in order to gain a better understanding of the limit value procedures and to provide an answer to the question, which fictitious radius should be used within the Neuber approach to the microstructural support effect. The notch stress situation is considered hereafter in the order of elliptical notches, parabolic notches and keyhole notches. The V-notches are dealt with further below.

1.4.4 Elliptical Notch Subjected to In-Plane Shear Loading

Surprisingly enough, no solution for the elliptical hole in the infinite plate subjected to pure in-plane shear loading was originally available from the literature. Neuber had solved the more complicated case of shear-bending loading of an infinite plate strip with an elliptical hole, Fig. 1.23, using the 'three functions approach'. There is a pure shear force acting in the midsection of plate strip and ellipse, but the shear force is superimposed by a bending moment in the cross-sections beside the midsection where σ'_{max} occurs.

The solution for the elliptical hole in the infinite plate subjected to pure in-plane shear loading, Fig. 1.24, has been developed on the basis of the complex stress function method (Radaj and Zhang 1993[(2)]). The solution for inclined tension loading can be found in the literature (Muskhelishvili 1963, ibid. pp. 343–345). Pure shear loading is derived by superimposition of plus 45° inclined tension and minus 45° inclined compression. The resulting stress functions for the elliptical hole under pure in-plane shear loading are the following:

$$\varphi(\zeta) = \frac{i\tau_{\mathrm{n}}R}{\zeta} \tag{1.52}$$

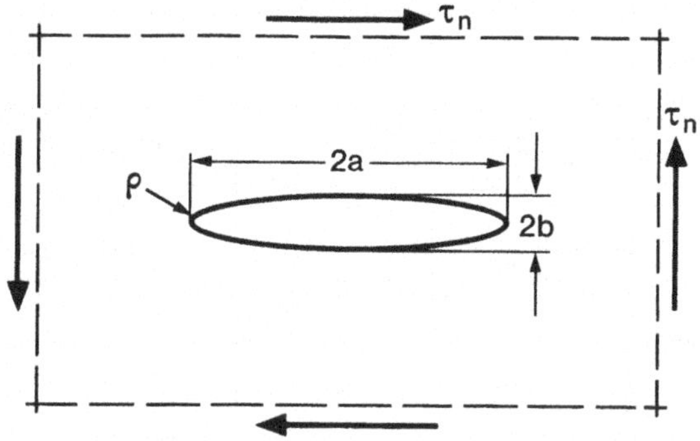

Fig. 1.24 Pure shear loading of infinite plate with elliptical hole; (Muskhelishvili 1963)

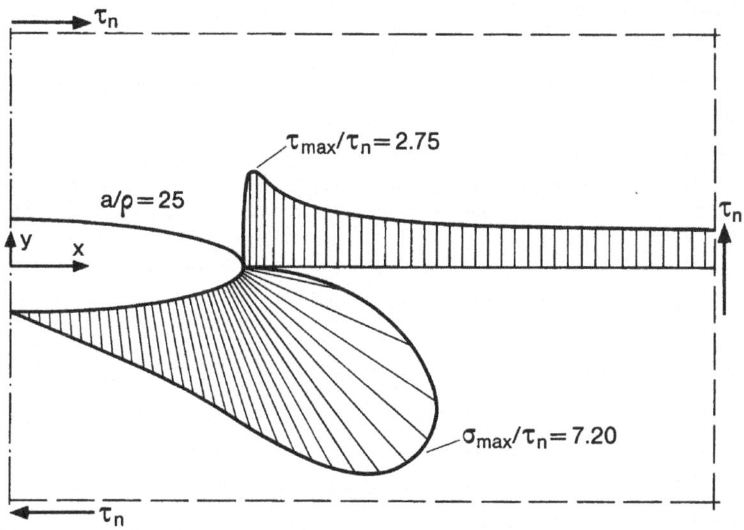

Fig. 1.25 Notch stresses at elliptical hole in in-plane shear-loaded plate (symmetry half); tangential stress at notch edge and shear stress along ligament; (Radaj and Zhang 1993[(2)])

$$\psi(\zeta) = -i\tau_n R\left(-\zeta + \frac{1}{m\zeta} - \frac{1+m^2}{m}\frac{\zeta}{\zeta^2 - m}\right) \tag{1.53}$$

with remote shear stress τ_n in infinity and with $R > 0$, $0 < m < 1$ designating size and diameter ratio a/b of the elliptical hole.

The evaluation of the two notch stresses which are of relevance to the corresponding crack problem under mode 2 loading, i.e. the maximum tangential stress σ'_{max} at the boundary of the hole and the maximum shear stress τ_{max} in the bisector plane (the ligament in the crack problem), had the following result, Fig. 1.25:

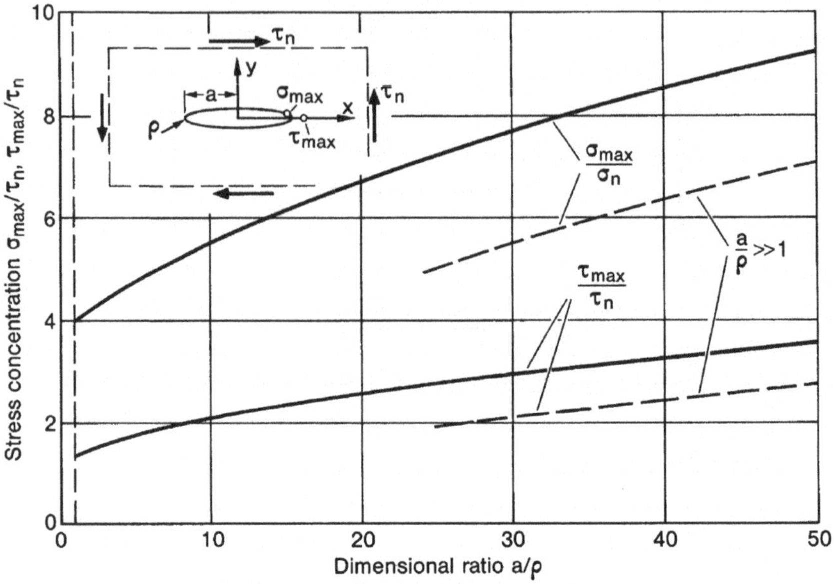

Fig. 1.26 Stress concentration dependent on notch curvature of elliptical hole in in-plane shear and shear-bending loaded infinite plate; sharp notch approximation ($a/\rho \gg 1$) for comparison; (Radaj and Zhang 1993[(2)])

$$\frac{\sigma'_{\text{max}}}{\tau_{\text{n}}} = 2 + \sqrt{\frac{a}{\rho}} + \sqrt{\frac{\rho}{a}} \tag{1.54}$$

$$\frac{\tau_{\text{max}}}{\tau_{\text{n}}} = \frac{2\sqrt{(a/3\rho + 2/3)^3} - 2\sqrt{a/\rho}}{\left(\sqrt{a/\rho} - 1\right)^2} \tag{1.55}$$

Whereas the expressions for the stress concentration are the same in the solution above for pure shear loading, Fig. 1.26, and in Neuber's solution for shear-bending loading (the factor 1.5 occurs if the reference stress is the averaged value of a parabolic distribution, as in Neuber's solution), the positions of σ'_{max} are slightly different (but not those of τ_{max}), Fig. 1.27. The different positions of σ'_{max} express the influence of superimposed bending.

The Eqs. (1.54) and (1.55) are simplified for $a/\rho \gg 1$, i.e. for the elliptical hole approximating a slit or crack (broken line curves in Fig. 1.26 deviating substantially from solid line curves, because $a/\rho \gg 1$ is met only insufficiently):

$$\frac{\sigma'_{\text{max}}}{\tau_{\text{n}}} = \sqrt{\frac{a}{\rho}} \tag{1.56}$$

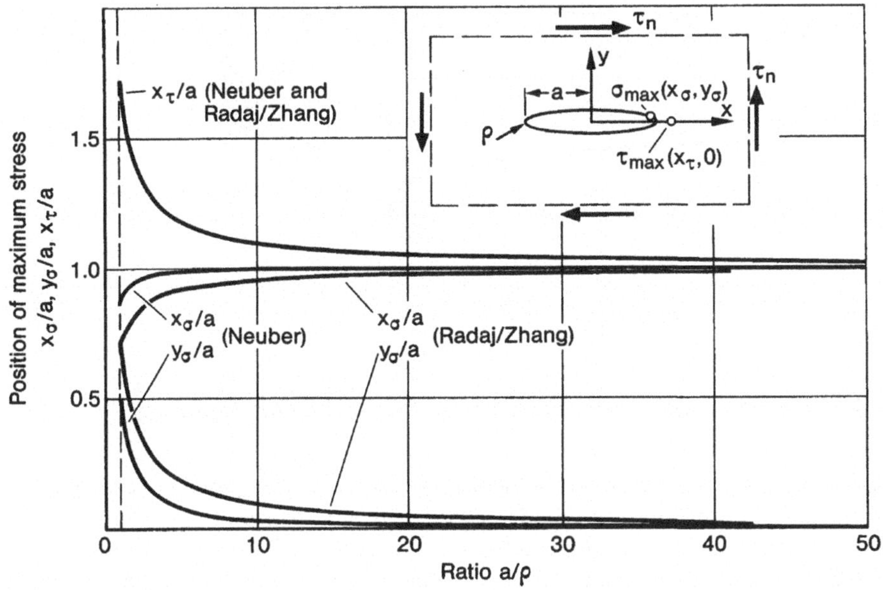

Fig. 1.27 Position of maximum stress $\sigma_{\max}$ and $\tau_{\max}$ at elliptical hole in in-plane shear and shear-bending loaded infinite plate; (Radaj and Zhang 1993[(2)])

$$\frac{\tau_{\max}}{\tau_{\mathrm{n}}} = \frac{2}{3\sqrt{3}}\sqrt{\frac{a}{\rho}} \tag{1.57}$$

The ratio of stress concentration $\sigma'_{\max}/\tau_{\max}$ converges rapidly to the value $1.5\sqrt{3} = 2.598$ having started with 3.0 for the circular hole ($a/\rho = 1.0$).

The maximum stress $\sigma'_{\max}$ occurs close to the apex point of the slender ellipse (after having been in the diagonal section of the circular hole in pure shear loading). It can be shown for pure in-plane shear loading that the inclination angle of $\sigma'_{\max}$ relative to the horizontal axis remains $\pi/4$ independent of the ratio a/ρ.

In order to assess sharp notches of circular root shape, the radius of curvature ρ' at the position of $\sigma'_{\max}$ was evaluated (because the stress concentration is supposed to be more dependent on the radius of curvature at the point where it acts than at the nearby apex point). Introducing the expression for $a/\rho \gg 1$, $\rho' = 2\rho\sqrt{2}$ or $\rho = \sqrt{2}\rho'/4$, into Eq. (1.56), the result is:

$$\frac{\sigma'_{\max}}{\tau_{\mathrm{n}}} = \frac{2}{\sqrt[4]{2}}\sqrt{\frac{a}{\rho'}} = 1.682\sqrt{\frac{a}{\rho'}} \tag{1.58}$$

It will be shown in Sect. 1.4.7 that Eq. (1.58) is inadequate.

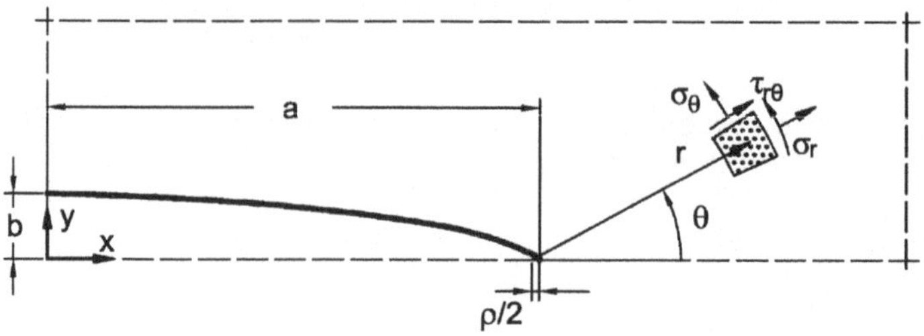

Fig. 1.28 Local polar r–θ coordinate system for notch stress analysis at blunt crack according to Creager; (Radaj and Zhang 1993[(2)])

1.4.5 Blunt Crack Subjected to In-Plane Shear Loading

The notch root and crack tip subjected to in-plane shear loading can also be investigated proceeding from the Creager stress field equations for 'blunt cracks' (Creager and Paris 1967). The condition $r/a \ll 1$ is used after introducing a local $r-\theta$ polar coordinate system at the apex point of an elliptical notch with radius of curvature ρ, Fig. 1.28. A power series expansion of the stress field is performed, neglecting the higher order terms. The resulting equations for in-plane shear loading (mode 2) read as follows:

$$\begin{Bmatrix} \sigma_x \\ \sigma_y \\ \tau_{xy} \end{Bmatrix} = \frac{K_{\mathrm{II}}}{\sqrt{2\pi r}} \begin{Bmatrix} \sin\theta/2[2+\cos\theta/2\cos 3\theta/2] \\ \sin\theta/2\cos\theta/2\cos 3\theta/2 \\ \cos\theta/2[1-\sin\theta/2\sin 3\theta/2] \end{Bmatrix} + \frac{K_{\mathrm{II}}}{\sqrt{2\pi r}}\frac{\rho}{2r} \begin{Bmatrix} \sin 3\theta/2 \\ -\sin 3\theta/2 \\ -\cos 3\theta/2 \end{Bmatrix} \tag{1.59}$$

$$\begin{Bmatrix} \sigma_r \\ \sigma_\theta \\ \tau_{r\theta} \end{Bmatrix} = \frac{K_{\mathrm{II}}}{4\sqrt{2\pi r}} \begin{Bmatrix} -5\sin\theta/2+3\sin 3\theta/2 \\ -3\sin\theta/2-3\sin 3\theta/2 \\ \cos\theta/2+3\cos 3\theta/2 \end{Bmatrix} + \frac{K_{\mathrm{II}}}{\sqrt{2\pi r}}\frac{\rho}{2r} \begin{Bmatrix} -\sin\theta/2 \\ \sin\theta/2 \\ -\cos\theta/2 \end{Bmatrix} \tag{1.60}$$

The shear stress in the bisector plane ($\theta = 0°$) follows from Eq. (1.60):

$$\tau_{r\theta} = \frac{K_{\mathrm{II}}}{\sqrt{2\pi r}}\left(1 - \frac{\rho}{2r}\right) \tag{1.61}$$

Inserting $r = 1.5\rho$ at the position of $\tau_{\max}$ according to the Neuber solution for the parabolic notch, the following relationship is gained:

$$\tau_{\max} = 0.3849\frac{K_{\mathrm{II}}}{\sqrt{\pi\rho}} \quad (r = 1.5\rho) \tag{1.62}$$

Using the ratio $\sigma'_{\max}/\tau_{\max} = 2.5981$ from the Neuber solutions for the parabolic, elliptical and hyperbolic notch, one gets:

$$\sigma'_{\max} = 2.5981\tau_{\max} = \frac{K_{\mathrm{II}}}{\sqrt{\pi\rho}} \tag{1.63}$$

in agreement with Eq. (1.50). The stress $\sigma'_{\max}$ in terms of $K_{\mathrm{II}}/(\pi\rho)^{1/2}$ is also gained from evaluating the maximum tangential stress $\sigma_{\mathrm{t\,max}}$ at ($r = \rho$, $\theta = 90°$) from σ_r, σ_θ and $\tau_{r\theta}$ according to Eq. (1.60). It can be shown (Lazzarin and Tovo 1996) that the Creager blunt crack equations above may also be interpreted as solutions for the parabolic notch.

The relationships between K_{II} and $\tau_{\max}$ or $\sigma'_{\max}$ are expressed by the stress concentration factors K_τ and $K_{\sigma'}$ with introduction of the geometry factors κ_τ and κ_σ, which depend on the notch geometry and the definition of the nominal stress. The in-plane shear loaded infinite plate with an internal crack or notch of semilength a is considered:

$$K_{\mathrm{II}} = \tau_{\mathrm{n}}\sqrt{\pi a} = \kappa_\tau \tau_{\max}\sqrt{\pi\rho} \quad (\rho \ll a) \tag{1.64}$$

$$K_{\mathrm{II}} = \tau_{\mathrm{n}}\sqrt{\pi a} = \kappa_{\sigma'}\sigma'_{\max}\sqrt{\pi\rho} \quad (\rho \ll a) \tag{1.65}$$

The stress concentration factors are defined as follows:

$$K_\tau = \frac{\tau_{\max}}{\tau_{\mathrm{n}}} \tag{1.66}$$

$$K_{\sigma'} = \frac{\sigma'_{\max}}{\tau_{\mathrm{n}}} \tag{1.67}$$

Then one gets the geometry factors from Eqs. (1.64) and (1.65):

$$\kappa_\tau = \frac{1}{K_\tau}\sqrt{\frac{a}{\rho}} \tag{1.68}$$

$$\kappa_{\sigma'} = \frac{1}{K_{\sigma'}}\sqrt{\frac{a}{\rho}} \tag{1.69}$$

The geometry factor Y in fracture mechanics, which describes the effect of nearby boundaries, has to be supplemented to $\tau_{\mathrm{n}}(\pi a)^{1/2}$ in Eqs. (1.64) and (1.65), when such conditions occur whereas this effect is already contained in the evaluated maximum stresses.

The Neuber solution for the elliptical notch with $\rho/a \ll 1$ in the in-plane shear-loaded infinite plate gives:

$$K_\tau = 0.3849\sqrt{\frac{a}{\rho}}, \quad \kappa_\tau = 2.5981 \tag{1.70}$$

$$K_{\sigma'} = \sqrt{\frac{a}{\rho}}, \quad \kappa_{\sigma'} = 1.0 \tag{1.71}$$

$$\frac{K_{\sigma'}}{K_\tau} = 2.5981 \tag{1.72}$$

Identical results are gained from the Creager solution with parabolic interpretation, but with K_{II} related to the internal elliptical notch. In this interpretation, σ'_{max} is determined on a parabolic path describing the notch geometry. Using a circular path, as in the Erdogan–Sih maximum tangential stress criterion for pointed slit tips, another relationship is derived for $\mathrm{K}_{\sigma'}$, evaluating σ_θ according to Eq. (1.60) for $r = \rho$ and $\theta = 70.56°$:

$$K_{\sigma'} = 0.6123\sqrt{\frac{a}{\rho}}, \qquad \kappa_{\sigma'} = 1.6332 \tag{1.73}$$

1.4.6 Parabolic and Hyperbolic Notch Subjected to In-Plane Shear Loading

The derivations above for the elliptical notch can also be performed for parabolic and hyperbolic notches based on the closed-form expressions for the stress concentration factors (Neuber 1958 or 1985). These notches are external notches as opposed to the elliptical notch, which is treated as an internal notch (in the infinite plate). Therefore, the nominal stress is now defined relating the shear force F to the net section width $2b$:

$$\tau_{\mathrm{n}} = \frac{F}{2b} \tag{1.74}$$

All parameters are referenced directly or indirectly to the parameter b instead of a. Thus their numerical values may not be compared without proper argumentation. It has to be noted that, in Neuber's solutions, the length a is indicated by t and the width b by a.

The following expressions for ($\rho \ll b$) are derived for the two-sided parabolic notch (Neuber 1985, p. 111):

$$K_\tau = 0.2722\sqrt{\frac{b}{\rho}}, \qquad \kappa_\tau = 3.6742 \tag{1.75}$$

$$K_{\sigma'} = 0.7071\sqrt{\frac{b}{\rho}}, \qquad \kappa_{\sigma'} = 1.4142 \quad (\theta = 90°) \tag{1.76}$$

$$\frac{K_{\sigma'}}{K_\tau} = 2.5981 \tag{1.77}$$

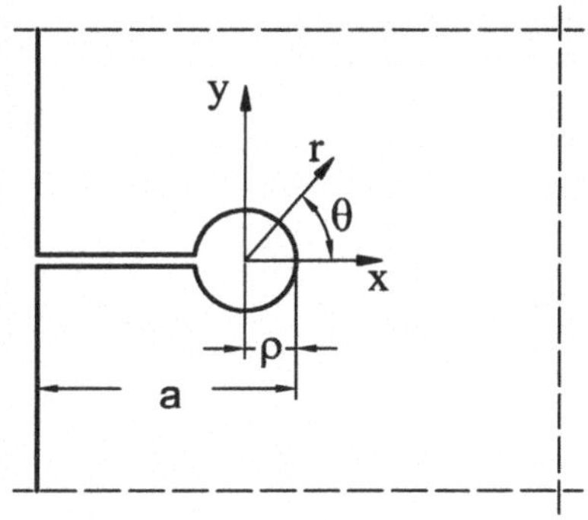

Fig. 1.29 Coordinate systems and geometrical parameters at external keyhole

Comparing the relationships for elliptical internal and parabolic external notches, theses are different by the parameters a and b and by the factor $2^{1/2}$ or $1/2^{1/2}$. Using gross cross-sectional nominal stresses τ_{ng} and the notch depth a for the parabolic external notches, the relationships are identical.

Slightly different expressions for ($\rho \ll b$) are derived for the two-sided hyperbolic notch (Neuber 1985, pp. 122–123):

$$K_\tau = 0.2450\sqrt{\frac{b}{\rho}}, \quad \kappa_\tau = 4.0816 \tag{1.78}$$

$$K_{\sigma'} = 0.6366\sqrt{\frac{b}{\rho}}, \quad \kappa_{\sigma'} = 1.5708 \tag{1.79}$$

$$\frac{K_{\sigma'}}{K_\tau} = 2.5981 \tag{1.80}$$

The relation $K_{\sigma'}/K_\tau$ is exactly the same for parabolic and hyperbolic notches, but the stress concentration and geometry factors are different by the factor 0.9.

1.4.7 Keyhole Notch Subjected to In-Plane Shear Loading

The stress field at the keyhole notch subjected to in-plane shear loading (Kullmer 1992; Radaj et al. 2001) has been derived using the Airy stress function method. The stress function is described by a six-term power series of the radial distance r combined with first and second order cyclic functions of the angle θ, Fig. 1.29. The boundary conditions enforce load-free edges of the hole ($r = \rho$) and of the slit ($\theta = \pm\pi$). The condition $\rho \ll a$ or $\rho \ll b$ with notch depth a or cross-sectional width b is introduced.

The boundary conditions cannot be completely satisfied simultaneously. There are two solutions available (Radaj et al. 2001). In one solution the stress σ_θ at the

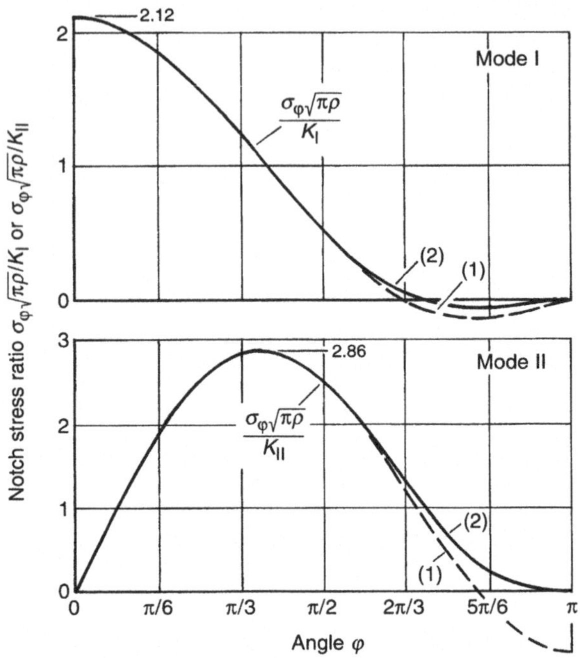

Fig. 1.30 Notch stress distribution at keyhole subjected to tensile loading (mode 1) and in-plane shear loading (mode 2), solution variants (1) and (2); polar angle $\varphi = \theta$; (Radaj et al. 2001)

slit edges is set to zero in the quadratic mean. In the other solution, the least square method is applied to the stress σ_r at the hole edge. The result of the analysis is shown in Fig. 1.30. The stress σ_θ at the hole edge according to the solution variants (1) and (2) is plotted over the polar angle θ for mode 1 and mode 2 loading. The maximum values are not dependent on the solution variant. In mode 2 loading, the necessary condition $\sigma_\theta = 0$ at $\theta = \pi$ (end point of the slit) is only met by the second solution variant.

The tangential stress on the boundary of the circular hole according to the first solution variant is given by the following expression:

$$\sigma_\theta = \frac{K_{II}}{\sqrt{\pi\rho}} \frac{1}{\sqrt{2}} \left(2\sin\frac{\theta}{2} + 3\sin\frac{3\theta}{2} \right) \quad (r = \rho) \tag{1.81}$$

The maximum tangential stress is resulting from the above expression:

$$\sigma_{\theta\,\max} = 2.8664 \frac{K_{II}}{\sqrt{\pi\rho}} \quad (r = \rho, \quad \theta = 67.115°) \tag{1.82}$$

The shear stress in the bisector plane of the keyhole according to the first solution variant is given by the following expression:

$$\tau_{r\theta} = \frac{K_{II}}{4\sqrt{2\pi r}} \left[-4 - \frac{13}{2}\frac{\rho}{r} + 3\left(\frac{\rho}{r}\right)^2 + \frac{15}{2}\left(\frac{\rho}{r}\right)^3 \right] \quad (\theta = 0°) \tag{1.83}$$

The maximum shear stress is resulting from the above expression:

$$\tau_{r\theta\,\max} = 0.7132\frac{K_{\mathrm{II}}}{\sqrt{\pi\rho}} \quad (r = 1.723\rho,\ \ \theta = 0^{\circ}) \tag{1.84}$$

Inserting the maximum notch stresses in the formal frame considered in the preceding subsections related to other notch shapes, the stress concentration and geometry factors for internal two-sided keyholes with ($\rho \ll a$) are:

$$K_{\tau} = 0.7132\sqrt{\frac{a}{\rho}}, \quad \kappa_{\tau} = 1.4021 \tag{1.85}$$

$$K_{\sigma'} = 2.8664\sqrt{\frac{a}{\rho}}, \quad \kappa_{\sigma'} = 0.3489 \tag{1.86}$$

$$\frac{K_{\sigma'}}{K_{\tau}} = 4.0191 \tag{1.87}$$

Identical equations result for external two-sided keyholes under the condition specified in the paragraph after Eq. (1.77).

Comparing these results with the Neuber solutions for the elliptical, parabolic and hyperbolic notches, it is found that the stress concentration factors K_{τ} and $K_{\sigma'}$ are substantially enlarged. Also the ratio $K_{\sigma'}/K_{\tau}$ is larger. These enlargements result from the constant curvature of the circular hole in comparison to the decrease in curvature proceeding from the apex point of the elliptical, parabolic or hyperbolic notch.

Comparing $K_{\sigma'}$ according to Eqs. (1.86) and (1.58), the conclusion is that Eq. (1.58) does not approximate the stress concentration factor of the keyhole sufficiently well. It remains by the factor 1.682/2.866 = 0.587 below the actual value. Considering the U-notch instead of the keyhole, the factor is 0.660. Therefore, those derivations by Radaj and Zhang which refer to the radius of curvature at the position of $\sigma_{\max}$ (Radaj and Zhang 1993[(2)]), ibid. Eqs. (11), (20–23) and (27), will be substituted below by the keyhole relationships.

Comparing the maximum stresses in mode 2 and mode 1 (Eq. 1.12) loading, the result is:

$$\frac{\sigma'_{\max}(\theta = 67.115^{\circ})}{\sigma_{\max}(\theta = 0^{\circ})} = 1.35\frac{K_{\mathrm{II}}}{K_{\mathrm{I}}} \tag{1.88}$$

The corresponding ratio according to the Erdogan–Sih maximum tangential stress criterion for pointed slit tips reads as follows:

$$\frac{\sigma'_{\max}(\theta = 70.56^{\circ})}{\sigma'_{\max}(\theta = 0^{\circ})} = 1.15\frac{K_{\mathrm{II}}}{K_{\mathrm{I}}}. \tag{1.89}$$

1.4.8 Limit Value Formulae for the Mode Two Stress Intensity Factor

The stress intensity factor K_{II} can be determined from a slender comparison notch (elliptical, parabolic, hyperbolic, keyhole or U-shaped) as the limit value of the maximum notch stress times the square root on the notch radius ρ when the latter becomes smaller and smaller ($\rho \to 0$ equivalent to $\rho/a \ll 1$ or $\rho/b \ll 1$). Based on the equations above for κ_τ and $\kappa_{\sigma'}$, the following relationships, valid as limit value formulae, can be used:

Elliptical (internal) notch:

$$K_{II} = 2.5981\tau_{max}\sqrt{\pi\rho}, \quad K_{II} = 1.0\sigma'_{max}\sqrt{\pi\rho} \tag{1.90}$$

Parabolic (external) notch:

$$K_{II} = 3.6742\tau_{max}\sqrt{\pi\rho}, \quad K_{II} = 1.4142\sigma'_{max}\sqrt{\pi\rho} \tag{1.91}$$

Hyperbolic (external) notch:

$$K_{II} = 4.0816\tau_{max}\sqrt{\pi\rho}, \quad K_{II} = 1.5708\sigma'_{max}\sqrt{\pi\rho} \tag{1.92}$$

Keyhole (internal) notch:

$$K_{II} = 1.4021\tau_{max}\sqrt{\pi\rho}, \quad K_{II} = 0.3489\sigma'_{max}\sqrt{\pi\rho} \tag{1.93}$$

U-shaped (internal) notch (FE analysis by F. Berto):

$$K_{II} \approx 1.5759\tau_{max}\sqrt{\pi\rho}\ , \qquad K_{II} \approx 0.3922\sigma'_{max}\sqrt{\pi\rho} \tag{1.94}$$

The maximum shear stress τ_{max} occurs in the bisector plane ($\theta = 0°$) ahead of the notch root (at $r > \rho/2$, for the keyhole at $r > \rho$). The maximum tangential stress σ'_{max}, designating the antimetric part of the notch stresses in the case of superimposed mode 2 and mode 1 loading, occurs at angles $\theta = 60$–$90°$. The Eqs. (1.90–1.94) above contain different geometry factors κ_τ or $\kappa_{\sigma'}$, but the numerical values of K_{II} for $\rho \to 0$ in actual evaluations are the same.

The relationships with τ_{max} have a physical meaning insofar as the maximum shear stress leads to the shear stress singularity for $\rho = 0$ referring to the ligament. A similar behaviour related to the normal stress singularity can be attributed to σ'_{max} if one considers the section normal to σ'_{max}.

The convergence behaviour producing the limit values according to Eqs. (1.90), (1.93) and (1.94), elliptical and keyhole notch supplemented by the U-shaped notch, is now considered in more detail.

The investigation has been performed for the in-plane shear-loaded (remote nominal shear stress τ_n) rectangular plate (width $2w$, length $2l$, $l/w = 0.5$) with a central crack or slit (length $2a$, $a = l$) in comparison to an elliptical hole (length $2a$, width $2b$, notch radius $\rho = b^2/a$), an U-shaped hole (length $2a$, gap width $2b = 2\rho$) and a keyhole (length $2a$, gap width zero), Fig. 1.31.

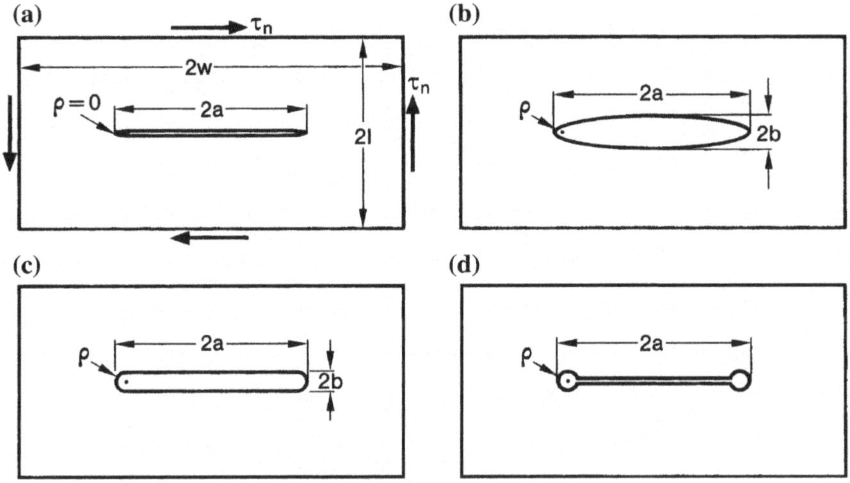

Fig. 1.31 In-plane shear-loaded rectangular plate with central crack or slit (**a**), elliptical hole (**b**), U-shaped hole (**c**) and keyhole (**d**); (Radaj and Zhang 1993$^{(2)}$)

At first, the normalised stress intensity factor of the crack, $k_{II} = K_{II}/\tau_n(\pi a)^{1/2}$, is determined using the boundary element method, evaluating K_{II} from crack edge displacements and ligament stresses in the immediate vicinity of the crack tip. The result was $k_{II0} = 1.360$.

Then the maximum notch stresses at the slender elliptical hole, U-shaped hole and keyhole with very small radii ($\rho/a = 0.1, 0.01, 0.001, 0.0001$) are determined using the boundary element method. The stress intensity factor K_{II} is evaluated for the elliptical and keyhole notches based on τ_{max} resulting in $k_{II\tau} = K_{II}/\tau_n(\pi a)^{1/2}$ and based on σ'_{max} resulting in $k_{II\sigma} = K_{II}/\tau_n(\pi a)^{1/2}$. The results are plotted over ρ/a indicating the convergence behaviour with decreasing ρ/a, Fig. 1.32. This graph substitutes the corresponding figure in the original publication (Radaj and Zhang 1993$^{(2)}$), where the keyhole equations are insufficiently approximated by Eq. (1.58), resulting in too high values of k_{II}. The values $k_{II\tau}$ and $k_{II\sigma}$ of the U-shaped notch coincide with the keyhole curve.

The conclusion from the figure is that the k_{II} curves for both types of notches converge to the same limit value at $\rho/a = 0$ which seems to be slightly lower than k_{II0}. The keyhole notch allows an acceptable accuracy already with $\rho/a = 0.01$.

A comparison has been made of the convergence behaviour of k_I, k_{II} and k_{III} from mode I, mode II and mode III loading of the rectangular plate with an elliptical hole, Fig. 1.33. The poor convergence of k_{II} in comparison to k_I and k_{III} can clearly be seen.

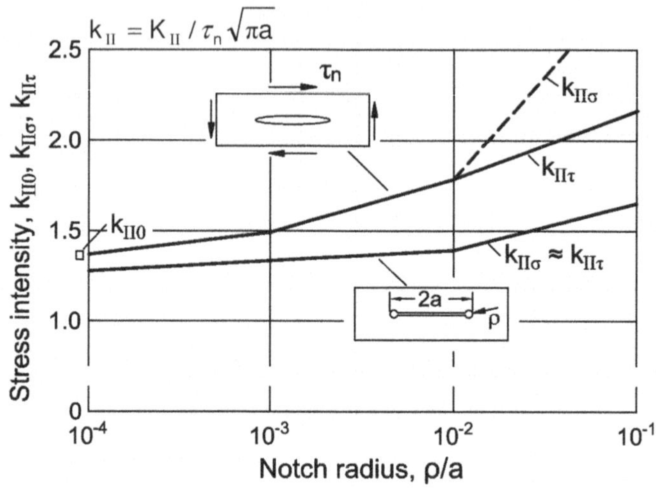

Fig. 1.32 Stress intensities (mode 2) derived for crack ($k_{II0} = 1.360$), elliptical hole and keyhole (inclusive of U-shaped hole) in rectangular plate under in-plane shear loading; convergence of limit value formulae for $\rho/a \to 0$; (Radaj and Zhang 1993[(2)], updated)

Fig. 1.33 Stress intensities (modes I, II and III) derived for elliptical hole in rectangular plate subjected to tensile loading, in-plane shear loading and out-of-plane shear loading, convergence behaviour of limit value formulae for $\rho/a \to 0$; (Radaj and Zhang 1993[(2)])

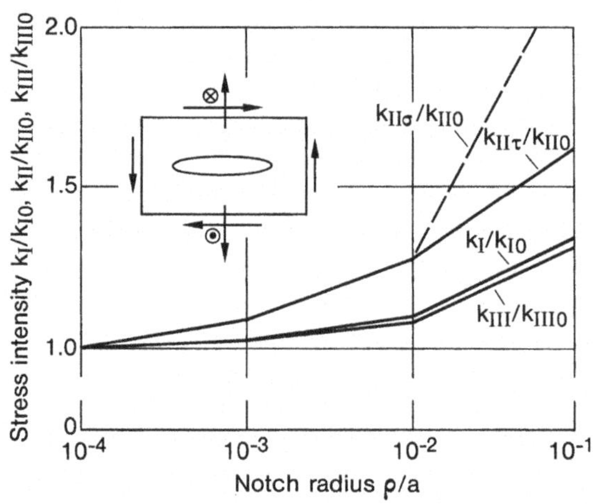

1.4.9 *Fictitious Notch Rounding Related to In-Plane Shear Loading with In-Bisector Crack Propagation*

As already mentioned, fictitious notch rounding according to Neuber, performed in order to take the microstructural support effect into account, has so far only been applied in mode 1 and mode 3 loading of pointed and sharply rounded notches and cracks propagating into the bisector or ligament, i.e. without a kink (with support factor $s = 2.0$–3.0 in tensile loading and $s = 0.5$ in out-of-plane shear loading).

The support factor s for non-kinking crack propagation under mode 2 loading of crack-like notches is determined by equating the shear stress $\bar{\tau}$ at the crack tip averaged over the length ρ^* along the ligament to the maximum shear stress $\tau_{\max}$ of the fictitiously rounded crack tip ($\rho_f = s\rho^*$, $\rho = 0$):

$$\bar{\tau} = K_{\mathrm{II}}\sqrt{\frac{2}{\pi\rho^*}} = \tau_{\mathrm{n}}\sqrt{\frac{2a}{\rho^*}} \tag{1.95}$$

The relationship between $\tau_{\max}$ and $a/\rho = a/\rho_f = a/s\rho^*$ depends on the type of notch (geometry factor κ_τ):

$$\tau_{\max} = \frac{1}{\kappa_\tau}\tau_{\mathrm{n}}\sqrt{\frac{a}{s\rho^*}} \quad (\rho = 0) \tag{1.96}$$

Equating $\bar{\tau}$ and $\tau_{\max}$, the factor s results in the following form:

$$s = \frac{1}{2\kappa_\tau^2} \tag{1.97}$$

The following values of s are determined with the relevant values of κ_τ given by Eqs. (1.70), (1.75), (1.78) and (1.85): for the elliptical notch $s = 0.0741$, for the parabolic notch $s = 0.0370$, for the hyperbolic notch $s = 0.0300$ and for the keyhole notch $s = 0.2543$.

These are the base point values ($\rho = 0$) of the $s(\rho)$ curve which have been considered above in mode 1 and mode 3 loading. There is no jump in multiaxiality between the pointed crack tip and the rounded notch root when considering the shear stresses in the bisector or ligament. This is true for any multiaxial failure criterion (normal stress, shear stress, von Mises, Beltrami), but the critical stresses will depend on the criterion.

For finding the dependency of the support factor on the notch radius, $s = s(\rho)$, the procedure corresponding to Eqs. (1.23–1.27) or (1.44–1.46) in mode 1 or mode 3 loading, respectively, cannot be used, because $\bar{\tau}(\rho_f, \rho^* \to 0)$ does not give the maximum stress of the fictitiously rounded notch. The procedure is applied in a modified form, proceeding from Creager's Eq. (1.61) for $\tau_{r\theta}$ ($\theta = 0°$) for determining the averaged stress $\bar{\tau}$ along the bisector line by integration and also the maximum stress $\tau_{\max}$ of the fictitiously rounded (parabolic) notch at the distance $r = 1.5\rho_f$:

$$\int \tau_{r\theta}\,\mathrm{d}r = \frac{K_{\mathrm{II}}}{\sqrt{2\pi r}}(2r + \rho) \tag{1.98}$$

$$\bar{\tau} = \frac{1}{\rho^*}\int\limits_{\rho/2}^{\rho/2+\rho^*} \tau_{r\theta}\,\mathrm{d}r = \frac{K_{\mathrm{II}}}{\rho^*\sqrt{2\pi}}\left[\frac{2(\rho/2+\rho^*)+\rho}{\sqrt{\rho/2+\rho^*}} - \frac{2(\rho/2)+\rho}{\sqrt{\rho/2}}\right] \tag{1.99}$$

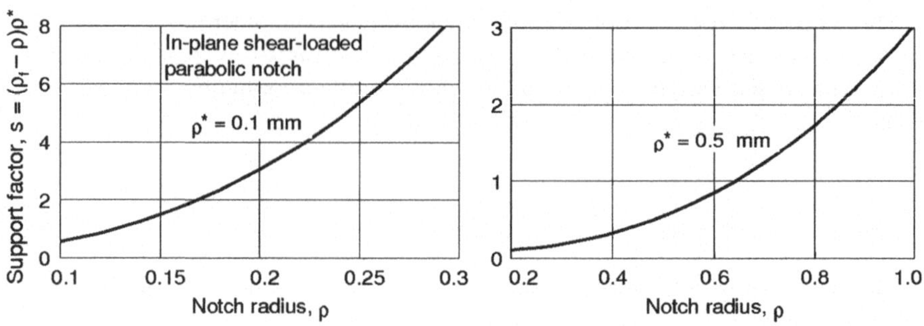

Fig. 1.34 Support factor s as function of the real notch radius ρ for in-plane shear-loaded parabolic (internal) notch with in-bisector crack propagation; $s = 0.0741$ for $\rho = 0$; (Berto 2010, private communication)

$$\tau_{\max} = \frac{K_{\mathrm{II}}}{\sqrt{2\pi}}\left[\frac{1}{\sqrt{1.5\rho_{\mathrm{f}}}} - \frac{\rho_{\mathrm{f}}}{2(1.5\rho_{\mathrm{f}})\sqrt{1.5\rho_{\mathrm{f}}}}\right] = \frac{2K_{\mathrm{II}}}{3\sqrt{3\pi\rho_{\mathrm{f}}}} \quad (1.100)$$

The following closed form expression results from $\tau_{\max} = \bar{\tau}$ (Berto 2010, private communication):

$$\rho_{\mathrm{f}} = \frac{0.296296(\rho^*)^3 + 0.148148(\rho^*)^2\rho}{4(\rho^*)^2 + 16\rho^*\rho + 8\rho^*\sqrt{\rho(2\rho^* + \rho)} - 8\rho^{3/2}\sqrt{2\rho^* + \rho}} \quad (1.101)$$

Two typical curve plots $s = s(\rho) = (\rho_{\mathrm{f}} - \rho)/\rho^*$ are shown in Fig. 1.34. The support factor s is progressively rising with ρ and highly dependent on ρ^*. No plateau values of s occur. Therefore, the concept of fictitious notch rounding is applicable only with crack-like notches ($\rho = 0$, $s = 0.0741$) in the case of mode 2 loading when evaluating the shear stresses in the bisector plane (non-kinking crack propagation).

1.4.10 V-Notch Subjected to In-Plane Shear Loading with Out-of-Bisector Crack Propagation, Blunt Notch Solution

Out-of-bisector crack propagation is the normal case at notches subjected to pure mode 2 loading. The crack propagates into a direction which allows pure mode 1 tensile loading of the propagating crack. This direction deviates substantially from the bisector plane. Only under special conditions (listed in Sect. 1.4.2), the crack will propagate in the bisector plane.

A parametric investigation considering a pointed V-notch specimen subjected to in-plane shear loading has been performed on a purely numerical basis using the FE method (Berto and Lazzarin 2010). The aim was to extend the Neuber concept of fictitious notch rounding to mode 2 loading with out-of-bisector crack propagation.

The following algorithmic procedure is applied. First, the propagation direction of the crack is determined by applying the maximum tangential stress criterion (Erdogan and Sih 1963). The criterion indicates the direction, in which pure mode 1 loading of the crack will occur. The minimum strain energy density criterion (Sih 1974) is applied in comparison, resulting in a slightly different direction. Then, stress averaging over the support length ρ^* along the crack propagation direction starting at the crack tip is performed, considering the equivalent stress according to different failure criteria (Rankine, von Mises and Beltrami). Extremely fine FE meshes around the notch tip are used for this purpose. The averaged (equivalent) notch stress $\overline{\sigma}$ is set equal to the maximum (equivalent) notch stress $\sigma'_{\max}$ of a fictitiously rounded substitute notch. These stresses occur in approximately the same propagation direction. The fictitiously enlarged radius is found by iterated FE calculations up to the point where the relative deviation between $\overline{\sigma}$ and $\sigma'_{\max}$ is less than 1 %.

The stress field of pointed V-notches subjected to mode 2 loading is described as follows (Lazzarin and Tovo 1996):

$$\begin{Bmatrix} \sigma_\theta \\ \sigma_r \\ \tau_{r\theta} \end{Bmatrix} = \frac{1}{\sqrt{2\pi}} \frac{r^{\lambda_2 - 1} K_2}{(1-\lambda_2) + \chi_2(1+\lambda_2)} \left[\begin{Bmatrix} -(1+\lambda_2)\sin(1-\lambda_2)\theta \\ -(3-\lambda_2)\sin(1-\lambda_2)\theta \\ (1-\lambda_2)\cos(1-\lambda_2)\theta \end{Bmatrix} + \chi_2(1+\lambda_2) \begin{Bmatrix} -\sin(1+\lambda_2)\theta \\ \sin(1+\lambda_2)\theta \\ \cos(1+\lambda_2)\theta \end{Bmatrix} \right] \tag{1.102}$$

where K_2 is the mode 2 stress intensity factor of pointed V-notches (Gross and Mendelson 1972), λ_2 is Williams' eigenvalue in mode 2 (Williams 1952) and χ_2 is a parameter related to the notch opening angle, 2α.

The crack propagation angle $\overline{\theta}$ in mode 2 loading according to the maximum tangential stress (MTS) criterion is obtained from Eq. (1.102) by imposing the condition $\partial\sigma_\theta/\partial\theta = \tau_{r\theta} = 0$ on the tangential stress σ_θ at the boundary of the circular 'core region' of the notch tip (small radius r^*):

$$(1+\lambda_2)\chi_2\cos(1+\lambda_2)\theta - (\lambda_2 - 1)\chi_2\cos(1-\lambda_2)\theta = 0 \tag{1.103}$$

The crack propagation angle $\overline{\theta}$ is identical for plane stress and plane strain conditions.

Two similar equations are derived in the case of the minimum strain energy density (MSED) criterion resulting in somewhat larger values of $\overline{\theta}$. This criterion predicts crack growth in the direction of minimum strain energy density at the boundary of the core region expressed by Sih's MSED factor S. The crack propagation angle $\overline{\theta}$ for plane stress and plane strain conditions, respectively, is obtained from the following equations:

Table 1.8 Pointed V-notches subjected to in-plane shear loading; eigenvalue λ_2, parameter χ_2 and crack propagation angle $\overline{\theta}$ according to MTS and MSED criteria, plane stress (ps) and plane strain (pn) conditions; different notch opening angles 2α; (Berto and Lazzarin 2010)

2α	λ_2	χ_2	$\overline{\theta}$(MTS)	$\overline{\theta}$(MSED, ps)	$\overline{\theta}$(MSED, pn)
0°	0.500	1.000	70.6°	79.7°	82.4°
30°	0.598	0.921	65.2°	80.4°	82.8°
45°	0.660	0.814	62.4°	80.5°	82.9°
60°	0.731	0.658	59.6°	80.6°	82.9°

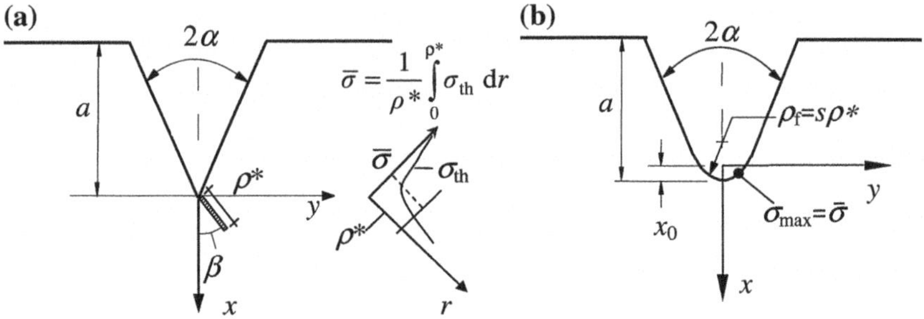

Fig. 1.35 Fictitious notch rounding concept applied to in-plane shear loaded pointed V-notches; real notch with stress $\overline{\sigma}$ averaged over ρ^* at angle $\overline{\theta}$ (**a**) and substitute blunt notch with fictitious notch radius ρ_f producing $\sigma_{max} = \overline{\sigma}$ (**b**); (Berto et al. 2012)

$$(1 + \lambda_2)(1 + \nu)\chi_2 \sin 2\theta + (\nu - 1)\sin(1 - \lambda_2)2\theta = 0 \tag{1.104}$$

$$(1 + \lambda_2)\chi_2 \sin 2\theta + (2\nu - 1)\sin(1 - \lambda_2)2\theta = 0 \tag{1.105}$$

The parameters λ_2 and χ_2 and the angles $\overline{\theta}$ are recorded in Table 1.8 for different notch opening angles 2α, crack direction criteria and multiaxiality conditions (plane stress compared with plane strain).

Stress averaging over the support length ρ^* in the crack propagation direction, Fig. 1.35a is performed on the theoretical (equivalent) notch stress σ_{th} resulting in the effective stress $\overline{\sigma}$ and in the effective stress concentration factor $\overline{K}_t$ of the pointed V-notch:

$$\overline{\sigma} = \frac{1}{\rho^*}\int_0^{\rho^*} \sigma_{th}(\overline{\theta}, r)\mathrm{d}r \tag{1.106}$$

$$\overline{K}_t = \frac{\overline{\sigma}}{\tau_0} \tag{1.107}$$

where τ_0 is the remote reference shear stress of the plate specimen without the rhombic hole (compare Fig. 1.36).

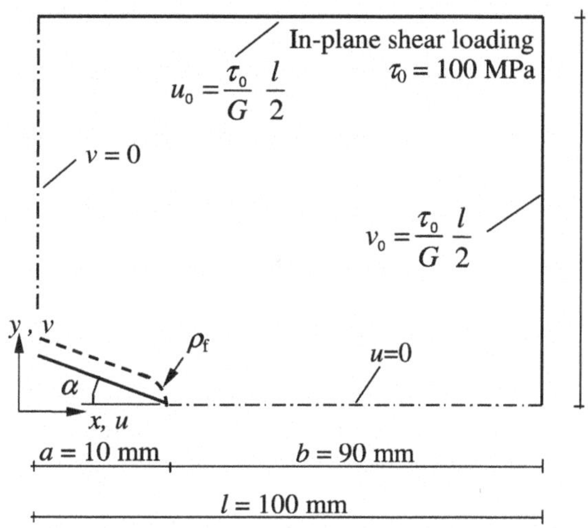

Fig. 1.36 Geometry and dimensions of the V-notched plate specimen (symmetry quarter); real pointed notch ($\rho = 0$) compared with fictitious blunt notch ($\rho = \rho_f$); remote loading by prescribed edge displacements; (Berto et al. 2012)

Considering the normal stress failure criterion in combination with the maximum tangential stress criterion for $\overline{\theta}$, it holds $\sigma_{th} = \sigma_\theta$ because $\sigma_r = 0$ and $\tau_{r\theta} = 0$ in the direction of $\overline{\theta}$, but when combined with the minimum strain energy criterion for $\overline{\theta}$, it holds $\sigma_{th} = \sigma_1$ (first principal stress).

The fictitiously rounded substitute notch ($\rho_f = s\rho^*$), Fig. 1.35b, presents the following stress concentration factor:

$$K_t(\rho_f) = \frac{\sigma_{max}(\rho^*, s)}{\tau_0} \tag{1.108}$$

The condition $K_t(\rho_f) = \overline{K}_t$ is approximated within a relative deviation Δ of ± 1 % by tentatively choosing values of ρ_f which might be appropriate:

$$\Delta = \frac{K_t(\rho_f) - \overline{K}_t}{\overline{K}_t} \tag{1.109}$$

The support factor s is subsequently derived from $s = \rho_f/\rho^*$.

There is no 'jump' in the multiaxiality of stress between the pointed and the fictitiously rounded notch in the crack propagation direction, angle $\overline{\theta}$, because $\sigma_r(0, \overline{\theta}) = 0$ for the pointed notch, and this is identical with the condition at the edge of the fictitiously rounded notch. But this is only true when applying the maximum tangential stress criterion for determining $\overline{\theta}$. There is a minor jump effect to be expected in the case of the minimum strain energy density criterion.

The parametric investigation based on FE analysis is performed for the V-notched specimen shown in Fig. 1.36, a quadratic plate with a pointed internal rhombic hole, the plate subjected to in-plane shear loading by prescribed remote boundary displacements. These are derived from the reference shear stress τ_0 of

Table 1.9 Microstructural support factor s of in-plane shear loaded V-notch dependent on notch opening angle 2α for different failure criteria under plane stress (ps) and plane strain (pn) conditions; crack propagation angle $\overline{\theta}(MTS)$, $\rho^* = 0.05$ mm; (Berto and Lazzarin 2010)

Angles		Support factor s					
		Rankine		von Mises		Beltrami	
2α	$\overline{\theta}$	ps	pn	ps	pn	ps	pn
0°	70.6°	2.47	2.47	2.52	3.22	2.52	2.76
30°	65.2°	4.00	4.00	3.20	3.77	3.45	3.72
45°	62.4°	5.90	5.90	3.94	4.55	4.43	4.72
60°	59.6°	10.90	10.90	5.60	6.48	6.68	7.15

the specimen without the hole. The following variations are considered in the numerical investigation:

- notch opening angle $2\alpha = 0$, 30, 45 and 60°,
- microstructural support length $\rho^* = 0.01$, 0.05, 0.1, 0.2, 0.3 and 0.5 mm,
- plane stress compared with plane strain conditions,
- normal stress criterion (Rankine), distortional strain energy criterion (von Mises) and total strain energy criterion (Beltrami).

First, the support factor s is determined for a particular support length, $\rho^* = 0.05$ mm, dependent on the notch opening angle 2α for different failure criteria under plane stress and plane strain conditions, Table 1.9. The values of s for the crack-like notch ($2\alpha = 0°$), $s = 2.47$–3.22 in mode 2 are slightly larger than $s = 2.00$–2.90 in mode 1 (Table 1.1). They are substantially larger for $2\alpha = 60°$.

The values of s determined for $\rho^* = 0.05$ mm are applied to cases with other values of ρ^* within the parametric investigation in order to show that these values of s are generally applicable, similar to the plateau values derived for mode 1 and mode 3 loading. An extract from the results is listed in Table 1.10. The relative deviation Δ is within the ± 1 % range for $\rho^* = 0.05$ mm, as prescribed in the iterated calculations. It is slightly enlarged for $\rho^* = 0.1$ mm, but it is substantially enlarged up to 10 % for $\rho^* = 0.5$ mm (conservative for strength assessments).

Comparing the Δ values in mode 2 loading (Table 1.10) with those in mode 1 loading (Table 1.2), it is evident, that the tendencies are similar and that the absolute values are in the same range. The conclusion by way of analogy is that plateau values of s should exist for mode 2 loading with out-of-bisector crack propagation, which can be used. As shown in Sect. 1.4.11, the plateau values are confirmed, at least for root hole notches.

The similar values of s for pointed crack-like notches ($2\alpha = 0°$, $\rho = 0$ mm) under mode 1 and under mode 2 loading conditions makes it possible to use the same fictitious radius in the two loading cases and especially too under mixed mode loading conditions. Then, the maximum tangential stress of the fictitiously rounded notch tip gives directly the crack propagation direction and the effective

Table 1.10 Fictitious notch rounding applied to pointed V-notch specimen ($\rho = 0$) subjected to in-plane shear loading using the value s for $\rho^* = 0.05$ mm; comparison of $K_t(\rho_f)$ with $\overline{K}_t$, both obtained from FE analysis (relative deviation Δ); variation of notch opening angle 2α, support length ρ^*, failure criterion and plane stress (ps) versus plane strain (pn); normal stress (ns), von Mises (vM) and Beltrami (B) criterion; (Berto and Lazzarin 2010)

		ρ^*= 0.05 mm				ρ^* = 0.1 mm				ρ^* = 0.5 mm			
	s	ρ_f [mm]	$K_t(\rho_f)$	$\overline{K}_t$	Δ %	ρ_f [mm]	$K_t(\rho_f)$	$\overline{K}_t$	Δ %	ρ_f [mm]	$K_t(\rho_f)$	$\overline{K}_t$	Δ %
$2\alpha = 0°$													
ns, ps, pn	2.47	0.124	23.13	23.07	0.26	0.247	16.4	16.31	0.55	1.235	8.02	7.29	10.01
vM, ps	2.52	0.126	23.02	23.07	−0.22	0.252	15.93	16.31	−2.33	1.260	7.95	7.29	9.05
vM, pn	3.22	0.161	20.48	20.51	−0.15	0.322	14.72	14.5	1.52	1.610	7.17	6.48	10.65
B, ps	2.52	0.126	23.02	23.07	−0.22	0.252	15.93	16.31	−2.33	1.260	7.95	7.29	9.05
B, pn	2.76	0.138	22.04	22.01	0.14	0.276	15.73	15.56	1.09	1.380	7.65	6.96	9.91
$2\alpha = 30°$													
ns, ps, pn	4.00	0.200	14.69	14.63	0.38	0.400	11.19	11.07	1.04	2.000	6.27	5.799	8.12
vM, ps	3.20	0.160	15.96	16.04	−0.49	0.320	12.11	12.14	−0.23	1.600	6.77	6.355	6.52
vM, pn	3.77	0.189	15.04	15.04	−0.03	0.377	11.45	11.39	0.56	1.885	6.38	5.962	7.01
B, ps	3.45	0.173	15.51	15.57	−0.40	0.345	11.91	11.79	1.06	1.725	6.59	6.171	6.79
B, pn	3.72	0.186	15.11	15.14	−0.21	0.372	11.48	11.46	0.18	1.860	6.41	6.000	6.82
$2\alpha = 60°$													
ns, ps, pn	10.90	0.545	7.42	7.45	−0.40	1.090	6.26	6.18	1.29	5.450	4.34	4.01	8.23
vM, ps	5.60	0.280	8.82	8.81	0.11	0.560	7.37	7.31	0.82	2.800	5.01	4.74	5.70
vM, pn	6.48	0.324	8.48	8.48	0.00	0.648	7.1	7.04	0.85	3.240	4.82	4.56	5.70
B, ps	6.68	0.334	8.44	8.41	0.32	0.668	7.06	6.98	1.15	3.340	4.83	4.53	6.62
B, pn	7.15	0.358	8.27	8.27	0.00	0.715	6.9	6.86	0.58	3.575	4.75	4.45	6.74

stress. Otherwise, the analysis is much more complicated. First, the crack tip stresses of the two loading modes must be superimposed. Then, the crack propagation direction must be determined from the maximum tangential stress. Finally, stress averaging must be performed in this direction.

Using the values of s of the mode 1 loading case also for mode 2 loading, where they may be substantially larger (V-notches with $2\alpha \gg 0°$) produces conservative results in engineering applications.

1.4.11 V-Notch Subjected to In-Plane Shear Loading with Out-of-Bisector Crack Propagation, Root Hole Solution

The purely numerical solution in the preceding subsection referring to blunt V-notches subjected to in-plane shear loading with out-of-bisector crack propagation has the character of a 'numerical experiment'. The values of s are determined for a large variety of geometrical configurations, but the more basic questions after convergence behaviour, limit values and parameter dependencies remain unanswered.

A sufficiently convenient function-analytical solution for blunt V-notches is not possible, but it is possible for V-notches with root hole. A general solution for the in-plane stress field (modes 1 and 2) has been derived, and the analytically simpler out-of-plane stress problem (mode 3) has also been solved (Zappalorto and Lazzarin 2011).

Referring to the in-plane stress field, the solution is based on Kolosov–Muskhelishvili's complex stress function theory (Muskhelishvili 1963). The eigenvalue problem presented by Williams for 'angular corners' (Williams 1952) is part of the solution. The stresses in the bisector plane are well approximated in general, whereas the stresses at the hole edge may show some deviation when approaching the flank edges.

A comprehensive function-analytical investigation combined with a purely numerical validation of results has been conducted for the in-plane shear loading case (mode 2) of V-notches with root hole based on the above analytical frame (Berto et al. 2012). Two analytical and one numerical method have been developed for determining the fictitious notch radius ρ_f and from it the support factor s dependent on the notch opening angle 2α. In all three methods, the crack propagation angle is determined based on the Erdogan–Sih maximum tangential stress (MTS) criterion:

- Method M1: The notch stress averaged over the inclined microstructural support length at pointed V-notches is expressed by the maximum notch stress of fictitiously rounded V-notches (with root hole) using a newly developed analytical solution.

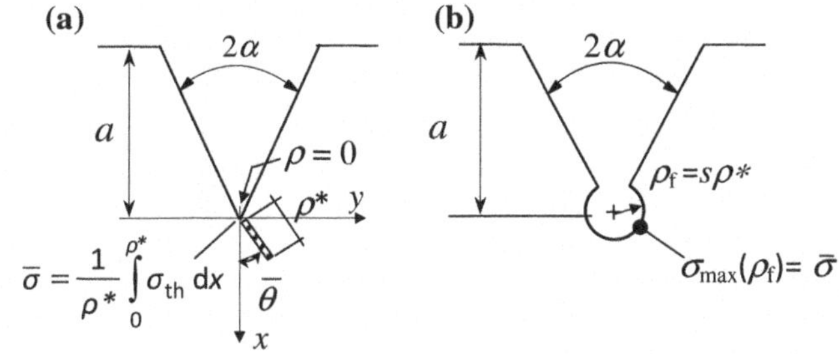

Fig. 1.37 Fictitious notch rounding concept applied to in-plane shear-loaded pointed V-notches; real notch with stress $\overline{\sigma}$ averaged over ρ^* at angle $\overline{\theta}$ (**a**) and substitute root hole notch with fictitious notch radius ρ_f producing $\sigma_{max} = \overline{\sigma}$ (**b**); (Berto et al. 2012)

- Method M2: By this new method, V-notches with root hole both in the real and in the fictitious configuration are considered. A procedure given by Neuber for determining the support factor *s* (Neuber 1958 or 1985) is applied in areas outside the bisector plane. This is an extension of what has been done in the cases of mode 1 and mode 3 loading of V-notches (Berto et al. 2008, 2009) now to mode 2 loading.
- Method M3: The support factor *s* is determined for pointed notches on a purely numerical basis by iteration of FE models (Berto and Lazzarin 2010). These data are also used for validating the results gained from the analytical methods.

Finally, using method M1, a more comprehensive evaluation is performed. Two additional multiaxial failure criteria are considered, von Mises (vM) and Beltrami (B), besides normal stress (ns), in combination with the maximum tangential stress (MTS) and minimum strain energy density (MSED) criteria applied to the hole edge stresses for the determination of the crack propagation angle. Plane stress and plane strain conditions, respectively, are evaluated.

The procedure of substituting notch stress averaging by fictitious notch rounding in method M1 and method M3 is demonstrated by Fig. 1.37. The pointed V-notch is considered on the one side, the V-notch with root hole on the other. The angle $\overline{\theta}$ denotes the out-of-bisector crack propagation direction. The corresponding procedure in method M2 is shown in Fig. 1.38. Here, real notch and fictitious notch have a root hole, and, once more, the crack propagation angle is $\overline{\theta}$.

In method M1, the stresses at the pointed V-notch are given by Eq. (1.102). The crack propagation angle $\overline{\theta}$ is obtained from Eq. (1.103), Table 1.8. The fictitious notch radius ρ_f results in the following form:

$$\rho_f = \left(-\frac{(1+\lambda_2)\left[\chi_2 \sin(1+\lambda_2)\overline{\theta} + \sin(1-\lambda_2)\overline{\theta}\right]}{\xi_\theta \lambda_2 [\chi_2(1+\lambda_2) + 1 - \lambda_2]}\right)^{\frac{1}{-1+\lambda_2}} \rho^* \qquad (1.110)$$

where the first term with the parenthesis on the right hand side is obviously the support factor *s* ($\rho_f = s\,\rho^*$). The parameters χ_2 and ξ_θ depend on 2α.

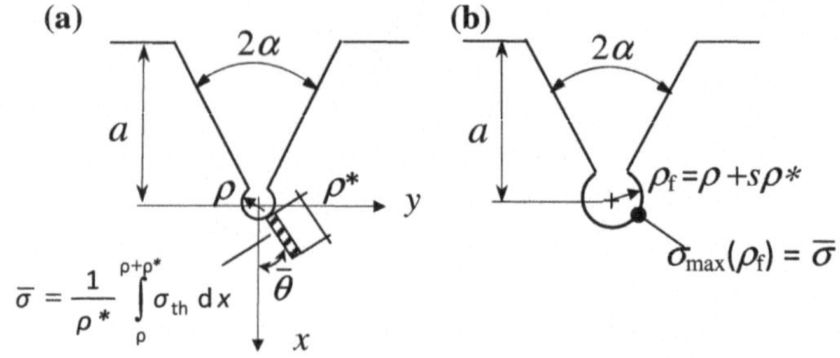

Fig. 1.38 Fictitious notch rounding concept applied to in-plane shear-loaded root hole V-notches; real notch with stress $\overline{\sigma}$ averaged over ρ^* at angle $\overline{\theta}$ (**a**) and substitute root hole notch with fictitious notch radius ρ_f producing $\sigma_{max} = \overline{\sigma}$ (**b**); (Berto et al. 2012)

In method M2, which follows Neuber's procedure, the averaged (equivalent) stress $\overline{\sigma}(\rho, \rho^*)$ at the real notch has to be equated to its limit value $\overline{\sigma}(\rho, \rho^* \to 0)$ at the fictitiously rounded notch, Eq. (1.23). From this equation, $\rho_f(\rho, \rho^*)$ is gained and from it the support factor s, Eq. (1.24). The crack propagation angle $\overline{\theta}$ (MTS) is 70.5, 61.9, 59.2 and 56.4° for $2\alpha = 0$, 30, 45 and 60°. It is compared with the corresponding data for the pointed notch in Table 1.8.

Leaving out the expressions for $\sigma_{th} = \sigma_\theta$, $\overline{\sigma}(\rho, \rho^*)$ and $\overline{\sigma}(\rho, \rho^* \to 0)$ here, one ends up with the following expression:

$$\begin{aligned}\rho_f = 4^{\frac{1}{1-\lambda}}\Bigg\{\Bigg[&(1+\lambda)(\rho^*+\rho)^\lambda\Bigg((\rho^*+\rho)^2+\left(\frac{\rho}{\rho^*+\rho}\right)^{2\lambda}(\rho^{*2}(-1+\lambda)\\&+2\rho^*(-1+\lambda)\rho-\rho^2)\Bigg)\phi\sin(1+\lambda)\overline{\theta}+(\rho^*+\rho)\big[-\rho^\lambda(\rho^*+\rho)[\lambda(-1+2\lambda)\\&+(-2+\lambda)\mathrm{cosec}\overline{\theta}\sin(1-2\lambda)\overline{\theta}]+(\rho^*+\rho)^\lambda\Big((1+\lambda)^2(\rho^*+\rho)\\&+\left(\frac{\rho}{\rho^*+\rho}\right)^{2\lambda}[-\rho^*(1+\lambda)-\rho+(-3+\lambda)\lambda\rho\\&+(\lambda-2)(\rho^*+\rho^*\lambda+\rho)\mathrm{cosec}\theta\sin(1-2\lambda)\overline{\theta})\big]\sin(1-\lambda)\overline{\theta}\\&/\Big[\lambda\rho^*(1+\lambda)(\rho^*+\rho)^2(\varphi\sin(1+\lambda)\overline{\theta}+\sin(1-\lambda)\overline{\theta})\Big]\Bigg\}^{\frac{1}{-1+\lambda}}\end{aligned} \tag{1.111}$$

A plot of $s = (\rho_f - \rho)/\rho^*$, Eq. (1.24), over ρ for $\rho^* = 0.1$ mm and different notch opening angles 2α is shown in Fig. 1.39, and a plot over ρ for different values of ρ^* and $2\alpha = 60°$ in Fig. 1.40.

The expression for ρ_f, Eq. (1.111), for pointed real notches ($\rho = 0$) reads:

$$\rho_f = \left[4^{\frac{1}{1-\lambda}}\left(\frac{1}{\lambda}+\frac{1}{1+\phi\,\mathrm{cosec}(1-\lambda)\overline{\theta}\sin(1+\lambda)\overline{\theta}}\right)^{\frac{1}{-1+\lambda}}\right]\rho^* \tag{1.112}$$

The term within the square brackets is the support factor s ($\rho_f = s\rho^*$).

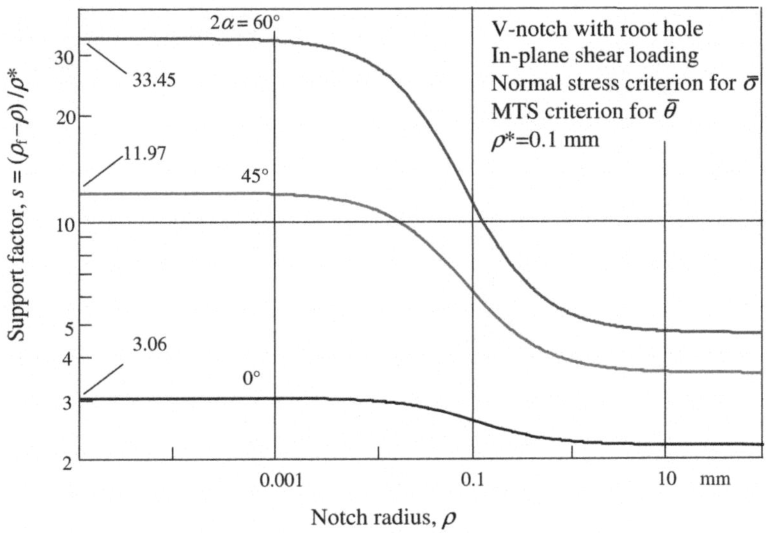

Fig. 1.39 Support factor s as function of the real notch radius ρ for different notch opening angles 2α for $\rho^* = 0.1$ mm; in-plane shear-loaded V-notch with root hole and out-of-bisector crack propagation; (Berto et al. 2012)

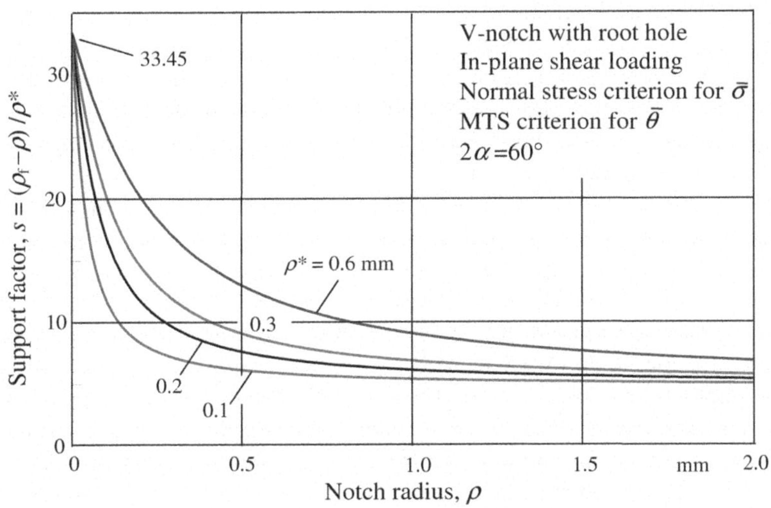

Fig. 1.40 Support factor s as function of the real notch radius ρ for different values of ρ^* considering $2\alpha = 60°$; in-plane shear-loaded V-notch with root hole and out-of-bisector crack propagation; (Berto et al. 2012)

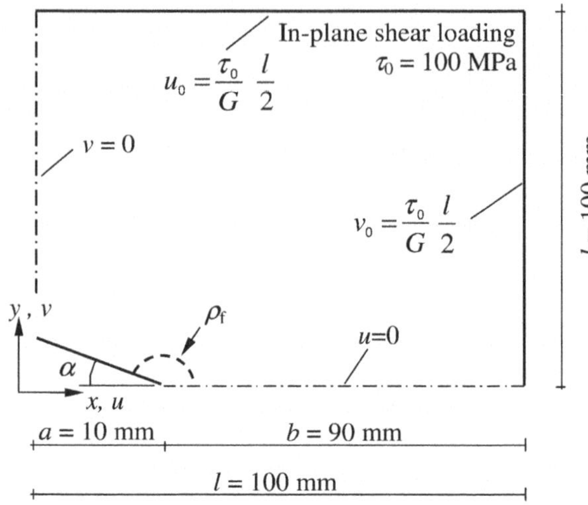

Fig. 1.41 Geometry and dimensions of the V-notched plate specimen (symmetry quarter); real pointed notch ($\rho = 0$) compared with fictitious root hole V-notch ($\rho = \rho_f$); remote loading by prescribed edge displacements; (Berto et al. 2012)

The logarithmic plot in Fig. 1.39 indicates that constant values of s occur both for $\rho \ll \rho^*$ (only in logarithmic scale of ρ) and $\rho \gg \rho^*$ with a sloping transition between the two plateau values. Only the values for $\rho \ll \rho^*$ (pointed notches) are further evaluated and used below. The linear plot in Fig. 1.40 shows a uniform value of s for $\rho = 0$ (pointed notch) and lower plateau values for $\rho \gg \rho^*$ which depend on ρ^* to some extent.

In method M3, the support factor s for in-plane shear loading is evaluated based on the purely numerical procedure already applied in Sect. 1.4.10. Whereas pointed real V-notches are considered in comparison to rounded fictitious V-notches (blunt notches) there, here the pointed real notches are contrasted with fictitious root hole notches, Fig. 1.41.

Before reporting on the validation program on the basis of method M3, a comparison of the values of s gained with the different methods is presented in Table 1.11. The values of s according to the normal stress criterion referring to the root hole notches evaluated by methods M1, M2 and M3 show excellent correspondence. They are more or less independent on the criterion which controls the crack propagation angle (MTS or MSED) and the multiaxiality condition (plane stress or plane strain). They are steeply rising with the notch opening angle 2α. The values of s referring to blunt notches evaluated by method M3 in the preceding subsection are stated in comparison. They are substantially smaller because the stress rising effect of blunt notches is much smaller than the stress rising effect of root holes. This is enhanced by the fact that the notch depth is slightly enlarged for root holes with $\rho_f > 0$. It has to be kept in mind, that a smaller value of s produces a stronger notch effect.

Further values of s for pointed V-notches according to method M1 combined with the MTS criterion have been derived for different failure criteria under plane stress and plane strain conditions, Table 1.12 (a similar table is available with values of s

Table 1.11 Support factor s obtained according to the normal stress criterion; pointed V-notches related to V-notches with root holes and to blunt circular V-notches; in-plane shear loading; plane stress (ps) and plane strain (pn); analysis methods M1, M2, M3; (Berto et al. 2011)

	Factor s for root hole V-notches					Factor s for blunt V-notches		
2α	MTS ps, pn M1	MTS ps, pn M2	MTS ps, pn M3	MSED ps M1	MSED pn M1	MTS ps, pn M3	MSED ps M3	MSED pn M3
0°	3.06	3.06	3.13	3.03	2.99	2.47	2.46	2.45
30°	6.34	6.40	6.35	6.15	6.12	4.00	3.90	3.85
45°	11.87	11.97	11.90	11.32	11.31	5.90	5.59	5.40
60°	33.03	33.45	33.00	31.00	31.29	10.90	10.00	9.46

Table 1.12 Support factor s for pointed V-notches according to analysis method M1; different failure criteria, plane stress (ps) and plane strain (pn) conditions; $\overline{\theta}$ values according to MTS criterion; in-plane shear loading; (Berto et al. 2011)

	Support factor s				
2α	Normal stress, ps, pn	von Mises, ps	von Mises, pn	Beltrami, ps	Beltrami, pn
0°	3.06	3.06	3.88	3.06	3.37
30°	6.34	5.05	5.93	5.44	5.83
45°	11.87	7.60	9.27	9.02	9.62
60°	33.03	17.75	20.44	21.03	22.43

from the combination with the MSED criterion). The values of s depend on the applied failure criterion, but the general trend in the dependency on 2α is the same.

The fictitious notch rounding approach applied to in-plane shear loading condition has been validated by comparisons based on the FE method considering pointed V-notches with different notch opening angles. The V-notched plate specimen shown in Fig. 1.41 is used. The $\overline{K}_t$ values of the pointed notches are determined from the relevant K_2 values found by the FE method for the selected values of the notch depth and the notch opening angle. The values of $K_t(\rho_f)$ are determined by the FE method, using the values of s for $\rho = 0$ from method M2.

The following parameters have been varied in the investigation with the normal stress criterion for $\overline{K}_t$ and the MTS criterion for $\overline{\theta}$:

- notch opening angle $2\alpha = 0$, 30, 45 and 60°,
- microstructural support length ρ^* between 0.05 and 0.4 mm,
- notch depth $a = 5$, 10 and 50 mm.

The table with the results inclusive of the relative deviations Δ between $K_t(\rho_f)$ and $\overline{K}_t$, Eq. (1.109), in the original paper (Berto et al. 2011) shows that for a large notch depth a relative to ρ_f, the Δ values rise with smaller a values and larger ρ^* values. This tendency can easily be explained by the fact, that a too large root hole destroys the dominance of the infinite plate conditions.

The results of another parametric investigation are recorded in Table 1.13. Here, different failure criteria, multiaxiality conditions and notch opening angles

Table 1.13 Fictitious notch rounding applied to pointed V-notch specimen ($\rho = 0$, $\rho^* = 0.2$ mm, $a = w = 50$ mm) subjected to in-plane shear loading; MTS criterion for $\overline{\theta}$, different failure criteria for $\overline{K}_t$: normal stress (ns), von Mises (vM) and Beltrami (B), plane stress (ps) and plane strain (pn); FE analysis results with values of s according to method M2; (Berto et al. 2011)

	s	ρ_f	$K_t(\rho_f)$	$\overline{K}_t$	Δ %	s	ρ_f	$K_t(\rho_f)$	$\overline{K}_t$	Δ %
	$2\alpha = 0°$					$2\alpha = 30°$				
ns, ps, pn	3.06	0.61	25.34	25.13	0.82	6.40	1.28	15.49	15.52	−0.17
vM, ps	3.06	0.61	25.60	25.14	1.80	5.05	1.01	16.97	17.00	−0.21
vM, pn	3.88	0.78	22.75	22.35	1.76	5.93	1.19	15.92	15.95	−0.19
B, ps	3.06	0.61	25.60	25.14	1.80	5.44	1.09	16.46	16.51	−0.30
B, pn	3.37	0.67	24.39	23.98	1.68	5.83	1.17	16.03	16.05	−0.13
	$2\alpha = 45°$					$2\alpha = 60°$				
ns, ps, pn	11.97	2.39	11.15	11.25	−0.85	33.45	6.69	7.43	7.59	−2.11
vM, ps	7.60	1.52	13.02	12.84	1.38	17.75	3.55	8.90	8.97	−0.81
vM, pn	9.27	1.85	12.16	12.23	−0.58	20.44	4.09	8.55	8.64	−1.07
B, ps	9.02	1.80	12.26	12.35	−0.73	21.03	4.21	8.48	8.57	−1.07
B, pn	9.62	1.92	12.00	12.08	−0.67	22.43	4.49	8.32	8.42	−1.15

are considered together with $\rho^* = 0.2$, $a = 50$, $w = 50$ mm using the MTS criterion for $\overline{\theta}$ (a similar table is available using the MSED criterion). The correspondence between $K_t(\rho_f)$ and $\overline{K}_t$ is satisfactory in all cases, the maximum relative deviation amounting to 1.8 %.

1.4.12 Conclusions

Based on a new function-analytical solution for rounded V-notches (with root hole) subjected to in-plane shear loading, the microstructural support factor s is evaluated for different notch opening angles 2α as function of the real notch radius ρ under the condition of out-of-bisector crack propagation. The plateau values of s for $\rho > \rho^*$ depend slightly on the microstructural length ρ^*. The maximum values of s for $\rho = 0$ are independent of it. Both values of s depend heavily on the notch opening angle 2α and to some extent on the applied failure criterion. The fictitious notch rounding concept for $\rho > 0$ fails in the case of in-bisector crack propagation.

As demonstrated by the parametric numerical investigation of pointed V-notch specimens ($\rho = 0$) subjected to in-plane shear loading with out-of-bisector crack propagation, the maximum values of s are well suited for engineering usage in structural strength assessments with the worst case assumption $\rho = 0$. Theses values of s are ready for incorporation in those codes and design recommendations for structural components under static or cyclic loading conditions (brittle fracture or fatigue failure) which comprise local approaches such as the notch stress or notch strain approach.

1.5 Application-Relevant Issues of the Fictitious Notch Rounding Concept

1.5.1 Survey of Section Contents

As explained in more detail in Sect. 1.1, the continuing efforts to improve Neuber's fictitious notch rounding concept originate both from academic and application-relevant motivations. The more academic questions have been dealt with in the preceding Sects. 1.2–1.4. Issues of application are now brought into focus.

In Sect. 1.5.2, fictitious notch rounding is considered in respect of mixed mode loading of crack or slit tips.

In Sect. 1.5.3, the peculiarities of fictitious notch rounding of short cracks or slits are presented.

In Sect. 1.5.4, non-singular and free-surface singular stress effects of pointed notch tips are dealt with.

In Sect. 1.5.5, the effect of cross-sectional weakening by the 'real fictitious' notch is addressed.

In Sect. 1.5.6, the effect of mean stress and residual stress is discussed in respect of the fictitious notch rounding concept.

In Sect. 1.5.7, the theoretical concept of fictitious notch rounding is compared with the application-relevant concept of a uniform reference notch.

In Sect. 1.5.8, the common design *S–N* curves referring to reference notches are reviewed.

In Sect. 1.5.9, the peculiar effects of a reference notch at the weld root in cruciform joints are discussed.

In Sect. 1.5.10, the peculiar effects of a reference notch at the weld root of thin-sheet welded lap joints are considered.

In Sect. 1.5.11, the theoretical *S–N* curve of a spot-welded lap joint is determined based on Neuber's microsupport and macrosupport concepts.

The references are found in a list at the end of this chapter, proceded by a list of symbols.

1.5.2 Fictitious Notch Rounding in Case of Mixed Mode Loading

Mixed mode loading of notches inclusive of cracks or slits is the normal case in engineering applications. There are two topics which can be dealt with in respect of mixed mode loading of crack tips using the stress averaging concept. One is the question what equivalent stress intensity factor K_{eq} in terms of K_{I}, K_{II}, K_{III} results from the assumption of the same value of ρ^* in the three loading modes. The other is directed to the engineer's demand to use only one fictitious notch radius despite different values of s for assessing the structural strength. To begin with, in-bisector crack propagation is assumed.

The equivalent stress intensity factor K_{eq} is derived below by inserting $\overline{\sigma} = K_{I}(2/\pi\rho^*)^{1/2}$, $\overline{\tau} = K_{II}(2/\pi\rho^*)^{1/2}$, $\overline{\tau}^* = K_{III}(2/\pi\rho^*)^{1/2}$ and $\overline{\sigma}_{eq} = K_{eq}(2/\pi\rho^*)^{1/2}$ into the relevant equivalent stress equation, compare Eqs. (1.11), (1.95) and (1.40).

The normal stress criterion under plane stress or plane strain conditions results in:

$$\overline{\sigma}_{eq} = \overline{\sigma} + \sqrt{\overline{\tau}^2 + \overline{\tau}^{*2}} \tag{1.113}$$

$$K_{eq} = K_{I} + \sqrt{K_{II}^2 + K_{III}^2} \tag{1.114}$$

The von Mises distortional strain energy criterion under plane stress conditions results in:

$$\overline{\sigma}_{eq} = \sqrt{\overline{\sigma}^2 + 3(\overline{\tau}^2 + \overline{\tau}^{*2})} \tag{1.115}$$

$$K_{eq} = \sqrt{K_{I}^2 + 3(K_{II}^2 + K_{III}^2)} \tag{1.116}$$

whereas under plane strain conditions, the coefficient of $\overline{\sigma}^2$ or K_{I}^2 is $(1 - 2\nu)^2$.

The Beltrami total strain energy criterion under plane stress conditions results in:

$$\overline{\sigma}_{eq} = \sqrt{2(1-\nu)\overline{\sigma}^2 + 2(1+\nu)(\overline{\tau}^2 + \overline{\tau}^{*2})} \tag{1.117}$$

$$K_{eq} = \sqrt{2(1-\nu)K_{I}^2 + 2(1+\nu)(K_{II}^2 + K_{III}^2)} \tag{1.118}$$

whereas under plane strain conditions, the coefficient of $\overline{\sigma}^2$ or of K_{I}^2 is $2(1-2\nu)(1+\nu)$.

The demand for a uniform fictitious notch radius in mixed mode loading is satisfied by transforming the maximum notch stresses in the equivalent stress formula to a mode-independent fictitious notch radius using the square root on a/ρ dependency of these stresses (notch depth a, notch radius ρ). Only the von Mises criterion combined with the keyhole notch is considered. Crack or slit-parallel loading is assumed to be negligible.

The maximum notch stresses are determined based on the relevant values of s appertaining to the keyhole notch. These values are indicated by the subscripts:

$$\sigma_{eq\,max} = \sqrt{\sigma^2_{max\,2.25} + 3\tau^2_{max\,0.2543} + 3\tau^{*2}_{max\,1.0}} \tag{1.119}$$

Converting the squared notch stresses to the fictitious radius $\rho_f = 2.25\rho^*$ of mode 1 loading, one gets:

$$\sigma_{eq\,max} = \sqrt{\sigma^2_{max2.25} + 26.48\tau^2_{max2.25} + 6.75\tau^{*2}_{max2.25}} \tag{1.120}$$

where the maximum notch stresses have to be determined in the bisector plane of the keyhole notch of radius $\rho = 2.25\rho^*$.

In addition, out-of-bisector crack propagation is investigated. Considering the values of s in the three loading modes referring to fictitious blunt circular notches, these are the relevant data, using the normal stress failure criterion:

- $2\alpha = 0°$: $s = 2.0$ (mode 1), $s \approx 2.5$ (mode 2), $s = 1.0$ (mode 3)
- $2\alpha = 60°$: $s = 2.4$ (mode 1), $s \approx 10.9$ (mode 2), $s = 1.2$ (mode 3)

Considering $2\alpha = 0°$, the fictitious notch radius of mode 1 loading may also be used in mixed mode 1 and 2 loading (resulting in conservative assessments), but mode 3 loading demands a substantially smaller radius. For $2\alpha \geq 60°$, the s values in mode 2 loading become so large that the effect of mode 2 loading can generally be neglected in mixed mode 1 and 2 loading. Once more the value of s in mode 3 loading is substantially smaller (about one half of the value of s in mode 1 loading).

Evaluating the stresses at fictitiously rounded notches is not superior to an evaluation of the stress intensity factors of the original slit or crack. Cross-sectional weakening by the 'real fictitious' notch is avoided in the latter case and also the disturbing notch effect of slit-parallel loading.

1.5.3 Fictitious Notch Rounding of Short Cracks or Slits

When applying the fictitious notch rounding concept to short cracks or slits in welded joints, in the latter case with the aim to assess weld root fatigue, a circular hole (instead of a keyhole) may substitute the crack or slit, Fig. 1.42. This occurs under the condition $a = \rho_f = s\rho^*$, which means that the restriction to the derivations above for fictitious notch rounding, $\rho = \rho_f \ll a$ (with crack or slit length a), is not met. An estimate is provided by the following considerations as to whether such a circular notch will give conservative results in structural strength assessments, i.e. maximum notch stresses which are higher than the averaged stresses at the crack or slit tip (Radaj et al. 2006).

For short cracks or slits, the averaged stresses over the microstructural support length ρ^* can be derived as follows (Sähn and Göldner 1989):

$$\frac{\bar{\sigma}}{\sigma_n} = \sqrt{1 + 2\frac{a}{\rho^*}} \tag{1.121}$$

$$\frac{\bar{\tau}}{\tau_n} = \sqrt{1 + 2\frac{a}{\rho^*}} \tag{1.122}$$

$$\frac{\bar{\tau}^*}{\tau_n^*} = \sqrt{1 + 2\frac{a}{\rho^*}} \tag{1.123}$$

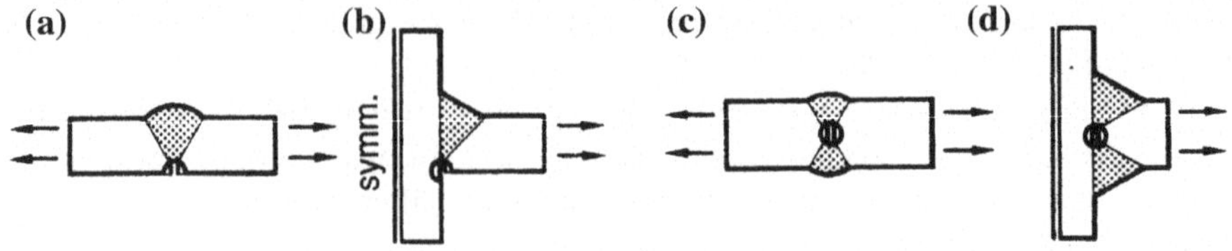

Fig. 1.42 Short slits fictitiously notch-rounded: single-V and double-V butt weld with root face (**a**, **c**); single-bevel and double-bevel butt weld with root face (**b**, **d**); (Radaj 1997)

These stresses have to be equated with the maximum stresses according to the unreduced equations for the elliptical notch:

$$\frac{\sigma_{\max}}{\sigma_n} = 1 + 2\sqrt{\frac{a}{s\rho^*}} \tag{1.124}$$

$$\frac{\tau^*_{\max}}{\tau^*_n} = 1 + \sqrt{\frac{a}{s\rho^*}} \tag{1.125}$$

The stress concentration $\tau_{\max}/\tau_n$ missing above is given by Eq. (1.55), inserting $\rho = s\rho^*$.

The question can now be answered, to what extent fictitious notch rounding of short cracks (or slits) in welded joints is conservative in structural strength assessments. Obviously the averaged stresses are enlarged for short cracks because of the first term in Eqs. (1.121–1.123). On the other hand, the values of s used in Eqs. (1.124), (1.125) and (1.55) are derived for long cracks. They are different from the value of s for short cracks which might be given in terms of a/ρ^* by equating the relevant unreduced equations above for the averaged and maximum notch stresses.

As a numerical example, a short crack or slit which is fictitiously rounded to a full circle using the value of s of long cracks is considered ($\rho_f = s\rho^*$). For tensile loading with $s = 2.0$, $\rho^* = 0.5$ mm, $a = \rho_f = 1$ mm, the result is:

$$\frac{\bar{\sigma}}{\sigma_n} = 2.236, \quad \frac{\sigma_{\max}}{\sigma_n} = 3.0, \quad \frac{\sigma_{\max}}{\bar{\sigma}} = 1.341 \tag{1.126}$$

For in-plane shear loading with $s = 0.0741$, $\rho^* = 13.495$ mm, $a = \rho_f = 1$ mm, the result is:

$$\frac{\bar{\tau}}{\tau_n} = 1.0715, \quad \frac{\tau_{\max}}{\tau_n} = 1.333, \quad \frac{\tau_{\max}}{\bar{\tau}} = 1.2443 \tag{1.127}$$

For out-of-plane shear loading with $s = 1.0$, $\rho^* = 1$ mm, $a = \rho_f = 1$ mm, the result is:

$$\frac{\bar{\tau}^*}{\tau^*_n} = 1.732, \quad \frac{\tau^*_{\max}}{\tau^*_n} = 2.0, \quad \frac{\tau^*_{\max}}{\bar{\tau}^*} = 1.155 \tag{1.128}$$

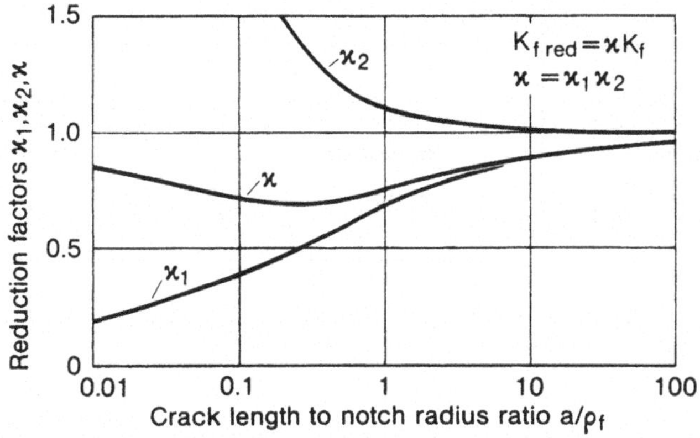

Fig. 1.43 Reduction of the fatigue notch factor of short cracks by the combined effects of a mild instead of a sharp notch (κ_1) and of closely opposed crack tips (κ_2); with $s = 2.0$; (Radaj 1997)

The conclusion from $\sigma_{max}/\overline{\sigma} > 1.0$, $\tau_{max}/\overline{\tau} > 1.0$ and $\tau^*_{max}/\overline{\tau}^* > 1.0$ is that conservative results are gained in all three cases. The value of s for the short crack is larger than for the long crack, $s = 5.236$ in the first case and $s = 1.309$ in the third case.

The short crack notch rounding option has also been addressed in a slightly different manner (Radaj 1997). Substituting the simplified equations valid for $a/\rho_f \gg 1.0$ by the original equations without this simplification establishes a rise both in the averaged stresses according to Eqs. (1.121–1.123) and in the stress concentration factors according to Eqs. (1.124), (1.55) and (1.125), but the rise in the latter is stronger than in the former. Therefore, the fatigue-effective stress concentration factor or fatigue notch factor K_f, evaluated at the fictitious notch, is reduced:

$$K_{f\,red} = \kappa_1\kappa_2 K_f = \frac{1}{1 + 2\sqrt{a/\rho_f}}\sqrt{1 + \frac{4a}{\rho_f}}\,K_f \qquad (1.129)$$

when considering tensile loading with $s = 2.0$. The reduction factors κ_1, κ_2 and $\kappa = \kappa_1\kappa_2$ are plotted dependent on a/ρ_f in Fig. 1.43 ($a/\rho_f < 1.0$ designates an elliptical hole turned into the loading direction). The fatigue notch factor is reduced in the whole range of a/ρ_f. The reduction at $a/\rho_f = 1.0$ is conforming with Eq. (1.126). The reduction expressed by Eq. (1.129) is only slightly changed when considering the two other loading cases with the other relevant values of s.

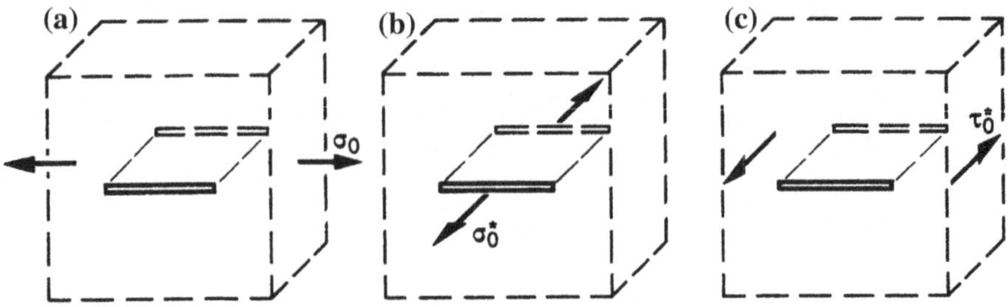

Fig. 1.44 Basic loading modes of slit or crack tips associated with non-singular stresses; mode 0I (**a**), mode 0II (**b**) mode 0III (**c**); (Radaj and Zhang 1993[(1)])

1.5.4 Non-Singular and Free-Surface Singular Stress Effects

The fictitious notch rounding concept implicitly assumes that the strength-effective loading conditions at pointed or sharp notches can sufficiently well be described by the superposition of the three loading modes 1, 2 and 3 at the notch tip, which are connected with singular stresses in the case of pointed notch tips (with the exception of mode 2 for $2\alpha > 102.6°$). But this is not always the case. There are further effects which may be important. One group of effects is related to three non-singular loading modes, Fig. 1.44: slit-parallel stresses ($2\alpha = 0°$), slit-front-parallel normal stresses and slit-front-parallel shear stresses (i.e. out-of-plane). Singular and non-singular stress components together describe the complete stress state at pointed notch tips under plane and antiplane stress field conditions. The effect of the three non-singular loading modes on fictitious notch rounding procedures is considered in the following.

Most important is slit-parallel loading which corresponds to the *T*-stress in the ligament of the slit tip and along the slit flanks. No stress concentration occurs with the ideal slit, but a substantial notch stress concentration may be produced by an actual hole at the slit tip representing fictitious notch rounding in a FE model. The corresponding stress concentration factor, $K_t \approx 3.0$, is unrealistic for strength assessments. Therefore, the slit-parallel load should be sufficiently low in applications of the method. It is recommended, not to apply actual notch rounding in cases of severe slit-parallel loading, but to determine the stresses at the fictitious notch from the stress intensity factors at the slit tip.

The notch stresses for the keyhole subjected to slit-parallel tensile (or compressive) loading according to the Airy stress function method (Radaj et al. 2001) are plotted in Fig. 1.45 (upper diagram). The more accurate solution results in

$$\sigma_{\theta\,\max} = 3.06\sigma_0 \quad (\theta = 90°) \tag{1.130}$$

as opposed to the simpler single hole solution, $\sigma_{\theta\ \max} = 3.0\sigma_0$ where σ_0 is the slit-parallel tensile or compressive basic stress. The slit-transverse stress at the apex point of the keyhole is given by:

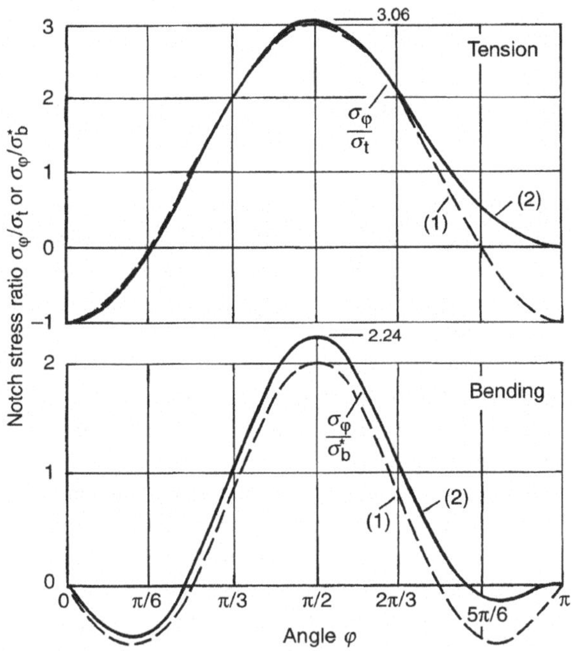

Fig. 1.45 Notch stress distribution at keyhole subjected to slit-parallel tensile and bending loading, solution variants (1) and (2); polar angle $\varphi = \theta$; (Radaj et al. 2001)

$$\sigma_\theta = -\sigma_0 \quad (\theta = 0°) \tag{1.131}$$

This stress is a disturbing element when evaluating the maximum notch stress produced by mode 1 loading of the slit.

The notch stresses for the keyhole subjected to a slit-parallel bending load are plotted in Fig. 1.45 (lower diagram). The more accurate solution (Radaj et al. 2001) results in

$$\sigma_{\theta\,\max} = 2.24\sigma_b^* \quad (\theta = 90°) \tag{1.132}$$

as opposed to the simpler single hole solution, $\sigma_{\theta\,\max} = 2.0\sigma_b^*$ where σ_b^* is the basic bending stress at the level of the hole radius.

Predominant slit-parallel loading combined with in-plane shear loading occurs in butt-welded joints with backing plate, in fillet-welded attachment joints, cover-plate joints and lap joints. These types of joint are analysed to some extent in respect of fatigue by the fictitious notch rounding concept in Sects. 1.5.9 and 1.5.10 as well as in Sect. 3.5.3.

Another non-singular stress state at pointed notches is out-of-plane shear loading with equally directed shear forces on the two slit sides, in contrast to mode 3 loading where these forces have opposite directions. The maximum notch stress at the keyhole results in:

$$\tau^*_{z\,\max} = 2\tau^*_0 \quad (\theta = 90^\circ) \tag{1.133}$$

with the non-singular basic shear stress τ^*_0. There is no problem with fictitious notch rounding in this non-singular loading case, because the maximum notch stress of the fictitious notch in mode 3 loading occurs at $\theta = 0^\circ$ where the effect of τ^*_0 disappears.

The third non-singular stress state at pointed notches is the tensile or compressive stress σ^*_0 in the direction of the slit front, which may be superimposed on the singular notch tip stresses in this direction, originating from the constraint condition of plane strain. These stresses do not disturb fictitious notch evaluations and can easily be taken into account within the relevant equivalent stress criterion.

For the sake of completeness, the free-surface singular stress effects are also addressed. The singular and non-singular loading modes 1, 2 and 3 refer to the in-plane and out-of-plane stress field conditions occurring in an infinitely extended continuum. In reality, plate-type specimens or components occur which are of finite thickness with load-free surfaces. Special stress effects originate at these surfaces, when the slit front is butting on it. Under mode 2 loading a transverse shear stress singularity occurs, which is described by a special corner or crack stress intensity factor K_O. The maximum value of K_O is found closely underneath the lateral surfaces of the specimen. Further details are given in Sect. 2.3.3. In mode 1 loading, K_1 is substantially reduced near the lateral surfaces. In mode 3 loading, K_3 must be reduced to zero at the lateral surfaces. These local stress intensity factor effects at lateral surfaces are not taken into account in the fictitious notch rounding concept. Corresponding stress effects are found to be substantially diminished in the case of finite notch radii.

1.5.5 Effect of Cross-Sectional Weakening

In applications of the fictitious notch rounding concept to welded joints, the real small notch radius at the weld toe or weld root is substituted by the larger fictitious notch radius. In the FE models of the welded joints, this fictitious notch is introduced as a real notch, termed 'reference notch'. Therefore, cross-sectional weakening may be produced by the enlarged notch radius. Its stress raising effect should be subtracted from the resulting notch stresses in order to separate the support effect represented by fictitious notch rounding.

In the following subsections, the symbol ρ_f for the 'real fictitious notch radius' is only occasionally substituted by the symbol ρ_r of the 'reference notch radius', despite the greater relevance of the latter symbol. This has historical reasons. The identity $\rho_f = \rho_r = 1$ mm is valid for welded joints in the following expositions.

The problem of cross-sectional weakening is further discussed with reference to the plane model of a tensile loaded transverse stiffener joint, Fig. 1.46. Pointed weld toe and weld root notches are assumed as the worst case of the original model (a).

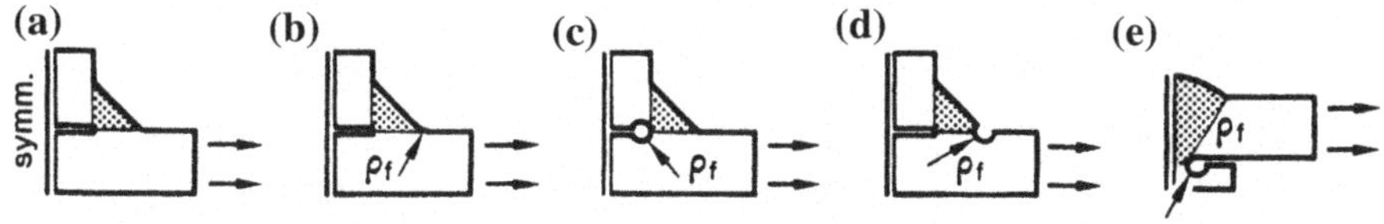

Fig. 1.46 Undercut by real fictitious notch rounding: pointed notches (**a**), weld toe notch rounded without undercut (**b**), weld root notch rounded with undercut (**c**), weld toe notch rounded with undercut (**d**), weld root notch rounded without undercut of the load-carrying cross-section (**e**); (Radaj et al. 2006)

Weld toe rounding with radius ρ_f without undercut, i.e. without cross-sectional weakening, is considered in general (b). Weld root rounding with radius ρ_f produces cross-sectional weakening both in the base plate and in the stiffener plate (c). With the hole centred by the theoretical root point, the cross-section of the fillet-weld would additionally be weakened. Centreing the toe by the theoretical weld toe point, produces severe undercut with corresponding cross-sectional weakening (d). Simultaneously, the force flow into the stiffener plate is reduced. Finally, weld root rounding with radius ρ_f for a butt weld with backing plate (e) demonstrates, that undercut can be avoided in the load-carrying plate whereas the notch in the non-load-carrying backing plate is without influence on the maximum notch stresses in the base plate.

The weakening of the cross-section by the notches can be taken into account in the fatigue-effective notch stresses or fatigue notch factors by way of a reduction factor which results from the tensile and bending stress increase in the equally weakened unnotched plate according to the plate bending theory (Radaj 1990; Radaj and Helmers 1997; Radaj et al. 2006). The reduction factor depends on the degree of cross-sectional weakening and on the type of loading (tensile and bending loads superimposed). Different reduction terms are introduced for the single-sided and double-sided notch, respectively, in comparison to the unnotched plate, Fig. 1.47:

$$K_f^* = \frac{(1-\rho_f^*)^2}{1+\rho_f^*(1+\sigma^*)}K_f \tag{1.134}$$

$$K_f^* = \frac{2(1-\rho_f^*)^2}{2-\rho_f^*(1+\sigma^*)}K_f \tag{1.135}$$

where K_f and K_f^* are the original and the reduced fatigue notch factor, respectively, ρ_f^* is the degree of cross-sectional weakening, $\rho_f^* = \rho_f/t$, σ^* is the surface stress ratio, $\sigma^* = \sigma_l/\sigma_u$, ρ_f is the fictitious or reference notch radius, t is the plate thickness and σ_l and σ_u are the remote stresses at the lower and upper plate surface. The condition $|\sigma_u| \geq |\sigma_l|$ or $|\sigma^*| \leq 1$ is introduced. The result will be accurate only for $\rho_f \ll t/2$.

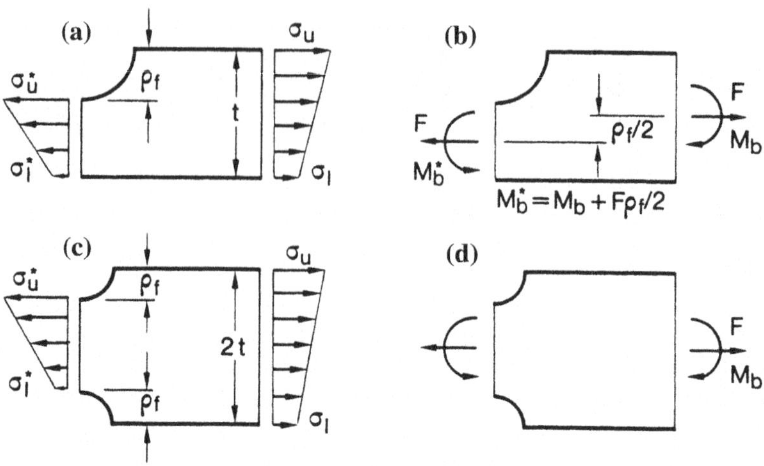

Fig. 1.47 Modification of the superimposed membrane and bending stresses at the upper and lower side of a structural component by real fictitious notch rounding with undercut on one side (**a**, **b**) or on both sides (**c**, **d**) of the cross-sectional model as the basis for the fatigue notch factor reduction: force and moment equilibrium (**b**, **d**) and associated membrane and bending stresses (**a**, **c**); (Radaj and Helmers 1997, Radaj et al. 2006)

Similarly, the reduction in the case of torsional-loaded circular shafts is given by the following formula:

$$K_f^* = (1 - \rho_f^*)^3 K_f \tag{1.136}$$

Another possibility for reducing the problem of cross-sectional weakening is to perform the notch stress analysis with extremely small fictitious notch radii and to use the generalised structure of the stress concentration formulae for welded joints according to Lawrence, $K_t = 1 + \alpha(t/\rho_f)^{1/2}$ or $K_t = \beta(t/\rho_f)^{\lambda}$ (Radaj et al. 2006), to convert the calculated stress concentration factors from the smaller to the larger fictitious notch radius. A first approximation may use the simple square root dependency on the ratio of the values of ρ_f.

A special modification of the fatigue notch factor is necessary in the case of two-sided cover plate joints if the cover plates are allowed to bend freely in the model whereas gap closure occurs in reality. The bending effect together with the constriction of the cross-section above the fictitious keyhole lowers the longitudinal stiffness of the cover plate. Therefore, the structural membrane stresses in the centre cross-section of the joint are lowered in the cover plates and increased in the base plate. Without the bending effect and keyhole, the longitudinal stresses in the slit faces of the two plates would be identical. A modification is therefore necessary to the fatigue notch factors calculated on the upper and lower semicircle of the keyhole and reduced according to the cross-sectional weakening, Eqs. (1.134) and (1.135). Note that the bending moment in the cover plate is

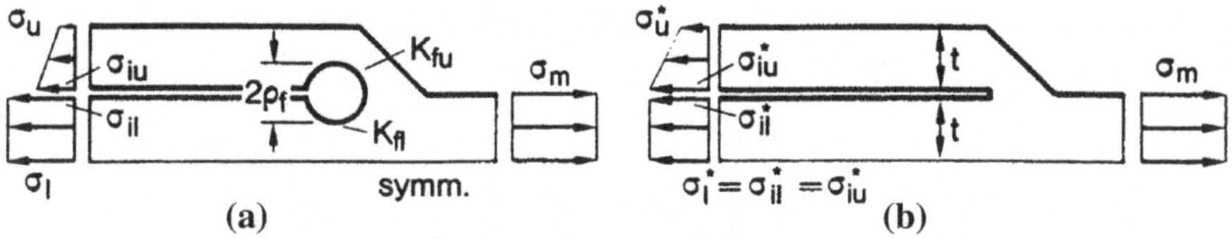

Fig. 1.48 Modification of the membrane plus bending stresses in the midsection of a double-sided cover plate joint (cross-sectional symmetry quarter) by real fictitious rounding of the root notch (**a**) in relation to the ideal slit (**b**); (Radaj and Helmers 1997; Radaj et al. 2006)

assumed to be unchanged within this modification. Different correction terms are introduced for the fatigue notch factors of the upper and lower semicircle, respectively, of the notch, Fig. 1.48 (Radaj and Helmers 1997; Radaj et al. 2006):

$$K_{\mathrm{f,u}}^{**} = \frac{1}{2}\left(1 + \frac{\sigma_{\mathrm{i,u}}}{\sigma_{\mathrm{i,l}}}\right) K_{\mathrm{f,u}}^{*} \tag{1.137}$$

$$K_{\mathrm{f,l}}^{**} = \frac{1}{2}\left(1 + \frac{\sigma_{\mathrm{i,l}}}{\sigma_{\mathrm{i,u}}}\right) K_{\mathrm{f,l}}^{*} \tag{1.138}$$

where K_{f}^{*} and K_{f}^{**} are the original reduced and the modified reduced fatigue notch factors, $\sigma_{\mathrm{i,u}}$ is the stress at the inner surface of the (upper) cover plate and $\sigma_{\mathrm{i,l}}$ is the stress at the inner surface of the (lower) base plate.

1.5.6 Effect of Mean Stress and Residual Stress

The fatigue limit is governed by the stress amplitude σ_{a} or stress range $\Delta\sigma = 2\sigma_{\mathrm{a}}$. The mean stress σ_{m} or its substitute, the stress ratio $R = \sigma_{\mathrm{l}}/\sigma_{\mathrm{u}}$ of lower to upper stress, is a secondary effect. The endurable stresses decrease with tensile mean stresses and rise with compressive mean stresses. The details are given in the author's books on fatigue (Radaj and Vormwald 2007; Radaj et al. 2006).

Only a minor complementary question is addressed here. To what extent is the fictitious notch radius or are the parameters s and ρ^* dependent on the mean stress or stress ratio? The microstructural length ρ^* is given by Neuber as a material parameter for $R = -1$, but it is generally accepted that identical values are valid for $R = -1$ and $R = 0$. Therefore, ρ^* may be introduced as approximately independent on σ_{m} or R. On the other hand, the support factor s was shown to be independent of ρ^* in the previous sections.

Residual stresses may be considered as a special case of mean stress. Their influence on fatigue is strong for mildly notched components where crack initiation processes are predominant, but minor for sharply notched members, where crack propagation prevails. There is no major influence on s or ρ^* according to the statements given above. The influence of the residual stresses is mainly expressed in different endurable stress amplitudes (literature as quoted above).

1.5.7 Fictitious Notch Compared with Reference Notch

The notch rounding concept has been developed on two different lines. The first development line comprises theoretical developments referring to the relationship between locally averaged and maximum substitute notch stresses for simple notch shapes without the interference of further dimensional or shape parameters such as nearby boundaries. The purely elastic theoretical frame of microstructural support may be extended to the locally elastic-plastic frame of macrostructural support. When applied to fatigue failure, the purely elastic concept refers to the high-cycle fatigue endurance limit, whereas the elastic-plastic concept includes the finite life range. A theoretical *S–N* curve may be found by elastic-plastic notch analysis combined with the strain *S–N* curve of the material.

The second development line aims at an empirically founded notch stress *S–N* curve for welded joints (Radaj et al. 2009[(2)]) based on a uniform reference notch radius. Two variants of the concept are in use. The conventional variant codified by the International Institute of Welding (IIW) applies a reference notch radius of 1 mm. It refers mainly to weld toe fatigue of fillet-welded and butt-welded joints. The endurance limit data are extrapolated into the finite life range on an empirical basis, approximating the scatter band of notch stress related *S–N* data of various welded specimens and structures (Hobbacher 2009; Fricke 2008).

The reference radius $\rho_r = 1$ mm was originally proposed based on a worst case consideration within the frame of fictitious notch rounding (Radaj 1990). The worst case assumption for the weld toe and weld root notches was $\rho = 0$ mm, that means pointed notch conditions. The support factor $s = 2.5$ (according to Neuber) was introduced referring to plane strain conditions and the von Mises equivalent stress. The relatively large microstructural support length $\rho^* = 0.4$ mm was related to the cast low-strength steel in the weld seam. These assumptions resulted in $\rho_f = 2.5 \times 0.4 = 1.0$ mm. This radius was used for the reference notch in later times, first for welded joints of steels, and then also for welded joints of aluminium alloys and of magnesium. The worst case within the fictitious notch rounding approach ($\rho = 0$) is not the worst case in the reference notch concept as far as slit-parallel basic stresses have to be taken into account.

A substantially smaller reference notch radius is considered more appropriate for thin-sheet ($t = 0.7$–3.0 mm) spot-welded and laser-beam-welded joints, in order to keep cross-sectional weakening low. The radius $\rho = 0.05$ mm is mainly used, which is slightly above the coarsened grain size of low-carbon steels. The common evaluation of the appertaining maximum notch stresses without averaging for the microstructural support effect is obviously a step-back from the available state of the art.

The relation between the theoretically founded fictitious notch rounding concept and the empirically based reference notch concept is a rather loose one. The theoretical developments are not directly transferable to engineering applications, but need proper scaling. This may be achieved by statistically evaluating fatigue test results in terms of the reference notch stresses, either by considering the total number of available fatigue test data or by analysing definite single joints (Olivier et al. 1989, 1994).

In engineering applications, not only the notch shape and notch radius have an influence on the notch stresses, but also the nearby boundaries as well as the remote boundary conditions. A typical geometrical multiparameter problem has to be solved. Closed-form analytical solutions are not possible. The FE method is generally used. A correction has to be applied to the FE results in order to take the effect of the finite reference notch size into account.

The procedure is explained by the example of seam-welded joints, the main application of the notch rounding approach. The actual small notch radius at the weld toe or weld root is substituted by a uniform reference notch radius. In the FE models of the welded joints, this reference notch is treated as a real notch, not as a fictitious notch. Thereby, cross-sectional weakening and also other effects produced by the enlarged notch radius within the nearby boundaries will take place. These effects should be removed from the results, because they do not occur with the original fictitious notch in the infinite domain.

The conclusion from this is, that analysing the actual notch and averaging the notch stresses along the probable crack propagation direction is the better methodological choice. Working with reference notches in FE models, which deviate substantially from the actual notch, can only give rough approximations.

The results from theoretical investigations into the fictitious notch rounding concept can be used to indicate trends of parametrical effects which should also be evident in the empirical findings with the reference notch stress evaluations. These can thus be better substantiated. The following trends can be predicted from the theoretical results presented in the preceding sections (see the support factor s in Tables 1.1, 1.3, 1.6 and 1.9, with the von Mises equivalent stress and plane strain conditions being relevant).

The main results from these theoretical investigations as compared with former knowledge is the existence of plateau values of the support factor s which depend heavily on the notch opening angle. The support factors are roughly twice as large for $2\alpha = 135°$ (weld toe) than for $2\alpha = 0°$ (weld root), with the exception of mode 2 loading for $2\alpha = 135°$ where they are undefined because of a non-singular notch stress behaviour. The conclusion is that the radius of the reference notch at the weld root should be approximately half as large as the radius at the weld toe, i.e. $\rho_r = 0.5$ mm instead of $\rho_r = 1.0$ mm, the reference radius in the IIW design recommendations. On the other hand, an enlargement of the smaller ρ_r value at the weld root may be substantiated by the effect of compressive residual stresses at this location.

The following trend can be stated with respect to the dependency of the support factor s on the loading mode. In mode 1 and mode 2 loading, identical reference notch radii, $\rho_r = 0.5$ mm at the weld root and $\rho_r = 1.0$ mm at the weld toe, may be used, but mode 3 loading demands roughly $0.4 \times \rho_r$ in order to remain conservative. On the other hand, the stress concentration in mode 3 is roughly one half of the stress concentration in mode 1, so that the effect of the smaller radius is more or less compensated in the maximum stresses.

From the point of view of microstructural crack initiation, the total strain energy criterion (after Beltrami) seems to be better suited than the other criteria for describing local fatigue failure at the slit or notch tip, because it comprises both the deviatoric and dilatoric stress and strain components, which are of influence on crack initiation. The plane strain condition approximates the actual condition in seam-welded or spot-welded joints. Plane stress, on the other hand, is only an academic device for this application. The jump factor $\kappa_B = 0.94$ for $2\alpha = 0°$ and $\kappa_B = 1.10$ for $2\alpha = 135°$ under plane strain conditions can certainly be neglected.

Because of the plateau behaviour of the support factor s with increasing real notch radius ρ, it should be possible to assess the fatigue limit of welded joints with post-weld enlarged weld toe radius by enlarging the reference radius correspondingly.

Major uncertainties into the fatigue assessment do not come in via the support factor s, but via the insufficient knowledge on the values of ρ^* for the material in the heat-treated state occurring at weld notches. The values of ρ^* of welded joints actually determined in the finite life range comprise elastic-plastic and crack propagation effects. The details of Neuber's evaluation of ρ^*, referring to non-welded materials, are not recorded in the literature. Also, the influence of static prestress either from the loading condition or from residual stresses on the values of ρ^* is uncertain. Neuber's values of ρ^* refer to alternating loading ($R = -1$).

1.5.8 Design S–N Curves Based on Reference Notches

The fictitious notch rounding concept for welded joints in its original version referred to the engineering high-cycle endurance limit at $N = 2 \times 10^6$ loading cycles, i.e. $\Delta\sigma_E = 240$ MPa (survival probability $P_s = 90$ %) for mild steel in the non-machined condition (Radaj 1990). The details of the various application-related development lines of this concept can be found in the first author's standard work on fatigue assessment of welded joints by local approaches (Radaj et al. 2006). In the following expositions, some central issues of the reference notch concept in applications to welded joints are discussed on an updated basis.

The endurance limit approach above was later on converted to an *S–N* curve approach. The *S–N* curve of the maximum notch stress σ_k at the reference notch ($\rho_r = 1$ mm) is described by:

$$(\Delta\sigma_{\mathrm{k}})^{k}N = C \tag{1.139}$$

$$C = (\Delta\sigma_{\mathrm{E}})^{k}(2 \times 10^{6}) \tag{1.140}$$

where the inverse slope exponent $k = 3$ is introduced, conforming with the generally accepted choice for the nominal and structural stress *S–N* curves of welded joints. The $\Delta\sigma_{\mathrm{E}}$ value, also named 'FAT class' (derived from 'fatigue strength'), refers to $P_{\mathrm{s}} = 97.7\ \%$. The basis are statistical evaluations of fatigue test data in terms of the maximum notch stresses at the reference notch. Fatigue test data obtained with high stress ratios (e.g. $R = 0.4$) are preferred, because they simulate the high tensile residual stresses in large welded components as distinguished from the smaller welded specimens. The notch stress approach to the fatigue assessment of welded joints is further specified in IIW design recommendations (Hobbacher 2009; Fricke 2008, 2011).

The notch stress *S–N* curve for welded joints of mild steel is characterised by $\Delta\sigma_{\mathrm{E}} = 225$ MPa ($P_{\mathrm{s}} = 97.7\ \%$). The corresponding value for aluminium alloys is $\Delta\sigma_{\mathrm{E}} = 71$ MPa and for magnesium $\Delta\sigma_{\mathrm{E}} = 28$ MPa. The inverse slope exponent $k = 3$ is generally introduced. The values of $\Delta\sigma_{\mathrm{E}}$ refer to the maximum principal stress range $\Delta\sigma_{1}$ at the notch root. When considering the von Mises equivalent stresses at the notch root, a lower FAT class may be appropriate because $\sigma_{\mathrm{vM}} \leq \sigma_{1}$. The plate thickness effect is already contained in the notch stresses.

The notch stress *S–N* curve above ($\Delta\sigma_{\mathrm{E}} = 225$ MPa) refers to sharp weld notches, i.e. to reference notches ($\rho_{\mathrm{r}} = 1$ mm) with a notch stress σ_{k} substantially above the structural stress σ_{s} in the base plate. A problem arises for relatively mild weld notches, i.e. to reference notches with a low stress concentration factor, e.g. welds with a low weld toe angle or welds in thin plates. It was decided in the IIW design recommendations, to assume a weld notch stress concentration factor $K_{\mathrm{w}} = \sigma_{\mathrm{k}}/\sigma_{\mathrm{s}} \geq 1.6$ for application of the curve FAT 225. Simultaneously, it has to be proven that the structural stress close to the weld remains below the structural stress curve FAT 160 of the parent material. The relevant design *S–N* curves are shown in Fig. 1.49. The parent material curve FAT 160 multiplied by $K_{\mathrm{w}} = 1.6$ intersects the notch stress curve FAT 225 at approximately 2×10^{5} cycles. The fatigue strength of the parent material thus governs the design in the medium to low-cycle fatigue range if $K_{\mathrm{w}} \leq 1.6$.

Another mild notch problem occurs with shape-improved weld toes with a real toe radius ρ, which has to be added to the worst case reference radius of 1 mm according to Neuber's fictitious notch rounding concept. It is proposed to use the curve FAT 200 in this case, once more restricted by a minimum notch stress concentration factor $K_{\mathrm{w}} = 1.6$ (Fricke 2008).

A re-analysis of fatigue data for welded joints in terms of the notch stress approach according to the IIW design recommendations (reference radius $\rho_{\mathrm{r}} = 1$ mm at the weld toe) has recently been published (Pedersen et al. 2010). A large number of experimental results are evaluated, 767 specimens in total:

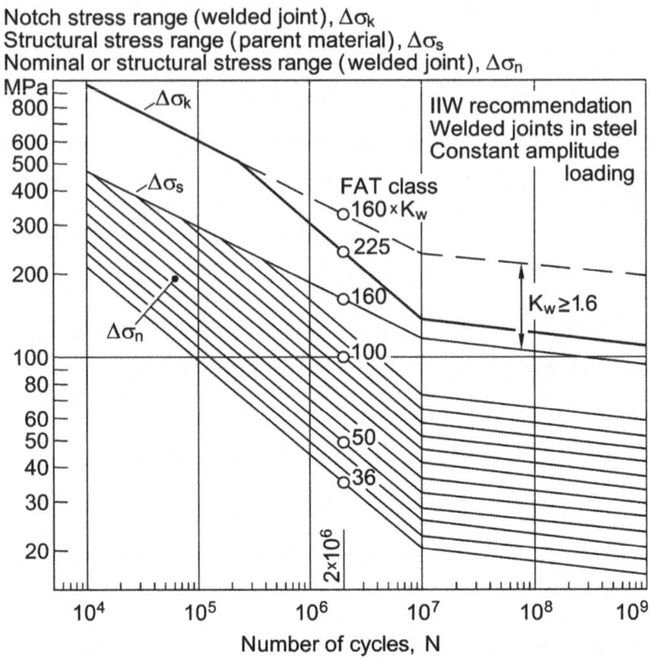

Fig. 1.49 Design *S–N* curves according to the IIW design recommendations; notch stress curve FAT 225 (relating to a reference notch radius $\rho_r = 1.0$ mm) truncated by FAT 160 × K_w with weld notch stress concentration factor $K_w \geq 1.6$; (Radaj et al. 2009[(2)])

T-joints, transverse attachments, butt joints and longitudinal attachments; plate thickness $t = 5$–25 mm, stress ratio $R = 0$–0.5, as-welded condition, steel grades S235 to S1110. It is found, that the butt joints with the relatively mild assigned notches (nominal weld toe angle 30°, $t = 5$–10 mm, $K_w = 1.6$–2.0) are assessed non-conservatively, when applying the FAT 200 curve. Therefore, a notch factor $K_w \geq 2.0$ is recommended instead of $K_w \geq 1.6$ for conservative assessments based on the FAT 225 curve without any truncation, Fig. 1.50. However, a reduction to FAT 200 is suggested in order to achieve approximately the same safety against failure as with the nominal stress approach ($P_s = 97.7$ %).

The general penalty resulting from $K_w \geq 1.6$ or 2.0 for well designed welded joints is not an incentive to optimise the shape of welded joints. It stands against well proven design recommendations.

Design *S–N* curves for thin-sheet welded joints referring to a reference micro-notch ($\rho_r = 0.05$ mm, $t \leq 5$ mm), defined without incorporating the microstructural support effect, are proposed as follows: FAT 630 for steels, FAT 180 for aluminium alloys and FAT 71 for magnesium (Fricke 2008).

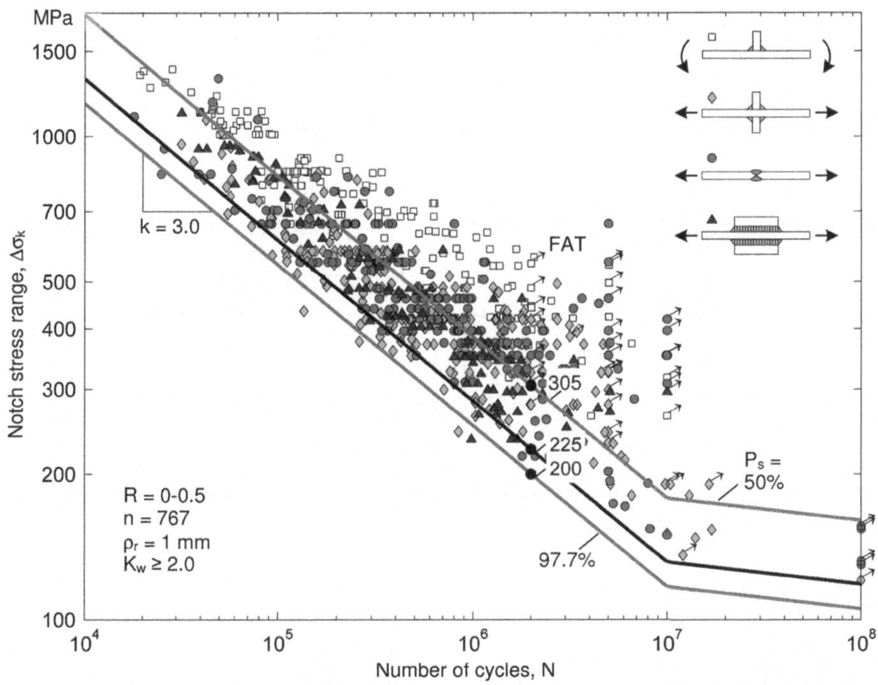

Fig. 1.50 Re-analysed fatigue test data of various welded joints and plate thicknesses in terms of endurable weld toe notch stresses for the reference notch radius $\rho_r = 1$ mm, compared with design *S–N* curves FAT 200 and FAT 225; (Pedersen et al. 2010)

1.5.9 Peculiarities of Reference Notches at the Weld Root in Cruciform Joints

The use of reference notches at weld toes for the fatigue assessment of welded joints based on fatigue-effective notch stresses is well established. The appropriate procedure at weld roots is still under discussion (Fricke 2011). Some peculiarities of the stress evaluation at reference notches at the weld root in fillet-welded cruciform joints are now considered. The basis is an early investigation which is not up-to-date anymore in all aspects (Radaj and Zhang 1993$^{(2)}$) and a more recent topical paper (Lazzarin et al. 2009).

The first joint considered is a tensile loaded cruciform joint with non-load-carrying fillet welds, the slits in the direction of the tensile load (transverse stiffener joint), Fig. 1.51a. The reference notch radius is $\rho_r = 1$ mm, the plate thickness $2t = 10$ mm, the weld throat thickness $t_w = 6$ mm, the overall slit length $2a = 10$ mm. The slit tip is rounded to a keyhole with its apex point coinciding with the root of the fillet weld. Cross-sectional weakening of the weld (not of the base plate) is thus avoided. The results, gained from a stress analysis using the boundary element (BE) method, are shown in Fig. 1.52a. The fatigue-effective

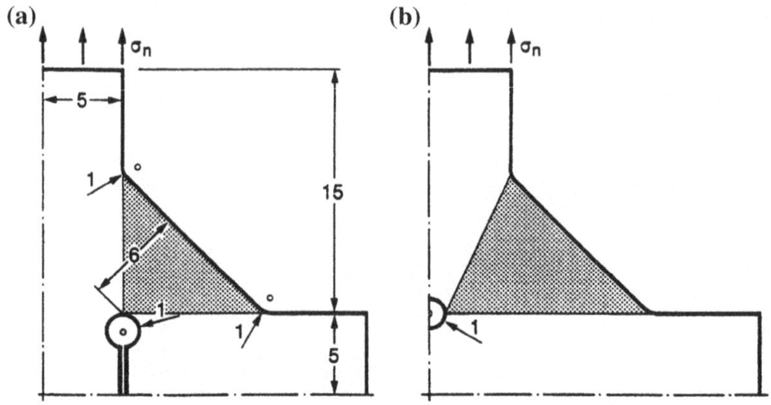

Fig. 1.51 Tensile loaded cruciform joints: transverse stiffener joint with non-load-carrying fillet welds (**a**) and cruciform joint with load-carrying double bevel butt weld with fillet welds and residual slit face (**b**); cross-sectional models of symmetry quarter; reference notch radius $\rho_r = 1.0$ mm; (Radaj and Zhang 1993(2))

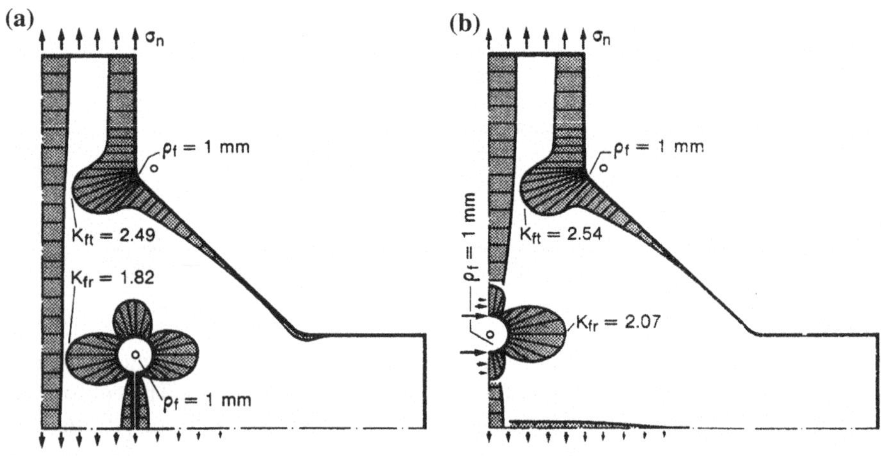

Fig. 1.52 Notch stresses at transverse stiffener joint with fillet welds (**a**) and cruciform joint with double bevel butt welds and residual slit face (**b**) subjected to tensile loading; fatigue notch factors K_f at weld toe (index t) and weld root (index r); (Radaj and Zhang 1993(2))

stress concentration factors (SCFs) at the keyhole are $K_f = 1.82$ ($\theta \approx -90°$, base plate side), $K_f \approx -1.3$ ($\theta = 0°$) and $K_f \approx 1.6$ ($\theta \approx +90°$, stiffener side).

For comparison, the SCF values are also calculated on the basis of the stress intensity factors (SIFs), K_{I} and K_{II}, of the pointed slit tip and the T-stress characterising slit-parallel loading in an attempt to separate the effect of nearby boundaries on the notch stresses as well as the effect of slit-parallel loading. The normalised SIFs resulted in $k_{\mathrm{I}} = K_{\mathrm{I}}/\sigma_n(\pi a)^{1/2} = -0.119$ (minus sign for slit closure) and

$k_{II} = K_{II}/\sigma_n(\pi a)^{1/2} = 0.032$. The normalised slit-parallel stress or T-stress is $T/\sigma_n = 0.592$. Referring to K_I, K_{II} and T, the SCFs for the keyhole, see Eqs. (1.12), (1.82), (1.130) and (1.131), are $K_{f,I} = -0.564$ ($\theta = 0°$), $K_{f,II} = \pm 0.205$ ($\theta = \pm 67.1°$), $K_{f,T} = -0.592$ ($\theta = 0°$) and $K_{f,T} \approx 1.812$ ($\theta = \pm 90°$).

The following comparisons between the actual SCF values in the model and the SIF-based keyhole SCF values above properly superimposed are made:

- hole at base plate side: $K_f = 1.82$ compared with $K_f \approx 1.82 - 0.20 = 1.61$,
- hole at ligament: $K_f \approx -1.30$ compared with $K_f \approx -0.56 - 0.59 = -1.15$,
- hole at stiffener side: $K_f \approx 1.60$ compared with $K_f \approx 1.81 + 0.20 = 2.01$.

There is only a rough correspondence between the actual and approximated K_f values because of nearby boundary effects not taken into account in the SCF formulae for the keyhole. Most striking is the change in position of the maximum value of K_f from the stiffener side to the base plate side. Highly remarkable is the fact that the T-stress or slit-parallel stress contributes most to these maxima despite its irrelevance in respect of the pointed slit.

The conclusion is that fictitious notch rounding simulated by a real reference notch is not appropriate for the pointed slit tips in cruciform and comparable non-load-carrying stiffener joints. There is some relevance only to slit tips with keyhole-similar manufacturing imperfections.

The second joint considered is a tensile loaded cruciform joint with load-carrying double bevel butt weld with fillet welds containing a residual short slit transverse to the tensile load, Fig. 1.51b. The short slit is rounded to a full circle ($a = \rho_f = 1$ mm). The result of the BE notch stress analysis is shown in Fig. 1.52b. The slit hole SCF is $K_f = 2.07$. The short-slit reduced SCF is $K_{f\ red} = \kappa K_f = 0.745 \times 2.07 = 1.544$, see Eq. (1.129).

The SIF analysis for pointed slit tips resulted in $k_I = 0.658$ and $k_{II} = 0$. The SCF derived therefrom using Eq. (1.10) is $K_f = 2k_I(a/\rho)^{1/2} = 1.316$. This value corresponds roughly with $K_{f\ red}$ above. The conclusion is that the value of K_f from the BE model is too high, but can be corrected to a more realistic value as shown above.

The third joint considered is a tensile loaded cruciform joint with load-carrying fillet welds (throat thickness t_w), the slits transverse to the direction of the tensile load ($\rho_r = 1$ mm, $2t = 10$ mm, $t_w = 10$ mm, $2a = 10$ mm), Fig. 1.53. The maximum tensile notch stress, SCF $K_f = 2.80$, occurs at $\theta = 0°$. Lower maximum compressive notch stresses are found at $\theta > \pm 90°$ on both sides of the hole. Slit-parallel compressive stresses, substantially below the nominal tensile stress σ_n and of different magnitude at the two slit faces are determined.

A further analysis of this type of joint using the FE method refers to an extremely small keyhole ('microkeyhole') in combination with a weld throat thickness substantially smaller than above ($\rho_r = 0.05$ mm, $t_w = 4$ mm), (Lazzarin et al. 2009). The SIF analysis for the joint with pointed slit tips resulted in

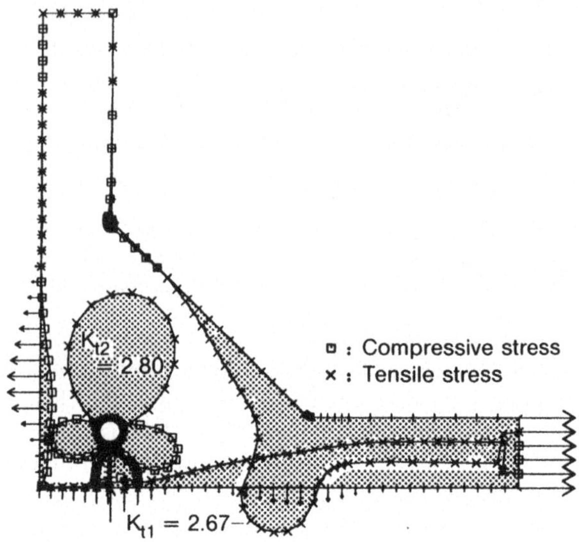

Fig. 1.53 Notch stresses at cruciform joint with load-carrying fillet welds subjected to tensile loading; fatigue notch factors $K_{t1} = K_{ft}$ at weld toe and $K_{t2} = K_{fr}$ at weld root; (Radaj and Zhang 1990, 1991)

$k_I = K_I/\sigma_n(\pi a)^{1/2} = 0.855$, $k_{II} = K_{II}/\sigma_n(\pi a)^{1/2} = -0.141$ and $T/\sigma_n = -0.865$. Based on these data, the SCF values for the keyhole are $K_{f,I} = 18.118$ ($\theta = 0°$), $K_{f,II} = \pm 4.043$ ($\theta = \pm 67.1°$), $K_{f,T} = 0.865$ ($\theta = 0°$) and $K_{f,T} = -2.647$ ($\theta = \pm 90°$). Thus, at the apex point of the keyhole, $K_f = K_{f,I} + K_{f,T} = 18.983$ ($\theta = 0°$). Slightly larger values $K_{1,\rho} > K_I$ and $K_{2,\rho} > K_{II}$ combined with $T = 0$ are recorded in the original paper (Lazzarin et al. 2009), resulting in a slightly larger $K_f = 20.17$ which is confirmed by FE analysis, $K_f = 19.97$.

The available FE analysis data do not allow final conclusions because they are restricted to $-90° \leq \theta \leq +90°$. The stresses at microkeyholes should be described very well based on K_I, K_{II} and T, but larger keyholes, e.g. $\rho_r = 1$ mm, will certainly produce major deviations by the effect of nearby boundaries, cross-sectional weakening among them. Also, the contribution of the T-stress on the notch stresses in relation to the SIFs K_I and K_{II} rises with increasing notch radii.

The tensile loaded cruciform joint with non-load-carrying fillet welds (stiffener joint), which was considered as the first example, is characterised by relatively high slit-parallel stresses or its substitute parameter, the T-stress. The stress rising effect with respect to fatigue is extremely overrated in this case. A similar behaviour occurs with other stiffener joints with through-running tensile loads in the main plate and no loading of the stiffener. Examples are the butt weld joint with backing plate, the attachment with one-sided fillet weld and the fillet-welded cover plate joint (Radaj et al. 2009[(1)]).

It is essential for a valid fatigue assessment at weld roots based on the reference notch concept that the ambiguity of notch positioning relative to the slit tip has to be overcome. The wide range of possible notch positions is demonstrated for the butt weld joint with backing plate, Fig. 1.54. The different positions are

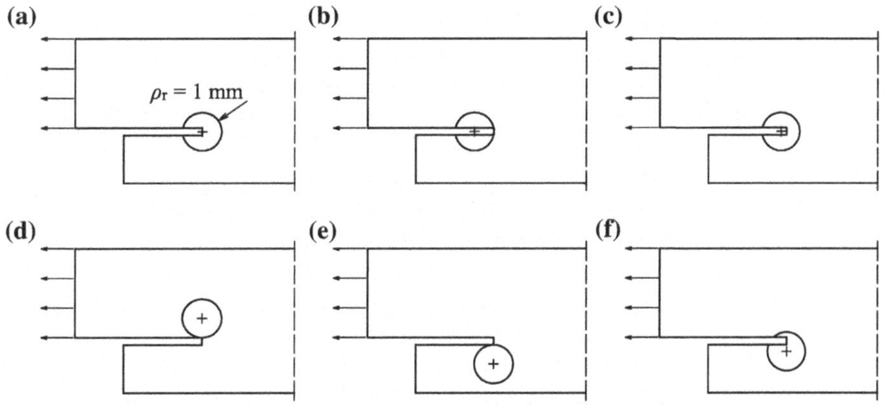

Fig. 1.54 Ambiguity of reference notch positioning within the notch rounding concept, when applied to the weld root of a butt weld joint with backing plate (symmetry half of cross-sectional model): concentric circle (**a**), eccentric circle horizontally displaced to different extents (**b**, **c**) and eccentric circle vertically displaced to different extents (**d**, **e**, **f**); (Radaj et al. 2009[(1)])

permissible options within the assessment procedure. Any notch in the tensile loaded main plate will generate too high notch stresses in respect of fatigue. Positioning of the notch completely in the non-loaded backing plate using an initially tangential notch contour will avoid a major stress-raising effect, but will produce too low notch stresses in respect of fatigue. Comparisons are available from the literature (Radaj et al. 2009[(1)]).

1.5.10 Peculiarities of Reference Notches at the Weld Root of Thin-Sheet Lap Joints

Thin sheet lap joints ($t = 0.7$–5.0 mm), either resistance spot-welded or laser beam seam-welded, are subject to special fatigue assessment procedures, especially in automotive engineering. These procedures deviate substantially from the procedures in areas where code regulations are mandatory, e.g. structural engineering or ship and offshore engineering. A variety of different approaches is available: hot spot structural stress or strain, stress intensity factors or *J*-integral, notch stress or strain and finally crack propagation concepts. Any procedure is acceptable if properly scaled by fatigue testing and confined to narrow areas of application (Radaj et al. 2006).

In the following, the results of two advanced, mainly theoretical investigations into the notch stress and strain approach applied to seam and spot welded lap joints are summarised. The first investigation refers to simple formulae for the calculation of the notch stresses at microkeyholes at the root of thin-sheet lap joints. The second investigation proceeds to the theoretical determination of the *S*–*N* curve of

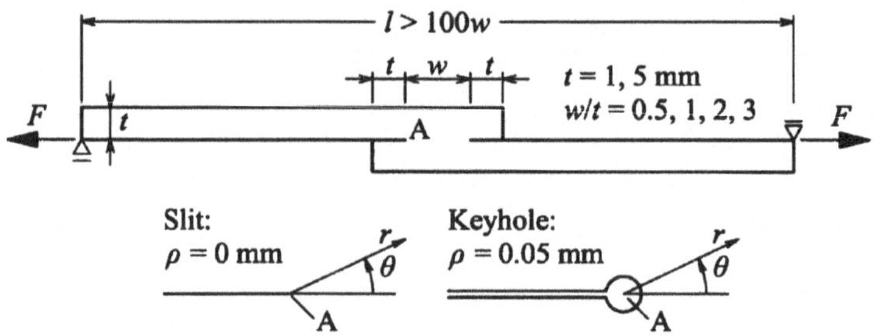

Fig. 1.55 Geometrical, loading and support conditions of tensile-shear loaded lap joint model, with enlargement of slit tip or keyhole; (Lazzarin et al. 2009)

a spot welded specimen based on the microstructural and macrostructural support concepts proposed by Neuber (Sect. 1.5.11).

The issue of the first investigation is to determine the notch stresses at the keyhole of a plane tensile-shear loaded lap joint model, Fig. 1.55. The historical details of the theoretical development of the notch stress formulae (two original and two final versions) are not addressed. Only the final amended equations are presented (Radaj et al. 2001; Lazzarin et al. 2009; Radaj 2010).

The notch stress formula for the microkeyhole in welded lap joint cross-sectional models (not only tensile-shear loaded) based on the structural stresses (linearised over the plate thickness) at the slit tip reads as follows (Radaj 2010):

$$\sigma_\theta = \frac{K_\mathrm{I}}{\sqrt{2\pi\rho}}\left(2\cos\frac{\theta}{2} + \cos\frac{3\theta}{2}\right) - \frac{K_\mathrm{II}}{\sqrt{2\pi\rho}}\left(2\sin\frac{\theta}{2} + 3\sin\frac{3\theta}{2}\right) + \overline{T}(1 - 2\cos 2\theta) \tag{1.141}$$

with the SIFs K_I and K_II of the corresponding pointed slit tip model, the keyhole radius ρ and the polar angle θ against the ligament. The T-stress $\overline{T}$ is the arithmetic mean between the slit flank stress $\sigma_{0\mathrm{s}}$ and the nominal membrane stress σ_m in the centre section of the lap joint model for $w/t \gg 1$ (with jointing face width w):

$$\overline{T} = \frac{1}{2}(\sigma_{0\mathrm{s}} + \sigma_\mathrm{m}) \tag{1.142}$$

The Eq. (1.141) above should separately be evaluated for $\theta \geq 0$ and $\theta \leq 0$ because the term $\overline{T}$ is different in the two angular ranges. The equation is also generally applicable by substituting $\overline{T}$ by the actual T-stress T.

The basic stress parameters K_I, K_II, $\sigma_{0\mathrm{s}}$ and σ_m are derived according to simple formulae from the linearised structural stresses in the cross-sectional model of the seam-welded lap joint or in the FE model of the spot-welded lap joint (Radaj et al. 2006). There is no FE analysis needed in the case of the cross-sectional model with $w/t \gg 1$ for determining the SIF values and the T-stress:

Table 1.14 SIFs, NSIFs and T-stresses in the tensile-shear loaded lap joint cross-sectional model dependent on plate thickness t and joint face width ratio w/t; nominal membrane stress $F/t = 10$ MPa; $K^*_{1\rho}$ defined with T-stress correction; signs of K_{II} and $K_{2\rho}$ inverted; (Radaj 2010; Lazzarin et al. 2009)

Geometry		SIFs K_{I}, K_{II}, NSIFs $K^*_{1\rho}$, $K_{2\rho}$ for $\rho = 0.05$ mm, [MPa mm$^{1/2}$]; T-stress T [MPa]								
t [mm]	w/t	K_{I}	$K^*_{1\rho}$	$K^*_{1\rho}/K_{\mathrm{I}}$	K_{II}	$K_{2\rho}$	$K_{2\rho}/K_{\mathrm{II}}$	$K_{\mathrm{II}}/K_{\mathrm{I}}$	$K_{2\rho}/K^*_{1\rho}$	T
1.0	0.5	1.62	1.78	1.10	12.22	17.05	1.40	7.54	9.56	20.0
1.0	1.0	4.35	4.38	1.01	10.28	11.06	1.08	2.36	2.52	19.5
1.0	2.0	8.08	8.75	1.08	9.95	10.05	1.01	1.23	1.15	18.0
1.0	3.0	8.61	9.51	1.10	9.95	10.05	1.01	1.16	1.06	17.6
5.0	0.5	3.67	3.66	1.00	27.30	28.73	1.05	7.44	7.84	20.0
5.0	1.0	9.80	9.82	1.00	23.02	23.25	1.01	2.35	2.37	19.6
5.0	2.0	18.14	18.4	1.02	22.27	22.30	1.00	1.23	1.21	18.0
5.0	3.0	19.33	19.7	1.02	22.27	22.29	1.00	1.15	1.13	17.6

$$K_{\mathrm{I}} = \frac{\sqrt{3}}{8}\sigma_{0\mathrm{s}}\sqrt{t} \tag{1.143}$$

$$K_{\mathrm{II}} = -\frac{1}{4}\sigma_{0\mathrm{s}}\sqrt{t} \tag{1.144}$$

$$\overline{T} = \frac{9}{16}\sigma_{0\mathrm{s}} \tag{1.145}$$

Another notch stress formula for the microkeyhole in the cross-sectional model has been proposed based on an accurate FE model of the rounded slit tip using the generalised NSIFs $K^*_{1,\rho}$ and $K_{2,\rho}$ together with the T-stress T (Lazzarin et al. 2009), which is given here in a slightly modified form (Radaj 2010):

$$\sigma_\theta = \frac{K^*_{1,\rho}}{\sqrt{2\pi\rho}}\left(2\cos\frac{\theta}{2} + \cos\frac{3\theta}{2}\right) - \frac{K_{2,\rho}}{\sqrt{2\pi\rho}}\left(2\sin\frac{\theta}{2} + 3\sin\frac{3\theta}{2}\right) + \frac{3}{2}T(1 - \cos 2\theta) \tag{1.146}$$

with the generalised NSIFs $K^*_{1,\rho}$ and $K_{2,\rho}$ and the T-stress T determined from the stresses in the ligament ($\theta = 0°$) of the keyhole in the cross-sectional FE model. $K^*_{1,\rho}$ follows from $\sigma_{\theta\ \max}$ at $r = \rho$, Eq. (2.167), after subtraction of the T-stress, $\sigma_{\theta,T} = -T$ (which means adding T). $K_{2,\rho}$ results from $\tau_{\theta r\ \max}$ at $r = 1.723\rho$, Eq. (2.155). The T-stress T results from the difference in the radial and tangential stresses in the ligament (Lazzarin et al. 2009):

$$T = \sigma_r - \sigma_\theta \quad (\theta = 0,\ r \to 0) \tag{1.147}$$

The Eq. (1.146) is valid in the full angular range ($-180° < \theta < +180°$). It is generally applicable, not only to lap joints under tensile-shear loading.

Table 1.15 Notch stress σ_θ [MPa] at keyhole in cross-sectional model of tensile-shear loaded lap joint, $F/t = 10$ MPa, evaluation in different angular directions; Radaj's Eq. (1.141), Lazzarin's Eq. (1.146), FE analysis and Radaj's Eqs. (1.141–1.144); (Radaj 2010)

	σ_θ (0°)	$\sigma_{\theta,\ \max}$	Δ %	$\bar{\theta}$	σ_θ (90°)	σ_θ (0°)	$\sigma_{\theta,\ \max}$	Δ %	$\bar{\theta}$	σ_θ (90°)
	$\rho = 0.05$ mm, $t = 1$ mm, $w/t = 1$					$\rho = 0.05$ mm, $t = 5$ mm, $w/t = 1$				
Radaj	3.9	141.4	−2.4	76°	133.8	30.0	241.6	−0.6	70°	217.8
Lazzarin	0.8	144.3	−0.4	78°	137.8	30.0	247.3	1.8	71°	225.1
FE analysis	2.8	144.8	0.0	79°	140.9	31.8	243.0	0.0	70°	223.9
	$\rho = 0.05$ mm, $t = 1$ mm, $w/t \gg 1$					$\rho = 0.05$ mm, $t = 5$ mm, $w/t \gg 1$				
Radaj	23.9	151.2		74°	141.5	81.2	266.8		66°	233.0

For the comparison below of the notch stresses, an FE model of the tensile-shear loaded lap joint, at first with pointed and then with rounded slit tips, has been analysed in respect of the parameters K_1, $K^*_{1,\rho}$, K_{II}, $K_{2,\rho}$ and T for two plate thicknesses and four ratios w/t. The results are listed in Table 1.14. K_{I} and K_{II} rise with the square root on t. K_{I} rises also with w/t, but not so K_{II}. The parameters $K^*_{1,\rho}$ and $K_{2,\rho}$ are slightly larger than K_{I} and K_{II} (up to 10 %).

The two Eqs. (1.141) and (1.146) for the notch stress σ_θ at the keyhole edge in the cross-sectional model of the tensile-shear loaded lap joint have been evaluated for the geometrical parameters $t = 1$ and 5 mm, $w/t = 1$ and $\rho = 0.05$ mm, Table 1.15. In order to allow direct comparisons, Eq. (1.141) is applied with the values of K_{I}, K_{II} and T from the FE analysis instead of using the simple structural stress formulae valid for $w/t \gg 1$. The reduction of w/t changes the values of K_{I}, K_{II} and T to some extent, see Table 1.14. The influence on K_{I} and thereby on $\sigma_\theta(0°)$ is rather strong. On the other hand, Eq. (1.146) is evaluated with $K^*_{1,\rho}$, $K_{2,\rho}$ and T from FE analysis according to the definition.

1.5.11 Theoretical S–N Curve of Spot-Welded Lap Joint Based on Neuber's Concepts

In this last subsection, the theoretical determination of the *S–N* curve of a tensile-shear loaded spot-welded lap joint is demonstrated, using Neuber's microstructural (elastic) and macrostructural (elastic-plastic) support concepts. A microkeyhole notch at the critical site of the weld spot edge is used in the model. The notch strain approach is applied which refers to crack initiation. Two formal variants exist which produce identical final results. In the one variant, the fatigue-effective (locally averaged) cyclic stresses and strains occurring at the elastic-plastic notch root are compared with the endurable cyclic stresses and strains in unnotched specimens. In the other variant, the actual (non-averaged) cyclic stresses and strains occurring at the elastic-plastic notch root are compared with endurable cyclic stresses and strains in unnotched specimens increased by the notch support index.

The elastic notch support index *n* related to the microstructural support effect is defined as follows:

$$n = \frac{\sigma_k}{\overline{\sigma}_k} = \frac{K_t}{K_f} \tag{1.148}$$

with the theoretical (maximum) notch stress σ_k, the fatigue-effective (maximum) notch stress $\overline{\sigma}_k$, the stress concentration factor K_t and the fatigue notch factor K_f. Using Neuber's fictitious notch rounding concept, Eq. (1.1), together with the sharp notch approximations $\sigma_k \propto (1/\rho)^{1/2}$ and $\overline{\sigma}_k \propto (1/\rho_f)^{1/2}$ results in:

$$n = \sqrt{\frac{\rho_f}{\rho}} = \sqrt{1 + s\frac{\rho^*}{\rho}} \tag{1.149}$$

with the support factor s, the microstructural length ρ^*, the fictitious notch radius ρ_f and the real notch radius ρ.

The two formal variants mentioned above of the notch strain approach, when confined to the (elastic) endurance limit, can be formulated as follows:

$$\overline{\sigma}_k \leq \overline{\sigma}_{kE} \quad (\overline{\sigma}_k = K_f \sigma_n,\ \overline{\sigma}_{kE} = \sigma_E) \tag{1.150}$$

$$\sigma_k \leq \sigma_{kE} \quad (\sigma_k = K_t \sigma_n,\ \sigma_{kE} = n\sigma_E) \tag{1.151}$$

with the nominal stress σ_n, the endurable notch stresses σ_{kE} and $\overline{\sigma}_{kE}$, the endurance limit σ_E and the notch support index *n*.

The basic formulae and material data of the notch strain approach comprise the following components (Radaj and Vormwald 2007):

- the cyclic stress-strain relationship according to Ramberg and Osgood,
- the strain *S–N* relationship according to Manson, Coffin and Morrow,
- the stress-strain hysteresis loop relationship after Masing,
- the elastic-plastic notch stresses and strains approximated by Neuber's macro-support formula, by other approximation formulae or by non-linear FE analysis,
- the initial residual stresses taken into account by a corresponding variant of Neuber's formula or by FE analysis,
- the damage parameter relationship taking the effects of mean stress and stress multiaxiality into account,
- the cyclic material parameters estimated on the basis of Seeger's 'uniform material law' (initial crack length $a_i = 0.25$ mm).

The notch strain approach has been applied to a tensile-shear loaded spot-welded lap joint specimen (Seeger et al. 2005). The geometrical and material data correspond to the investigation of McMahon in order to compare the analysis results with the published experimental data referring to a spot diameter

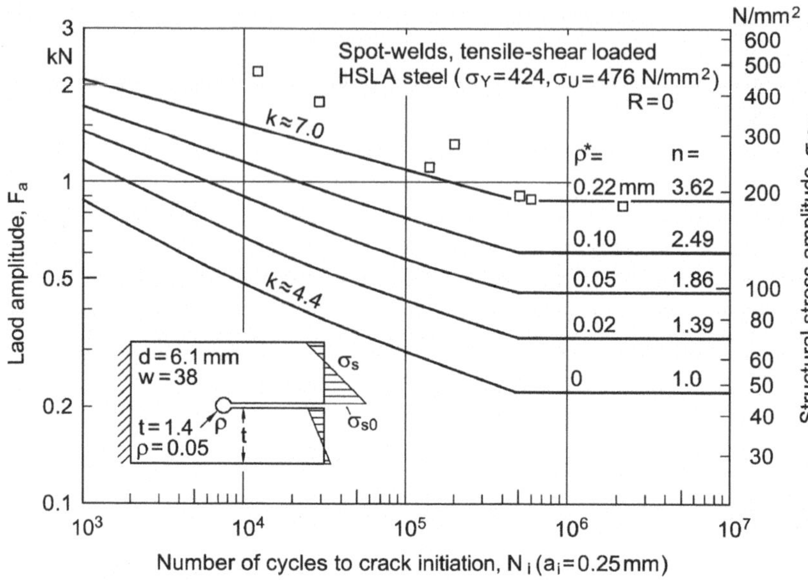

Fig. 1.56 *S-N* curves characterising crack initiation ($a_i = 0.25$ mm) in tensile-shear loaded spot-welded lap joint specimen; analysis results based on the notch strain approach (solid line curves) in comparison to test results (point symbols) by McMahon; structural stresses σ_s in FE model transferred to cross-sectional model with keyhole; notch support index n determined with microstructural length ρ^*; (Seeger et al. 2005)

$d = 6.1$ mm, a sheet thickness $t = 1.4$ mm, a specimen width $w = 38$ mm and a high-strength low-alloy steel (McMahon et al. 1990).

The elastic structural stresses in the specimen which are the basis for the elastic-plastic notch stresses and strains, result from an accurate FE model. The elastic notch stresses at the front side of the weld spot, considered as the crack initiation site, are determined for the relevant cross-sectional model (keyhole radius $\rho = 0.05$ mm) according to Eq. (1.141) proceeding from the structural stresses (reference value σ_{s0} at the inner side of the tensile-loaded plate). The (elastic) notch support index n is derived as the ratio of the theoretical and fatigue-effective (maximum) notch stresses, Eq. (1.148), the latter being averaged over the microstructural length ρ^* normal to the keyhole edge.

Different *S–N* curves result from the above-mentioned formulae of the notch strain approach, depending on the choice of the material parameter ρ^*, Fig. 1.56. Referring to Neuber's ρ^* over σ_Y diagram, $\rho^* = 0.07$ mm has to be chosen with the considered yield limit of ferritic steels (Neuber 1968). Another evaluation by Kuhn results in $\rho^* = 0.1$ mm with the ultimate tensile strength of wrought steel being considerd (Radaj and Vormwald 2007, ibid. Fig. 4.35). The available fatigue strength data for spot-welded joints suggest a fictitious notch radius $\rho_f \approx 0.6$ mm leading to $\rho^* = 0.22$ mm when introducing a real notch radius $\rho = 0.05$ mm (Radaj et al. 2006, ibid. pp. 524–527). The two limit values of the inverse slope exponent k entered into the diagram are supplementary approximations for the calculated curves.

The comparison of the theoretical with the experimental results confirms that $\rho^* = 0.22$ mm should be used instead of $\rho^* = 0.07$ mm. The theoretical endurance limit occurs at $N_E = 5 \times 10^5$ cycles (corresponding to the base material) whereas the test results indicate a slightly higher value of N_E. The slope in the medium-cycle fatigue range seems to be insufficiently reproduced by the theoretical results corresponding to $\rho^* = 0.22$ mm, but the number of experimental data is too small for a definite statement. Additionally, deviations have to be expected because the crack length $a_i = 0.25$ mm is not adequate for defining crack initiation at microkeyholes with radius $\rho = 0.05$ mm.

The assessment procedure based on notch strains described above has been used to investigate the effect of further influencing parameters (Seeger et al. 2005). Introducing the material parameters of the heat-affected zone (hardness increased by factor 1.5) reduces ρ^* and n with the consequence that the increase in fatigue strength is compensated by the reduction in notch support. Reducing the notch radius from $\rho = 0.05$ to 0.01 mm increases the high-cycle fatigue strength, an effect which is well known from non-damaging microholes in homogeneous stress fields.

The conclusion is that the theoretical *S–N* curves, not only of common seam-welded joints (Radaj et al. 2006), but also of thin-sheet resistance spot-welded or laser beam seam-welded joints, can be determined based on the notch stresses and strains at a reference notch. An extremely small notch radius may be appropriate in the case of thin-sheet material. The application of Neuber's microstructural and macrostructural support concepts on this notch allows a realistic solution after proper scaling in the sense discussed above.

List of Symbols

a	Notch depth, semilength of slit or crack
a_i	Initiated crack length
a, b	Semiaxes of elliptical hole
b	Net section semiwidth
C	Auxiliary parameter of *S–N* curve
C_1, C_2	Constants in Neuber's stress field equations
d	Weld spot diameter
F	Shear force
F_a	Load amplitude
K_1, K_2, K_3	Notch stress intensity factors, mode 1, 2, 3
$K_{1,\rho}$, $K_{2,\rho}$	Generalised notch stress intensity factors, mode 1, 2
$K^*_{1,\rho}$	*T*-stress-corrected $K_{1,\rho}$
K_f, $K_{f\ red}$	Theoretical fatigue notch factor, reduced K_f

K_f^*, K_f^{**}	Fatigue notch factor, reduced and modified
$K_{f,l}$, $K_{f,u}$	Fatigue notch factor, lower and upper slit face
K_O	Stress intensity factor, transverse singular mode
K_t	Theoretical (equivalent) stress concentration factor
$\overline{K}_t$	Stress concentration factor of locally averaged notch stresses
$K_t(\rho_f)$	Stress concentration factor after fictitious notch rounding
$K_{\sigma'}, K_\tau$	Stress concentration factor relating to σ'_{max}, τ_{max}
K_I, K_{II}, K_{III}	Stress intensity factor, mode I, II, III
K_{Ic}	Fracture toughness
K_{eq}	Equivalent stress intensity factor
K_w	Stress concentration factor of weld notch
k	Inverse slope exponent, *S–N* curve
k_I, k_{II}, k_{III}	Normalised stress intensity factors of modes I, II and III
l	Semilength of plate
m	Parameter in complex stress function
N, N_i	Endurable number of cycles, the same up to crack initiation
n	Elastic notch support index
q, r	Stress field parameters linked to V-notch opening angle
P_s	Survival probability
R	Ratio of lower to upper stress
R	Parameter in complex stress function
r	Polar coordinate, radial distance
r^*	Radius of circular core region
r_0	Distance between notch tip and origin of polar coordinate system
S	Endurable nominal stress
S	Sih's strain energy density factor
s, $\overline{s}$	Microstructural support factor, plateau value
s_{vM}, s_B	Support factor, von Mises and Beltrami equivalent stress
T	*T*-Stress
t	Notch depth, semiaxis of ellipse, plate thickness
t_w	Weld throat thickness
u, u_0	Horizontal displacement, remote value
v, v_0	Vertical displacement, remote value
w	Plate semiwidth, joint face width, specimen width
x_0	Distance between notch tip and origin of polar coordinate system
x, y, z	Cartesian coordinates
α	Notch opening semi-angle
Δ	Relative deviation
ζ	Complex coordinate
$\theta, \overline{\theta}$	Polar angle, value for σ'_{max}
κ	Multiaxiality jump factor
κ_0	Free surface factor
$\kappa_{\sigma'}$, κ_τ	Geometry factor, relating to σ' and τ
κ_{vM}, κ_B	Jump factor, von Mises and Beltrami equivalent stress
κ_1, κ_2, κ_3	Short crack reduction factor on K_f, mode I, II, III
λ_1, λ_2	Eigenvalue of stress distribution at V-notch, mode 1, 2

μ	Exponent in Filippi's stress equations
ν	Poisson's ratio
ρ	Real notch radius
ρ'	Radius of curvature at point of maximum stress σ'_{max}
ρ^*	Microstructural support length
ρ_f, ρ_r	Fictitious and reference notch radius
ρ_f^*	Degree of cross-sectional weakening, ρ_f/t
σ^*	Lower to upper surface stress ratio, σ_l/σ_u
$\overline{\sigma}$	Locally averaged notch stress
σ_a	Stress amplitude
σ_b^*	Structural bending stress at level of hole radius
$\overline{\sigma}_c$	Locally averaged notch stress at fracture
σ_E, σ_{kE}	Endurance limit, notch stress endurance limit
$\overline{\sigma}_{eq}$	Locally averaged equivalent stress
$\sigma_{eq\,max}$	Maximum equivalent stress
$\sigma_{i,l}, \sigma_{i,u}$	Lower and upper plate inner side stresses
σ_k	Notch stress
σ_l, σ_u	Stresses on lower and upper plate surface
σ_n, σ_{ng}	Nominal stress, value in gross cross-section
σ_m	Static mean stress or membrane stress
σ_{max}, σ'_{max}	Maximum notch stress, symmetric and antimetric component
σ_r, σ_θ	Stresses in the polar coordinate system
σ_s	Structural stress
σ_{th}	Theoretical (equivalent) notch stress
σ_{vM}, σ_B	Equivalent stress, von Mises and Beltrami
σ_U	Ultimate tensile strength
σ_Y	Yield limit
σ_0^*	Structural stress parallel to slit front
σ_0	Structural basic stress parallel to slit
σ_{0s}	Structural stress at slit flank
σ_1, σ_2, σ_3	Principal stresses
$\sigma_{\theta\ max}$	Maximum tangential stress
$\overline{\tau}$, $\overline{\tau}^*$	Locally averaged notch shear stress, in-plane and out-of-plane
τ_0	Reference shear stress
τ_{max}, τ^*_{max}	Maximum notch shear stress, in-plane and out-of-plane
τ_n, τ_n^*	Nominal shear stress, in-plane and out-of-plane
τ_{ng}, τ^*_{ng}	Nominal shear stress, gross cross-section, in-plane, out-of-plane
$\tau_{r\theta}$	Shear stress in polar coordinate system
τ_{th}, τ^*_{th}	Theoretical notch shear stress, in-plane and out-of-plane
τ_{yz}	Shear stress along notch bisector
$\tau^*_{z\,max}$	Maximum out-of-plane shear stress produced by τ_0^*
τ_0^*	Non-singular out-of-plane structural shear stress
ϕ, ψ	Complex stress function
$\tilde{\omega}_1$	Auxiliary parameter in Filippi's stress equations
B	Beltrami (criterion)
BE	Boundary element

FAT	Fatigue strength class in IIW design recommendations
FE	Finite element
IIW	International Institute of Welding
NSIF	Notch stress intensity factor
MSED	Minimum strain energy density (criterion)
MTS	Maximum tangential stress (criterion)
SCF	Stress concentration factor
SIF	Stress intensity factor
ns	Normal stress (criterion)
ps	Plane stress
pn	Plane strain
vM	von Mises (criterion)

References

Berto F (2012) Fictitious notch rounding concept applied to V-notches with end holes under mode 3 loading. Int J Fatigue 38:188–193

Berto F, Lazzarin P (2010) Fictitious notch rounding approach of pointed V-notches under in-plane shear. Theor Appl Fract Mech 53:127–135

Berto F, Zappalorto M (2011) Fictitious notch rounding concept applied to V-notches with endholes under mode 1 loading. Int J Fract 171:91–98

Berto F, Zappalorto M (2012) The fictitious notch rounding approach applied to V-notches with root holes subjected to mode 1 loading, J Strain Anal, available online: doi 10.1177/0309324712437106

Berto F, Lazzarin P, Radaj D (2008) Fictitious notch rounding concept applied to sharp V-notches: Evaluation of the microstructural support factor. Part I: Basic stress equations. Engng Fract Mech 75:3060–3072

Berto F, Lazzarin P, Radaj D (2009) Fictitious notch rounding concept applied to sharp V-notches: Evaluation of the microstructural support factor. Part II: Microstructural support analysis. Engng Fract Mech 76:1151–1175

Berto F, Lazzarin P, Radaj D (2012) Fictitious notch rounding concept applied to V-notches with root holes subjected to in-plane shear loading. Engng Fract Mech 79:281–294

Creager M, Paris PC (1967) Elastic field equations for blunt cracks with reference to stress corrosion cracking. Int J Fract Mech 3:247–252

Erdogan F, Sih GC (1963) On the crack extension in plates under plane loading and transverse shear. J Basic Engng 85:519–525

Filippi S, Lazzarin P, Tovo R (2002) Developments of some explicit formulas useful to describe elastic stress fields ahead of notches in plates. Int J Solids Struct 39:4543–4565

Fricke W (2008) Guideline for the fatigue assessment by notch stress analysis for welded structures. IIW Doc XIII-2240r1-08/XV-1289r1-08

Fricke W (2011) Guideline for the assessment of weld root fatigue. IIW Doc XIII-2380r1-11/XV-1383r1-11

Gross R, Mendelson A (1972) Plane elastic static analysis of V-notched plates. Int J Fract Mech 8:267–427

Hasebe N, Nakamura T, Iida J (1990) Notch mechanics for plane and thin plate bending problems. Engng Fract Mech 37:87–99

Hobbacher A (ed) (2009) Fatigue design of welded joints and components. IIW-Doc XIII-2151-07/XV-1254-07. Welding Research Council, Bulletin 520, New York

Inglis CE (1913) Stresses in a plate due to the pressure of cracks and sharp corners. Trans Inst Naval Arch 60:219–230

Irwin GR (1957) Analysis of stresses and strains near the end of a crack traversing a plate. J Appl Mech 24:361–364

Kullmer G (1992) Elastic stress fields in the vicinity of a narrow notch with circular root. In: Proc Eur Conf on Reliability and Structural Integrity of Advanced Materials, ECF9, pp 905–910

Lazzarin P, Filippi S (2006) A generalised stress intensity factor to be applied to rounded V-shaped notches. Int J Solids Struct 43:2461–2478

Lazzarin P, Tovo R (1996) A unified approach to the evaluation of linear-elastic fields in the neighbourhood of cracks and notches. Int J Fract 78:3–19

Lazzarin P, Berto F, Radaj D (2009) Fatigue-relevant stress field parameters of welded lap joints: pointed slit tip compared with keyhole notch. Fatigue Fract Engng Mater Struct 32:713–735

McMahon JC, Smith GA, Lawrence FV (1990) Fatigue crack initiation and growth in tensile-shear spot weldments. ASTM STP 1058:47–77

Muskhelishvili NI (1963) Some basic problems of the mathematical theory of elasticity. P. Noordhoff, Groningen

Neuber H (1936) Zur Theorie der technischen Formzahl. Forschung Ing Wes 7:271–274

Neuber H ([1]1937, [2]1958, [3]1985) Kerbspannungslehre. Springer, Berlin

Neuber H (1968) Über die Berücksichtigung der Spannungskonzentration bei Festigkeitsberechnungen. Konstruktion 20:245–251

Olivier R, Köttgen VB, Seeger T (1989) Schweißverbindung I – Schwingfestigkeitsnachweise für Schweißverbindungen auf der Grundlage örtlicher Beanspruchungen. FKM-Forschungsheft 143, FKM, Frankfurt/M

Olivier R, Köttgen VB, Seeger T (1994) Schweißverbindung II – Untersuchung zur Einbindung eines neuartigen Zeit- und Dauerfestigkeitsnachweises von Schweißverbindungen aus Stahl in Regelwerke. FKM-Forschungsheft 180, FKM, Frankfurt/M

Pedersen MM, Mouritsen OØ, Hansen MR, Andersen JG, Wenderby J (2010) Re-analysis of fatigue data for welded joints using the notch stress approach. Int J Fatigue 32:1620–1626

Peterson RE (1950) Relation between stress analysis and fatigue of metals. Proc SESA 11: 199–206

Radaj D (1969) Näherungsweise Berechnung der Formzahl von Schweißnähten. Schweißen u Schneiden 21:97-103, 151–158

Radaj D (1990) Design and analysis of fatigue resistant welded structures. Abington Publ, Cambridge

Radaj D (1997) Fatigue notch factor of gaps in welded joints reconsidered. Engng Fract Mech 57:405–407

Radaj D (2010) *T*-stress corrected notch stress intensity factors with application to welded lap joints. Fatigue Fract Engng Mat Struct 33:378–389

Radaj D, Helmers K (1997) Bewertung von Schweißverbindungen hinsichtlich Schwingfestigkeit nach dem Kerbspannungskonzept. Konstruktion 49:21–27

Radaj D, Vormwald M ([3]2007) Ermüdungsfestigkeit – Grundlagen für Ingenieure. Springer, Berlin

Radaj D, Zhang S (1990) Mehrparametrige Strukturoptimierung hinsichtlich Spannungserhöhungen. Konstruktion 42:289–292

Radaj D, Zhang S (1991) Multiparameter design optimisation in respect of stress concentrations. In: Eschenauer HA et al. (eds) Engineering optimization in design processes. Springer, Berlin, pp. 181–189

Radaj D, Zhang S (1992) Notch effect of welded joints subjected to antiplane shear loading. Engng Fract Mech 43:663–669

Radaj D, Zhang S (1993[(1)]) Analogies between crack tip and rigid line tip stresses and displacements. Engng Fract Mech 44:913–919

Radaj D, Zhang S (1993[(2)]) On the relations between notch stress and crack stress intensity in plane shear and mixed mode loading. Engng Fract Mech 44:691–704

Radaj D, Lehrke HP, Greuling S (2001) Theoretical fatigue-effective notch stresses at weld spots. Fatigue Fract Engng Mater Struct 24:293–308

Radaj D, Sonsino CM, Fricke W ([2]2006) Fatigue assessment of welded joints by local approaches. Woodhead Publ, Cambridge and CRC Press, Boca Raton Fla

Radaj D, Lazzarin P, Berto F (2009[(1)]) Fatigue assessment of welded joints under slit-parallel loading based on strain energy density or notch rounding. Int J Fatigue 31:1490–1504

Radaj D, Sonsino CM, Fricke W (2009[(2)]) Recent developments in local concepts of fatigue assessment of welded joints. Int J Fatigue 31:2–11

Radaj D, Lazzarin P, Berto F (2013) Generalised Neuber concept of fictitious notch rounding. Int J Fatigue 35 (accepted for publication)

Sähn S, Göldner H (1989) Bruch- und Beurteilungskriterien in der Festigkeitslehre. VEB Fachbuchverlag, Leipzig

Seeger T, Greuling S, Brüning J, Leis P, Radaj D, Sonsino CM (2005) Bewertung lokaler Berechnungskonzepte zur Ermüdungsfestigkeit von Punktschweißverbindungen. FAT-Schriftenreihe 196, Frankfurt

Sih GC (1974) Strain energy density factor applied to mixed mode crack problems. Int J Fract 10:305–321

Taylor D (1999) Geometrical effects in fatigue – A unifying theoretical model. Int J Fatigue 21:413–420

Weiss V (1971) Eine Bruchmechanik für Kerben. Schweiz Arch Angew Wiss Techn 37:1–7

Wieghardt K (1907) Über das Spalten und Zerreißen elastischer Körper. Z Math Phys 55:60–103

Williams ML (1952) Stress singularities resulting from various boundary conditions in angular corners on plates in tension. J Appl Mech 19:526–528

Zappalorto M, Lazzarin P (2011) In-plane and out-of-plane stress field solutions for V-notches with end holes. Int J Fract 167:168–180

Chapter 2
Extended Stress Intensity Factor Concepts

D. Radaj

Abstract The stress intensity factor concept for describing the stress field at pointed crack or slit tips is well known from fracture mechanics. It is substantially extended here in two directions. One extension refers to pointed V-notches with stress intensities depending on the notch opening angle. The loading mode related notch stress intensity factors K_1, K_2 and K_3 are introduced. Another extension refers to rounded notches with crack shape or V-notch shape in two variants: parabolic, elliptic or hyperbolic notches ('blunt notches') on the one hand and root hole notches ('keyholes' when considering crack shapes) on the other hand. Here, the loading mode related generalised notch stress intensity factors $K_{1\rho}$, $K_{2\rho}$ and $K_{3\rho}$ are defined. The concepts of elastic stress intensity factor, notch stress intensity factor and generalised notch stress intensity factor are extended into the range of elastic-plastic (work-hardening) or perfectly plastic notch tip or notch root behaviour. Here, the plastic notch stress intensity factors $K_{1\mathrm{p}}$, $K_{2\mathrm{p}}$ and $K_{3\mathrm{p}}$ are of relevance. The original stress intensity factor concept is also transferred from cracks or slits to rigid thin inclusions.

Contents

D. Radaj and M. Vormwald, *Advanced Methods of Fatigue Assessment*,
DOI: 10.1007/978-3-642-30740-9_2,

2.1 Survey of Chapter Contents

The stress intensity factor concept has originally been developed in the context of fracture mechanics. For fracture phenomena, both brittle fracture and fatigue failure, the asymptotic singular stress field at the pointed crack or slit tip is most important. The stress level around the singularity is described by the stress intensity factor (SIF), possibly superimposed by the crack-parallel non-singular T-stress. In certain cases (e.g. thin-sheet lap joints), it is necessary, also to take higher order terms of the stress field approximation into account.

The local three-dimensional stress singularity at a definite point of the crack or slit front can generally be described by superimposition of three two-dimensional stress singularities corresponding to three independent loading or opening modes of the crack tip: transverse tensile loading (mode I), in-plane shear loading (mode II) and out-of-plane shear loading (mode III). The appertaining SIFs are K_{I}, K_{II} and K_{III}. Non-singular stresses may be superimposed: crack-parallel stresses (the T-stress) as well as normal stresses and symmetric shear stresses in the crack front direction. In special cases, the separation of the elementary crack tip loading modes is not possible, coupling effects together with transverse singular effects are occurring, e.g. where the crack front butts on a free surface.

The SIF concept has been substantially extended in two directions. One extension refers to pointed V-notches with stress intensities depending on the notch opening angle. Here, the loading mode related notch stress intensity factors (NSIFs) are K_1, K_2 and K_3. Another extension refers to rounded notches with crack shape or V-notch shape in two variants: parabolic, elliptic or hyperbolic notches ('blunt notches') on the one hand and root hole notches ('keyholes' when considering crack shapes) on the other hand. The notch stresses remain finite in this case, but their angular and radial distribution is linked to that of the corresponding pointed notches. Here, the loading mode related generalised NSIFs are $K_{1\rho}$, $K_{2\rho}$ and $K_{3\rho}$. They are slightly larger than K_1, K_2 and K_3, their value increases with reduced notch severity.

The elastic SIF, NSIF and generalised NSIF concepts may be extended into the range of elastic-plastic (work-hardening) or perfectly plastic notch tip or notch root behaviour. Here, the plastic NSIFs $K_{1\mathrm{p}}$, $K_{2\mathrm{p}}$ and $K_{3\mathrm{p}}$ are of relevance.

Further extensions of the SIF concepts are documented in the literature. The original SIF concept may be transferred from cracks or slits to rigid thin inclusions, from pointed V-notches to pointed rigid inclusions with V-notch shape and from blunt V-notches to the corresponding rounded rigid inclusions. Another extension is the bimaterial wedge problem with inclusion of interface crack configurations.

In Sect. 2.2, the original stress intensity factor concept is reviewed, referring to the stress field near crack tips, with application to slit tips in welded joints.

In Sect. 2.3, the notch stress intensity factor concept is summarised, referring to the stress field near pointed V-notch tips, with application to fillet-welded joints.

In Sect. 2.4, the generalised notch stress intensity factor concept is described, referring to the stress field at blunt V-notch tips and at root hole V-notch tips, with application to round bars with shoulder fillets.

In Sect. 2.5, plastic notch stress intensity factors are introduced for describing the stress and strain field at pointed V-notch tips and at parabolic notch tips. A new version of the Neuber rule accounting for the influence of the notch opening angle is presented.

In Sect. 2.6, the stress intensity factor concept is applied to rigid thin inclusions, with one example of SIF evaluation.

The bimaterial wedge problem with inclusion of interface crack configurations cannot be presented within the available space of this book. Reference is made instead to the early basic solutions (Hein and Erdogan 1971; Bogy 1971) and to a recent comprehensive study on asymptotic near-field analyses of plane multi-material joints (Sator 2010). Typical applications are interface cracks or cracks butting on the interface, the interface itself butting on a unnotched surface, on a corner-notched surface or on a more generally V-notched surface. A well known special problem is the bearing pressure between a plane rigid stamp and a plane elastic continuum.

2.2 Original Stress Intensity Factor Concept

2.2.1 Survey of Section Contents

The basic formulation of the SIF concept is reviewed. Basic is the description of the singular and non-singular stress field in the vicinity of a crack tip or slit tip. The three-dimensional stress state at any point of the crack or slit front is separated into the three two-dimensional singular loading modes I, II and III and into three non-singular modes (*T*-stress, *S*-stress and front-parallel normal stress). A coupling of singular modes together with transverse singular effects may occur where the crack or slit front butts on a free surface.

The stress field equations in the vicinity of crack or slit tips are presented and the corresponding SIFs, *T*-stress and *S*-stress are defined. These characteristic parameters are specified for the joint face edge of seam-welded and spot-welded lap joints proceeding from the linearised structural stresses which suppress the singularity completely. Lap joints with equal and unequal plate thickness, respectively, are considered.

Equivalent SIFs are defined, referring to failure under mixed-mode loading conditions. Endurable SIFs for spot-welded lap joints are finally given, both in the form of a threshold value and in the form of a *K–N* curve with a defined, sufficiently low scatter range.

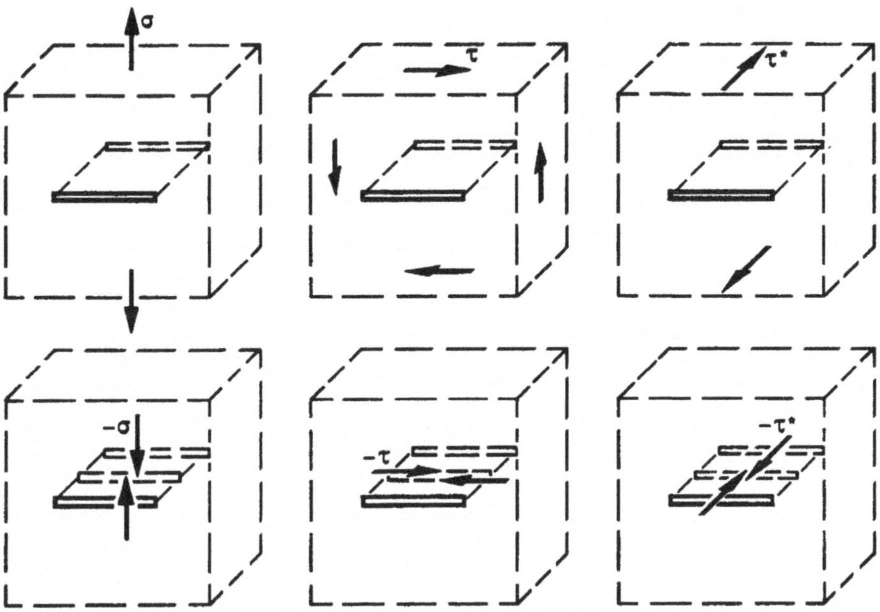

Fig. 2.1 Basic singular loading modes I (**a**), II (**b**) and III (**c**) of cracks; planar crack in the infinite plate; plane and antiplane strain fields; arrows designate remote boundary or crack flank stresses; (Radaj and Zhang 1993[(1)])

The expositions are based on readily available text books on fracture mechanics and on special publications. A list of symbols and a list of references are found at the end of Chap. 2.

2.2.2 Basic Crack Tip Loading Modes

The stress intensity factor concept originates from fracture mechanics. For fracture phenomena, both brittle fracture and fatigue failure, the local stress field close to the crack tip is decisive. The stresses rise to a singularity at the pointed crack tip. The intensity of this singularity is dependent on the magnitude of the relevant structural stresses.

The local three-dimensional stress field at a definite point of the crack front may generally be separated into two-dimensional stress fields which are superimposed: three loading modes with singular stresses and three loading modes with non-singular stresses. The singular loading modes comprise transverse tensile loading (mode I), in-plane shear loading (mode II) and out-of-plane shear loading (mode III), Fig. 2.1. The non-singular loading modes consist of crack-parallel tensile or compressive loading (mode 0I), crack-front-parallel tensile or

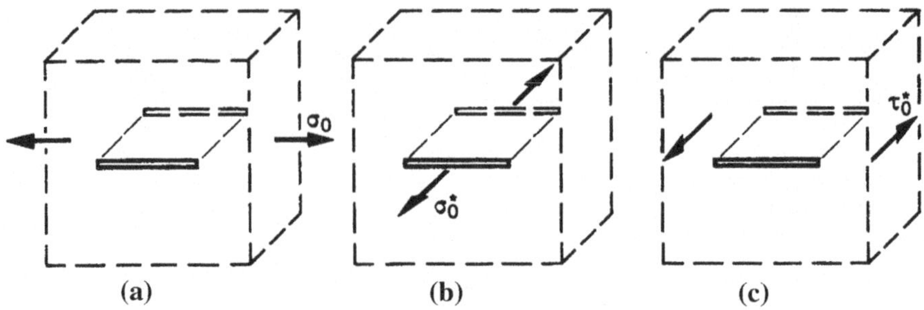

Fig. 2.2 Basic non-singular loading modes 0I (**a**), 0II (**b**) and 0III (**c**) of cracks; planar crack in the infinite plate; plane and antiplane strain fields; arrows designate remote boundary stresses; (Radaj and Zhang 1993[1])

compressive loading (mode 0II) and crack-front-parallel shear loading (mode 0III), Fig. 2.2.

In exceptional cases, coupling effects may occur locally between the mode-related singularities within locally three-dimensional configurations. One typical example is the crack tip in a plate of finite thickness subjected to in-plane shear loading. A transverse shear stress singularity is generated locally underneath the plate surface. Further details can be found in Sect. 2.3.3.

2.2.3 *Asymptotic Stress Field near Crack Tips*

The three basic loading modes with singular stresses at the crack tip produce the following asymptotic stress distribution around the crack tip (Westergaard 1939; Sneddon 1946; Irwin 1957; Williams 1957), Fig. 2.3:

$$\sigma_{\mathrm{ij}} = \frac{1}{\sqrt{2\pi r}}\left[K_{\mathrm{I}} f_{\mathrm{I,ij}}(\theta) + K_{\mathrm{II}} f_{\mathrm{II,ij}}(\theta) + K_{\mathrm{III}} f_{\mathrm{III,kz}}(\theta)\right] \qquad (2.1)$$
$$(\mathrm{i, j} = x, y \text{ and } \mathrm{k} = x, y \text{ or } \mathrm{i, j} = r, \theta \text{ and } \mathrm{k} = r, \theta)$$

The mode-related stress intensity factors (SIFs) K_{I}, K_{II} and K_{III} depend on the magnitude of the load, the crack length and further geometrical parameters of the considered configuration. The mode-related angle-dependent functions $f_{\mathrm{I,ij}}$, $f_{\mathrm{II,ij}}$, $f_{\mathrm{III,iz}}$ describe the angular distribution of the stresses at the crack tip. The relationship above is strictly valid for $r \to 0$ and approximately valid for values of r which are small in relation to the crack length and other geometrical parameters of the configuration.

These are the asymptotic stresses in the three singular loading modes, given in polar coordinates:

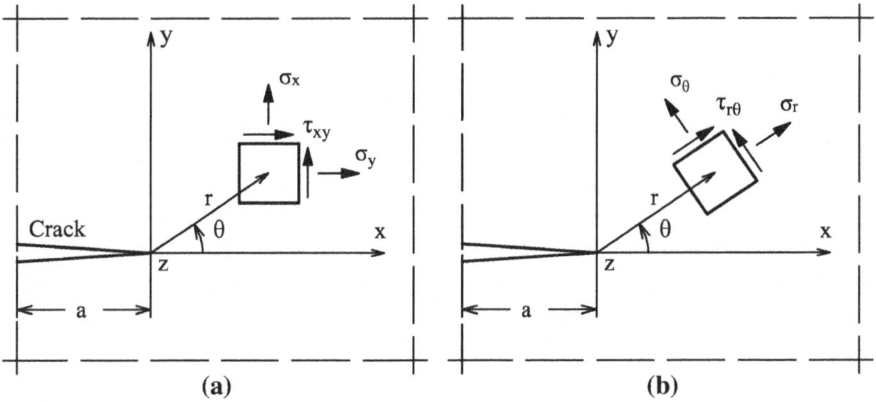

Fig. 2.3 Cartesian (**a**) and polar (**b**) coordinate systems with correspondingly defined stresses at the crack tip; (Lazzarin and Tovo 1998)

Mode I (transverse tensile loading):

$$\begin{Bmatrix} \sigma_r \\ \sigma_\theta \\ \tau_{r\theta} \end{Bmatrix} = \frac{K_\mathrm{I}}{4\sqrt{2\pi r}} \begin{Bmatrix} 5\cos\theta/2 - \cos 3\theta/2 \\ 3\cos\theta/2 + \cos 3\theta/2 \\ \sin\theta/2 + \sin 3\theta/2 \end{Bmatrix} \tag{2.2}$$

$$\begin{aligned} &\sigma_z = 0 \quad \text{(plane stress)} \\ &\sigma_z = \nu(\sigma_r + \sigma_\theta) = \frac{2\nu K_\mathrm{I}}{\sqrt{2\pi r}} \cos\frac{\theta}{2} \quad \text{(plane strain)} \end{aligned} \tag{2.3}$$

Mode II (in-plane shear loading):

$$\begin{Bmatrix} \sigma_r \\ \sigma_\theta \\ \tau_{r\theta} \end{Bmatrix} = \frac{K_\mathrm{II}}{4\sqrt{2\pi r}} \begin{Bmatrix} -5\sin\theta/2 + 3\sin 3\theta/2 \\ -3\sin\theta/2 - 3\sin 3\theta/2 \\ \cos\theta/2 + 3\cos 3\theta/2 \end{Bmatrix} \tag{2.4}$$

$$\begin{aligned} &\sigma_z = 0 \quad \text{(plane stress)} \\ &\sigma_z = \nu(\sigma_r + \sigma_\theta) = -\frac{2\nu K_\mathrm{II}}{\sqrt{2\pi r}} \sin\frac{\theta}{2} \quad \text{(plane strain)} \end{aligned} \tag{2.5}$$

Mode III (out-of-plane shear loading):

$$\begin{Bmatrix} \tau_{rz} \\ \tau_{\theta z} \end{Bmatrix} = \frac{K_\mathrm{III}}{\sqrt{2\pi r}} \begin{Bmatrix} \sin\theta/2 \\ \cos\theta/2 \end{Bmatrix} \tag{2.6}$$

2.2.4 Higher Order Non-Singular Stress Terms

The Eqs. (2.1–2.6) above represent asymptotic or first order approximations of the stress field around the singularity at the crack tip. They deviate from the exact solution with increasing distance from the crack tip. The complete solution for the stress field is represented by an infinite series expansion based on eigenvalues and eigenfunctions resulting in non-singular higher order stress terms (Williams 1952, 1957).

The boundary condition of load-free crack or slit flanks ($\theta = \pm\pi$), introduced into the general complex stress functions, results in a homogeneous equation system. Nontrivial solutions exist for the coefficient determinant being zero, resulting in identical eigenvalues in the three loading modes:

$$\sin 2\lambda\pi = 0 \quad \rightarrow \quad \lambda = \frac{n}{2} \quad (n = 1, 2, 3, \ldots) \tag{2.7}$$

The stresses are expressed by an infinite series of the corresponding eigenfunctions (nomenclature after Gross and Seelig 2001):

$$\sigma_{\mathrm{ij}} = r^{-1/2}\hat{\sigma}_{\mathrm{ij}}^{(1)}(\theta) + \hat{\sigma}_{\mathrm{ij}}^{(2)}(\theta) + r^{1/2}\hat{\sigma}_{\mathrm{ij}}^{(3)}(\theta) + r\hat{\sigma}_{\mathrm{ij}}^{(4)}(\theta) + \ldots \quad (\mathrm{i, j} = r, \theta) \tag{2.8}$$

$$\tau_{\mathrm{kz}} = r^{-1/2}\hat{\tau}_{\mathrm{kz}}^{(1)}(\theta) + \hat{\tau}_{\mathrm{kz}}^{(2)}(\theta) + r^{1/2}\hat{\tau}_{\mathrm{kz}}^{(3)}(\theta) + r\hat{\tau}_{\mathrm{kz}}^{(4)}(\theta) + \ldots \quad (\mathrm{k} = r, \theta) \tag{2.9}$$

The stress field equation σ_{ij} can be split into symmetric and antimetric parts in respect of θ, the symmetric part connected with mode I loading, the antimetric part with mode II loading. The stress field equation τ_{kz} is related to mode III loading.

The first order terms define the stress singularity at the crack tip ($r = 0$). They contain the SIFs K_{I}, K_{II} and K_{III}. The second order terms define a uniform non-singular stress at the crack tip, the tensile or compressive stress directed in crack direction named '*T*-stress' (Larsson and Carlsson 1973) and the symmetric shear stress in crack front direction which may be named '*S*-stress' in conformity with the *T*-stress.

These parameters control the stress field close to the crack tip (the 'near-field'), now written in the usual nomenclature up to the second term:

$$\sigma_{\mathrm{I,ij}}^{(1,2)} = \frac{K_{\mathrm{I}}}{\sqrt{2\pi r}} f_{\mathrm{I,ij}}^{(1)}(\theta) + T f_{\mathrm{T,ij}}^{(2)}(\theta) \tag{2.10}$$

$$\sigma_{\mathrm{II,ij}}^{(1,2)} = \frac{K_{\mathrm{II}}}{\sqrt{2\pi r}} f_{\mathrm{II,ij}}^{(1)}(\theta) \tag{2.11}$$

$$\tau_{\mathrm{III,kz}}^{(1,2)} = \frac{K_{\mathrm{III}}}{\sqrt{2\pi r}} f_{\mathrm{III,kz}}^{(1)}(\theta) + S f_{\mathrm{S,kz}}^{(2)}(\theta) \tag{2.12}$$

The stress terms of higher than second order converge to zero for $r \rightarrow 0$, but may be important for larger values of r. Under certain conditions, e.g. a low sheet thickness in lap joints, the range of r, where the second order approximation is appropriate, may be very small (e.g. $r = 0.01$ mm in the tensile-shear specimen with a plate thickness of 1 mm). Then, higher order approximations are needed. The stress equations for mode I and II loading have been derived up to the seventh term represented by $r^{5/2}$ (Berto and Lazzarin 2010).

A possible misunderstanding should immediately be removed. The higher order terms do not describe the stress field in the total structure. The series developments in Eqs. (2.8) and (2.9) are valid only within a certain distance from the slit tip, smaller than the nearest external boundary.

The T-stress σ_0 corresponds to the loading mode 0I in Fig. 2.2, the S-stress τ_0^* to the loading mode 0III. The uniform normal stress σ_0^* in the direction of the crack front produced by mode 0II loading needs no special comment.

The T-stress results in the following polar stress components:

$$\begin{Bmatrix} \sigma_r \\ \sigma_\theta \\ \tau_{r\theta} \end{Bmatrix}_T = T \begin{Bmatrix} \cos^2\theta \\ \sin^2\theta \\ -0.5\sin 2\theta \end{Bmatrix} \tag{2.13}$$

$$\sigma_z = \nu T \quad \text{(plane strain)} \tag{2.14}$$

The corresponding expression for the S-stress is:

$$\begin{Bmatrix} \tau_{rz} \\ \tau_{\theta z} \end{Bmatrix}_S = S \begin{Bmatrix} \cos\theta \\ -\sin\theta \end{Bmatrix} \tag{2.15}$$

2.2.5 Limit Value Formulae for SIFs, T-Stress and S-Stress

The SIFs K_{I}, K_{II} and K_{III} may be determined based on Eqs. (2.10–2.12) from the following limit value formulae which consider the predominant stresses in the ligament ($\theta = 0$):

$$K_{\mathrm{I}} = \lim_{r \to 0} \sqrt{2\pi r}\, \sigma_\theta \tag{2.16}$$

$$K_{\mathrm{II}} = \lim_{r \to 0} \sqrt{2\pi r}\, \tau_{r\theta} \tag{2.17}$$

$$K_{\mathrm{III}} = \lim_{r \to 0} \sqrt{2\pi r}\, \tau_{\theta z} \tag{2.18}$$

Alternatively, limit values of the predominant displacements of the crack flanks may be evaluated. Any other non-vanishing stress or displacement component for the same or other values of θ may be used in the limit value formulae.

The limit conditions above may be applied to analytical stress field solutions and their series expansion or to numerical solutions based on the finite element or boundary element method. In the latter case, the limit value evaluation is facilitated by plotting the stress or displacement components over the distance r from the crack tip in log–log scales. The first order stress approximations appear as decreasing straight lines with the gradient corresponding to $1/r^{1/2}$. The SIFs can easily be calculated from these approximations.

T-stress and S-stress may also be determined from analytical solutions after series expansion of the stress field. In the case of numerical modelling, the T-stress is usually determined as the medium value of the tensile or compressive stresses acting in the two flank sides of the crack. In general, this crack-parallel stress is evaluated in the midsection of the (internal) crack. It has the same sign and value on both sides of the crack, whereas the corresponding stresses from mode II loading have opposite signs (Williams 1957).

Using another method, the T-stress results from the difference in the stresses σ_r and σ_θ acting in the ligament close to the crack tip (Lazzarin et al. 2009):

$$T = \sigma_r - \sigma_\theta \quad (\theta = 0,\ r \to 0) \tag{2.19}$$

whereas Mode II loading has no effect on these stresses.

The S-stress is determined as the medium of the shear stresses acting along the crack front in the plane normal to the ligament at $r \to 0$. It is the symmetric mode 0III shear stress component.

The parameters T and S are defined for $r \to 0$, but may be extended to the near field $r > 0$, $T(r)$ and $S(r)$, with a linear dependency on r.

2.2.6 J-Integral Expressed by SIFs and T-Stress

The line integral J (Rice 1968), designed for describing the fracture behaviour in nonlinear-elastic materials, can successfully be applied with certain restrictions to the elastic-plastic fracture behaviour. Rice has shown that the nonlinear energy release rate can be written as a path-independent line integral J around the crack tip. It has also been shown that the J-integral uniquely characterises the stress and strain field close to the crack tip (Hutchinson 1968; Rice and Rosengren 1968). Thus the J-integral is both an energy parameter and a stress intensity parameter (Anderson 1995).

In its simplified linear-elastic form, the J-integral may be used to determine SIFs on the basis of remote boundary conditions chosen on the integration path. Vice versa, the J-integral may be determined based on the available SIFs and T-stress.

It can be shown that, under linear-elastic conditions, the J-integral is equal to the energy release rate, provided coplanar crack propagation can be assumed. Then, the total J is the sum of the mode-related J-values, J_{I}, J_{II} and J_{III}:

$$\begin{aligned} &J = J_{\mathrm{I}} + J_{\mathrm{II}} + J_{\mathrm{III}} = \frac{K_{\mathrm{I}}^2}{E'} + \frac{K_{\mathrm{II}}^2}{E'} + \frac{K_{\mathrm{III}}^2}{2G} \\ &E' = E \quad (\text{plane stress}), \quad E' = \frac{E}{1-\nu^2} \quad (\text{plane strain}), \quad G = \frac{E}{2(1+\nu)} \end{aligned} \tag{2.20}$$

where E is the modulus of elasticity and G the shear modulus.

In the case of kinking crack propagation, additional terms occur in the expression for J. It is assumed that the crack propagates in the direction of the maximum energy release rate $G_{\max}$. One additional term J_K is related to mixed mode I and II loading, the other J_T to the T-stress (Erdogan and Sih 1963; Williams and Ewing 1972; Cottrell and Rice 1980; Lazzarin et al. 2009):

$$J_2 = J_K + J_T = -\frac{2K_{\mathrm{II}}}{E'}\left(K_{\mathrm{I}} + T\sqrt{\frac{8}{\pi}R}\right) \tag{2.21a}$$

In Eq. (2.21a), R is the radius of the selected circular path, i.e. J_T is path-dependent. The total J-value has the following square root form (Gdoutos 1990):

$$J_{\mathrm{tot}} = \sqrt{J^2 + J_2^2} \tag{2.21b}$$

2.2.7 General SIF Formulae

The SIFs are needed for assessing the brittle fracture or fatigue strength of structural members and specimens containing cracks, slits or flaws. The mode I SIF K_{I} is most important, because crack propagation is mainly controlled by this parameter. The crack takes a path where mode I conditions are prevailing. The mode II SIF K_{II} is important as a superimposed, mostly secondary loading effect. The mode III SIF K_{III} is of relevance to engineering problems such as torsional loaded round bars.

The SIFs depend on the geometry and position of the crack as well as on the geometrical and loading parameters of the considered structural member or specimen. Dominating parameters are the crack length and the load level. The dimension of the SIFs is $\mathrm{N/mm^{3/2}} = \mathrm{MPa\ mm^{1/2}}$, compare Eqs. (2.16–2.18).

The SIFs K_{I}, K_{II} and K_{III} in combination with the T-stress, S-stress and non-singular crack-front-parallel stress σ_0^* describe the linear-elastic stress field close to the crack tip at any point of the crack front completely. Today, the generally available numerical tools (FE and BE) allow to analyse any special crack configuration and to determine the stress field parameters just mentioned. On the other hand, formulae for SIFs are indispensable for assessing the influence of the geometrical, loading and support conditions. Any parametrical investigation and any integration into more general calculation methods is based on such formulae.

The analytical and numerical methods for determining the SIFs have been systematically reviewed (Radaj and Vormwald 2007, ibid. pp. 356–363). Formulae for SIFs are available for a large number of crack configurations in specimens and structural members (Murakami 1987; Rooke and Cartwright 1976; Sih 1973; Tada et al. 1999). Also, the T-stress solutions have been compiled (Sherry et al. 1995).

A review on the SIFs of cracks and slits in welded joints with reference to fatigue assessments is available (Radaj et al. 2006, ibid. pp. 250–259). A summary of the contents is given below.

The primary concern of the fatigue crack propagation approach applied to welded joints is directed towards surface cracks in areas of notch stress concentration such as weld toes. These cracks are assumed to be oblong and mostly continuous in the case of transverse welds or to be elliptical in the case of ends of longitudinal welds. The semi-ellipse is defined as circumscribing the more complex actual shape of several simultaneously initiated but coalescing cracks or of other planar defects.

The maximum SIF occurring at the deepest point of the semi-elliptical surface crack in a plate with superimposed notch effect is approximated by the following engineering formulae:

$$K_{\mathrm{I}} = (M_{\mathrm{km}} Y_{\mathrm{m}} \sigma_{\mathrm{m}} + M_{\mathrm{kb}} Y_{\mathrm{b}} \sigma_{\mathrm{b}}) \frac{\sqrt{\pi a}}{\Phi} \tag{2.22}$$

with the membrane and bending structural stresses σ_{m} and σ_{b}, with the geometry factors Y_{m} and Y_{b} due to the free plate surfaces (dependent on a/t and $a/2c$, with crack depth a, crack length $2c$ and plate thickness t), with the magnification factors M_{km} and M_{kb} on membrane and bending stresses due to the notch stress concentration factor K_{t} (dependent on the geometrical notch parameters and the ratio a/t, $M_{\mathrm{k}} = K_{\mathrm{t}}$ for $a/t = 0$) and with Φ the complete elliptical integral of the second kind due to the curved crack front (dependent on the aspect ratio a/c). The factors M_{k} and Y may be supplemented by a factor denoting the influence of a finite plate width. The parameter $1/\Phi$ characterises the decrease in the SIF of an internal elliptical or external semi-elliptical crack relative to a crack with a straight crack front penetrating the plate thickness ($1/\Phi = 0.64$ for $a/c = 1.0$).

A large body of data characterising the dependency of $M_{\mathrm{k}} Y_{\mathrm{m}}$ on a/t and a/c is available for plane models of one-sided transverse attachment joints which are considered as representative for the local situation at tubular welded joints. A semi-elliptical surface crack at the toe of the one-sided fillet weld is analysed for membrane and bending stresses in the base plate. Easy-to-use formulae for the factors M_{k} and Y of cracks at the weld toe of various other welded joints are available.

Another application of the SIF concept is directed to the crack-like slit ends at the root of fillet welds in cruciform joints, attachment joints, lap joints and cover plate joints. Various degrees of weld penetration are analysed. Corresponding SIF formulae are available in the open literature.

The stress intensity factors referenced above presume force-controlled loading conditions, i.e. the basic structural stresses at the crack site remain unchanged while the crack propagates with increasing SIF. As far as crack growth in structural components is considered, deformation-controlled loading conditions are often more appropriate. The dependency of the SIF on crack depth is then fundamentally changed, it is decreasing. This effect is termed 'load shedding'.

2.2.8 SIFs at Spot-Welded Lap Joints

The singular and non-singular slit tip stresses at the edge of spot welds in lap joints may be determined in a two-step analysis procedure (Radaj et al. 2006, ibid. pp. 366–512). In a first step, the structural stresses around the weld spot edge are calculated based on engineering formulae or based on a FE model consisting of plane thin-shell elements. Bernoulli's condition that plane cross sections of the thin-shell elements remain plane is valid. This results in a linear distribution over the plate thickness of superimposed membrane and bending tensile or compressive stresses and also of membrane and bending spot-edge-parallel shear stresses. The spot-edge-transverse shear stresses are defined as the medium of a parabolic distribution over the plate thickness. In a second step, these internal structural stresses which vary along the edge line of the weld spot are applied to the plane cross-sectional model of the weld spot edge containing the slit tip. Simple but accurate formulae are available for the SIFs, the T-stress and the S-stress dependent on the decomposed structural stresses and on the square-root on plate thickness t. The assumption behind the cross-sectional model of only one half of the weld spot is, that the weld spot diameter d is substantial larger than the plate thickness t, so that a direct influence of the opposed weld spot side is excluded. It has been found that in the SIF formulae for the slit tips in thin-sheet lap joints, the sheet thickness t is the dominating square-root parameter, substituting the crack length in the conventional formulae for crack tip SIFs.

The originally proposed decomposition procedure for the structural stresses at the weld spot edge in joints of equal plate thickness, Fig. 2.4, is based on symmetric and antimetric (indices ++ and +−) stresses or forces in the upper and lower plate (indices u and l) of the cross-sectional model, resulting in the different singular and non-singular loading modes (Radaj 1989). This procedure attributes primary concern to self-equilibrating force groups. The idea behind this is that forces which are transmitted from the upper to the lower plate may produce other slit tip stresses than when transmitted to a remote support within the weld spot. A stress is termed 'symmetric', if it has the same value and sign (or direction) at corresponding points above and below the slit tip. For 'antimetric' stresses, the sign is reversed.

The procedure is presented and discussed in detail in the author's book (Radaj et al. 2006, ibid. pp. 447–453), so that the presentation here can be restricted to the derived SIF formulae.

The original SIF formulae with the symmetric and antimetric stress portions in joints of equal plate thickness t read as follows:

$$K_{\mathrm{I}} = \left(\frac{1}{\sqrt{3}}\sigma_{\mathrm{b}}^{++} + 2.23\tau_{\perp}^{+-}\right)\sqrt{t} \tag{2.23}$$

$$K_{\mathrm{II}} = \left(\frac{1}{2}\sigma_{\mathrm{m}}^{+-} + \frac{1}{2}\sigma_{\mathrm{b}}^{+-} + 0.55\tau_{\perp}^{++}\right)\sqrt{t} \tag{2.24}$$

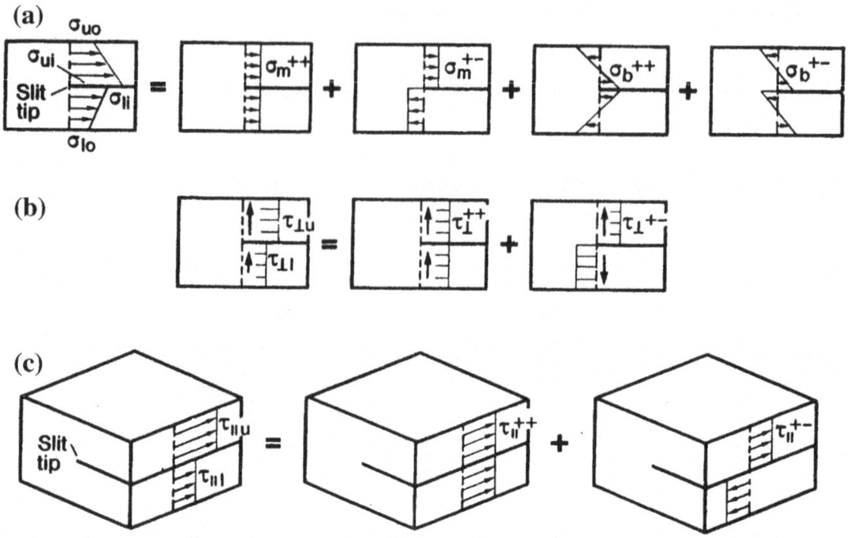

Fig. 2.4 Decomposition of the total structural stress state at the slit tip into symmetrical and antimetrical stress portions; membrane and bending stresses (**a**), transverse shear stresses (**b**) and longitudinal shear stresses (**c**); (Radaj 1989; Radaj and Zhang 1991[(1)], 1991[(2)])

$$K_{\mathrm{III}} = \sqrt{2}\tau_{\|}^{+-}\sqrt{t} \tag{2.25}$$

The factors 2.23 and 0.55 have been determined by an accurate boundary element analysis (Radaj 1989). Slightly different values, 2.36 and 0.60, have been derived based on a *J*-integral formulation (Zhang 1999). It is obvious from Eqs. (2.23) and (2.24) that the edge-transverse shear stresses have a remarkable influence on the SIFs K_{I} and K_{II}.

The SIFs may also be expressed in the non-decomposed stresses:

$$K_{\mathrm{I}} = [0.144(\sigma_{\mathrm{ui}} - \sigma_{\mathrm{uo}} + \sigma_{\mathrm{li}} - \sigma_{\mathrm{lo}}) + 1.115(\tau_{\perp\mathrm{u}} - \tau_{\perp\mathrm{l}})]\sqrt{t} \tag{2.26}$$

$$K_{\mathrm{II}} = [0.25(\sigma_{\mathrm{ui}} - \sigma_{\mathrm{li}}) + 0.275(\tau_{\perp\mathrm{u}} + \tau_{\perp\mathrm{l}})]\sqrt{t} \tag{2.27}$$

$$K_{\mathrm{III}} = 0.707(\tau_{\|\mathrm{ui}} - \tau_{\|\mathrm{li}})\sqrt{t} \tag{2.28}$$

By introduction of the edge-parallel shear stresses on the inner plate side (index i) in Eq. (2.28), equilibrated bending shear stresses can be assumed superimposed on the membrane shear stresses. The K_{III} value thus determined should be conservative (Radaj et al. 2006, ibid. pp. 449–450).

The original decomposition procedure above has two deficiencies. The non-singular bending stresses are not separated and the SIF K_{II} is derived from two decomposed normal stresses, Eq. (2.24), whereas one such stress should be sufficient. These deficiencies were soon realised and removed (Radaj and Zhang 1991[(1)], 1991[(2)]). A strictly mode-related decomposition was the result, with the

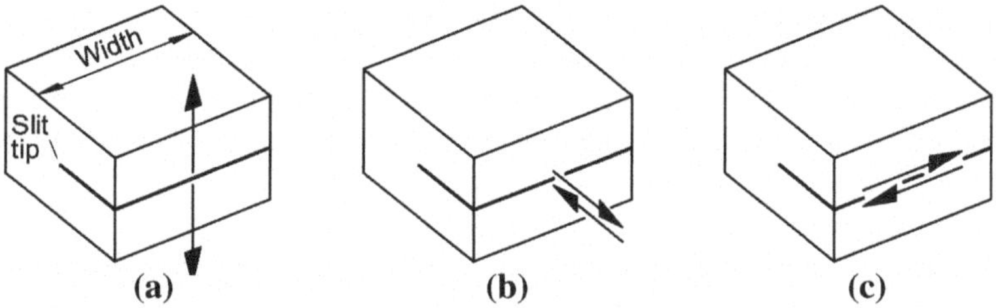

Fig. 2.5 Self-equilibrating loading states generating a stress singularity at the slit tip; pure mode I (**a**), pure mode II (**b**) and pure mode III (**c**); the pairs of resultant forces are assumed to act on the slit flanks in opposite directions in the same line each; they are evenly distributed over the specimen width; (Radaj et al. 2006)

definition that the pure modes I, II and III are produced by self-equilibrated remote boundary forces, Fig. 2.5. Non-singular modes 0 may be superimposed (in-plane membrane and bending stresses, out-of-plane membrane shear stress), which correspond to the support forces.

The characteristic stress of pure mode I loading is the counter-bending stress σ_{cb}, and that of pure mode II loading the tensile-bending stress σ_{tb}, compare Eqs. (2.26) and (2.27):

$$\sigma_{\mathrm{cb}} = \frac{1}{4}(\sigma_{\mathrm{ui}} - \sigma_{\mathrm{uo}} + \sigma_{\mathrm{li}} - \sigma_{\mathrm{lo}}) \tag{2.29}$$

$$\sigma_{\mathrm{tb}} = \frac{1}{2}(\sigma_{\mathrm{ui}} - \sigma_{\mathrm{li}}) \tag{2.30}$$

Neglecting the edge-transverse shear stress $\tau_{\perp}$, the SIFs according to Eqs. (2.26) and (2.27) have the following simple form:

$$K_{\mathrm{I}} = \frac{1}{\sqrt{3}}\sigma_{\mathrm{cb}}\sqrt{t} \tag{2.31}$$

$$K_{\mathrm{II}} = \frac{1}{2}\sigma_{\mathrm{tb}}\sqrt{t} \tag{2.32}$$

Other mode-related decompositions (mode I, mode II, mode 0 membrane and mode 0 bending) of the in-plane normal stresses are also possible. A principal statement is elucidating in this context (Seeger et al. 2005). The four independent normal stresses at the outer and inner plate surfaces in the cross-sectional model can always be substituted by or decomposed in four other independent stresses characterising definite linearised stress groups. It is straightforward to select the following mode-related stresses as such: the mode II related antimetric membrane stress $\sigma_{\mathrm{m}}^{\mathrm{II}}$, the mode I related symmetric bending stress $\sigma_{\mathrm{b}}^{\mathrm{I}}$ and the mode 0 related membrane and bending stresses, σ_{m}^{0} and σ_{b}^{0}. Preference is given to these decomposed stress groups, because any stress state in the cross-sectional model can thus uniquely be described by simple expressions, especially too when considering unequal plate thickness. Self-equilibrating stress groups are not a necessary condition, because the non-equilibrated stress portions are compensated by mode 0 stress distributions. Above, the pure modes I, II and III are defined by membrane and bending stresses which are not self-equilibrated in general. Mode 0 stresses and support forces are not identical as a consequence.

Another argument supports the decomposition of the shear stresses $\tau_{\perp u}, \tau_{\perp l}$, $\tau_{\| u}$, $\tau_{\| l}$ into symmetric and antimetric stress groups without consideration of the equilibrium conditions. The resulting decomposed stresses are identical with those from a force-equilibrated decomposition also in the case of unequal plate thickness. Once more, self-equilibration is not a necessary condition.

The revised decomposition into the mode-related stresses just mentioned for joints of equal plate thickness is presented in Fig. 2.6. The SIFs are derived therefrom by the following expressions:

$$K_{\mathrm{I}} = \left(\frac{1}{\sqrt{3}}\sigma_{\mathrm{b}}^{\mathrm{I}} + 2.23\tau_{\perp}^{\mathrm{I}}\right)\sqrt{t} \tag{2.33}$$

$$K_{\mathrm{II}} = \left(\frac{1}{2}\sigma_{\mathrm{m}}^{\mathrm{II}} + 0.55\tau_{\perp}^{\mathrm{II}}\right)\sqrt{t} \tag{2.34}$$

$$K_{\mathrm{III}} = \sqrt{2}\tau_{\|}^{\mathrm{III}}\sqrt{t} \tag{2.35}$$

It should be noted that $\sigma_{\mathrm{b}}^{\mathrm{I}} = \sigma_{\mathrm{cb}}$ and $\sigma_{\mathrm{m}}^{\mathrm{II}} = \sigma_{\mathrm{tb}}$, which are the characteristic self-equilibrated stresses. This is an indication that self-equilibrated loading is the primary condition of the pure loading modes.

The expression for the non-singular slit tip stresses are:

$$T = \sigma_{\mathrm{m}}^{0} + \sigma_{\mathrm{b}}^{\mathrm{I}} = \frac{1}{2}(\sigma_{\mathrm{ui}} + \sigma_{\mathrm{li}}) \tag{2.36}$$

$$S = \tau_{\|}^{0} = \frac{1}{2}(\tau_{\| \mathrm{u}} + \tau_{\| \mathrm{l}}) \tag{2.37}$$

The following remark may be added. The non-singular bending stress σ_{b}^{0} characterises a third order term in the series expansion of the stress field, whereas T-stress and S-stress are second order terms.

The original decomposition procedure for the structural stresses at the weld spot edge in joints of unequal plate thickness (Radaj and Zhang 1991$^{(1)}$, 1991$^{(2)}$) is related to the self-equilibrated pure modes I, II and III. The mode 0 bending stresses were not separated in the first version. The correspondingly amended procedure is visualised for the in-plane normal-stresses in Fig. 2.7.

At first, some notes are necessary on the status of the relevant publications, which present two slightly different versions of the method. The original version (Radaj and Zhang 1991$^{(1)}$) does not separate the mode 0 bending stress. This version has been taken over into the author's book (Radaj et al. 2006), because the coefficients for the K_{I} and K_{II} expressions are available only for this deficient

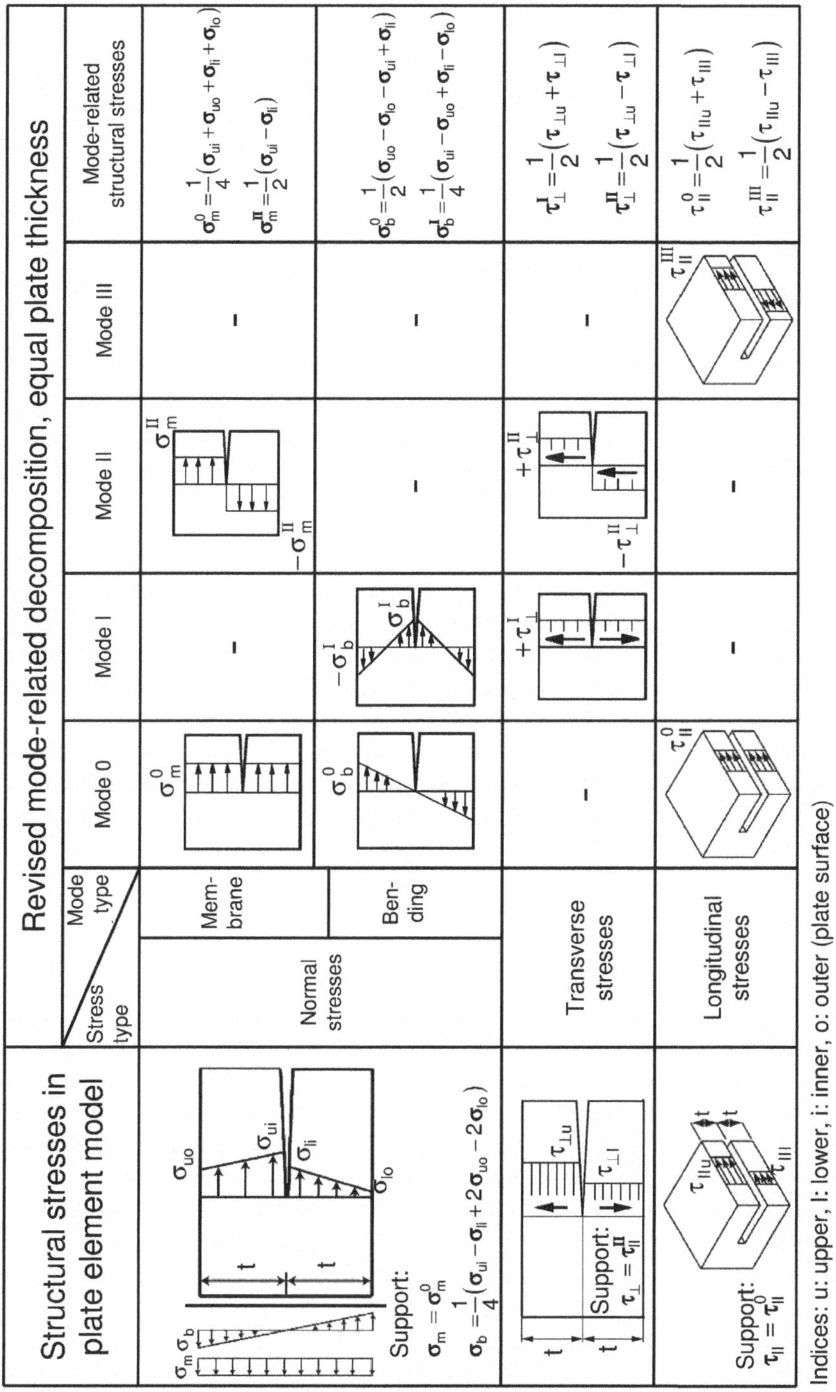

Fig. 2.6 Revised mode-related decomposition with equal plate thickness; (Seeger et al. 2005)

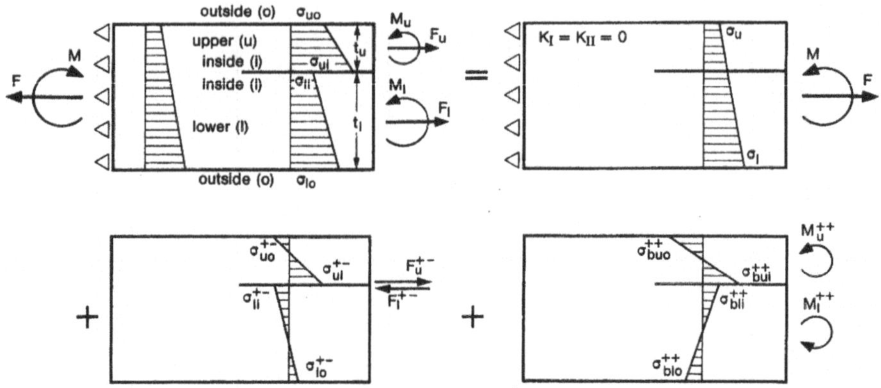

Fig. 2.7 Decomposition of the in-plane structural normal stress state at the slit tip of a lap joint cross-sectional model with unequal plate thickness into antimetrical (tension-bending) force and symmetrical moment (counterbending) portions after separation of the non-singular membrane and bending stress portion; (Radaj and Zhang 1991[(2)])

method (without effect on the numerical SIF results). An inconsistency in the expressions for the decomposed edge-parallel shear stresses in the book, ibid. Eqs. (10.29) and (10.30), has to be mentioned. These stresses are not self-equilibrating for thickness ratios $\delta \neq 1.0$. They should have the same form as the preceding transverse shear expressions. The coefficients K_{IIIu}^{+-} are related to the self-equilibrating stresses (Radaj and Zhang 1991[(1)]).

The revised decomposition into mode-related membrane and bending stresses for joints of unequal plate thickness without the condition of self-equilibration is presented in Fig. 2.8. The SIFs are derived therefrom by the following expressions:

$$K_{\mathrm{I}} = (k_{\mathrm{I,b}}\sigma_{\mathrm{b}}^{\mathrm{I+II}} + k_{\mathrm{I,m}}\sigma_{\mathrm{m}}^{\mathrm{II+I}} + k_{\mathrm{I},\perp}\tau_{\perp}^{\mathrm{I}} + k_{\mathrm{I},\perp}^{*}\tau_{\perp}^{\mathrm{II+I}})\sqrt{t_{\mathrm{u}}} \tag{2.38}$$

$$K_{\mathrm{II}} = (k_{\mathrm{II,b}}\sigma_{\mathrm{b}}^{\mathrm{I+II}} + k_{\mathrm{II,m}}\sigma_{\mathrm{m}}^{\mathrm{II+I}} + k_{\mathrm{II},\perp}\tau_{\perp}^{\mathrm{II+I}})\sqrt{t_{\mathrm{u}}} \tag{2.39}$$

$$K_{\mathrm{III}} = k_{\mathrm{III},\,\|}\tau_{\|}^{\mathrm{III}}\sqrt{t_{\mathrm{u}}} \tag{2.40}$$

The situation above is more complicated with unequal compared with equal plate thickness. The decomposition procedure is the same in principle, but not all decomposed stresses can be connected with a single loading mode. Some are related to two modes, but one of the two modes is dominant in general.

The stresses with a two-mode effect have an additional index related to the second mode: $\sigma_{\mathrm{b}}^{\mathrm{I}} \rightarrow \sigma_{\mathrm{b}}^{\mathrm{I+II}}$, $\sigma_{\mathrm{m}}^{\mathrm{II}} \rightarrow \sigma_{\mathrm{m}}^{\mathrm{II+I}}$, $\tau_{\perp}^{\mathrm{II}} \rightarrow \tau_{\perp}^{\mathrm{II+I}}$. The number of stress terms in the expressions for K_{I} and K_{II} is correspondingly enlarged. The coefficients of the stress terms have an index for the mode type (I, II, III), to which they contribute, and an index for the stress type (b, m, $\perp$, $\|$), to which they are connected. These coefficients depend on the ratio of the plate thicknesses, $\delta = t_{\mathrm{u}}/t_{\mathrm{l}}$ (indices for upper

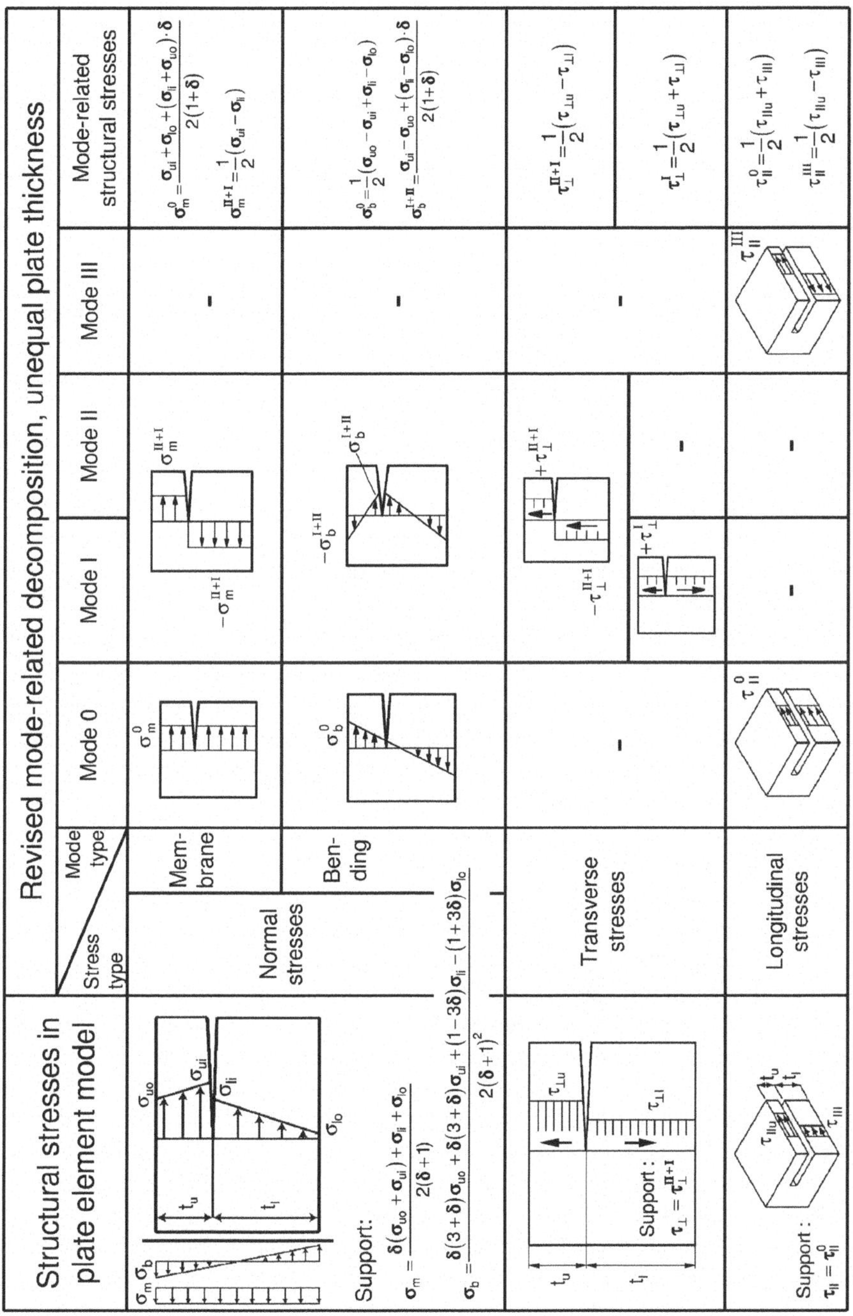

Fig. 2.8 Revised mode-related decomposition with unequal plate thickness; (Seeger et al. 2005)

Table 2.1 Coefficients of mode-related structural stresses for the SIFs K_{I}, K_{II} and K_{III} in joints of unequal plate thicknesses t_{u} (upper plate) and t_{l} (lower plate); (Radaj and Zhang 1999; Seeger et al. 2005)

	Mode I				Mode II			Mode III
$\delta = t_{\mathrm{u}}/t_{\mathrm{l}}$	$k_{\mathrm{I,b}}$	$k_{\mathrm{I,m}}$	$k_{\mathrm{I},\perp}$	$k^*_{\mathrm{I},\perp}$	$k_{\mathrm{II,b}}$	$k_{\mathrm{II,m}}$	$k_{\mathrm{II},\perp}$	$k_{\mathrm{III},\parallel}$
1.0	0.578	0	2.236	0	0	0.500	0.550	1.414
0.556	0.652	−0.218	2.025	0.876	−0.068	0.593	0.429	1.247
0.333	0.695	−0.409	1.911	1.245	−0.130	0.700	0.325	1.155
0.120	0.734	−0.671	1.794	1.656	−0.226	0.905	0.099	1.058

and lower plate). The upper plate thickness t_{u} is the square root parameter in the SIFs.

The coefficients of the accordingly decomposed stresses in the SIF formulae are listed in Table 2.1 for several values of δ. In the line with $\delta = 1.0$, the coefficients from Eqs. (2.23) and (2.24) appear. The coefficients $k_{\mathrm{I,b}}$, $k_{\mathrm{I,m}}$, $k_{\mathrm{II,b}}$, $k_{\mathrm{II,m}}$ for the other values of δ have been determined by application of the FE programme FRANC2D/L. The values of $k_{\mathrm{I},\perp}$, $k^*_{\mathrm{I},\perp}$, $k_{\mathrm{II},\perp}$, $k_{\mathrm{III},\parallel}$ are derived from the formulae in the original publication (Radaj and Zhang 1991[(1)], 1991[(2)]). The coefficient $k_{\mathrm{III},\parallel}$ is found for a membrane shear stress distribution. Bending shear stresses are not taken into account.

The coefficients of the originally decomposed stresses in the SIF formulae (inclusive of $K_{\mathrm{res}} = (K_{\mathrm{I}}^2 + K_{\mathrm{II}}^2)^{1/2}$) are available dependent on the plate thickness ratio δ as closed form expressions and also in graphical form (Radaj and Zhang 1991[(1)]). They are determined based on a combination of *J*-integral and boundary element method. The application to the spot-welded tensile-shear and cross tension specimens is also demonstrated.

Spot-welded lap joints have also been investigated with dissimilar plate materials (Radaj and Zhang 1992, 1994). Different elastic constants are assigned to the upper and lower plate. An oscillatory stress singularity occurs at the bimaterial slit tip. It is described by special unconventional SIFs which may be substituted by the conventional SIFs in an approximative sense for engineering applications.

Only the original mode-related version of the method, which is based on self-equilibrated stress groups, is reviewed below. A revised version based on possibly simpler, non-equilibrated stress groups has been proposed, but not elaborated (Seeger et al. 2005). Any version has to come up with two complications. First, the mode 0 membrane and bending loading states have to be based on a linear distribution of the membrane and bending strains (not of the stresses), because only these produce no stress singularity at the slit tip. The corresponding membrane and bending stresses show a step and kink in gradient in the slit and interface plane. Second, the conventional SIFs K_{I} and K_{II} (not K_{III}) describe the stresses at the slit tip only approximately. Especially, the step in the interface-parallel normal stress values is not reproduced, but the deviations from the exact values are not too large (Radaj and Zhang 1994).

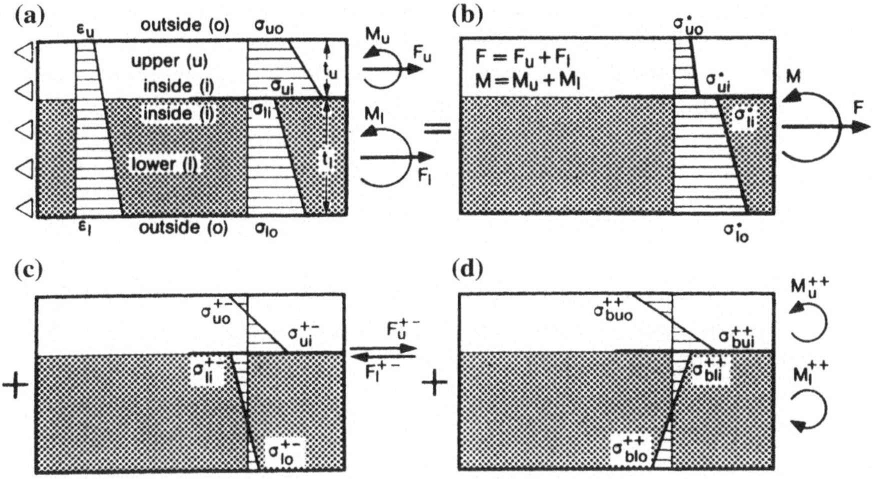

Fig. 2.9 Decomposition of the in-plane structural normal stress state at the slit tip of a bimaterial lap joint cross-sectional model (**a**), into a stress portion without singularity effect (**b**), and into tension-bending and counterbending portions with singularity effect (**c**, **d**); (Radaj and Zhang 1992)

The self-equilibrated decomposition of the membrane and bending stresses taking bimaterial conditions into account is based on Fig. 2.9. The condition of plane cross-sections is expressed by the linear distribution of the strains between ε_u and ε_l. The original plate stresses to be decomposed are σ_{uo}, σ_{ui}, σ_{li}, σ_{lo}. The mode 0 loading state is characterised by σ^*_{uo}, σ^*_{ui}, σ^*_{li}, σ^*_{lo}. The step in the stress values and in their gradient is clearly visible. Finally, the tension and counter-bending loading states are shown which are related to the stress singularity.

The SIFs K_I, K_{II} and K_{III} are given by the following expressions, all coefficients and stresses referring to the upper plate:

$$K_I = (k_{I,tb}\sigma_{tb} + k_{I,cb}\sigma_{cb} + k^{++}_{I,\perp}\tau^{++}_{\perp} + k^{+-}_{I,\perp}\tau^{+-}_{\perp})\sqrt{t_u} \tag{2.41}$$

$$K_{II} = (k_{II,tb}\sigma_{tb} + k_{II,cb}\sigma_{cb} + k^{++}_{II,\perp}\tau^{++}_{\perp} + k^{+-}_{II,\perp}\tau^{+-}_{\perp})\sqrt{t_u} \tag{2.42}$$

$$K_{III} = k_{III,\parallel}\tau^{+-}_{\parallel}\sqrt{t_u} \tag{2.43}$$

The coefficients k_μ (μ = I,tb, I,cb etc.) depend on the modulus of elasticity ratio of the two materials and on the plate thickness ratio $\delta = t_u/t_l$. They are available as closed form expressions for $\delta = 1.0$ (Radaj and Zhang 1992), gained by a combination of the boundary element and the J-integral method. The plane strain condition is appropriate when considering spot-welded joints. The linear-elastic behaviour of isotropic materials is described by two independent elastic constants, mostly the modulus of elasticity E and Poisson's ratio ν, but sometimes also the shear modulus G. The coefficients k_μ are then stated dependent on the shear modulus ratio $\beta = G_u/G_l$ supplemented by Poisson's ratios ν_u and ν_l.

The initial step in the normal stress decomposition procedure is the separation of the mode 0 non-singular membrane and bending stresses based on the condition of a linear strain distribution (plane cross-sections remaining plane). Rather complicated expressions are gained for the appertaining stresses σ^*_{uo}, σ^*_{ui}, σ^*_{li}, σ^*_{lo} (see Fig. 2.9). The characteristic stresses for mode I and mode II loading, counter-bending (index cb) and tensile-bending (index tb), are then expressed as follows:

$$\sigma_{\mathrm{cb,u}} = \sigma^*_{\mathrm{ui}} - \sigma_{\mathrm{ui}} + 2(\sigma^*_{\mathrm{uo}} - \sigma_{\mathrm{uo}}) \tag{2.44}$$

$$\sigma_{\mathrm{tb,u}} = 2(\sigma_{\mathrm{ui}} - \sigma^*_{\mathrm{ui}} + \sigma_{\mathrm{uo}} - \sigma^*_{\mathrm{uo}}) \tag{2.45}$$

These two equations are easily derived by considering the stresses in the upper plate after deduction of the mode 0 stresses:

$$\sigma_{\mathrm{ui}} - \sigma^*_{\mathrm{ui}} = \sigma_{\mathrm{tb,u}} + \sigma_{\mathrm{cb,u}} \tag{2.46}$$

$$\sigma_{\mathrm{uo}} - \sigma^*_{\mathrm{uo}} = -\frac{1}{2}\sigma_{\mathrm{tb,u}} - \sigma_{\mathrm{cb,u}} \tag{2.47}$$

They are valid for any plate thickness ratio and any material combination.

In the referenced publication (Radaj and Zhang 1992), ibid. Eq. (37), $\sigma_{\mathrm{tb,u}}$ is erroneously introduced with the factor 1/2 instead of 2 in Eq. (2.45) above, but the expressions for $k_{\mathrm{I,tb}}$ and $k_{\mathrm{II,tb}}$ are simultaneously changed, so that K_{I} and K_{II} remain unchanged. Actually, $k_{\mathrm{I,tb}} = 0$ for any modulus of elasticity ratio combined with $\delta = 1.0$ (see Fig. 6 in the referenced publication).

The decomposition of the edge-transverse and edge-parallel shear stresses based on self-equilibrated forces can be substituted, as already stated, by a decomposition into symmetric and antimetric stress groups without causing a change in the results:

$$\tau_{\perp}^{++} = \frac{1}{2}(\tau_{\perp\mathrm{u}} + \tau_{\perp\mathrm{l}}) \tag{2.48}$$

$$\tau_{\perp}^{+-} = \frac{1}{2}(\tau_{\perp\mathrm{u}} - \tau_{\perp\mathrm{l}}) \tag{2.49}$$

$$\tau_{\parallel}^{+-} = \frac{1}{2}(\tau_{\parallel\mathrm{u}} - \tau_{\parallel\mathrm{l}}) \tag{2.50}$$

The expression for $\tau_{\parallel}^{+-}$ is derived without considering the effect of superimposed bending shear stresses.

2.2.9 *Equivalent SIFs under Mixed Mode Loading Conditions*

Mixed mode loading conditions at crack or slit tips pose the question of how the SIFs of the different modes should be superimposed to give an equivalent SIF (also termed 'resultant SIF'), equivalent in respect of fatigue failure by crack initiation

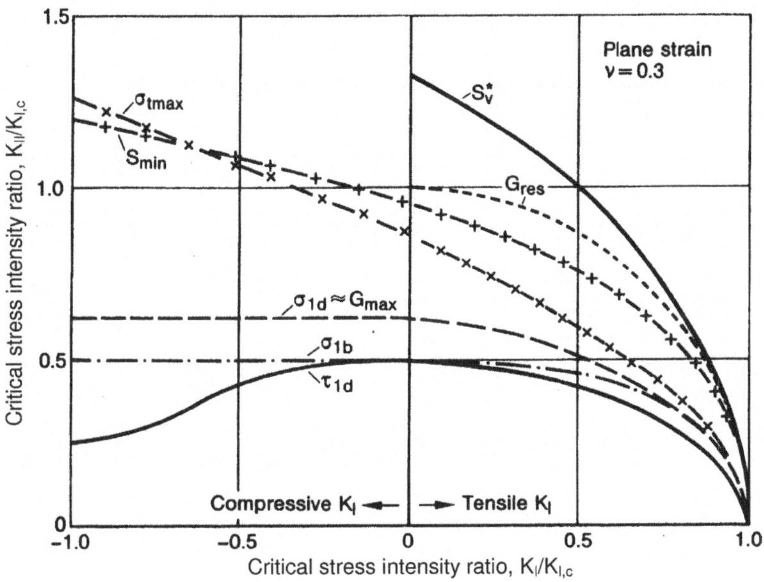

Fig. 2.10 Critical stress intensities (normalised by $K_{I,c}$) characterising crack propagation under mixed mode I and II loading conditions; various failure criteria; curve designations specified in the text; (Radaj and Zhang 1995$^{(1)}$)

and propagation. This includes the question of what critical value of K_{II} or K_{III} is equivalent to the critical value of K_I. The answer to these questions is complicated by the fact that the crack under mixed-mode loading conditions usually propagates in a non-coplanar manner, i.e. the crack changes its direction of propagation ('kinked crack'). The crack propagation angle ('kink angle') must be found first. Additional assumptions are necessary to do this. Several hypotheses and related criteria have been proposed to describe crack initiation and propagation under mixed-mode conditions. They are not generally verified by testing. It is therefore necessary to check the applicability of the criterion selected in each individual case.

The critical conditions for crack initiation and propagation under mixed mode I and II conditions, specified according to different failure criteria, are plotted in Fig. 2.10. The tensile K_I part of the diagram is mainly taken from the literature (Sih 1975). The compressive part is supplemented as far as the criteria are defined in this region. The critical SIF ratios $K_I/K_{I,c}$ and $K_{II}/K_{I,c}$ refer to the critical SIF $K_{I,c}$ in mode 1 loading: $K_{I,c} = K_{Ic}$ (fracture toughness) for brittle fractures, $K_{I,c} = K_{th}$ (threshold SIF) or $K_{I,c} = \overline{\sigma}_E(\pi\rho^*/2)^{1/2}$ (endurance limit stress $\overline{\sigma}_E$ averaged over microstructural support length ρ^*) for fatigue fractures. The following failure criteria are evaluated:

- The maximum tangential stress $\sigma_{t\ max}$ at the edge of a 'core region' around the crack tip (radius r^*) is decisive for crack propagation (Erdogan and Sih 1963).
- The minimum strain energy density at the edge of the core region expressed by the minimum strain energy density factor S_{min} is decisive for crack propagation (Sih 1973, 1974, 1975).
- The maximum dilatational strain energy density at the edge of the core region expressed by the corresponding factor S^*_{max} is decisive for crack propagation (Radaj and Heib 1978).
- The resultant energy release rate G_{max} is decisive for coplanar crack propagation (Irwin 1957).
- The maximum energy release rate G_{res} is decisive for non-coplanar crack propagation (Hussain et al. 1974).
- The process zone criteria: brittle initiation of fracture occurs at a critical value of the first principal stress σ_{1b} at distance r^* in the direction of maximum (tensile) dilatational strain energy density; 'ductile' initiation of fracture occurs at a critical value of the principal shear stress τ_{1d} (or first principal stress σ_{1d}) at distance r^* in the direction of maximum (tensile) distortional strain energy density (Radaj and Zhang 1995[(1)])

There are major differences in curve shape and curve position both in the tensile and the compressive range of K_I which are caused by the different formal and physical contents of the criteria. The process zone criteria differ from one another to a minor extent only. They are conservative compared with the conventional criteria. The curve for the well-known and usually preferred Erdogan–Sih criterion of maximum tangential stress, for example, runs considerably higher, especially in the compressive K_I range. The reason for this is the fact that the criterion takes only the tangential stresses around the crack tip into account and neglects the radial stresses which are rather high under mode II conditions.

The consideration above, as far as fatigue-relevant, refers to constant-amplitude loading of an ideal crack or slit tip with negligible mean stress effects. In reality further important influencing parameters have to be considered such as mean stress intensity, residual stress intensity, non-singular stresses, non-proportional loading, crack or slit tip microstructural conditions and crack closure effects among others. Whether or not a definite criterion is applicable in the special case under consideration should finally be decided on the basis of mixed mode fatigue testing results. Such data are not readily available from the literature.

The preferred criterion in the case of spot-welded or seam-welded joints is the maximum tangential stress criterion. It states that the crack or slit tip subjected to mixed mode I and II loading will propagate in the direction in which the tangential tensile stress at a small distance from the crack tip reaches its maximum. This results in the crack propagation angle $\overline{\theta}$ as a function of the SIF ratio K_{II}/K_I, Fig. 2.11. The equivalent SIF K_{eq} then follows from the condition of identical values of $\sigma_{t\ max}$ in mixed mode and pure mode I loading:

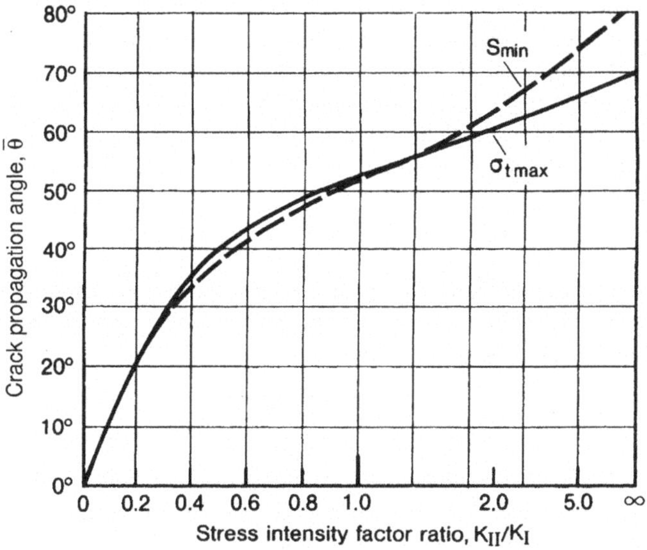

Fig. 2.11 Crack propagation angle $\overline{\theta}$ against ligament direction dependent on stress intensity factor ratio; maximum tangential stress criterion ($\sigma_{t\ max}$) compared with strain energy density criterion (S_{min}); (Yuuki et al. 1985[1], 1985[2]; Yuuki and Ohira 1986, 1989)

$$K_{eq} = \left(K_I \cos^2 \frac{\overline{\theta}}{2} - \frac{3}{2} K_{II} \sin \overline{\theta} \right) \cos \frac{\overline{\theta}}{2} \tag{2.51}$$

$$\tan \frac{\overline{\theta}}{2} = \frac{1}{4} \left(\frac{K_I}{K_{II}} \right) \pm \frac{1}{4} \sqrt{\left(\frac{K_I}{K_{II}} \right)^2 + 8} \tag{2.52}$$

where the positive root corresponds to a negative K_{II}, the negative root to a positive K_{II} (with positive K_I). Pure mode I loading results in $\overline{\theta} = 0°$ and $K_{eq} = K_I$, whereas pure mode II loading is associated with $\overline{\theta} = 70.5°$ and $K_{eq} = 1.15K_{II}$.

The strain energy density criterion (Sih 1973, 1974, 1975) which comprises not only K_I and K_{II} but also K_{III}, is open to more varied physical interpretations and adaptable to more complex test data. The criterion states that the crack propagates in the direction of the minimum total strain energy density at the edge of the core region (approximately equal to the direction of maximum dilatational strain energy density), and that a critical value of the strain energy densities mentioned must be reached. The equivalent stress intensity factor K_{eq} follows from:

$$K_{eq} = \sqrt{A_{11} K_I^2 + A_{12} K_I K_{II} + A_{22} K_{II}^2 + A_{33} K_{III}^2} \tag{2.53}$$

The coefficients A_{11} to A_{33} are mainly dependent on the crack propagation angle (besides Poisson's ratio) which in turn is defined by the stress intensity factor ratios

K_{II}/K_I and K_{III}/K_I, respectively. The coefficients are often introduced as constant in order to simplify the procedure. The following simplified version of Eq. (2.53) follows from the application of the strain energy density criterion to the single-mode loading states of a centre crack in the infinite plate. With Poisson's ratio $\nu = 0.28$, the ratios of critical SIFs are $K_{II,c}/K_{I,c} = 0.985$ and $K_{III,c}/K_{I,c} = 0.663$, thus yielding the following formula, (Sih 1973), ibid. Eqs. (16) and (18):

$$K_{eq} = \sqrt{K_I^2 + 1.03K_{II}^2 + 2.27K_{III}^3} \tag{2.54}$$

A similar equation is obtained proceeding from the resultant energy release rate criterion (Irwin 1957) which assumes coplanar crack propagation and ignores the influence of the crack propagation angle:

$$K_{eq} = \sqrt{K_I^2 + K_{II}^2 + 1.39K_{III}^2} \tag{2.55}$$

Here, $\nu = 0.28$ is introduced, resulting in $1/(1 - \nu) = 1.39$. The same expression is derived based on the J-integral. This relationship has occasionally been applied to spot-welded lap joints (Linder et al. 1998).

The process zone criteria (Radaj and Zhang 1995[(1)]) also yield the general form of Eq. (2.53) partly extended by the addition of linear K_I and K_{II} terms. The corresponding curves in Fig. 2.10 (σ_{1d} and τ_{1d} for 'ductile' behaviour, σ_{1b} for brittle behaviour) result from evaluating the crack propagation angle in the different mixed-mode (inclusive of single-mode) loading states. The following more conservative form, Eq. (2.56), is proposed on the basis of the lower plotted curves in Fig. 2.10 substituting the older, less conservative form, Eq. (2.57), chosen in several comparative evaluations related to spot-welded joints (Radaj et al. 1990, 2006; Radaj and Giering 1994, 1995; Radaj and Zhang 1995[(2)]):

$$K_{eq} = \sqrt{K_I^2 + 3K_{II}^2 + K_{III}^2} \tag{2.56}$$

$$K_{eq} = \sqrt{K_I^2 + K_{II}^2 + K_{III}^2} \tag{2.57}$$

Kurath has found that K_{II}^2 combined with the factor 3 is valid in the case of spot welds of low-carbon steels and combined with the factor 2 in the case of spot welds of high-strength steels (Kurath 1992). The factor 4.3 has been derived from fatigue test results gained from single-spot specimens with various ratios K_I/K_{II} (Lee and Kim 2004).

Similar formulae have been derived based on the assumption of coplanar crack propagation described by a critical averaged stress over the microstructural length ρ^* (Neuber's hypothesis). Different failure criteria (Rankine's normal stress, von Mises' distortional strain energy, Beltrami's total strain energy) are combined with plane stress or plane strain multiaxiality conditions, see Sect. 1.5.2.

A diagram derived for kinked crack propagation under mixed-mode loading conditions of the crack in non-welded steels (Pook 1989) is presented in Fig. 2.12.

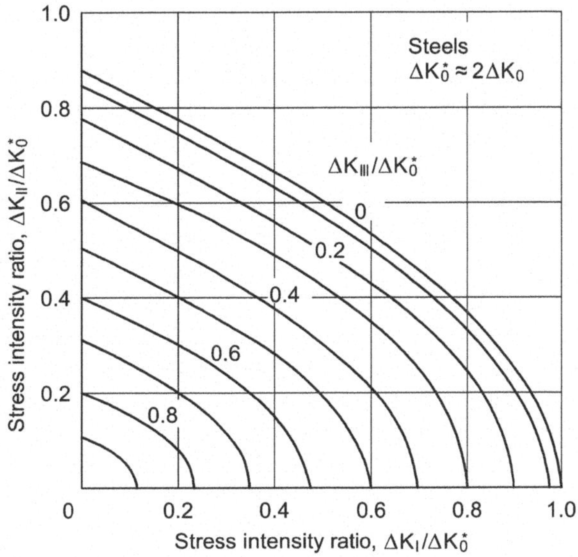

Fig. 2.12 Lower limit curves of the cyclic threshold stress intensity factors for kinked or branched macrocrack propagation under mixed mode loading conditions (modes I, II and III superimposed), based on test results for various steels; (Pook 1989)

The reference quantity ΔK_0^* is the cyclic threshold stress intensity factor for kinked or branched macro-crack propagation which may be about twice the threshold value for coplanar crack propagation in mode I.

The equivalent stress intensity factors according to Eqs. (2.51–2.57), when applied to cyclic loading, do not include the further influencing parameters governing fatigue, e.g. prestress or crack closure. Another shortcoming is the absence of the non-singular stress terms whose influence on fatigue crack initiation and propagation may be important (Haefele and Lee 1995).

2.2.10 Endurable Stress Intensity Factors

Endurable SIFs were originally defined in respect of static loading considering brittle fracture as limit condition. The fracture toughness $K_{\rm Ic}$ obtained in mode 1 fracture tests is the relevant material parameter. To ensure a sufficiently brittle specimen behaviour is an essential requirement on the testing procedure. Size and thickness of the specimen must be kept sufficiently large in order to maintain a predominantly elastic behaviour with three-dimensional tensile stresses ahead of the crack tip and with plastic deformations at the crack tip more or less avoided.

Endurable SIFs defined in respect of fatigue loading are the threshold SIFs $\Delta K_{\rm th}$, which are considered as material parameters. The parameter $\Delta K_{\rm th}$ denotes the minimum cyclic stress range at which propagation of a (sufficiently long) crack is observed. In general, mode I testing in the high-cycle fatigue range is performed for this purpose. A typical lower bound value for structural steels is $\Delta K_{\rm th} = 180$ MPa mm$^{1/2}$ (valid for $R = 0$).

The existence of a threshold value is explained by crack closure phenomena occurring in cyclic loading (Elber 1970, 1971). Under plane stress conditions, crack closure originating from plastic deformation at the crack tip is the predominant effect. Under plane strain conditions, other phenomena are prevailing: surface roughness, fracture particles and oxide formation.

A rough estimate of ΔK_{th} can also be gained on the basis of Neuber's microstructural support hypothesis, using the available data for structural steels (endurance limit for averaged stresses in polished surface, $\Delta\overline{\sigma}_E = 270\,\text{MPa}$, microstructural length, $\rho^* = 0.1$ mm) the result is $\Delta K_{th} = \Delta\overline{\sigma}_E(\pi\rho^*/2)^{1/2} = 107\,\text{MPa}\,\text{mm}^{1/2}$.

The use of endurable SIFs ΔK for the fatigue assessment of spot-welded joints is well established. Several investigations have been performed with tensile-shear loaded and peel-tension loaded specimens made of low-alloy steels, high-strength steels among them. The test results are plotted in the form of *K–N* curves (endured cycles *N* for different SIFs ΔK, interpreted as endurable SIFs ΔK over cycles *N*). These investigations and their results have been reviewed and compared (Radaj et al. 2006, ibid. pp. 471–476). The scatter of the *K–N* curve of a definite specimen type under investigation is generally small, but the deviations between the results of different authors are large and often unexplained.

Most influential on these deviations is a deficient determination of the SIFs in the spot-weld specimen under consideration. Factors 0.2–3.0 occur in the literature compared with the correct values. Thus, the corresponding *K–N* curves are by the same factors too low or too high. In the following, only one typical *K–N* curve for spot-welded joints (Yuuki et al. 1985$^{(1)}$, 1985$^{(2)}$; Yuuki and Ohira 1986, 1989) is further discussed. The SIF ΔK_{eq} is plotted over number of cycles *N* in logarithmic scales, Fig. 2.13, representing the *K–N* curve in the following form:

$$N = A_k(\Delta K_{eq})^{-k} \tag{2.58}$$

with the inverse slope exponent *k* and the coefficient A_k assumed to be material parameters. Similar investigations with comparable results can be found in the literature (Mizui et al. 1988; Linder et al. 1998).

The SIF ΔK_{eq} is determined according to the maximum tangential stress criterion. The inverse slope exponent results as $k = 4.5$. The calculated SIFs are by a factor of 0.7–0.8 too low. Taking this factor into account, the endurable SIFs for $N \geq 5 \times 10^6$ cycles roughly agree with the generally accepted lower bound value for non-welded structural steels, $\Delta K_{th} = 180$ MPa mm$^{1/2}$.

The width of the scatter band of the *K–N* curve is well in agreement with other local parameter curves which are used for fatigue assessment. Attributing the failure probabilities $P_f = 2.3$ and 97.7 % to the limiting curves in the figure, the following scatter range indices are derived:

$$T_N = N_{2.3}/N_{97.7} = 0.1, \quad T_\sigma = (T_N)^{1/k} = 0.6 \tag{2.59}$$

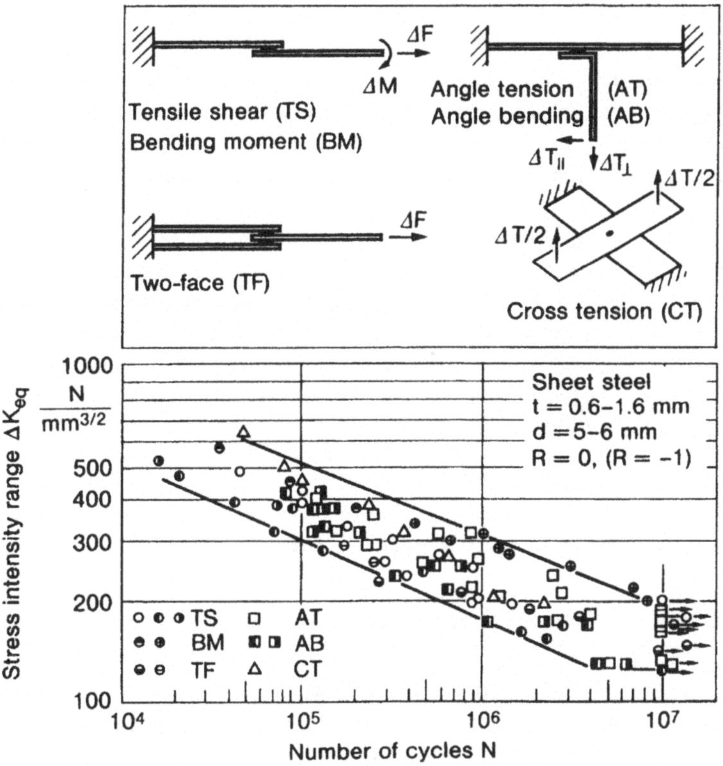

Fig. 2.13 Endurable stress intensity factor range at the weld spot of various specimens under different loading conditions according to maximum tangential stress criterion; inverse slope exponent $k = 4.5$; (Yuuki et al. 1985[(1)], 1985[(2)]; Yuuki and Ohira 1986, 1989)

Several deficiencies besides too low SIFs in the investigation leading to Fig. 2.13 have to be noted: no thickness adaption of ΔK_{eq}, the non-singular or T-stress neglected, the total life evaluated instead of the crack initiation life.

2.2.11 Conclusions

The definition and use of SIFs for describing the stress field at crack tips or pointed slit tips and their application to fatigue and brittle fracture phenomena is a well established procedure for more than 50 years. Both the theoretical concepts and the application-relevant details are generally known. In this section, the SIF concept has been reviewed, giving the basic stress field equations for areas close to the crack tip or slit tip, separated in singular and non-singular parts, the singular part subdivided into the three loading modes, the non-singular part consisting of

T-stress, *S*-stress and a crack-front-parallel normal stress. Coupled singular effects may additionally occur where the crack front butts on a free surface.

The presented stress field equations and their governing parameters (SIFs K_{I}, K_{II} and K_{III}, *T*-stress, *S*-stress) are made available for seam-welded and spot-welded lap joints based on a structural stress analysis which avoids modelling of the singular stresses. Lap joints with equal and unequal plate thickness, respectively, are considered.

Mixed mode loading conditions are taken into account by equivalent SIF formulae representing various fracture criteria. Endurable SIFs for spot-welded joints are given in the form of a *K–N* curve with defined scatter range. These data are unreliable because the underlying SIF analysis results may be rather inaccurate. There is an urgent demand for a better substantiated and more accurate design *K–N* curve. Only relative fatigue assessments are possible based on accurate SIF analyses without a reliable *K–N* curve.

2.3 Notch Stress Intensity Factor Concept

2.3.1 Survey of Section Contents

The stress intensity factor (SIF) concept referring to crack tips or slit tips is extended to V-notch tips or re-entrant corner tips in the form of the notch stress intensity factor (NSIF) concept. The stress singularities associated with such pointed notches are less marked and expressed by a smaller negative exponent of the radial distance from the notch tip. The stress field equations for the area close to the notch tip are given for the three singular loading modes 1, 2 and 3 and for the non-singular mode characterised by the *S*-stress. A transverse singular effect coupled with mode 2 loading at free surfaces is identified and described.

The NSIF concept is applied to the re-entrant corner notches at the toe of fillet welds and butt welds. The size effect is naturally included. A link between structural stress and NSIF is established. Equivalent NSIF formulae are presented, referring to failure under mixed mode loading conditions. Endurable NSIFs dependent on number of load cycles *N* are given for non-load-carrying fillet welds of attachment joints. In parallel, the *J*-integral concept for pointed V-notches is presented and applied.

The expositions are primarily based on the pioneering analytical developments and their practical applications (mainly to welded joints) carried out by Atzori, Lazzarin, Tovo, Livieri, Meneghetti and others. A list of symbols and a list of references is available at the end of Chap. 2

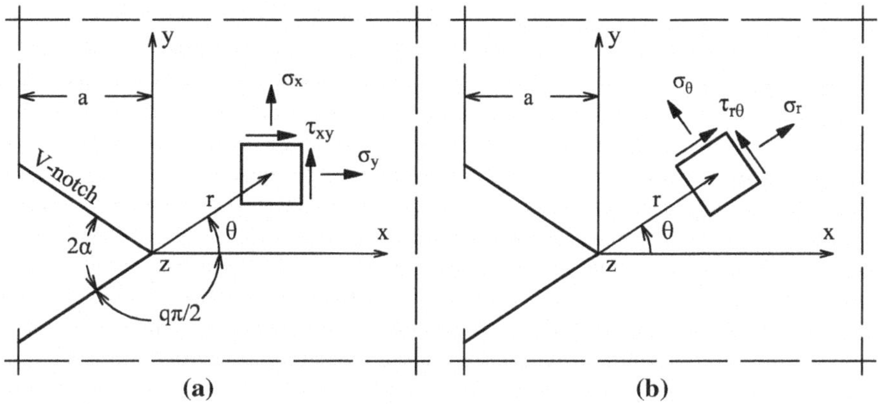

Fig. 2.14 Cartesian (**a**) and polar (**b**) coordinate systems with correspondingly defined stresses at the pointed V-notch tip; (Lazzarin and Tovo 1998)

2.3.2 Stress Field near Pointed V-Notches

The well-known concept of stress intensity factors describing the stress singularity at crack tips or slit tips under elastic material conditions can be transferred to pointed re-entrant corner notches (e.g. V-notches, stepped bars, weld toe notches). Whereas the asymptotic stress drop from the singularity at the crack tip is described by the inverse square root of the radial distance r from the crack tip (the exponent is minus 0.5), a smaller, notch angle dependent exponent occurs in the case of corner notches, which means that the degree of the singularity is reduced (Williams 1952). The stress field close to corner notches (just as the stress field close to crack tips) can be described by stress intensity factors (Gross and Mendelson 1972). These are named 'notch stress intensity factors' (NSIFs) as distinguished from the conventional stress intensity factors (SIFs) of crack tips.

The singular in-plane and out-of-plane stress fields at pointed corner notches can be specified by three notch loading modes (in analogy to the crack opening modes) related to the bisector plane of the notch: symmetric in-plane stresses (mode 1), antimetric in-plane stresses (mode 2) and out-of-plane shear stresses (mode 3). The corresponding notch loading modes are in-plane tensile loading, in-plane shear loading and out-of-plane shear loading.

The three basic loading modes with singular stresses at the notch tip produce the following asymptotic stress distribution (stress tensor σ_{ij}) around the notch tip, restricted to the first order terms (Williams 1952; Hasebe et al. 1990; Lazzarin and Tovo 1996, 1998; Quian and Hasebe 1997), Fig. 2.14:

$$\sigma_{\mathrm{ij}} = \frac{1}{\sqrt{2\pi}}\left[K_1 r^{\lambda_1 - 1} f_{1,\mathrm{ij}}(\theta) + K_2 r^{\lambda_2 - 1} f_{2,\mathrm{ij}}(\theta) + K_3 r^{\lambda_3 - 1} f_{3,\mathrm{kz}}(\theta)\right] \tag{2.60}$$
$$(\mathrm{i,j} = x, y \text{ and } \mathrm{k} = x, y \text{ or } \mathrm{i,j} = r, \theta \text{ and } \mathrm{k} = r, \theta)$$

The NSIFs K_1, K_2, K_3 depend on the magnitude of the load, the notch depth a, the notch opening angle 2α and further geometric parameters of the considered configuration. The angular functions $f_{1,\mathrm{ij}}, f_{2,\mathrm{ij}}, f_{3,\mathrm{kz}}$ describe the angular distribution of the stress close to the notch tip. The relationship above is strictly valid for $r \to 0$ and approximately valid for values of r which are small in relation to the notch depth and other geometrical parameters of the configuration.

Just as with the crack problem ($2\alpha = 0$), the complete solution comprises non-singular higher order terms, but only a finite number for rather small values of 2α and no such terms for about $2\alpha > 45°$ (Atzori et al. 1997, ibid. Fig. 3).

The original solution for the in-plane stress field is based on the Airy stress function in polar coordinates in the following form, which comprises a symmetrical and an antimetrical component (Williams 1952; Sanford 2003):

$$F(r,\theta) = r^{\lambda+1} f(\theta) \tag{2.61}$$

where the values of λ have to be determined as part of the solution. The angular functions have to comply with the boundary conditions on the load-free faces of the V-notch (actually Williams started with the wedge, extending the solution to V-notches by considering inscribed wedge angles larger than π).

The stress can now be expressed in terms of r, λ and $f(\theta)$. Application of the boundary conditions produces a system of four simultaneous equations for four unknown constants. This system can be separated into two independent sets of equations related to the symmetrical and antimetrical stress fields. A non-trivial solution can be obtained only if the determinants of the coefficient matrices are equal to zero each. From this requirement, the condition follows:

$$\sin 2\lambda\alpha = \pm\lambda \sin 2\alpha \tag{2.62}$$

Since 2α is a fixed parameter for a specific wedge or notch, Eq. (2.62) provides the values of λ, called 'eigenvalues', necessary to ensure a nontrivial solution. Williams' solution has been widely used under the name 'eigenfunction expansion method' (Hasebe et al. 1990).

Another function-analytical approach for solving the problem of the in-plane loaded blunt V-notch comprising the pointed V-notch has been applied by Lazzarin et al. while the principal mathematical steps remained the same as in the Airy stress function method (Lazzarin and Tovo 1996, 1998; Lazzarin et al. 1998; Atzori et al. 1997). According to the Kolosov–Muskhelishvili complex stress function method, the stress fields at symmetrical V-notches subjected to mode 1 and mode 2 loading can be derived from two analytical functions defined in the following form:

$$\varphi(z) = az^{\lambda}, \quad \psi(z) = bz^{\lambda} + cz^{\mu} \tag{2.63}$$

where the coefficients a, b, c are complex and the exponents λ, μ real ($\lambda > \mu$). Williams' results would be gained with the term cz^{μ} being neglected.

In order to impose the boundary conditions, an auxiliary system of curvilinear coordinates $w = u + \mathrm{i}v$ is introduced (Neuber 1958), which is related to the Cartesian coordinates $z = x + \mathrm{i}y$ by the conformal mapping function

$$z = w^q \tag{2.64}$$

The condition $u = 0$ describes the pointed V-notch with notch opening angle 2α in the z-plane:

$$2\alpha = 2\pi - q\pi, \quad q = \frac{2\pi - 2\alpha}{\pi} \tag{2.65}$$

where $q = 2.0\text{–}1.0$ for $2\alpha = 0\text{–}2\pi$.

The eigenvalues in the considered method and application result from the following condition, separated into the mode 1 and mode 2 parts:

$$\sin \lambda_1 q\pi + \lambda_1 \sin q\pi = 0 \tag{2.66}$$

$$\sin \lambda_2 q\pi - \lambda_2 \sin q\pi = 0 \tag{2.67}$$

The stress field analysis for the V-notch subjected to out-of-plane shear loading (mode 3) is easier to perform, because the governing equation in terms of the out-of-plane displacements is a potential function substituting the bipotential function in the case of the in-plane stresses. Performing similar mathematical steps as before, the following eigenvalue equation is found (Quian and Hasebe 1997):

$$\sin \lambda_3 (2\pi - 2\alpha) = 0 \tag{2.68}$$

The first non-zero eigenvalue is simply:

$$\lambda_3 = \frac{\pi}{2\pi - 2\alpha} \tag{2.69}$$

Actually, Quian and Hasebe have solved the more complex case of an unsymmetrical V-notch with a bimaterial interface ahead of the notch root, but with restriction to pointed notches.

The (first) eigenvalues λ_1, λ_2, λ_3 defining the degree of the stress singularity at the notch tip depend solely on the notch opening angle 2α (Williams 1952; Carpenter 1984), Fig. 2.15. The eigenvalue 0.5 is related to crack tips, $2\alpha = 0$, and the eigenvalue 1.0 (no singularity) to straight edges, $2\alpha = \pi$ (but only for mode 1 and mode 3). The eigenvalues for mode 1 loading are slightly smaller than those for mode 3 loading. They are substantially smaller in relation to mode 2 loading, thus designating severer singularities. The stress singularity in mode 2 loading is weaker and vanishes completely for $2\alpha \geq 102.6°$.

The in-plane stresses at the pointed V-notch ($\rho = 0$) are derived in the following form after introduction of the NSIFs K_1 and K_2 proposed by Gross and Mendelson (Lazzarin et al. 1998):

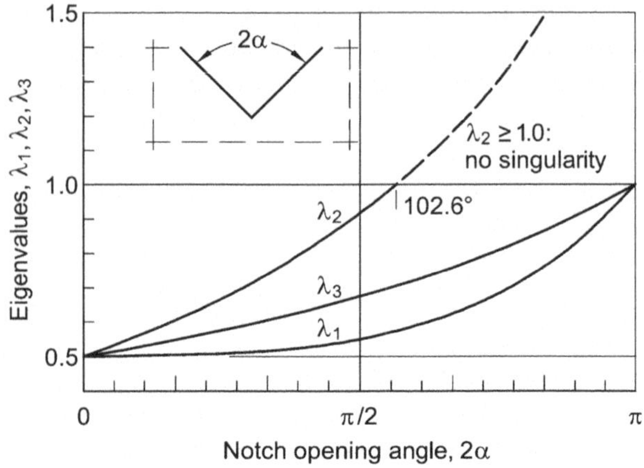

Fig. 2.15 Eigenvalues λ_1, λ_2, λ_3 defining the degree of stress singularity at pointed V-notches subjected to mode 1, 2, 3 loading conditions, dependent on notch opening angle 2α; (Lazzarin et al. 2004)

$$\begin{aligned}
\begin{Bmatrix} \sigma_\theta \\ \sigma_r \\ \tau_{r\theta} \end{Bmatrix}_1 &= \frac{1}{\sqrt{2\pi}} \frac{r^{\lambda_1 - 1} K_1}{(1+\lambda_1) + \chi_1(1-\lambda_1)} \\
&\times \left[\begin{Bmatrix} (1+\lambda_1)\cos(1-\lambda_1)\theta \\ (3-\lambda_1)\cos(1-\lambda_1)\theta \\ (1-\lambda_1)\sin(1-\lambda_1)\theta \end{Bmatrix} + \chi_1(1-\lambda_1) \begin{Bmatrix} \cos(1+\lambda_1)\theta \\ -\cos(1+\lambda_1)\theta \\ \sin(1+\lambda_1)\theta \end{Bmatrix} \right] \\
\sigma_{z,1} &= \nu(\sigma_{\theta,1} + \sigma_{r,1}) \quad \text{(plane strain)} \\
\chi_1 &= -\frac{\sin(1-\lambda_1)q\pi/2}{\sin(1+\lambda_1)q\pi/2}
\end{aligned} \tag{2.70}$$

$$\begin{aligned}
\begin{Bmatrix} \sigma_\theta \\ \sigma_r \\ \tau_{r\theta} \end{Bmatrix}_2 &= \frac{1}{\sqrt{2\pi}} \frac{r^{\lambda_2 - 1} K_2}{(1+\lambda_2) + \chi_2(1-\lambda_2)} \\
&\times \left[\begin{Bmatrix} -(1+\lambda_2)\cos(1-\lambda_2)\theta \\ -(3-\lambda_2)\cos(1-\lambda_2)\theta \\ (1-\lambda_2)\sin(1-\lambda_2)\theta \end{Bmatrix} + \chi_2(1-\lambda_2) \begin{Bmatrix} -\sin(1+\lambda_2)\theta \\ \sin(1+\lambda_2)\theta \\ \cos(1+\lambda_2)\theta \end{Bmatrix} \right] \\
\sigma_{z,2} &= \nu(\sigma_{\theta,2} + \sigma_{r,2}) \quad \text{(plane strain)} \\
\chi_2 &= -\frac{\sin(1-\lambda_2)q\pi/2}{\sin(1+\lambda_2)q\pi/2}
\end{aligned} \tag{2.71}$$

The antimetrical singular and symmetrical non-singular out-of-plane shear stresses have the following form (Zappalorto and Lazzarin 2011[(3)]) corresponding to Eq. (2.126):

$$\begin{Bmatrix} \tau_{rz} \\ \tau_{\theta z} \end{Bmatrix} = \frac{1}{\sqrt{2\pi}} K_3 r^{\lambda_3 - 1} \begin{Bmatrix} \sin \lambda_3 \theta \\ \cos \lambda_3 \theta \end{Bmatrix} \tag{2.72}$$

$$\begin{Bmatrix} \tau_{rz} \\ \tau_{\theta z} \end{Bmatrix}_S = S \begin{Bmatrix} \cos \theta\pi/(\pi - \alpha) \\ -\sin \theta\pi/(\pi - \alpha) \end{Bmatrix} \tag{2.73}$$

with $K_3^{(a)}$ and λ_a substituted by K_3 and λ_3, with $K_3^{(s)} r^{\lambda_s - 1}/(2\pi)^{1/2}$ substituted by the S-stress already introduced at crack tips ($2\alpha = 0$), Eq. (2.15), and with λ_s substituted by $\pi/(\pi - \alpha)$. The substitutions are introduced in order to make the equations above compatible with the hitherto used parameters.

The substitution related to the S-stress in only formally consistent with the existing analytical frame, because $S \to 0$ for $r \to 0$ according to $S = K_3^{(s)} r^{\lambda_s - 1}/(2\pi)^{1/2}$. Considering the S-stress to be a physical reality not only for $2\alpha = 0$, but also for $2\alpha > 0$ (similar to the situation with the T-stress), Eq. (2.73) is only a temporary expedient.

The characteristic stress components in the bisector plane ($\theta = 0$) have the following simple form:

$$\sigma_\theta(r, 0) = \frac{1}{\sqrt{2\pi}} K_1 r^{\lambda_1 - 1} \tag{2.74}$$

$$\tau_{r\theta}(r, 0) = \frac{1}{\sqrt{2\pi}} K_2 r^{\lambda_2 - 1} \tag{2.75}$$

$$\tau_{\theta z}(r, 0) = \frac{1}{\sqrt{2\pi}} K_3 r^{\lambda_3 - 1} \tag{2.76}$$

The NSIFs K_1, K_2, K_3 may be evaluated on the basis of the characteristic stress components above considering the limit values for $r \to 0$ (Gross and Mendelson 1972):

$$K_1 = \lim_{r \to 0} \sqrt{2\pi} r^{1-\lambda_1} \sigma_\theta(r, 0) \tag{2.77}$$

$$K_2 = \lim_{r \to 0} \sqrt{2\pi} r^{1-\lambda_2} \tau_{r\theta}(r, 0) \tag{2.78}$$

$$K_3 = \lim_{r \to 0} \sqrt{2\pi} r^{1-\lambda_3} \tau_{\theta z}(r, 0) \tag{2.79}$$

The dimensions of K_1, K_2, K_3 are $\mathrm{N/mm}^{1+\lambda_1}$, $\mathrm{N/mm}^{1+\lambda_2}$, $\mathrm{N/mm}^{1+\lambda_3}$. The numerical values of K_1, K_2, K_3 can be set into comparison, e.g. as failure criteria, provided their dimension is identical, i.e. only under the condition of an identical notch angle besides an identical loading mode (for example, the stress intensity factors of crack tips are used in this way). In mode 1 loading, the stress intensity factors of crack tips are directly comparable with those of sharp notches, provided the notch opening angle is not too large ($2\alpha \le 60°$, see Fig. 2.15), because the dimension does not change substantially. Mixed mode failure criteria must

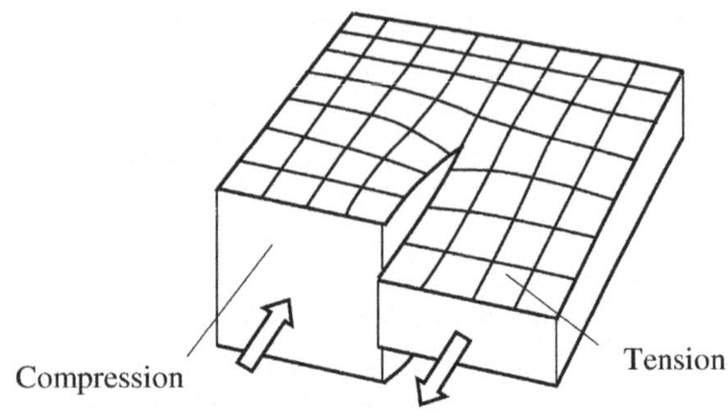

Fig. 2.16 Enlarged deformation of a mode II loaded plate specimen with a crack showing a transverse singular loading effect closely underneath the surfaces at the crack tip; (Harding et al. 2010)

therefore be based on self-normalised NSIFs (Sect. 2.3.6). Failure criteria based on averaged notch stresses (Chap. 1) or on averaged strain energy density (Chap. 3) are appropriate in other cases.

2.3.3 Transverse Singular Effect at In-Plane Shear-Loaded V-Notches

The elementary singularities in the loading modes 1, 2 and 3 may be locally coupled in three-dimensional configurations. One typical example is the crack front in a plate of finite thickness subjected to in-plane shear loading (mode 2), Fig. 2.16.

The exaggerated deformation plot of the crack tip region subjected to mode 2 loading shown in the figure illustrates the transverse expansion of the compressive loaded plate side and the transverse contraction of the tensile loaded plate side. At the crack front, the two opposed deformations lock each other. This produces out-of-plane shear stresses of opposite direction above and below the midplane. They vanish in the midplane and on the two plate surfaces and have a maximum closely underneath the surfaces. They constitute an out-of-plane ('transverse') shear stress singularity which resembles a mode 3 singularity, but remains symmetrical relative to the midplane. Its intensity is significantly affected by the Poisson's ratio ν. The singularity effect is largest for $\nu = 0.5$ and zero for $\nu = 0$. Simultaneously, the mode 2 intensity varies through the plate thickness, assuming Williams' original value only at a distance from the crack front of about half the plate thickness. Similar effects can be expected for non-zero notch opening angles.

The literature on coupled three-dimensional singular effects is shortly reviewed. Benthem studied these effects associated with the intersection of a wedge front of finite length with the free surface of the half space (Benthem 1977). Nakamura and Parks investigated the in-plane shear-loaded crack front intersecting a free surface as described above (Nakamura and Parks 1989). Kotousov analysed the tensile loaded V-notch in three dimensions (Kotousov and Wang 2002; Kotousov 2005;

Kotousov and Lew 2006). The following presentation refers to in-plane shear-loaded V-notches (Harding et al. 2010; Berto et al. 2011).

For describing the transverse singular effect at in-plane shear-loaded V-notches (with inclusion of cracks), the Kane and Mindlin assumptions for a flat plate of thickness $2h$ are adopted in the following form (Kane and Mindlin 1956):

$$u_x = u_x(x, y), \quad u_y = u_y(x, y), \quad u_z = \frac{z}{h} w\,(x, y) \tag{2.80}$$

The following two governing equations of a corresponding first order plate theory may be derived (Yang and Freund 1985):

$$\nabla^4 w - \kappa^2 \nabla^2 w = 0 \tag{2.81}$$

$$\nabla^6 \Phi - \kappa^2 \nabla^4 \Phi = 0 \tag{2.82}$$

$$\kappa^2 = \frac{6}{h^2(1 - \nu)} \tag{2.83}$$

where Φ is the stress resultant function and κ a stiffness parameter.

A further harmonic displacement function ψ is introduced for describing the in-plane and out-of-plane displacements. Assuming a general asymptotic behaviour of w, Φ and ψ at the notch tip and adopting the eigenfunction expansion method, the following eigenvalue equation is derived (Kotousov 2007; Harding et al. 2010):

$$[\sin 2\lambda_2(\pi - \alpha) - \lambda_2 \sin 2(\pi - \alpha)] \cos \lambda_{\mathrm{O}}(\pi - \alpha) = 0 \tag{2.84}$$

The expression in brackets set to zero corresponds to Williams' in-plane shear stress solution, Eq. (2.62). The term $\cos \lambda_{\mathrm{O}}(\pi - \alpha)$ set to zero describes an out-of-plane shear loading mode, Eq. (2.68). It has been shown that $0 < \lambda_2 < 1.0$ (singularity) occurs for $0 < 2\alpha < 102.6°$ whereas $\lambda_2 > 1.0$ (no singularity) results for $102.6° < 2\alpha < 180°$. On the other hand, $0 < \lambda_{\mathrm{O}} < 1.0$ is found for $0 < 2\alpha < 180°$. The eigenvalues λ_{O} and λ_3 are identical because the eigenvalue equations can be presented in the same form. The expression for $\lambda_{\mathrm{O}} = \lambda_3$ is given by Eq. (2.69).

However, there are major differences between mode O and mode 3. The out-of-plane singular mode O is a local mode coupled to the applied mode 2 due to a Poisson's ratio effect producing a symmetric shear stress distribution relative to the midplane of the plate. In contrast, the out-of-plane singular mode 3 is an independent loading mode associated with an antimetric shear stress distribution relative to the midplane that is totally independent of Poisson's ratio (Harding et al. 2010).

A numerical investigation using the FE method has been performed in order to study the dependence of the out-of-plane singular mode on notch opening angle and Poisson's ratio. The region near to the notch tip (radial distance $r \leq 5h$, with plate thickness $2h$) is discretised into a fine mesh of higher-order three-dimensional elements (15-node wedge-shaped and 20-node trapezoidal). The NSIFs are

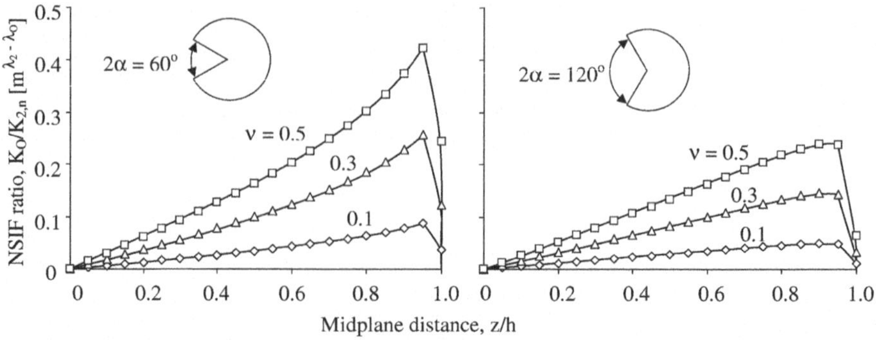

Fig. 2.17 NSIF ratio $K_O/K_{2,n}$ dependent on midplane distance z/h; out-of-plane singular mode O coupled with in-plane singular mode 2; plate thickness $2h$, Poisson's ratio ν and nominal NSIF $K_{2,n}$; (Harding et al. 2010)

found by the following limit values applied to the numerical data in the close neighbourhood of the notch tip (Seweryn and Zwolinski 1993):

$$K_2 = \lim_{r \to 0} (2\pi r)^{1-\lambda_2} \tau_{r\theta} \tag{2.85}$$

$$K_O = \lim_{r \to 0} (2\pi r)^{1-\lambda_O} \tau_{\theta z} \tag{2.86}$$

In the vicinity of the notch tip, the stress state is essentially three-dimensional. It comprises the singular out-of-plane loading mode O as well as the variability of the NSIF K_2 in the plate thickness direction. As a reference value, the nominal NSIF $K_{2,n}$ is introduced which corresponds to Williams' two-dimensional stress field solution.

Typical results of the investigation are shown in Fig. 2.17 where the NSIF ratio $K_O/K_{2,n}$ is plotted over the midplane distance z/h for three Poisson's ratios ν. The out-of-plane NSIF K_O is continuously rising to a maximum closely underneath the free surfaces of the plate. The rise is steeper for small notch opening angles. The largest values of K_O occur for $\nu = 0.5$, and K_O vanishes for $\nu = 0$. The NSIF K_2 is also much larger close to the free surface in comparison to its midplane value. Since the degree of the mode O singularity is greater than that of the mode 2 singularity (for $2\alpha = 0$), the effect of transverse mode O loading increases with the plate thickness.

The transverse singular effect in mode 2 loading is of relevance for application. Some interesting results are presented by Lazzarin and Berto (Harding et al. 2010; Berto et al. 2011). The welded lap joint models investigated using the FE method are shown in Fig. 2.18. The transverse singular effect is initiated at the stress-free lateral surface of the three-dimensional models and decreases in the direction of the midplane.

Based on the strain energy density approach, it is shown that the critical area in fatigue loading of slits ($2\alpha = 0$) is at the lateral surfaces for plate thicknesses greater than approximately 10 mm. It is shifted to the midplane for smaller thicknesses. The effect would be more pronounced if the ligament width were

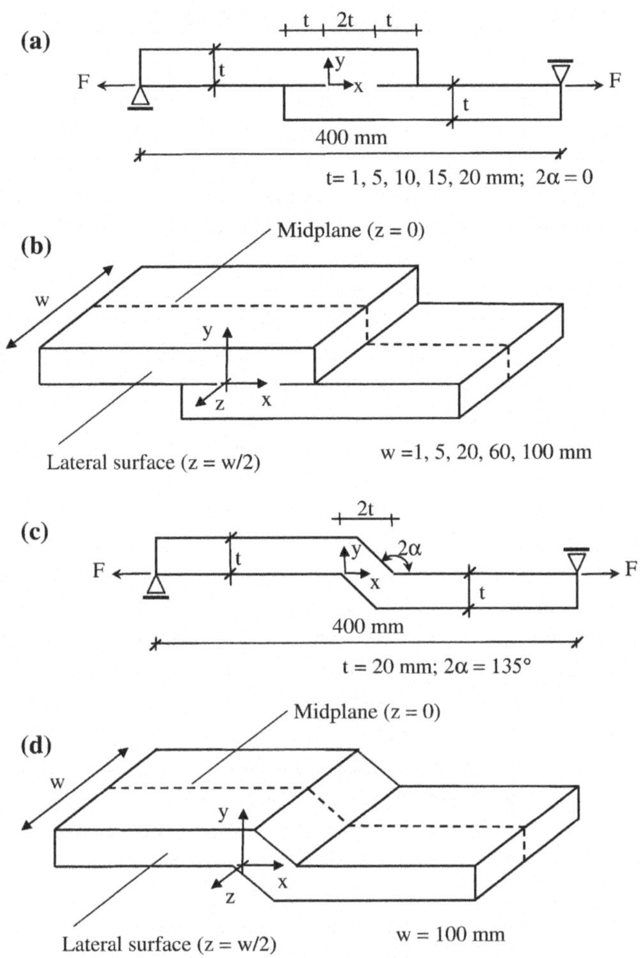

Fig. 2.18 Welded lap joint models subjected to tensile-shear loading: two-dimensional model with slit opening angle $2\alpha = 0°$ (**a**) and corresponding three-dimensional model (**b**); two-dimensional model with notch opening angle $2\alpha = 135°$ (**c**) and corresponding three-dimensional model (**d**); (Harding et al. 2010)

reduced relative to the plate thickness, resulting in an increase of mode 2 relative to mode 1 loading.

It is shown for notch opening angles $2\alpha \geq 102.6°$, where the in-plane shear stress singularity disappears, that the associated out-of-plane shear stress singularity still exists and may be dominant in the failure initiation phase.

2.3.4 NSIF Values of V-Notches and Weld Toe Notches

The NSIF values of V-notches and weld toe notches have been determined by a major number of authors. The weld toe notches are treated as a special case of V-notches ($2\alpha \approx 135°$) mainly by Italian researchers (Atzori et al. 1999(2); Lazzarin et al. 1998). Different methods are involved. The most general and convenient method is the eigenvalue expansion of the notch stress field according to Williams (also named 'asymptotic solution'), mostly combined with conventional FE analysis.

Asymptotic stress distributions at weld toe notches have been derived as early as 1985/86 without defining the NSIF parameter and without separating the mode 1 and mode 2 contributions (Atzori 1985; Lazzarin 1986). The clear analytical frame was established 1998 by Lazzarin and Tovo in the publication referenced above.

Various other methods have been used. Originally, the boundary collocation method was preferred (Gross and Mendelson 1972). Later on, special notch tip finite elements have been designed which incorporate the stress singularity: hybrid finite elements (Lin and Tong 1980), degenerate asymptotic finite elements (Akin 1976) and finite elements with constraints (Seweryn 2002). Three approaches are available for evaluating the NSIFs: direct evaluation at the singularity point, asymptotic evaluation outside the singularity point (comparison of numerical and analytical solutions) and energy methods (J-integral and strain energy density averaged over a control volume). The boundary element method is applicable with or without special singularity elements. Singular integral equations have also been successfully used (Theocaris and Ioakimidis 1979; Noda et al. 1996; Noda and Takase 2003; Savruk and Kazberuk 2006, 2007, 2010).

Useful NSIF data can be found in several publications (Gross and Mendelson 1972; Chen 1995; Seweryn and Molski 1996; Dunn et al. 1997(1), 1997(2); Lazzarin et al. 1998; Atzori et al. 1999(1); Noda and Takase 2003; Savruk and Kazberuk 2010). An extensive parametric investigation has been performed on the NSIFs at torsional loaded round bars with a circumferential hyperbolic notch of different notch opening angles 2α and different ratios of notch depth a to net cross-section radius R (Zappalorto et al. 2009). The closed form solution for the deep hyperbolic notch is combined with FE analysis results for notches of finite depth. The evaluated dimensionless NSIF is $k_3 = K_3/\tau_{\mathrm{n}}^{*} R^{1-\lambda_3}$.

The main area of application of the NSIFs are fillet-welded joints showing any degree of penetration, but also butt-welded joints with the typical weld reinforcement are analysed in this way. The weld toe may be considered as a re-entrant corner notch of opening angle $2\alpha = 135°$ resulting in a bisector plane inclination of 112.5° against the plate surface. The stress field at the weld toe may then be described by the NSIFs K_1, K_2 and K_3, possibly supplemented by non-singular stress components (the out-of-plane S-stress and a superimposed normal stress in the direction of the toe line). Based on the NSIFs, fatigue test data may be evaluated in a form which includes the size effect. Geometries scaled in geometrical proportion and subjected to the same nominal stress are characterised by different

NSIF values. The NSIFs are used independent of whether the weld notch is really pointed or only sharply rounded. Only excessively large toe notch radii (shape-optimised welds) must be excluded. The assumption of pointed toe notches corresponds to the worst-case procedure in the fatigue assessment codes.

Considering a cross-section of finite width ahead of the notch, this width may be set equal to the plate thickness t of fillet-welded joints, the NSIFs can be expressed in the following form (Lazzarin et al. 1998):

$$K_1 = k_1 \sigma_\mathrm{n} t^{1-\lambda_1} \tag{2.87}$$

$$K_2 = k_2 \sigma_\mathrm{n} t^{1-\lambda_2} \tag{2.88}$$

$$K_3 = k_3 \tau_\mathrm{n}^* t^{1-\lambda_3} \tag{2.89}$$

where σ_n and τ_n^* are the nominal stresses over the plate thickness t and k_1, k_2, k_3 are the geometry coefficients quantified further below. With regard to the application to fillet-welded joints, the reference stress of K_2 in Eq. (2.88) is not the nominal shear stress τ_n but the nominal tensile stress σ_n.

The NSIFs for fillet-welded joints (notch opening angle $2\alpha = 135°$) have the following form:

$$K_1 = k_1 \sigma_\mathrm{n} t^{0.326} \tag{2.90}$$

$$K_2 = k_2 \sigma_\mathrm{n} t^{-0.302} \tag{2.91}$$

$$K_3 = k_3 \tau_\mathrm{n}^* t^{0.200} \tag{2.92}$$

The negative exponent of t in the expression for K_2, Eq. (2.91), designates non-singular stresses at the pointed notch. These can be neglected in strength predictions in comparison to the singular stresses described by K_1 and K_3, but they have to be taken into account when evaluating the stresses in front of the weld toe, Eq. (2.95), where strain gauges may be applied.

The Eqs. (2.87–2.89) are well suited to catch the size effect on the static or fatigue strength of corner-notched members, fillet-welded joints among them, provided the strength can be characterised by the NSIF (identical notch opening angles are a precondition). The following relationship is derived from Eq. (2.87) for the ratio of endurable nominal stresses $\sigma_{\mathrm{n}1}$ and $\sigma_{\mathrm{n}2}$ acting on the plate thicknesses t_1 and t_2:

$$\frac{\sigma_{\mathrm{n}2}}{\sigma_{\mathrm{n}1}} = \frac{k_{11}}{k_{12}} \left(\frac{t_1}{t_2}\right)^{1-\lambda_1} \tag{2.93}$$

where k_{11} and k_{12} are the geometry coefficients k_1 of the members 1 and 2.

In cases of a large net cross-sectional width between the notch tips, the notch depth a is the governing parameter instead of the plate thickness t, leading to the following relationship (Dunn et al. 1997^(1)^, 1997^(2)^):

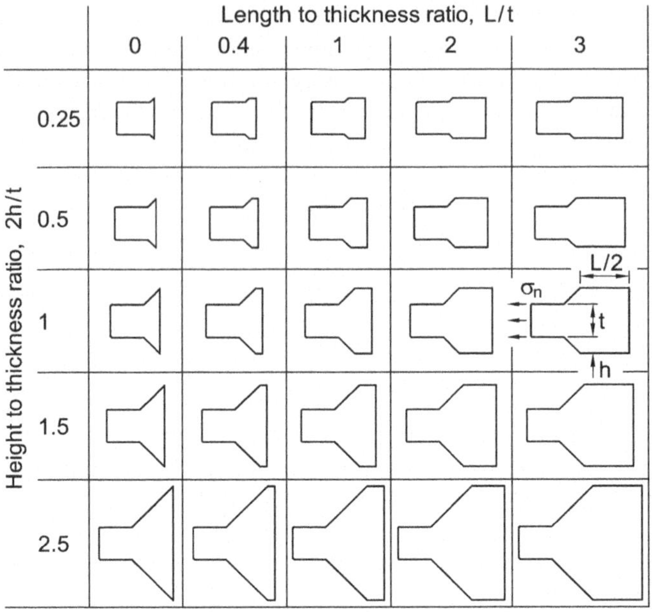

Fig. 2.19 Plane models with weld-like outer geometries (i.e. without the inner slits) subjected to tensile loading, investigated numerically using the FE method; (Lazzarin et al. 1998)

$$\frac{\sigma_{n2}}{\sigma_{n1}} = \frac{Y_1}{Y_2}\left(\frac{a_1}{a_2}\right)^{1-\lambda_1} \tag{2.94}$$

where Y_1 and Y_2 are the geometry coefficients of the members 1 and 2.

The dimensionless geometry coefficients k_1, k_2, k_3 in Eqs. (2.87–2.89) are determined based on extremely fine-meshed cross-sectional FE models of the considered welded joints evaluating the stresses in the bisector plane where they are decoupled for the three loading modes. Any stress value along the bisector plane, but sufficiently close to the notch tip, can be used to determine the NSIF by applying Eqs. (2.77), (2.78) or (2.79) and therefrom the geometry coefficient by evaluating Eqs. (2.90), (2.91) or (2.92). Systematic numerical investigations related to fillet-welded joints have been conducted (Lazzarin et al. 1998; Tovo and Lazzarin 1999).

A first investigation was related to plane models of weld-like outer geometries without the inner slits, Fig. 2.19. The geometry coefficients k_1 and k_2 dependent on the geometrical parameters (dimension ratios) are shown in Fig. 2.20. The numerical results are approximated by polynomials applying the least square method. The coefficient k_1 rises with L/t and $2h/t$, the coefficient k_2 does the opposite. Plastic notch stress intensity factors with appertaining stress distributions have also been determined for the considered joint-similar models (Lazzarin et al. 2001).

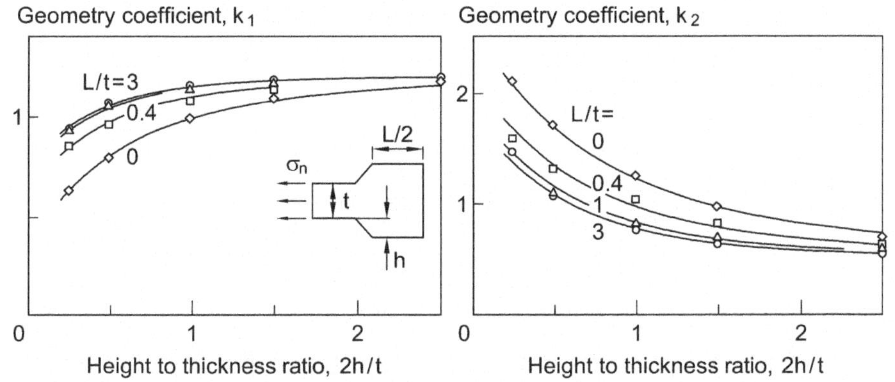

Fig. 2.20 Geometry coefficients k_1 and k_2 dependent on geometrical parameters for plane models with weld-like geometry subjected to tensile loading; FE analysis results; (Lazzarin et al. 1998)

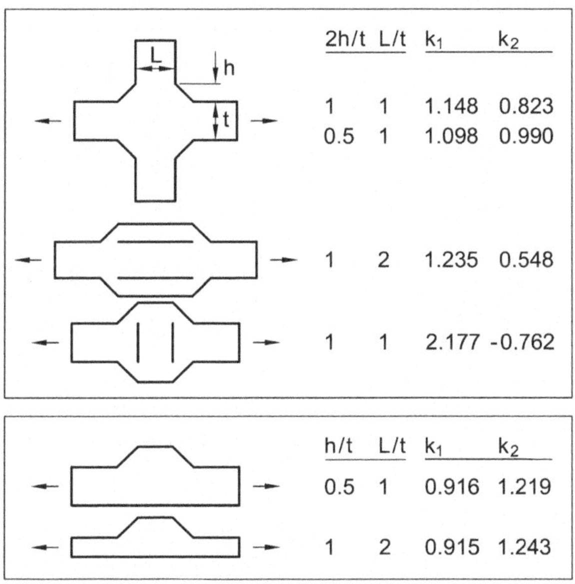

Fig. 2.21 Geometry coefficients k_1 and k_2 relating to cross-sectional models of various fillet-welded joints subjected to tensile loading; FE analysis results; (Lazzarin et al. 1998)

A second investigation was related to more realistic cross-sectional models of fillet-welded joints (through-welded cruciform joint, cover plate joint, top-cut cruciform joint) subjected to tensile loading, Fig. 2.21. Parametrical variations were considered for the cruciform attachment joint with non-load-carrying fillet welds resulting in the values of k_1 and k_2 plotted in Fig. 2.22. The curve shapes are similar to those referring to the weld-like plane models (Fig. 2.20). This means that the non-bevelled height of the transverse stiffener plate is of secondary influence.

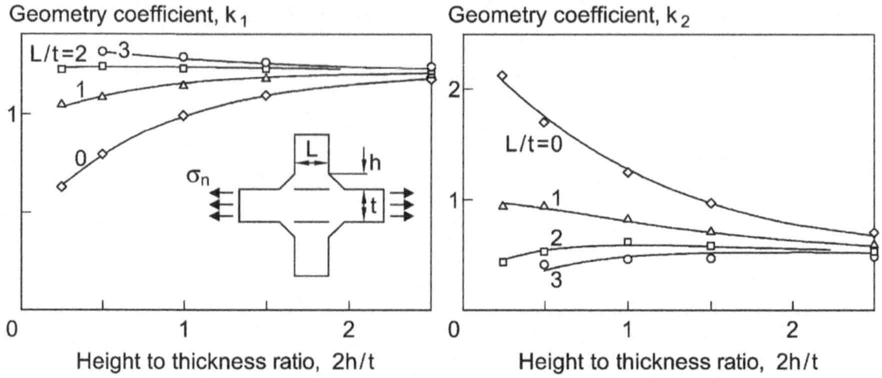

Fig. 2.22 Geometry coefficients k_1 and k_2 dependent on geometrical parameters for cruciform joint with non-load-carrying fillet welds (transverse attachment joint) subjected to tensile loading; FE analysis results; (Lazzarin et al. 1998)

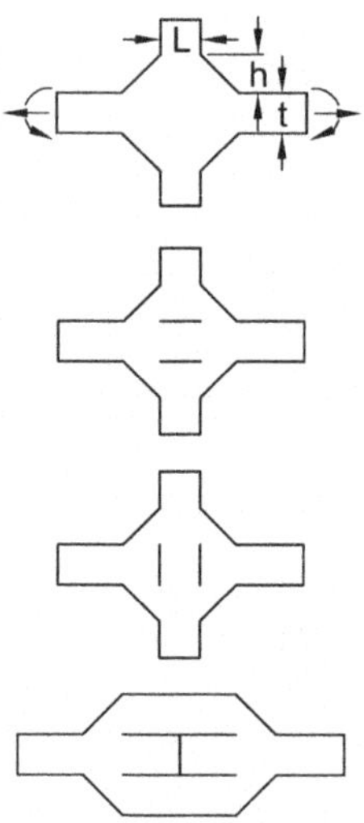

Ratios 2h/t	L/t	Geometry coefficients k_{1t} (Tension load)	k_{2t} (Tension load)	k_{1b} (Bending load)	k_{2b} (Bending load)
2	1	1.173	0.662	0.897	0.824
2	0.1	1.165	0.720	0.899	0.821
2	1	1.199	0.625	0.900	0.819
2	3	1.250	0.251	0.900	0.818
2	1	1.432	-0.018	0.904	0.790
2	5	1.543	-0.465	0.907	0.769

Fig. 2.23 Geometry coefficients k_{1t}, k_{2t} and k_{1b}, k_{2b} relating to typical cruciform and lap joints subjected to tensile force and bending moment, respectively; FE analysis results; (Tovo and Lazzarin 1999)

The plots in Fig. 2.22 refer to tensile loaded cruciform joints. Different diagrams are reported for cruciform attachment joints subjected to a bending load (Atzori et al. 1997). In the latter case, the values of k_1 and k_2 are lower, in general, explaining the higher fatigue strength of these joints under bending loading in comparison to tensile loading.

A third investigation with comparisons to the hot spot structural stress approach was related to typical cruciform and lap joints subjected to tension force or bending moment, respectively, Fig. 2.23. The values of k_1 are always higher for tensile than for bending loading. The opposite holds for the values of k_2 which are less influential on fatigue. A substantial rise in the values of k_1 occurs for the cruciform and lap joints with load-carrying fillet welds.

The coefficients k_1, k_2 and k_3 have also been determined for a fillet-welded circular tube penetrating a tensile loaded plate strip and for comparable cylindrical attachments (Susmel and Tovo 2004). Curve plots over peripheral angle indicate various crack initiation sites.

There remains an open demand for the notch stress intensity factors of butt-welded joints with different degrees of reinforcement height and width, comprising laser beam welds and gas metal-arc welds and referring to the top and root sides of the welds. The notch opening angles may vary between 90 and 180° in these cases. A tentative investigation of such joints has been presented (Lazzarin et al. 2006).

2.3.5 Relationship between Structural Stress and NSIF for Welded Joints

The stress rise in front of fillet welds is of major concern in respect of the hot spot structural stress approach which is widely used for the fatigue assessment of welded joints. This approach proceeds from the local structural stresses determined at the weld toe or closely in front of it by means of the FE method, of engineering formulae or of strain gauge measurements. The structural stress *S–N* curve can approximately be set equal to the nominal stress *S–N* curve of a locally comparable welded joint specimen.

The assumption behind the structural stress approach is that the fatigue-effective local parameter is correctly and generally described by the maximum structural stress. As far as the NSIF K_1 can also be considered to be fatigue-effective, the link between maximum structural stress and NSIF is of interest. Two application examples demonstrate this link in the following.

The stress rise in front of fillet welds has been determined based on the NSIFs in order to validate this approach in general by comparison with FE analysis results (Lazzarin et al. 1998). The radial stress along the free edge of the cross-sectional model, which corresponds to the surface stress in the base plate in a line normal to the fillet weld, is calculated by introducing the notch opening angle $2\alpha = 135°$ and the polar angle $\theta = 112.5°$ against the bisector plane into the general equation of stress, Eq. (2.70) and (2.71), thus yielding:

$$\sigma_r = 0.423 K_1 r^{-0.326} - 0.553 K_2 r^{0.302} \tag{2.95}$$

with the distance r from the pointed notch tip. In this equation, the sign of K_2, positive or negative, is chosen according to Williams' analytical frame. In the case

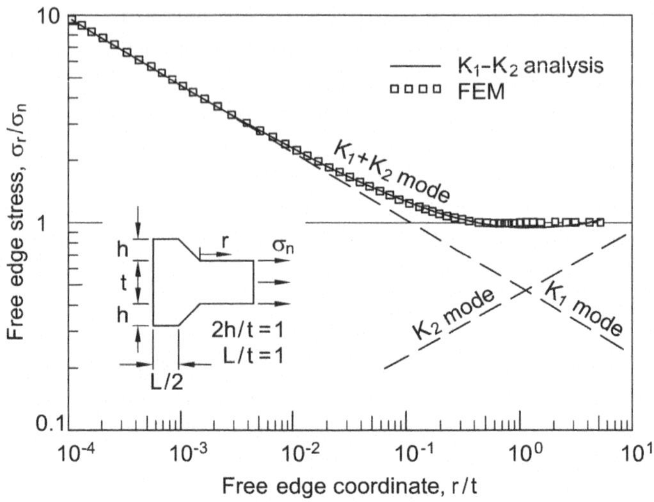

Fig. 2.24 Stress rise close to the pointed notch tip for weld-like model subjected to tensile loading; contribution of mode 1 and mode 2 loading to the free edge stress; K_1 and K_2 analysis results compared with FE analysis results; (Lazzarin et al. 1998)

considered below, K_2 is negative, so that the contribution due to mode 2 increases the contribution due to mode 1.

The stress rise close to the pointed notch tip according to Eq. (2.95) with FE-based values of K_1 and K_2 for a weld-like plane model subjected to tensile loading in comparison to the direct FE analysis results is plotted in Fig. 2.24. The logarithmic scale of the *r*/*t* axis is a means to illustrate the details of stress superposition (stresses related to K_1 and K_2) and to avoid plotting the stress singularity at $r = 0$. The stress values gained from the two methods are more or less identical up to the second crossing of the nominal stress level.

Another validation of the NSIF approach has been conducted for the same weld-like model as before, but with a blunt notch instead of a pointed notch. The NSIF stress equations extended by the notch radius term or its equivalent, Eqs. (2.116) and (2.117) reveal the basic structure, are used, Fig. 2.25. Once more, satisfactory agreement with the two methods (FE-based K_1 and K_2 as well as FE directly) is achieved.

Strain gauges applied to determine a fatigue-relevant local parameter at welds are generally placed on the free surface of the joint at a convenient distance from the notch tip of the weld toe. At this position, the effect of mode 2 loading is always present. Then, if the fatigue strength is thought of as mainly controlled by the mode 1 stress distribution, the contribution of mode 2 loading to the measured strains should be disregarded. This is possible only if one knows both NSIFs, K_1 and K_2, for a given geometry and remote loading condition. The greater the distance x from the notch tip is, the greater will be the perturbing effect due to mode 2 loading.

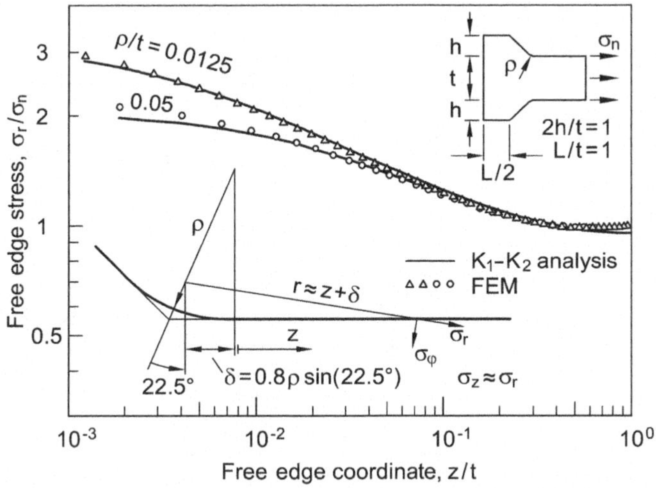

Fig. 2.25 Stress rise close to the rounded notch tip for weld-like model subjected to tensile loading; 'radial' stress σ_r along the straight free edge; K_1 and K_2 analysis results compared with FE analysis results; (Lazzarin et al. 1998)

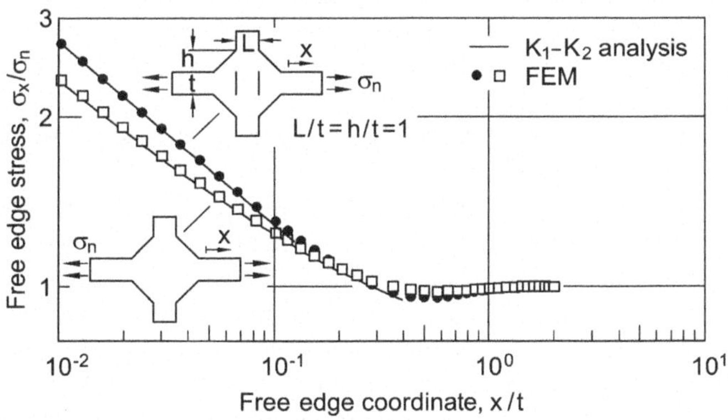

Fig. 2.26 Local stress rise above the structural stress level ($\sigma_s = \sigma_n$) plotted over distance from the pointed notch tip for the cross-sectional model of cruciform joints subjected to tensile loading; joint with slits in comparison to joint without slits; FE-based K_1 and K_2 analysis results compared with direct FE analysis results; (Tovo and Lazzarin 1999)

Two application examples are given which demonstrate deficiencies of the conventional structural stress approach and draw attention to the advantages of combining the NSIF approach with the structural stress approach (Tovo and Lazzarin 1999).

The first example refers to the different K_1 and K_2 values of the cruciform joint with full penetration welds compared with fillet welds without penetration (joint with internal slits), Fig. 2.26. The structural stresses are identical ($\sigma_x/\sigma_n = 1$), but the notch stress intensity factors, expressed by the gradients of the stress rise

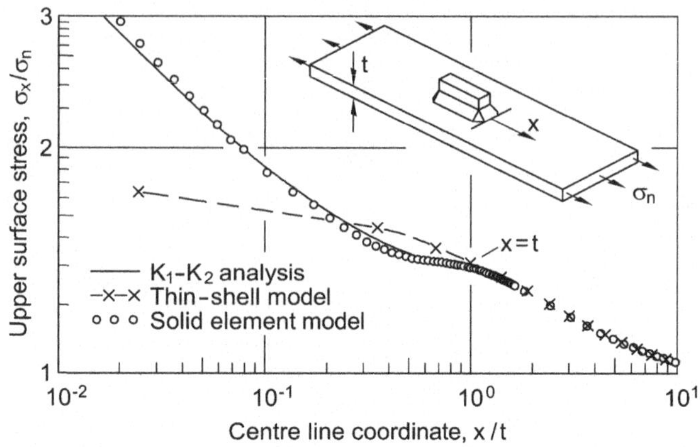

Fig. 2.27 Local stress rise plotted over distance from the weld toe for single-sided longitudinal attachment joint subjected to tensile loading; FE-based K_1 and K_2 analysis results based on structural stresses on thin-shell element model (coarse mesh) at distance $x = t$ from the weld toe (special extrapolation formula) compared with direct FE analysis results from solid element model (fine mesh); (Tovo and Lazzarin 1999)

towards the corner notch, are not. The K_1 factor ratio 1.432/1.173 = 1.221 is not too far from the IIW design recommendation, FAT class ratio 71/63 = 1.127 (Hobbacher 2009). Another conclusion from the stress plot is that direct measurement or calculation of the asymptotic stress rise towards the corner notch tip (for determining the fatigue-relevant NSIF) must be performed close to the notch tip, $x/t \leq 0.1$ (Atzori and Meneghetti 2001).

The second example of a longitudinal attachment joint refers to the fact that the conventional structural stress extrapolation to the hot spot at the weld toe may be rather inaccurate when using the FE analysis results from a thin-shell element model, Fig. 2.27. The stress rise to the weld toe in front of the attachment is well described by the NSIFs K_1 and K_2 determined on the basis of the relevant structural stresses in the thin-shell element model at distance $x = t$ from the weld toe (special extrapolation formula), as can be seen from the comparison with the FE analysis results using a fine-meshed three-dimensional solid element model. On the other hand, the structural stresses in the thin-shell element model in the region in front of the weld toe, where the conventional extrapolation to the hot spot is performed ($0.4 \leq x/t \leq 1.0$), are not sufficiently accurate, at least not with the applied element mesh.

Another link between structural stress FE analysis methods and the NSIF K_1 concept has been established (Meneghetti and Lazzarin 2007), see Sect. 3.7.3. The NSIF K_1 can be approximated by the 'peak stress' σ_p at the V-notch tip in a rather coarse globally even finite element mesh (element size typically $d = 1$ mm). The non-dimensional ratio $K_1/(\sigma_p d^{1-\lambda_1})$ is found to be independent of the notch opening angle. Once, the ratio has been determined for a definite configuration, it can be applied within a wide range of notch depths and notch angles.

2.3.6 Empirical Mixed-Mode Failure Criterion for Welded Joints

An empirical mixed-mode failure criterion for welded joints in terms of the NSIFs has been conceived in analogy to a conventional formulation in terms of the nominal stresses. The formula is based on test results for unnotched round bar specimens (ductile materials: unalloyed and alloyed steels) subjected to superimposed tensile and torsional in-phase high-cycle fatigue loading (Gough and Pollard 1935, 1937; Gough 1950). The endurable (nominal) stress ranges $\Delta\sigma$ and $\Delta\tau$ under the condition of combined loading follow an elliptical curve:

$$\left(\frac{\Delta\sigma}{\Delta\sigma_{\mathrm{E}}}\right)^2 + \left(\frac{\Delta\tau}{\Delta\tau_{\mathrm{E}}}\right)^2 = 1 \quad (R = -1) \tag{2.96}$$

where $\Delta\sigma_{\mathrm{E}}$ and $\Delta\tau_{\mathrm{E}}$ are the endurance limit values under the condition of single mode loading, with stress ratio $R = -1$ characterising reversed stressing. The relationship between $\Delta\tau_{\mathrm{E}}$ and $\Delta\sigma_{\mathrm{E}}$ may be chosen according to conventional failure criteria (Rankine: $\Delta\tau_{\mathrm{E}} = \Delta\sigma_{\mathrm{E}}$; von Mises: $\Delta\tau_{\mathrm{E}} = \Delta\sigma_{\mathrm{E}}/3^{1/2}$; Tresca: $\Delta\tau_{\mathrm{E}} = \Delta\sigma_{\mathrm{E}}/2$). For cast iron materials and for the round bar specimens with a sharp notch (brittle material behaviour), the curve shape deviates to some extent from the elliptical shape, necessitating a slightly more complex functional representation.

An elliptical failure curve has also been found for pointed V-notches evaluating averaged notch stresses in the most critical direction (Seweryn and Mróz 1995).

Converting the nominal stress based expression due to Gough, Eq. (2.96), into a NSIF based expression, neglecting the influence of mode 2 loading (actually the mode 2 singularity is weak and vanishes for $2\alpha > 102.6°$), the following relationship is gained (Lazzarin et al. 2004):

$$\left(\frac{\Delta K_1}{\Delta K_{1,\mathrm{E}}}\right)^2 + \left(\frac{\Delta K_3}{\Delta K_{3,\mathrm{E}}}\right)^2 = 1 \quad (R = -1) \tag{2.97}$$

where ΔK_1 and ΔK_3 are the endurable NSIFs in mixed mode loading compared with their endurance limit values $\Delta K_{1,\mathrm{E}}$ and $\Delta K_{3,\mathrm{E}}$ in single mode loading ($N_{\mathrm{E}} = 1 \times 10^5$ up to 2×10^6 cycles). It can be shown that Eq. (2.97) is substantiated by the criterion of strain energy density averaged over the circle section volume surrounding the V-notch tip.

In order to define an equivalent NSIF, Eq. (2.97) is rewritten in the following form (Lazzarin et al. 2004):

$$\left(\frac{\Delta K_1}{\Delta K_{1,\mathrm{E}}}\right)^2 + k_{1/3}^2\left(\frac{\Delta K_3}{\Delta K_{1,\mathrm{E}}}\right)^2 = 1 \tag{2.98}$$

$$k_{1/3} = \frac{\Delta K_{1,\mathrm{E}}}{\Delta K_{3,\mathrm{E}}} \tag{2.99}$$

The coefficient $k_{1/3}$ is the ratio of the fatigue strength in pure tensile (or bending) loading and pure torsional (or out-of-plane shear) loading, expressed by the endurable NSIF ranges at the endurance limit N_E. The coefficient is not a dimensionless parameter. The dimension depends via λ_1 and λ_3 on the notch opening angle 2α.

The equivalent NSIF can now be written in the following form:

$$\Delta K_{1,\mathrm{eq}} = \sqrt{(\Delta K_1)^2 + k_{1/3}^2 (\Delta K_3)^2} \tag{2.100}$$

where the units of $\Delta K_{1,\mathrm{eq}}$ coincide with those of ΔK_1, that is $\mathrm{MPa\,m}^{1-\lambda_1}$. Alternatively, the equivalent NSIF can be defined in units of ΔK_3, that is $\mathrm{MPa\,m}^{1-\lambda_3}$:

$$\Delta K_{3,\mathrm{eq}} = \sqrt{(\Delta K_3)^2 + k_{1/3}^{-2} (\Delta K_1)^2} \tag{2.101}$$

The coefficient $k_{1/3}$ has been determined based on the local strain energy density approach (see Sect. 3.2.1), evaluating the distortional strain energy density averaged in a sector volume with control radius R_0 around the notch tip (Lazzarin et al. 2004). The same radius R_0 is introduced for mode 1 and mode 3 loading conditions. The high-cycle fatigue data for a V-notch with $2\alpha = 135°$ (weld toe) made of fine-grained high-strength steel FeE460 resulted in $R_0 = 0.36$ mm, or when made of the age-hardened aluminium alloy AlSi1MgMn in $R_0 = 0.11$ mm. The following expression is then derived:

$$k_{1/3} = \sqrt{\frac{e_{\mathrm{d3}}}{e_{\mathrm{d1}}} R_0^{\lambda_3 - \lambda_1}} \tag{2.102}$$

where e_{d1} and e_{d3} are distortional strain energy density coefficients depending on the notch opening angle 2α (besides on Poisson's ratio ν).

The derivations up to here refer to high-cycle fatigue conditions ($N_\mathrm{E} = 1 \times 10^5$ up to 2×10^6 cycles) where the plastic deformations at the notch tip can be neglected. In the low-cycle and medium-cycle fatigue range, plastic deformations are an important effect and an accurate determination of R_0 would require an elastic-plastic analysis. The latter is avoided by considering the *S–N* curves in terms of the endurable nominal stresses $\Delta\sigma_\mathrm{E}$ and $\Delta\tau_\mathrm{E}$:

$$\Delta\sigma(N) = \Delta\sigma_\mathrm{E} \left(\frac{N_\mathrm{E}}{N} \right)^{1/k_\sigma} \tag{2.103}$$

$$\Delta\tau(N) = \Delta\tau_\mathrm{E} \left(\frac{N_\mathrm{E}}{N} \right)^{1/k_\tau} \tag{2.104}$$

where k_σ and k_τ are the inverse slope exponents of the *S–N* curves for tensile and shear loading.

In the case of identical values of k_σ and k_τ (same slope of the *S–N* curves), the coefficient $k_{1/3}$ is independent of the considered endurable number of cycles N_E. Then, the Eqs. (2.100) and 2.101) are also valid in the low-cycle and medium-cycle fatigue range.

In the conventional codes for the fatigue assessment of structural members, especially welded joints, different values of k_σ and k_τ are introduced, e.g. $k_\sigma = 3.0$ and $k_\tau = 5.0$ (Eurocode 3, 2005). In analogy to Eqs. (2.103) and (2.104), the *K–N* curves of the endurable NSIFs have the following form:

$$\Delta K_1(N) = \Delta K_{1,\mathrm{E}}\left(\frac{N_\mathrm{E}}{N}\right)^{1/k_\sigma} \tag{2.105}$$

$$\Delta K_3(N) = \Delta K_{3,\mathrm{E}}\left(\frac{N_\mathrm{E}}{N}\right)^{1/k_\tau} \tag{2.106}$$

and the mixed mode relationship in Eq. (2.97) is now formulated for the number of cycles $N_\mathrm{C} < N_\mathrm{E}$ in the low-cycle or medium-cycle fatigue range:

$$\left(\frac{\Delta K_1}{\Delta K_{1,\mathrm{C}}}\right)^2 + \left(\frac{\Delta K_3}{\Delta K_{3,\mathrm{C}}}\right)^2 = 1 \quad (R = -1) \tag{2.107}$$

In order to define an equivalent NSIF, Eq. (2.107) is rewritten in the following form:

$$\left(\frac{\Delta K_1}{\Delta K_{1,\mathrm{C}}}\right)^2 + k_{1/3,\mathrm{C}}^2\left(\frac{\Delta K_3}{\Delta K_{1,\mathrm{C}}}\right)^2 = 1 \tag{2.108}$$

The coefficient $k_{1/3,\mathrm{C}}$ is the ratio of the fatigue strength in pure tensile (or bending) loading and pure torsional (or out-of-plane shear) loading, expressed by the endurable NSIF ranges at the number of cycles, $N_\mathrm{C} < N_\mathrm{E}$:

$$k_{1/3,\mathrm{C}} = \frac{\Delta K_{1,\mathrm{C}}}{\Delta K_{3,\mathrm{C}}} \tag{2.109}$$

Based on the *K–N* curve relationships according to Eqs. (2.105) and (2.106), introducing $N = N_\mathrm{C}$, one gets:

$$k_{1/3,\mathrm{C}} = \frac{\Delta K_{1,\mathrm{E}}(N_\mathrm{E}/N_\mathrm{C})^{1/k_\sigma}}{\Delta K_{3,\mathrm{E}}(N_\mathrm{E}/N_\mathrm{C})^{1/k_\tau}} = k_{1/3}\left(\frac{N_\mathrm{E}}{N_\mathrm{C}}\right)^{1/k_\sigma - 1/k_\tau} \tag{2.110}$$

where $k_{1/3}$ is given by Eq. (2.99).

The equivalent NSIFs can now be written in analogy to Eqs. (2.100) and (2.101) in the following form:

$$\Delta K_{1,\mathrm{eq}} = \sqrt{(\Delta K_1)^2 + k_{1/3}^2(\Delta K_3)^2\left(\frac{N_\mathrm{E}}{N_\mathrm{C}}\right)^{2(1/k_\sigma - 1/k_\tau)}} \tag{2.111}$$

$$\Delta K_{3,\mathrm{eq}} = \sqrt{(\Delta K_3)^2 + k_{1/3}^{-2}(\Delta K_1)^2\left(\frac{N_\mathrm{E}}{N_\mathrm{C}}\right)^{2(1/k_\sigma - 1/k_\tau)}} \tag{2.112}$$

These relationships presume that the Gough ellipse is valid also in the low-cycle and medium-cycle fatigue range. An experimental proof of this presumption is not available.

2.3.7 Endurable NSIFs of Fatigue-Loaded Welded Joints

The fatigue strength or life of fillet-welded joints in the medium-cycle to high-cycle fatigue range ($N \geq 10^4$) can advantageously be described by endurable (elastic) NSIFs (Boukharouba et al. 1995; Verreman and Nie 1996, 1997). In contrast to the endurable nominal stresses, the endurable NSIFs are independent of the plate thickness, i.e. the size effect is already included. Fatigue test data from the literature referring to non-load-carrying fillet welds at tensile loaded transverse attachment joints of steel (mainly cruciform joints) in the as-welded condition have been evaluated to demonstrate this, Fig. 2.28 (Lazzarin and Tovo 1998). Large variations in the geometrical data are covered by the fatigue test data: plate thickness $t = 13$–100 mm, fillet weld leg length $h = 5$–16 mm, attachment length $L = 3$–220 mm. Only the NSIF K_1 is used to summarise the fatigue data, whereas the effect of K_2 is considered negligible in this type of joint ($2\alpha = 135°$), because no stress singularity occurs in the latter case. The large scatter range of the original nominal stress *S–N* curve is substantially reduced to conventional values experienced with geometrically uniform specimens.

An explanation for the fact that not only the crack initiation life but actually the evaluated total life is correctly described by endurable values of ΔK_1 is given in the following form (Atzori et al. 1999[(1)]): The crack initiation life (initiated crack length $a_\mathrm{i} = 0.3$ mm) and the residual crack propagation life (critical crack length $a_\mathrm{cr} = t/3$), occurring in the ratio approximately 2:1 in the investigation just quoted, are both controlled by the NSIF ΔK_1 as the dominant parameter. The SIF ΔK_I for a developing crack in the bisector plane is proportional to the NSIF ΔK_1.

The evaluation above based on ΔK_1 has been extended to tensile and bending loaded fillet-welded joints of further steels as well as of aluminium alloys (thickness $t = 3$–24 mm), Fig. 2.29 (Lazzarin and Livieri 2001). Different relationships are used to correlate ΔK_1 and $\Delta\sigma_\mathrm{n}$ in the tension and bending load cases. In the high-cycle range, the fatigue strength of the steel joints is about twice the

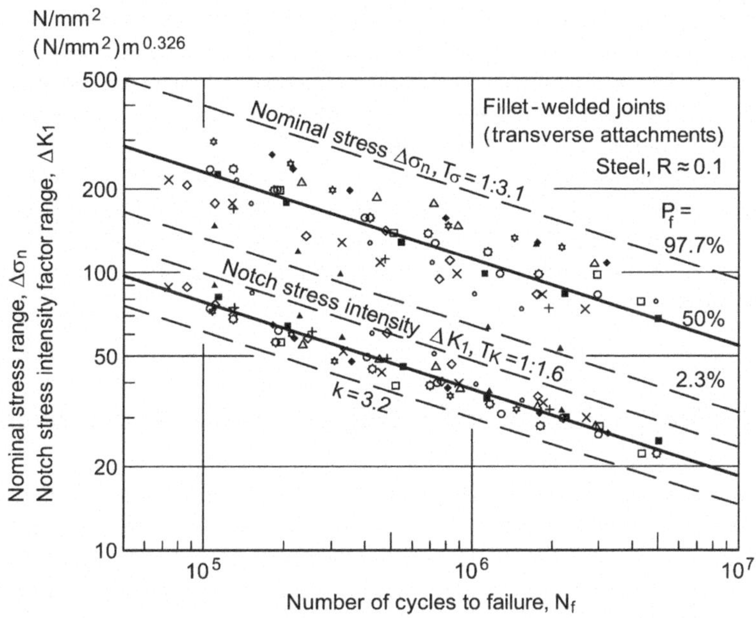

Fig. 2.28 Fatigue test data (toe failures) for fillet-welded transverse attachment joints of steel subjected to tensile loading; *S–N* curves contrasted with *K–N* curves; T_σ and T_K refer to $P_f = 2.3$ and 97.7 %; (Lazzarin and Tovo 1998)

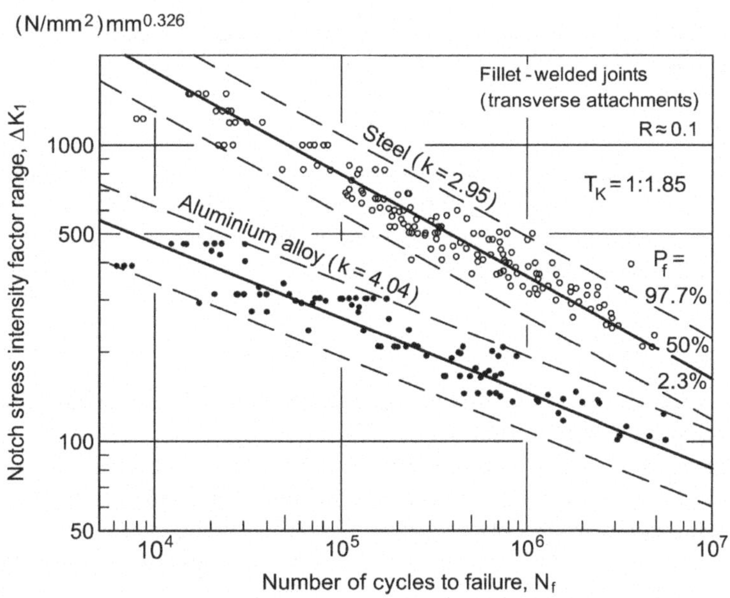

Fig. 2.29 Fatigue test data (toe failures) for fillet-welded transverse attachment joints of steel and aluminium alloy subjected to tensile and bending loads in terms of *K–N* curves; T_K refers to $P_f = 2.3$ and 97.7 %; (Lazzarin and Livieri 2001)

fatigue strength of the aluminium alloy joints. The size effect on the endurable nominal stresses is characterised by Eq. (2.93), leading to an exponent $(1 - \lambda_1) = 0.326$ of the thickness ratio. The exponent 0.25 is recommended in the Eurocode 3 (2005), exponents 0.1–0.3 dependent on joint type are found in the IIW design recommendations (Hobbacher 2009).

It has also been proposed to use the cyclic plastic zone radius R_p at the notch tip, approximated based on the elastic NSIF, for assessing the fatigue strength of fillet-welded joints. The *S–N* curves for different types of fillet-welded specimens are then replaced by a uniform R_p–*N* or R_p/t–*N* curve (Lazzarin and Livieri 2000; Koibuchi et al. 1999; Tanaka et al. 2002).

2.3.8 Endurable J-Integral for Fatigue-Loaded Welded Joints

The path-independent line integral around crack tips, usually named '*J*-integral', is a well-defined parameter characterising the fracture behaviour of crack tips (Rice 1968), compare Sect. 2.2.6. It was originally derived for linear-elastic material behaviour, but later on extended to nonlinear-elastic or elastic-plastic material behaviour, provided the loading curve is monotonic in the latter case.

The *J*-integral concept has been extended to pointed V-notches, at first under linear-elastic and then under elastic-plastic conditions (Lazzarin et al. 2002). The integral is path-dependent for corner notches subjected to mode 1 and mode 2 loading.

The J_V-integral of V-notches in linear-elastic materials can be given in the following form:

$$J_V = 2(A_{11}K_1^2 + A_{22}K_2^2) \tag{2.113}$$

where A_{11} and A_{22} depend on the notch opening angle 2α and the path around the notch tip. The dimension of J_V is MPa mm. The above integral can be made path-independent under the restriction of pure mode 1 (or pure mode 2) loading conditions, leading to the parameter

$$J_L = \frac{J_V}{R^{2\lambda_1 - 1}} \tag{2.114}$$

where R is the radius of the circular integration path. The parameter J_L remains dependent on the notch opening angle 2α.

The integrals J_V and J_L can also be defined for elastic-plastic material behaviour under the restriction of pure mode 1 loading. Using the elastic J_V-integral approach, the contributions of mode 1 and mode 2 loading can be superimposed, i.e. a well-founded failure criterion for mixed-mode loading can be established.

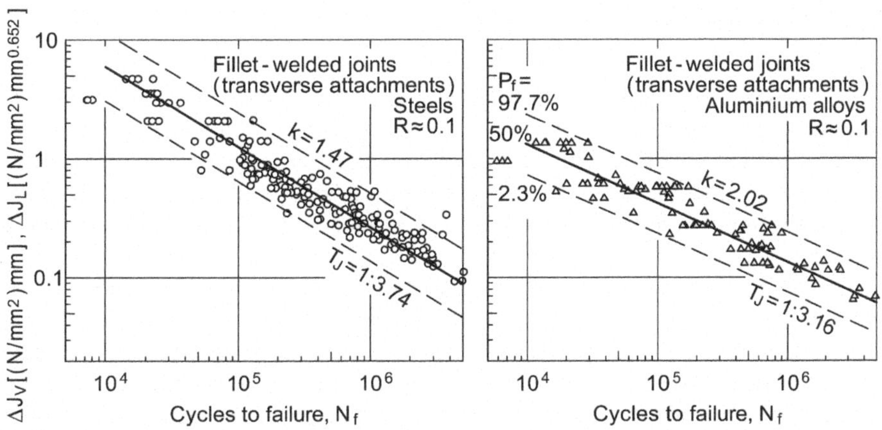

Fig. 2.30 Fatigue test data (toe failures) for fillet-welded transverse attachment joints of steels and aluminium alloys under tensile and bending loads in terms of *J–N* curves; scatter band characterises mean values plus minus two standard deviations; T_J refers to $P_f = 2.3$ and 97.7 %; (Lazzarin et al. 2002)

The integrals J_V and J_L have also been defined for pure mode 3 loading with inclusion of elastic-plastic material behaviour (Lazzarin and Zappalorto 2008). A different line integral around elastic corner notches subjected to mode 3 loading has independently been derived (Quian and Hasebe 1997). It evolves to be path-independent provided the notch flanks are free of loads.

The *J*-integral, defined above for V-notches, can be used as a fatigue-relevant parameter applicable to corner notches under mixed mode loading conditions, to varying notch opening angles (defined by the toe angle of the considered butt-welded or fillet-welded joints) and under conditions of local yielding at the notch tip. Evaluating the published fatigue test results for non-load-carrying transverse stiffener joints using a uniform circular path radius $R = 1$ mm, so that the values of J_V and J_L are identical, resulted in the $J_V - N$ curves or $J_L - N$ curves, respectively, plotted in Fig. 2.30 for fillet-welded-joints made of steels and aluminium alloys. The contribution of ΔK_2 is neglected with reference to the missing stress singularity in the case of large notch opening angles. When comparing the scatter range indices T_J and T_K, it has to be noted that a quadratic increase occurs solely by substituting a stress parameter by an energy parameter. The presented curves do not yet prove the applicability of the ΔJ-concept with regard to varying notch opening angles and marked mixed mode conditions.

More recently, J_V evaluations for welded joints have been performed based on a substantially extended data base of fatigue test results with failures proceeding from the weld toe and weld root (Livieri and Tovo 2009). The data base is almost identical with that previously used for local SED evaluations (Fig. 3.12).

2.3.9 Conclusions

The NSIFs K_1, K_2 and K_3 together with the non-singular stress components describe the stress field in the close vicinity of pointed V-notches or re-entrant corners. A transverse singular effect coupled with mode 2 loading at free surfaces is additionally identified. The numerical basis of NSIF evaluations in configurations with finite boundaries is the FE analysis method performed with extremely fine meshes.

The NSIF concept is well suited to describe the fatigue strength of fillet-welded and butt-welded joints with failure initiation at the weld toe. A uniform *K–N* curve with sufficiently low scatter range is derived which includes the size or plate thickness effect. Alternatively, a presentation in terms of the endurable *J*-integral is possible. Equivalent NSIFs are derived for cases of mixed mode loading conditions.

The NSIF parameters are well suited to constitute the basis of the endurable average strain energy density at weld notch tips which is under discussion to be implemented in the IIW recommendations for fatigue assessment.

2.4 Generalised Notch Stress Intensity Factor Concept

2.4.1 Survey of Section Contents

The notch stress intensity factor (NSIF) concept referring to pointed V-notch tips with inclusion of the conventional stress intensity factor (SIF) approach referring to pointed crack tips is extended to the generalised notch stress intensity factor concept referring to sharply rounded (blunt) or root-holed V-notches or crack tips. Notch rounding changes the stresses at the V-notch or crack tip substantially insofar as the stress singularity is removed. But the stress distribution connected with the singularity remains widely unchanged at distances from the notch root larger than one half of the notch radius ρ, provided the radius is sufficiently small in relation to the notch depth or crack length (sharp notches).

It is shown that the stress distribution at sharply rounded notches subjected to the loading modes 1, 2, 3 can be described by the generalised NSIFs $K_{1\rho}$, $K_{2\rho}$, $K_{3\rho}$ as the governing field parameters. These NSIFs are related to the maximum notch stresses σ_{max}, τ_{max}, τ^*_{max} which constitute the conventional stress concentration factors (SCFs) in engineering science. In contrast to the SCFs, the generalised NSIFs describe not only the maximum stress but the whole stress field in the vicinity of the notch root. The field information is needed for assessing failure processes such as crack initiation and propagation or microstructural damaging phenomena.

In the following, the stress field equations for blunt V-notches subjected to the three basic loading modes are given. The corresponding generalised NSIFs are defined and evaluation procedures described. Examples for notch stress distributions determined on the basis of generalised NSIFs are given. Furthermore, V-notches with root hole are analysed under the three loading modes, the stress field being likewise described by generalised NSIFs. The NSIFs of parabolic and keyhole notches are set into comparison. The *T*-stress effect at rounded notch configurations is clarified. The generalised NSIF concept is finally applied to a round bar with shoulder fillets. The fatigue limit of sharply rounded notches can only exceptionally be directly expressed by critical NSIF values.

The expositions are mainly based on the analytical development of the generalised NSIF concept by Lazzarin, Zappalorto, Filippi and Tovo. A list of symbols and a list of references are found at the end of Chap. 2.

2.4.2 Stress Field at Blunt V-Notches Subjected to Tensile and In-Plane Shear Loading

A general solution for the stress field at sharply rounded (blunt) V-notches, subjected to tensile and in-plane shear loading is available (Lazzarin and Tovo 1996; Filippi et al. 2002$^{(1)}$, 2002$^{(2)}$; Lazzarin and Filippi 2006). The nomenclature is shown in Fig. 2.31. The theoretical frame, Kolsov–Muskhelishvili's complex stress function method combined with Neuber's auxiliary system of curvilinear coordinates, has been reviewed in Sect. 2.3.2 with application to pointed V-notches. Now, the improved solution by Filippi is evaluated in respect to blunt V-notches. The solution is not exact, but represents the first term in a series perturbating the pointed notch solution.

The auxiliary system of curvilinear coordinates is shown in Fig. 2.32. The parametric curves u = const. and v = const. are plotted in the x–y plane using the conformal mapping function $z = w^q$ with $z = x + \mathrm{i}y$ and $w = u + \mathrm{i}v$. The factor q is related to the notch opening angle 2α of the pointed notch, curve $u = 0$, Eq. (2.65) and Fig. 2.31. The curve $u = u_0$ is assumed to describe the blunt notch. Its radius of curvature ρ at the intersection with the x-axis, where $r_0 = u_0^q$, is equal to

$$\rho = \frac{q r_0}{(q-1)} \tag{2.115}$$

This radius of curvature corresponds to the notch root radius. The distance r_0 between the origin of the x–y coordinate system and the notch root depends (via q) on the notch opening angle 2α. Only for the parabolic notch ($2\alpha = 0$, $q = 2$), the relationship $r_0 = \rho/2$ is valid.

The stress field equations for blunt V-notches subjected to tensile and in-plane shear loading (mode 1 and mode 2) are presented in extension of the equations for pointed V-notches, Eqs. (2.70) and (2.71):

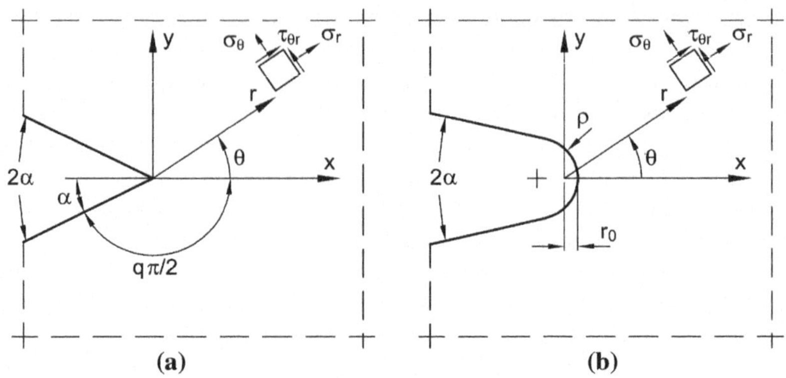

Fig. 2.31 Coordinate systems, symbols and in-plane notch stress components at pointed V-notch (**a**) and at the corresponding rounded V-notch (**b**); with $r_0 = \rho(q - 1)/q$; (Lazzarin and Tovo 1998)

Fig. 2.32 Auxiliary system of curvilinear coordinates applied to solve the stress field problem at V-notches or hyperbolic notches; (Lazzarin and Tovo 1996)

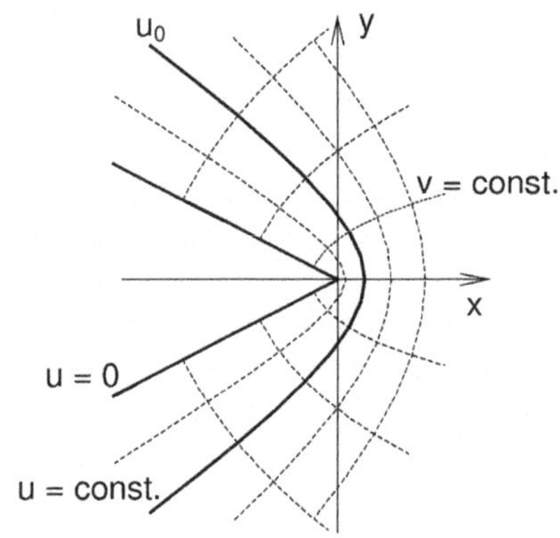

$$\begin{Bmatrix} \sigma_\theta \\ \sigma_r \\ \tau_{r\theta} \end{Bmatrix}_1 = \frac{1}{\sqrt{2\pi}} \frac{r^{\lambda_1 - 1} K_1}{(1+\lambda_1) + \chi_1(1-\lambda_1)} \times \left[\begin{Bmatrix} (1+\lambda_1)\cos(1-\lambda_1)\theta \\ (3-\lambda_1)\cos(1-\lambda_1)\theta \\ (1-\lambda_1)\sin(1-\lambda_1)\theta \end{Bmatrix} + \chi_1(1-\lambda_1) \begin{Bmatrix} \cos(1+\lambda_1)\theta \\ -\cos(1+\lambda_1)\theta \\ \sin(1+\lambda_1)\theta \end{Bmatrix} + \left(\frac{r}{r_0}\right)^{\mu_1 - \lambda_1} [(3-\lambda_1) - \chi_1(1-\lambda_1)] \begin{Bmatrix} \cos(1+\mu_1)\theta \\ -\cos(1+\mu_1)\theta \\ \sin(1+\mu_1)\theta \end{Bmatrix} \right] \tag{2.116}$$

$$\begin{Bmatrix} \sigma_\theta \\ \sigma_r \\ \tau_{r\theta} \end{Bmatrix}_2 = \frac{1}{\sqrt{2\pi}} \frac{r^{\lambda_2 - 1} K_2}{(1+\lambda_2) + \chi_2(1-\lambda_2)} \\ \times \left[\left[\begin{Bmatrix} -(1+\lambda_2)\cos(1-\lambda_2)\theta \\ -(3-\lambda_2)\cos(1-\lambda_2)\theta \\ (1-\lambda_2)\sin(1-\lambda_2)\theta \end{Bmatrix} + \chi_2(1-\lambda_2) \begin{Bmatrix} -\sin(1+\lambda_2)\theta \\ \sin(1+\lambda_2)\theta \\ \cos(1+\lambda_2)\theta \end{Bmatrix} \right] \right. \\ \left. - \left(\frac{r}{r_0}\right)^{\mu_2 - \lambda_2} [(1-\lambda_2) + \chi_2(1+\lambda_2)] \begin{Bmatrix} \sin(1+\mu_2)\theta \\ -\sin(1+\mu_2)\theta \\ \cos(1+\mu_2)\theta \end{Bmatrix} \right] \tag{2.117}$$

The NSIFs K_1 and K_2 are determined for $r \to 0$ according to Gross–Mendelson's limit value definition, Eqs. (2.77) and (2.78). Several well known solutions for crack or notch configurations (Williams 1952; Irwin 1957; Creager and Paris 1967; Glinka and Newport 1987) can be considered as special cases of the above general solution, in which the NSIFs K_1 and K_2 may be substituted by the parameters a_1 and a_2 or by the maximum stresses σ_{max} at the notch tip and τ_{max} ahead of the notch tip.

Williams' solution for pointed V-notches results with $r_0 = 0$ (because $\rho = 0$), compare Eqs. (2.70) and (2.71). Westergaard–Sneddon–Irwin's solution for crack tips is derived with $r_0 = 0$, $q = 2$ and $\lambda_1 = \lambda_2 = 0.5$ (because $2\alpha = 0$), compare Eqs. (2.2) and (2.4). Creager's blunt crack (or parabolic notch) solution is obtained by imposing $r_0 = \rho/2$, $q = 2$, $\lambda_1 = \lambda_2 = 0.5$ and $\mu_1 = \mu_2 = -0.5$ (Creager and Paris 1967):

$$\begin{Bmatrix} \sigma_\theta \\ \sigma_r \\ \tau_{r\theta} \end{Bmatrix}_1 = \frac{K_I}{\sqrt{2\pi r}} \left(\frac{1}{4} \left[\begin{Bmatrix} 3\cos\theta/2 \\ 5\cos\theta/2 \\ \sin\theta/2 \end{Bmatrix} + \begin{Bmatrix} \cos 3\theta/2 \\ -\cos 3\theta/2 \\ \sin 3\theta/2 \end{Bmatrix} \right] + \frac{\rho}{2r} \begin{Bmatrix} \cos\theta/2 \\ -\cos\theta/2 \\ \sin\theta/2 \end{Bmatrix} \right) \tag{2.118}$$

$$\begin{Bmatrix} \sigma_\theta \\ \sigma_r \\ \tau_{r\theta} \end{Bmatrix}_2 = \frac{K_{II}}{\sqrt{2\pi r}} \left(\frac{1}{4} \left[\begin{Bmatrix} 3\sin\theta/2 \\ 5\sin\theta/2 \\ \cos\theta/2 \end{Bmatrix} + \begin{Bmatrix} 3\sin 3\theta/2 \\ -3\sin 3\theta/2 \\ 3\cos 3\theta/2 \end{Bmatrix} \right] + \frac{\rho}{2r} \begin{Bmatrix} \sin\theta/2 \\ -\sin\theta/2 \\ \cos\theta/2 \end{Bmatrix} \right) \tag{2.119}$$

It can also be shown that Neuber's solution for the stress distribution along the notch bisector for tensile loaded V-notches is in agreement with Eq. (2.116) to a wide extent (Lazzarin and Tovo 1996).

The general stress field equations above, Eqs. (2.116) and (2.117), are approximations. They are deficient in accuracy, when applied to blunt V-notches with a large notch opening angle, e.g. $2\alpha = 135°$ referring to weld toes. With the aim of improving the accuracy of the analytical solution, the part of the equations with the exponent μ is extended. This is achieved by adding a second analytical function to the first expression in Eq. (2.63), in order to provide further free parameters for adjustment to the boundary conditions (Filippi et al. 2002[(2)]):

$$\varphi(z) = az^{\lambda} + dz^{\mu}, \quad \psi(z) = bz^{\lambda} + cz^{\mu} \tag{2.120}$$

The resulting stress field equations are more complex than before in the second order terms which contain the exponent μ, but the accuracy of the solution is substantially enhanced, especially for large notch opening angles. An additional improvement of the exponent μ_2 for mode 2 loading has been introduced (Lazzarin et al. 2011).

2.4.3 Stress Field at Blunt V-Notches Subjected to Out-of-Plane Shear Loading

A general solution for the stress field at sharply rounded (blunt) V-notches subjected to out-of-plane shear loading has been derived on a similar analytical basis as before in the case of tensile and in-plane shear loading (Smith 2004; Zappalorto and Lazzarin 2011[(3)]). Pointed V-notches are included.

Only one analytical function (consisting of two terms) is needed for the mathematical description of the shear stress field:

$$H(z) = az^{\lambda} + bz^{\mu} \tag{2.121}$$

where the coefficients a and b are complex whereas the exponents λ and μ are real. The coefficients are introduced as real if only the antimetric part of the solution is requested. The pointed notch problem is described by neglecting the second term with the exponent μ. The reason for one analytical function being sufficient is, that the out-of-plane stress field is governed by a potential function whereas the in-plane stress fields must obey a bipotential function.

The out-of-plane shear stresses are derived in the following form, separated into antimetrical (subscript a) and symmetrical (subscript s) parts in the r—θ coordinate system:

$$\begin{Bmatrix} \tau_{rz} \\ \tau_{\theta z} \end{Bmatrix} = \frac{K_{3\rho}^{(a)}}{\sqrt{2\pi}} r^{\lambda_a - 1} \begin{Bmatrix} \sin \lambda_a \theta \\ \cos \lambda_a \theta \end{Bmatrix} + \frac{K_{3\rho}^{(s)}}{\sqrt{2\pi}} r^{\lambda_s - 1} \begin{Bmatrix} \cos \lambda_s \theta - (r/r_0)^{\mu_s - \lambda_s} \cos \mu_s \theta \\ -\sin \lambda_s \theta + (r/r_0)^{\mu_s - \lambda_s} \sin \mu_s \theta \end{Bmatrix} \tag{2.122}$$

where the antimetric and symmetric eigenvalues are dependent on the notch opening angle 2α expressed by the parameter q ranging between 2.0 and 1.0, Eq. (2.65):

$$\lambda_a = \frac{1}{q}, \quad \lambda_s = \frac{2}{q}, \quad \mu_s = \frac{1}{q}, \quad q = \frac{2\pi - 2\alpha}{\pi} \tag{2.123}$$

The generalised NSIFs $K_{3\rho}^{(a)}$ and $K_{3\rho}^{(s)}$ are determined from the stresses in the bisector plane ($\theta = 0$) of the blunt V-notch (with $r_{0+} > r_0$ in the logarithmic plot):

$$K_{3\rho}^{(a)} = \lim_{r \to r_{0+}} \sqrt{2\pi}\left[r^{1-\lambda_a}\tau_{\theta z}(r, 0)\right] \tag{2.124}$$

$$K_{3\rho}^{(s)} = \lim_{r \to r_{0+}} \sqrt{2\pi}\left[\frac{r^{1-\lambda_s}\tau_{rz}(r, 0)}{1 - (r/r_0)^{\mu_s - \lambda_s}}\right] \tag{2.125}$$

The above NSIF $K_{3\rho}^{(a)}$ can be substituted by the maximum notch stress, $K_{3\rho}^{(a)} = \tau_{\max}^{(a)} r_0^{1-\lambda_a}(2\pi)^{1/2}$ with $r_0 = (1 - 1/q)\rho$ (Zappalorto et al. 2009), whereas this is not possible with the NSIF $K_{3\rho}^{(s)}$.

For the pointed V-notch, notch radius ρ and distance r_0 are equal to zero, so that the terms with r/r_0 disappear (the exponent, $\mu_s - \lambda_s$, is negative):

$$\begin{Bmatrix} \tau_{rz} \\ \tau_{\theta z} \end{Bmatrix} = \frac{K_3^{(a)}}{\sqrt{2\pi}} r^{\lambda_a - 1} \begin{Bmatrix} \sin \lambda_a\theta \\ \cos \lambda_a\theta \end{Bmatrix} + \frac{K_3^{(s)}}{\sqrt{2\pi}} r^{\lambda_s - 1} \begin{Bmatrix} \cos \lambda_s\theta \\ -\sin \lambda_s\theta \end{Bmatrix} \tag{2.126}$$

The exponent ($\lambda_a - 1$) is negative, thus characterising a singular stress distribution. The other exponent ($\lambda_s - 1$) is positive, thus denoting a non-singular distribution. In the case of a crack or slit ($2\alpha = 0$), the eigenvalues are $\lambda_a = 1/2$ and $\lambda_s = 1.0$, the exponents being ($\lambda_a - 1$) $= -1/2$ and ($\lambda_s - 1$) $= 0$, Eq. (2.123) with $q = 2$. Comparison with Eq. (2.73) gives the non-singular S-stress, $S = K_3^{(s)} r^{\lambda_s - 1}/(2\pi)^{1/2}$.

In order to improve the solution for U-notches and blunt V-notches subjected to out-of-plane shear loading, a modified mapping function has been introduced which complies better with the actual notch shape consisting of a circular arc tapering off into the rectilinear flanks of the V-notch (Zappalorto et al. 2010). The resulting, more complex stress field equations have been derived. It is found that the notch stresses are markedly influenced by the details of the notch root shape. A solution for semi-elliptical notches is also available (Lazzarin et al. 2007). In both investigations the NSIFs are found by FE analysis.

2.4.4 Generalised Notch Stress Intensity Factors

For evaluating the stress fields at blunt V-notches, generalised notch stress intensity factors have to be introduced which may be differently defined depending on the type of notch geometry. Also, the simple SIFs or NSIFs may be used. For example, in Eqs. (2.116) and (2.117) the NSIFs K_1 and K_2 of the corresponding pointed notch are introduced, or in Eqs. (2.118) and (2.119) the SIFs K_I and K_{II} of the corresponding pointed crack tip. It has to be noted, that the distance r_0 is measured between the origin of r and the pointed or rounded notch or crack tip.

In Eq. (2.122), the generalised NSIF $K_{3\rho}$ is introduced, which characterises the stress field at rounded V-notches directly. There are three different limit value procedures available for determining the SIFs, NSIFs or generalised NSIFs.

The first procedure evaluates the stresses in the bisector plane of the pointed notch or crack for determining K_1, K_2, K_3 according to Eqs. (2.77–2.79) or K_{I}, K_{II}, K_{III} according to Eqs. (2.16–2.19) as limit values for $r \to 0$ (Gross and Mendelson 1972).

The second procedure evaluates the maximum notch stresses in the bisector plane of the rounded crack or notch (in mode 2 loading also outside the bisector plane) for determining K_{I}, K_{II}, K_{III} or K_1, K_2, K_3 as limit values for $\rho \to 0$. The following formulae are available, first for cracks under mode I loading based on $\sigma_{\max} = \sigma_{\theta\ \max}$ in the bisector plane (Irwin 1957, 1958),

$$K_{\mathrm{I}} = \lim_{\rho \to 0} \frac{1}{2} \sigma_{\max} \sqrt{\pi \rho} \qquad (T = 0) \tag{2.127}$$

further for cracks under mode II loading based on $\tau_{\max} = \tau_{r\theta\ \max}$ in the bisector plane (Cheng 1988; Radaj and Zhang 1993[(3)]) or based on $\sigma_{\max} = \sigma_{\mathrm{t}\ \max}$ outside the bisector plane (Sih and Liebowitz 1968),

$$K_{\mathrm{II}} = \lim_{\rho \to 0} \frac{3\sqrt{3}}{2} \tau_{\max} \sqrt{\pi \rho} \tag{2.128}$$

$$K_{\mathrm{II}} = \lim_{\rho \to 0} \sigma_{\max} \sqrt{\pi \rho} \qquad (K_{\mathrm{I}} = 0,\ T = 0) \tag{2.129}$$

and finally for cracks under mode III loading based on $\tau^*_{\max} = \tau_{\theta z\ \max}$ (Hahn 1976; Hasebe and Kutanda 1978):

$$K_{\mathrm{III}} = \lim_{\rho \to 0} \tau^*_{\max} \sqrt{\pi \rho} \tag{2.130}$$

The limit value for K_{I}, Eq. (2.127), presumes that the crack-parallel or slit-parallel stress $\sigma_{\parallel}$ (or T-stress in the limit $\rho = 0$) is zero. This is further elaborated in Sect. 2.4.8.

Similar formulae are available for V-notches under mode 1 loading (Benthem 1987; Hasebe et al. 1990; Nui et al. 1994),

$$K_1 = \lim_{\rho \to 0} \frac{\sqrt{\pi}}{2} \sigma_{\max} \rho^{1-\lambda_1} \tag{2.131}$$

and also for hyperbolic V-notches under mode 2 loading (Benthem 1987), $a_{2\tau}$ and $a_{2\sigma}$ according to Table 2.2 (using a special mapping function):

$$K_2 = \lim_{\rho \to 0} \frac{\sqrt{\pi}}{a_{2\tau}} \tau_{\max} \rho^{1-\lambda_2} \tag{2.132}$$

$$K_2 = \lim_{\rho \to 0} \frac{\sqrt{\pi}}{a_{2\sigma}} \sigma_{\max} \rho^{1-\lambda_2} \tag{2.133}$$

Table 2.2 Parameter values for Benthem's NSIF K_2 of pointed V-notches, Eqs. (2.132) and (2.133); (Benthem 1987)

2α	λ_2	$a_{2\tau}$	$a_{2\sigma}$	2α	$\lambda_{2\perp}$	$a_{2\tau}$	$a_{2\sigma}$
0°	0.500	0.385	1.000	90°	0.909	0.620	1.213
30°	0.598	0.492	1.242	102.54°	1.000	0.707	1.179
60°	0.731	0.550	1.275				

Table 2.3 Parameter values for evaluation of $K_{1\rho}$ and $K_{2\rho}$ according to Eqs. (2.135) and (2.136); (Lazzarin and Filippi 2006)

2α	q	λ_1	λ_2	μ_1	μ_2	$\tilde{\omega}_1$	$\tilde{\omega}_2$
0°	2.0	0.500	0.500	−0.500	−0.500	1.000	−1.0
30°	1.833	0.501	0.598	−0.456	−0.447	1.034	−1.0
45°	1.750	0.505	0.660	−0.432	−0.412	1.014	−1.0
60°	1.667	0.512	0.731	−0.406	−0.373	0.970	−1.0
90°	1.500	0.545	0.909	−0.345	−0.288	0.810	−1.0
120°	1.333	0.616	1.149	−0.268	−0.198	0.570	−1.0
135°	1.250	0.674	1.302	−0.220	−0.151	0.432	−1.0

and finally for V-notches under mode 3 loading in analogy to Eq. (2.130):

$$K_3 = \lim_{\rho \to 0} \sqrt{\pi} \tau^*_{\max} \rho^{1-\lambda_3} \tag{2.134}$$

The Eqs. (2.127–2.134) above may be used without the limes with distinct finite values of ρ in an approximative manner, but the convergence behaviour must be checked (compare Figs. 1.32 and 1.33).

The third procedure evaluates the stresses in the bisector plane of the rounded notch or crack for determining $K_{1\rho}$, $K_{2\rho}$, $K_{3\rho}$ as the limit values for $r \to r_0$ (or $r \to r_{0+}$, the position of $\tau_{r\theta \max}$ in mode 2 loading) which may be combined with averaging these NSIFs over a small distance ahead of the notch root. The following formulae are derived for V-notches inclusive of cracks (Lazzarin and Filippi 2006; Zappalorto et al. 2008), Table 2.3:

$$K_{1\rho} = \lim_{r \to r_0} \frac{\sqrt{2\pi} r^{1-\lambda_1} \sigma_\theta(r, 0)}{1 + \tilde{\omega}_1 (r/r_0)^{\mu_1 - \lambda_1}} = \frac{\sqrt{2\pi} r_0^{1-\lambda_1} \sigma_\theta(r_0, 0)}{1 + \tilde{\omega}_1} \tag{2.135}$$

$$K_{2\rho} = \lim_{r \to r_{0+}} \frac{\sqrt{2\pi} r^{1-\lambda_2} \tau_{r\theta}(r, 0)}{1 + \tilde{\omega}_2 (r/r_0)^{\mu_2 - \lambda_2}} = \frac{\sqrt{2\pi} r_0^{1-\lambda_2} \tau_{r\theta}(r_{0+}, 0)}{1 - (r_{0+}/r_0)^{\mu_2 - \lambda_2}} \tag{2.136}$$

$$K_{3\rho} = \lim_{r \to r_0} \sqrt{2\pi} r^{1-\lambda_3} \tau_{\theta z}(r, 0) = \sqrt{2\pi} r_0^{1-\lambda_3} \tau_{\theta z}(r_0, 0) \tag{2.137}$$

The value $\tilde{\omega}_2 = -1$ in Eq. (2.136) is part of the stress field solution. The limit value $r \to r_{0+}$ instead of $r \to r_0$ in mode 2 loading is proposed here in order to avoid numerical problems when introducing $r = r_0$. An 'equivalent V-notch' may be defined for other notch types (e.g. semicircular) based on Eq. (2.135); (Filippi and Lazzarin 2004).

The NSIFs $K_{1\rho}$, $K_{2\rho}$, $K_{3\rho}$ differ from K_1, K_2, K_3 for finite values of ρ. They are generally enlarged for increasing values of ρ. The following relationship should be valid, because $r_0 \propto \rho$:

$$K_{\rm i} = \lim_{\rho \to 0} K_{{\rm i}\rho} \quad ({\rm i} = 1,\ 2,\ 3) \tag{2.138}$$

The condition according to Lazzarin, that the enlarged values of $K_{1\rho}$, $K_{2\rho}$, $K_{3\rho}$ can be used as characteristic stress field parameters at rounded notches ($\rho > 0$), is the constancy of $K_{1\rho}$, $K_{2\rho}$, $K_{3\rho}$ over a certain distance from the notch root. If this condition is met, the NSIF concept is applicable also to notches with a substantially increased notch radius.

In cases of a slightly (oscillating) dependence of $K_{1\rho}$, $K_{2\rho}$, $K_{3\rho}$ on $(r - r_0)$, the distance from the notch root, the following averaging procedure is proposed (Lazzarin and Filippi 2006):

$$\overline{K}_{{\rm i}\rho} = \frac{1}{\eta\rho} \int\limits_{r_0}^{r_0+\eta\rho} K_{{\rm i}\rho} dr \quad ({\rm i} = 1,\ 2,\ 3) \tag{2.139}$$

where $\eta = 0.4$ is a recommended choice. The averaging procedure is also preferred in cases of a very small notch radius where sufficiently fine FE meshing at the notch root is difficult to achieve.

Now the question arises, to what extent the stress field equations expressed by K_1, K_2, K_3 or $K_{\rm I}$, $K_{\rm II}$, $K_{\rm III}$ can be used with the generally enlarged values of $K_{1\rho}$, $K_{2\rho}$, $K_{3\rho}$. The answer depends on the considered case. As far as possible, the stress field equations should be used, from which the relationships for $K_{1\rho}$, $K_{2\rho}$, $K_{3\rho}$ were derived.

These relationships depend on the notch shape. Those mentioned above refer to V-notches or hyperbolic notches with the parabolic notch as a special case. The comparison of hyperbolic with circular arc V-notches reveals a slight difference even in mode 1 loading (Savruk and Kazberuk 2010). The formulae for elliptical notches are also different to some extent. The relationships for V-notches with root hole, the keyhole notch among them, deviate even more. They will be presented in Sect. 2.4.6.

Not only the notch shape exerts an influence on the values of $K_{1\rho}$, $K_{2\rho}$, $K_{3\rho}$, but also the geometrical and dimensional parameters of the notched members as well as its loading and support conditions. Cross-sectional weakening and notch-parallel loading (or T-stress) are especially influential.

The generalised NSIFs $K_{1\rho}$, $K_{2\rho}$, $K_{3\rho}$ have potential for application insofar as they define the stress field over a finite area (or volume), just as the simple NSIFs or SIFs do. The widely used stress concentration factors (SCFs) give only the maximum stress at one point. Additional information on the stress field is needed, in order to characterise the fatigue life or brittle fracture strength, e.g. notch stress averaging, notch stress gradient, critical distance among others. In contrast, the NSIFs (possibly supplemented by the T-stress and S-stress terms) represent the complete notch stress field without additional information being necessary. Whereas the SCFs are not applicable to pointed notches, the NSIFs proceed from pointed notches.

The NSIFs in contrast to the SCFs also include the scale or size effect on brittle fracture or fatigue strength. Considering a small notched component, the notch with radius ρ_1, in comparison to a self-similar enlarged notched component, the notch with radius ρ_2, both components subjected to the same nominal stresses, then the SCFs K_t are identical, but the NSIFs are related by $(\rho_2/\rho_1)^{1-\lambda}$ according to Eqs. (2.131–2.134). They define two stress profiles of different level ahead of the notch root starting with the same value of the maximum stress. One method to determine failure-effective notch stresses is to average the notch stresses over a definite, material-dependent microstructural support length. These averaged notch stresses will be higher for the larger component. Thus, in comparison with a constant endurable averaged notch stress, the larger component with the higher NSIF will come off worse.

2.4.5 Evaluation Examples for Stresses at Blunt V-Notches

Some typical evaluation examples for the stresses at sharply rounded (blunt) V-notches subjected to tensile and in-plane shear loading are reviewed in the following in order to achieve further clarity on the generalised NSIF concept.

In the first four figures, the determination of the NSIFs $K_{1\rho}$ and $K_{2\rho}$ (as far as relevant) according to Eqs. (2.135) and (2.136) is exemplified. The tensile loaded flat bar with two-sided U-notch is considered in Fig. 2.33. A slight oscillation of $K_{1\rho}$ over $(r - r_0)$ occurs for $\rho \neq 0$ in the evaluation range up to $(r - r_0) = 1$ mm. The value of $K_{1\rho}$ rises with the ratio ρ/a (with notch depth a), as specified further below. The oscillation is enlarged for the V-notch with $2\alpha = 135°$, while the influence of ρ/a is reduced, Fig. 2.34. Mixed mode loading conditions occur in the bisector plane of the shoulder fillet flat bar in Fig. 2.35 and of the weld-like flat bar in Fig. 2.36. Once more, some variability of $K_{1\rho}$ or $K_{2\rho}$ is detected in the evaluation range.

The stress distribution in the notch bisector ahead of the notch root in a tensile loaded V-notched flat bar is plotted in Fig. 2.37 for three notch opening angles. Original and $K_{1\rho}$-based analysis results are compared. A first curve segment up to 0.1 mm is mainly controlled by the maximum notch stress. A second segment up to 5 mm presents the gradient of the corresponding pointed notch configuration. The third segment correlates with the nominal stress, actually in the net cross-section. Another similar diagram, where the notch radius is varied, is shown in Fig. 2.38. The averaged solution for $K_{1\rho}$ is evaluated. The first curve segment is limited by $(r - r_0)/\rho \approx 0.1$. The nominal stress is related to the net cross-section.

The linearised rise of $K_{1\rho}/K_1$ over the ratio of notch radius to notch depth, ρ/a, is shown in Fig. 2.39 and also by Table 2.4. Actually, the curves should start at $K_{1\rho}/K_{1\rho} = 1.0$ for $\rho/a = 0$, which is suppressed by the linearisation, compare Eq. (2.138). The result is independent on b/a for sufficiently large values of this ratio.

Corresponding results for in-plane shear loading are recorded in subsequent graphs. The analysed configuration is a rectangular plate with a two-sided internal parabolic, hyperbolic, U-shaped or V-shaped notch, the plate subjected to

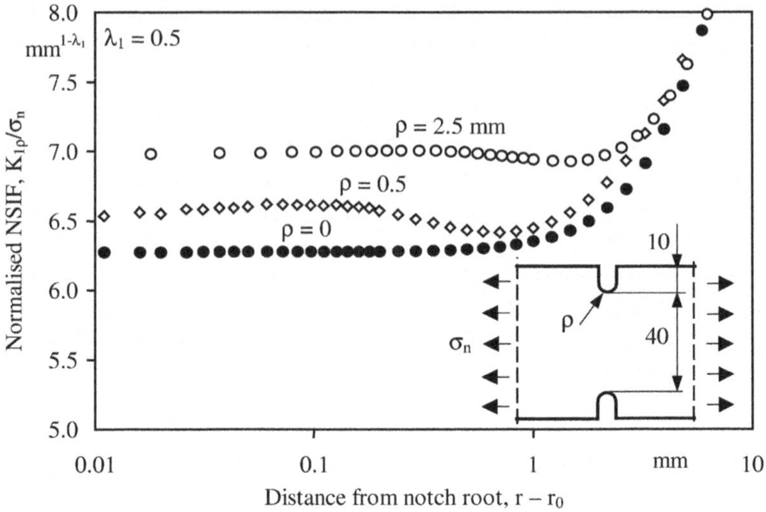

Fig. 2.33 Normalised NSIF $K_{1\rho}$ dependent on distance from U-notch root, $(r - r_0)$, Eq. (2.135) for $2\alpha = 0°$; evaluation of FE analysis results; (Lazzarin and Filippi 2006)

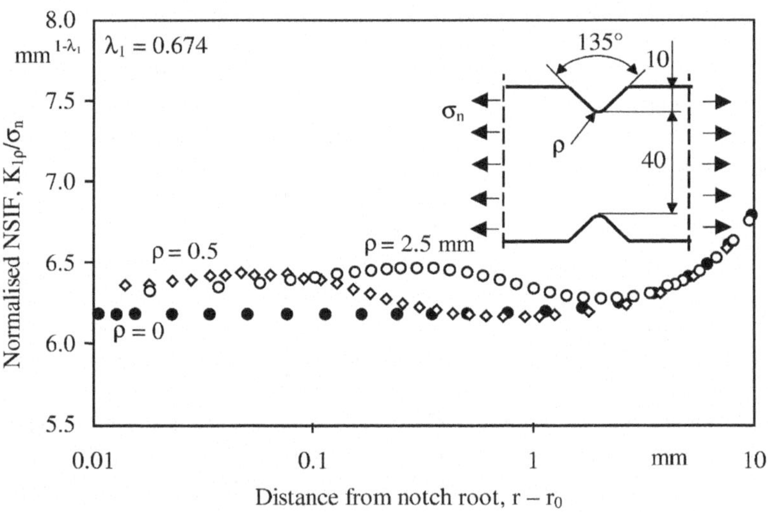

Fig. 2.34 Normalised NSIF $K_{1\rho}$ dependent on distance from V-notch root, $(r - r_0)$, Eq. (2.135) for $2\alpha = 135°$; evaluation of FE analysis results; (Lazzarin and Filippi 2006)

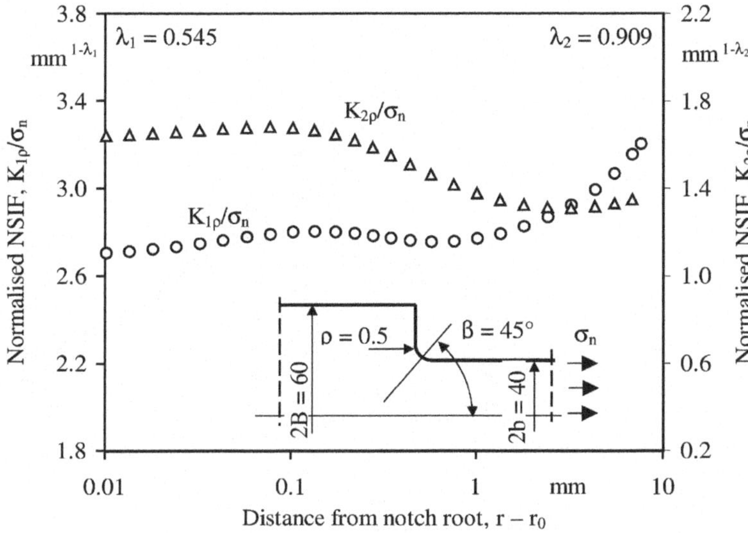

Fig. 2.35 Normalised NSIFs $K_{1\rho}$ and $K_{2\rho}$ dependent on distance from corner notch root, $(r - r_0)$, Eqs. (2.135) and (2.136) for $2\alpha = 90°$; evaluation of FE analysis results; (Lazzarin and Filippi 2006)

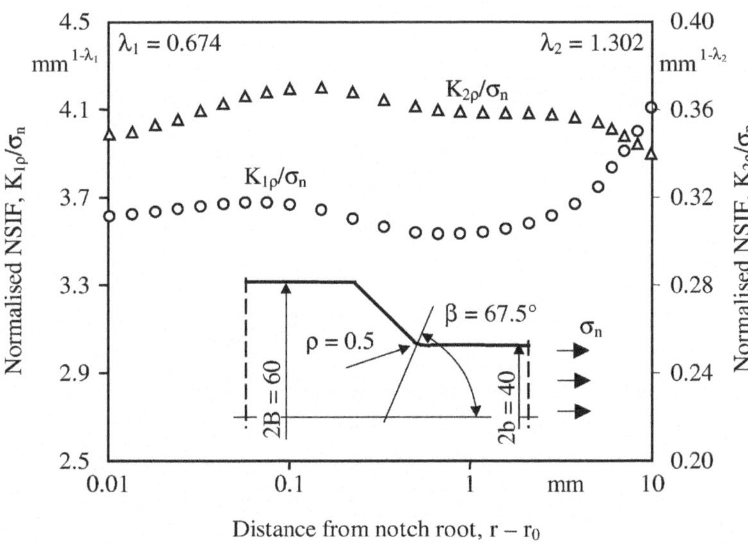

Fig. 2.36 Normalised NSIFs $K_{1\rho}$ and $K_{2\rho}$ dependent on distance from corner notch root, $(r - r_0)$, Eqs. (2.135) and (2.136) for $2\alpha = 135°$; evaluation of FE analysis results; (Lazzarin and Filippi 2006[1])

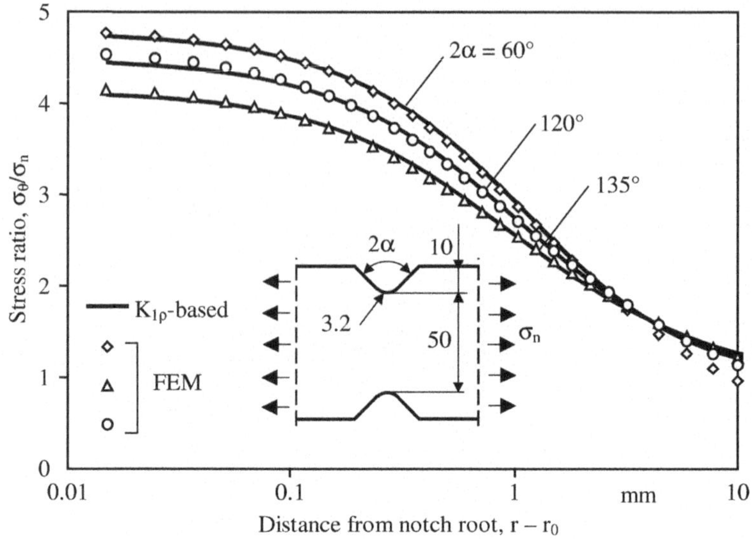

Fig. 2.37 Stress σ_θ in notch bisector ahead of the V-notch root, notch radius $\rho = 3.2$ mm, for three notch opening angles 2α; original compared with $K_{1\rho}$-based FE analysis results; $K_{1\rho}/K_1$ approximated according to Eq. (2.139) and Table 2.4; (Lazzarin and Filippi 2006)

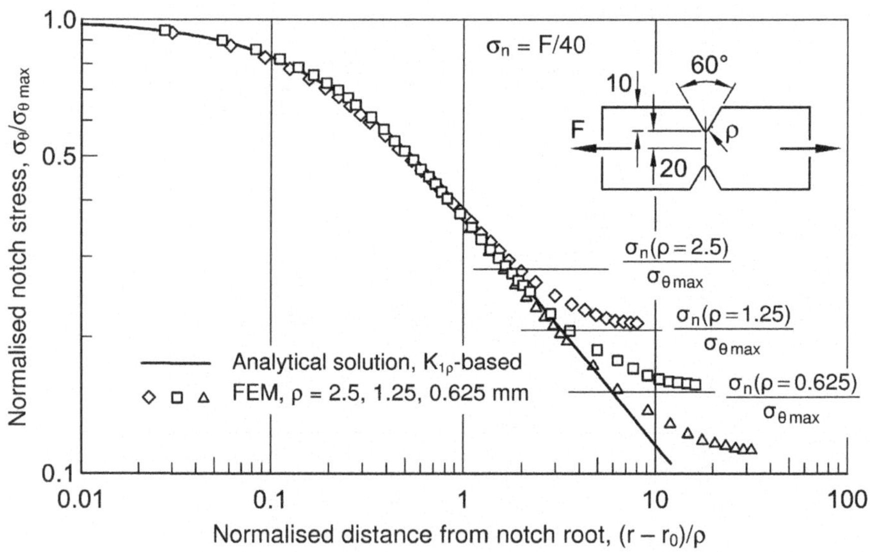

Fig. 2.38 Stress σ_θ in notch bisector over distance from V-notch root, $(r - r_0)$; direct compared with $K_{1\rho}$-based FE analysis results; $K_{1\rho}$ according to Eq. (2.139) with finite values of ρ; (Lazzarin and Tovo 1996)

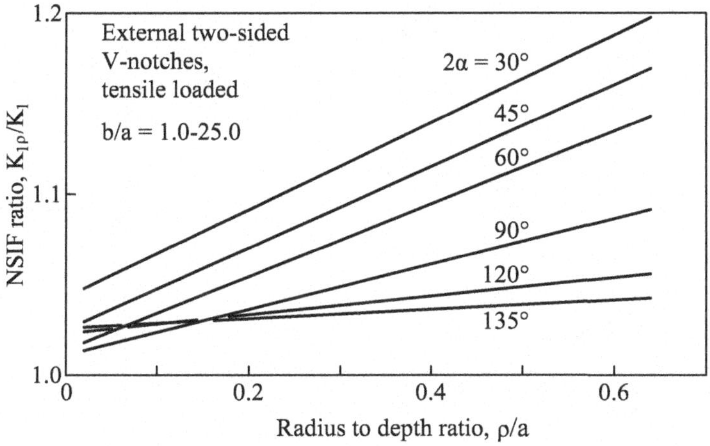

Fig. 2.39 NSIF ratio $K_{1\rho}/K_1$ dependent on the ratio of V-notch radius to V-notch depth, ρ/a, based on FE analysis results; various notch opening angles 2α; (Lazzarin and Filippi 2006)

Table 2.4 Parameter values for the approximation of $K_{1\rho}/K_1 = \varphi + \psi(\rho/a)$ for V-notched flat bar; different notch opening angles 2α; net cross-sectional width $2b$ to notch depth a ratios $2b/a = 2, 5, 10, 50$ (Lazzarin and Filippi 2006)

2α	30°	45°	60°	90°	120°	135°
φ	1.043	1.025	1.014	1.011	1.023	1.026
ψ	0.241	0.226	0.200	0.127	0.051	0.025

prescribed remote edge displacements u_0 and v_0 together with shear-compatible support conditions in the two geometrical symmetry planes, Fig. 2.40. The reference stress τ_0 is produced in the rectangular plate without the notch. It is smaller than the nominal stress in the gross cross-section of the plate with the internal notch.

At first, parabolic and hyperbolic notches are considered. The curves of $K_{2\rho}/\tau_0$ over distance from the notch root for the hyperbolic notch ($2\alpha = 60°$), Fig. 2.41, show that $K_{2\rho}$ is substantially enlarged even for small values of ρ, e.g. $\rho = 0.1$ mm. The corresponding shear stresses in the bisector are plotted in Fig. 2.42. The curves run asymptotically to the straight line of the pointed notch. The FE analysis results of the parabolic and hyperbolic notches have also been evaluated using $\overline{K}_{2\rho}$ according to Eq. (2.139). The parabolic notch results are set in comparison to $K_{2\rho}(\tau_{max})$ according to Eq. (2.128) with K_{II} substituted by $K_{2\rho}$ (Cheng 1988). Additionally, the maximum principal stresses at the notch edge corresponding to the maximum tangential stress are compared, Table 2.5. The correspondence is satisfactory.

The rise of $K_{1\rho}/K_2$ over the ratio of notch radius to notch depth, ρ/a, is shown in Fig. 2.43 for parabolic and hyperbolic notches. All the curves converge to $K_{2\rho}/K_2 = 1.0$ for $\rho/a = 0$, as demanded by Eq. (2.138). The rise is strongest for the parabolic notch.

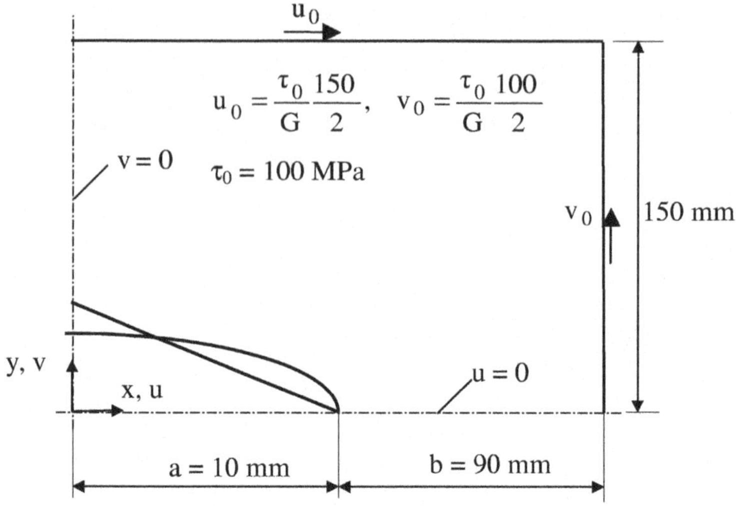

Fig. 2.40 One symmetry quarter of rectangular plate with parabolic or hyperbolic internal two-sided V-notch subjected to prescribed displacements at the remote plate edges producing in-plane shear loading conditions; same edge displacements applied to the plate with central crack or pointed V-notch of semidepth $a = 10$ mm; (Lazzarin et al. 2011)

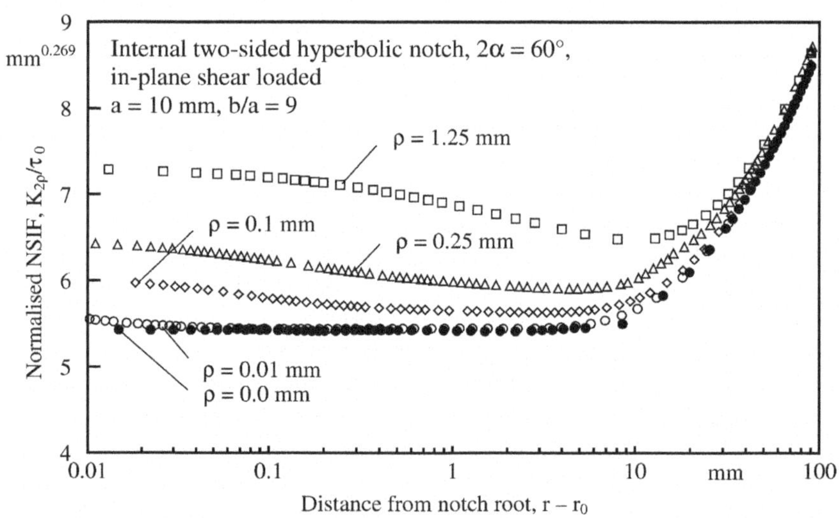

Fig. 2.41 Normalised NSIF $K_{2\rho}$ of internal two-sided hyperbolic notch ($2\alpha = 60°$, different notch radii) dependent on distance from notch root, $(r - r_0)$, $K_{2\rho}$ according to Eq. (2.136); evaluation of FE analysis results; (Lazzarin et al. 2011)

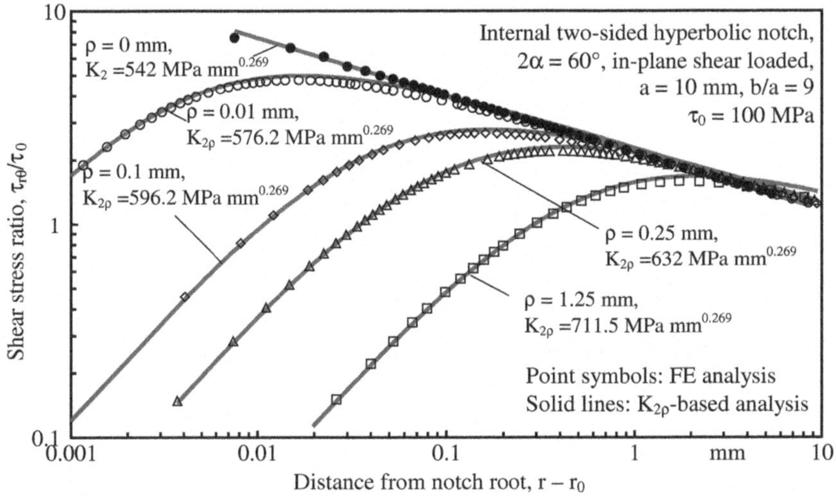

Fig. 2.42 Shear stress $\tau_{r\theta}$ in notch bisector over distance from notch root, $(r - r_0)$; internal two-sided hyperbolic notch ($2\alpha = 60°$, different notch radii), in-plane shear loaded; direct compared with $K_{2\rho}$-based FE analysis results; $K_{2\rho}$ according to Eq. (2.136) averaged by Eq. (2.139); (Lazzarin et al. 2011)

Table 2.5 Generalised NSIFs of two-sided parabolic notch ($\lambda_2 = 0.5$) in rectangular plate subjected to in-plane shear loading; $\overline{K}_{2\rho}$ averaged over 0.4ρ, Eq. (2.139), or based on $\tau_{\max}$, Eq. (2.128); additionally maximum principal (or tangential) stresses at notch edge; deviations $\Delta_K = (K_{2\rho,\tau}/\overline{K}_{2\rho} - 1) \times 100$ and $\Delta_\sigma = (\sigma_{1\,\max,\overline{K}}/\sigma_{1\,\max,\mathrm{FE}} - 1) \times 100$; (Lazzarin et al. 2011)

ρ [mm]	$\tau_{r\theta\ \max,\mathrm{FE}}$ [MPa]	$\overline{K}_{2\rho}$ [MPa mm$^{0.5}$]	$K_{2\rho,\tau}$ [MPa mm$^{0.5}$]	Δ_K [%]	$\sigma_{1\max,\mathrm{FE}}$ [MPa]	$\sigma_{1\max,\overline{K}}$ [MPa]	Δ_σ [%]
0.01	1306.1	596.0	601.5	0.92	3204.0	3211.0	0.22
0.10	465.8	677.0	678.2	0.18	1204.5	1207.9	0.28
0.50	249.3	809.6	811.8	0.27	639.2	646.0	1.05
1.25	190.3	977.7	979.8	0.21	477.8	493.4	3.26
2.50	159.5	1177.5	1161.3	−1.37	405.6	420.2	3.59
4.00	145.8	1376.2	1342.8	−2.43	370.6	388.2	4.75

At second, the U-notch and V-notches are analysed, using the $K_{2\rho}$ formula of root holes, Eq. (2.144). It was found that this formula is more accurate than the corresponding formula for hyperbolic (and parabolic) notches, Eq. (2.136), which shifts the position of $\tau_{r\theta\ \max}$ closer to the notch root. The curves $K_{2\rho}/\tau_0$ over distance $(r - r_0)$ from the notch root for the U-notch, Fig. 2.44, show that $K_{2\rho}$ is substantially enlarged only for larger values of ρ ($\rho/a \geq 0.1$). The corresponding shear stresses in the bisector are plotted in Fig. 2.45. The curves are identical up to the maximum stress and then deviate in approximation of the nominal stress in the net cross-section with the exception of the $K_{2\rho}$-based solid line which presumes an infinitely wide cross-section.

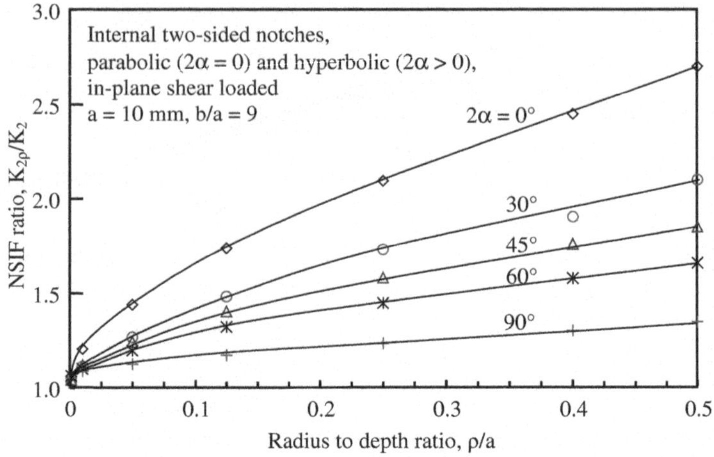

Fig. 2.43 Parabolic and hyperbolic notches: NSIF ratio $K_{2\rho}/K_2$ dependent on the ratio of notch radius to notch depth, ρ/a, based on FE analysis results; various notch opening angles 2α; (Lazzarin et al. 2011)

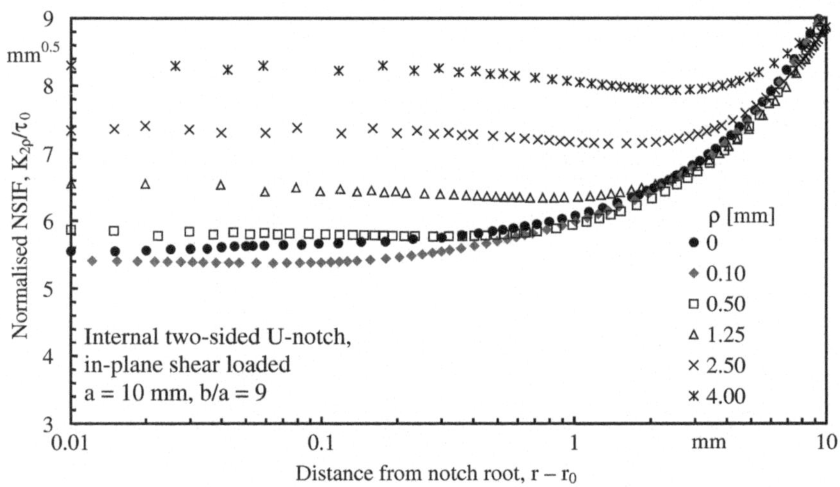

Fig. 2.44 Normalised NSIF $K_{2\rho}$ of internal two-sided U-notch (different notch radii) dependent on distance from notch root, $K_{2\rho}$ according to Eq. (2.144); evaluation of FE analysis results; (Lazzarin et al. 2011)

The rise of $K_{2\rho}/K_2$ over the ratio of notch radius to notch depth, ρ/a, is plotted in Fig. 2.46 for U-notch and V-notches. The deviation between actual and analytical notch shape has the effect that the limes condition, Eq. (2.138), is violated for larger values of 2α ($K_{2\rho}/K_2$ does not converge to 1.0 for $\rho \to 0$).

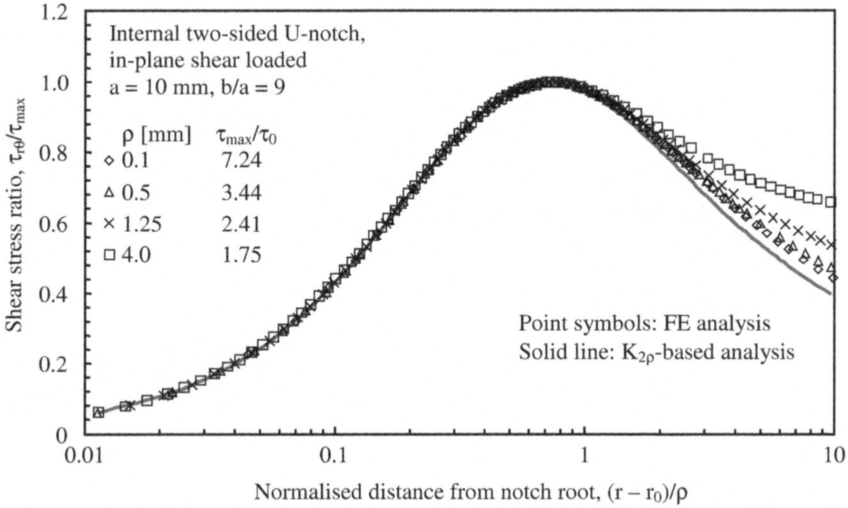

Fig. 2.45 Shear stress $\tau_{r\theta}$ in notch bisector over distance from notch root, $(r - r_0)$; internal two-sided U-notch (different notch radii), in-plane shear loaded; direct compared with $K_{2\rho}$-based FE analysis results; $K_{2\rho}$ according to Eq. (2.144); (Lazzarin et al. 2011)

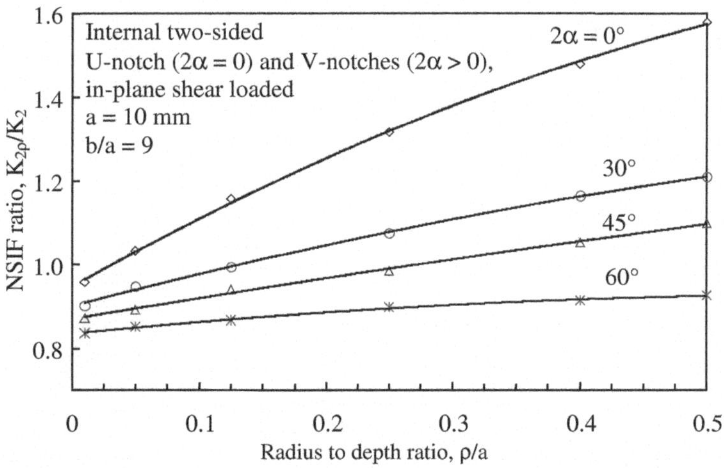

Fig. 2.46 U-notch and V-notches: NSIF ratio $K_{2\rho}/K_2$ dependent on the ratio of notch radius to notch depth, ρ/a, based on FE analysis results evaluated by Eq. (2.144); various notch opening angles 2α; (Lazzarin et al. 2011)

Finally, the notch stress curves for a quadratic plate specimen with one-sided V-notch, used for brittle fracture testing of brittle materials, are drawn in Fig. 2.47. The notch opening angle $2\alpha = 45°$ is fixed, the V-notch radius ρ is varied between 0 and 1 mm. The tensile forces F produce superimposed nominal tensile and

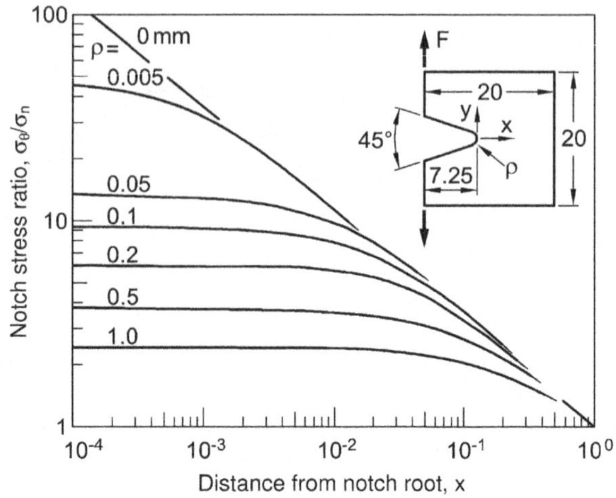

Fig. 2.47 Stress σ_θ in notch bisector over distance from notch root, x; one-sided rounded V-notches in comparison to pointed V-notch in square plate; notch opening angle $2\alpha = 45°$; (Nui et al. 1994)

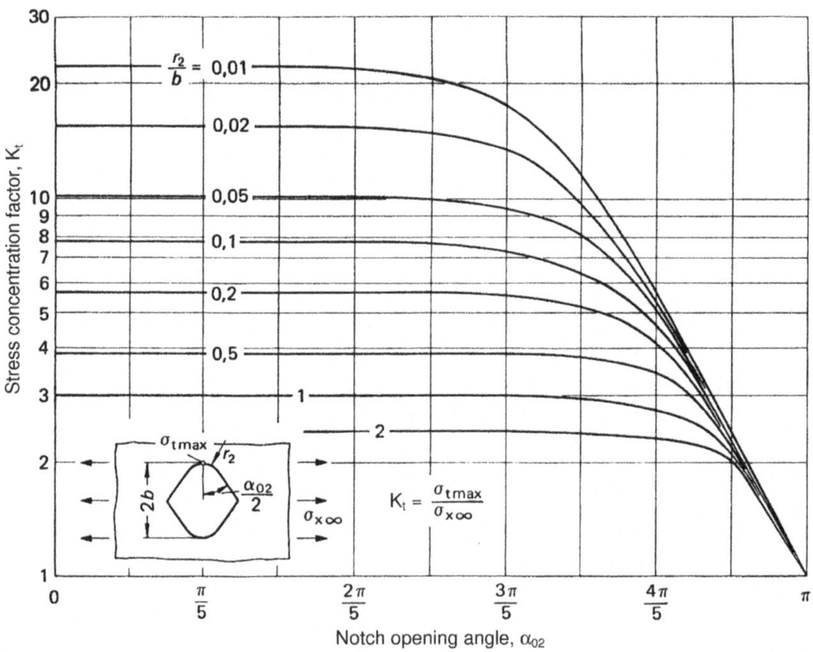

Fig. 2.48 SFCs of internal two-sided rounded V-notches ('rhombic cutouts') dependent on notch opening angle α_{02} for various ratios r_2/b of notch radius to notch depth; infinite plate subjected to remote tensile loading; results by functional analysis; (Radaj and Schilberth 1977)

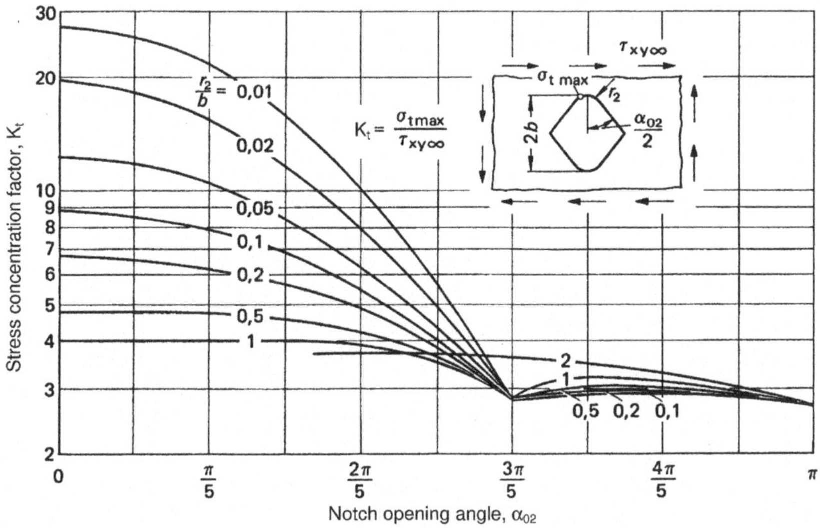

Fig. 2.49 SFCs of internal two-sided rounded V-notches ('rhombic cutouts') dependent on notch opening angle α_{02} for various ratios r_2/b of notch radius to notch depth; infinite plate subjected to remote in-plane shear loading; results by functional analysis; (Radaj and Schilberth 1977)

bending stresses in the net cross-section. The solution of the stress field problem has been achieved based on Kolosov–Muskhelishvili's complex stress function method combined with conformal mapping of the polygonal plate area onto the unit circle by applying the Schwartz–Christoffel transformation (Nui et al. 1994). For evaluating the notch stress at the pointed notch, Williams' singularity exponents are introduced. It is demonstrated that the maximum notch stresses in the close neighbourhood of the notch root, $(r - r_0)/\rho \leq 0.2$, depend predominantly on the notch radius, whereas outside this region, the notch opening angle is decisive, as given by the pointed notch solution. It is visible from the graph, that the rounded notch curves run slightly above the pointed notch curve, which indicates $K_{1\rho} > K_1$.

It is also shown more generally and in quantitative terms that the intermediate NSIF-governed stress field may be used to characterise the yielding or strength behaviour of sharply rounded notches (Nui et al. 1994; Dini and Hills 2004). This field is intermediate between the maximum notch stress governed near-field and the nominal stress governed far-field.

This subsection with the evaluation examples for the notch stresses at rounded V-notches is closed with two diagrams showing the SCFs of internal two-sided V-notches ('rhombic cutouts') in the infinite plate subjected to remote in-plane tensile and shear loading, Figs. 2.48 and 2.49; (Radaj and Schilberth 1977).

These SCFs have been determined very accurately based on complex analytical stress functions according to Kolosov–Muskhelishvili, combined with conformal mapping of the rhombic cutout with rounded V-notches according to Schwartz–Christoffel. A polygonal shape is considered, consisting of straight lines and

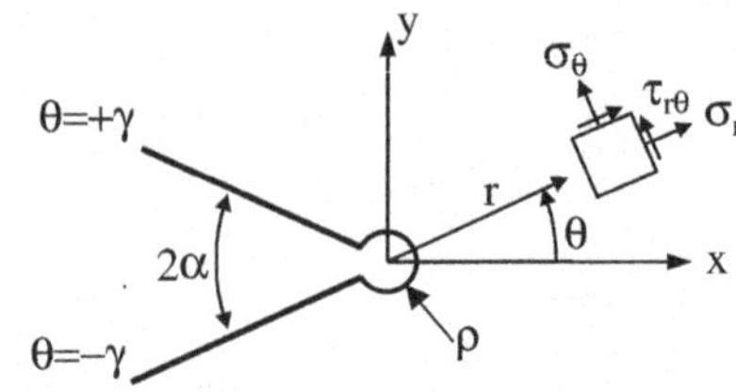

Fig. 2.50 V-notch with root hole, polar in-plane coordinate system with origin in centre of hole; (Zappalorto and Lazzarin 2011[(1)])

circular arcs. The circular arcs are approximated by multi-corner polygons. Their series expansion as a mapping function is truncated, so that the corners are smoothed. The NSIF concept is not applied.

The SCFs are primarily dependent on the ratio r_2/b of notch radius to notch depth. They may additionally depend on the notch opening angle α_{02}. In tensile loading, the latter dependency is marked only for $\alpha_{02} \geq 90°$. In shear loading, the SCFs decrease rapidly with increasing values of $\alpha_{02} > 0°$. They reach their maximum at $\alpha_{02} = 108°$. This angle corresponds to the limit value for singular stresses in mode 2 loading of pointed V-notches, $2\alpha = 102.6°$. Actually, the curve points were calculated in steps of $\Delta\alpha_{02} = 18°$, so that $\alpha_{02} = 108°$ is the curve point nearest to the no-singularity angle.

The NSIF concept needs the FE analysis as basis for evaluations in engineering applications. Parametric investigations, as just considered above, would necessitate variation of the geometrical parameters in the FE model before addressing the NSIF behaviour. To the author's knowledge, no such investigations have been performed up to now. Deviations from the SCF behaviour will predominantly be caused by a varying stress gradient normal to the notch root. This gradient is primarily dependent on the reciprocal value of ρ and only occasionally on the notch opening angle to a larger extent.

2.4.6 Stress Field at V-Notches with Root Hole

V-notches with root hole and the special case of the keyhole are rarely used in engineering designs. Here, blunt notches with a smooth transition from the rectilinear to the circular notch edge are preferred, because they produce lower notch stresses and are easier to produce. On the other hand, they are an important design within the theoretical frame of stress analysis. They allow an easier handling of solutions and they simulate the circular notch shape accurately, which is important in those cases where the maximum notch stress occurs outside the bisector. The advantage of their use in NSIF considerations is, that the variation of the notch radius does not affect the other geometrical parameters of the notch. This is not the case with elliptical, parabolic or hyperbolic notches.

First, tensile and in-plane shear loading is dealt with, then out-of-plane shear loading and finally the transverse (out-of-plane) shear loading effect produced by in-plane shear loading at load-free surfaces.

Table 2.6 Parameter values for evaluation of the stresses $\sigma_\theta(r, 0)$ normal to the bisector plane under mode 1 loading dependent on notch opening angle 2α; (Zappalorto and Lazzarin 2011[(1)])

2α	0°	30°	45°	60°	90°	120°	135°
λ_1	0.5000	0.5015	0.5050	0.5122	0.5445	0.6157	0.6736
φ_1	0.5000	0.5338	0.5770	0.6401	0.8388	1.1538	1.3556
g_1	2.0000	2.0352	2.0820	2.1523	2.3833	2.7696	3.0292
g_2	1.2500	1.2705	1.3006	1.3486	1.5116	1.7638	1.9030
g_3	1.5000	1.4942	1.4800	1.4514	1.3260	1.0639	0.8659
g_4	1.2500	1.3352	1.4453	1.6081	2.1342	3.0181	3.6243

For analysing the stress field under in-plane loading conditions for the V-notch with root hole shown in Fig. 2.50, the Kolosov–Muskhelishvili method is applied (Zappalorto and Lazzarin 2011[(1)]). The following analytical functions are used, with $\varphi(z)$ and $\psi(z)$ being interchanged in the article just mentioned (Muskhelishvili 1963):

$$\varphi(z) = az^{\lambda} + bz^{-\lambda} \tag{2.140}$$

$$\psi(z) = cz^{\lambda} + dz^{-\lambda} + ez^{-\lambda-1} + fz^{-\lambda-2} \tag{2.141}$$

where λ is real and positive whereas the coefficients a, b, c, d, e and f are complex.

The eigenvalues λ_1 and λ_2 result from the boundary condition of load-free notch flanks, as before. Further boundary conditions can be stated on the circular edge of the root hole for determining the free coefficients:

$$(\sigma_r)_{r=\rho} = (\tau_{r\theta})_{r=\rho} = 0 \tag{2.142}$$

The resulting stress field is separated into a symmetric (mode 1) and antimetric (mode 2) part. In the following, only the stress σ_θ (mode 1) or $\tau_{r\theta}$ (mode 2) in the bisector as well as the stress σ_θ at the hole edge are recorded and compared with FE analysis results for external or internal two-sided V-notches in a rectangular plate specimen. The nominal notch depth is $a = 10$ mm, the nominal semiwidth of the net cross-section is $b = 9a$. Taking the root hole into account, the actual notch depth is $(a + \rho)$ and the actual semiwidth $(b - \rho)$.

In the case of mode 1 loading, the stresses $\sigma_\theta(r, 0)$ normal to the bisector plane have the following form:

$$\sigma_\theta(r, 0) = \frac{K_{1\rho}}{\sqrt{2\pi}} \frac{r^{\lambda_1 - 1}}{(1 + \lambda_1) + \phi_1(\gamma)} \left[g_1 + g_2 \left(\frac{\rho}{r}\right)^{2\lambda_1} + g_3 \left(\frac{\rho}{r}\right)^{2\lambda_1 + 1} + g_4 \left(\frac{\rho}{r}\right)^{2\lambda_1 + 2} \right] \tag{2.143}$$

with the coefficients g_1 to g_4 and ϕ_1 depending on the notch opening angle 2α (or $2\gamma = 2\pi - 2\alpha$) according to Table 2.6. The NSIF $K_{1\rho}$ may be determined by evaluation of $\sigma_\theta(r, 0)$ from a FE analysis of the considered V-notch with root hole, solving Eq. (2.143) for $K_{1\rho}$ dependent on the distance from the notch root.

Table 2.7 Parameter values for evaluation of the stresses $\tau_{r\theta}(r, 0)$ in the bisector plane under mode 2 loading dependent on notch opening angle 2α; (Zappalorto and Lazzarin 2011[(1)])

2α	0°	30°	45°	60°	90°	120°	135°
λ_2	0.5000	0.5982	0.6597	0.7309	0.9085	1.1489	1.3021
h_1	1.6250	1.6422	1.6639	1.6941	1.7782	1.8873	1.9451
h_2	−0.7500	−0.6011	−0.5393	−0.4849	−0.3921	−0.3079	−0.2614
h_3	−1.8750	−2.0411	−2.1246	−2.2092	−2.3861	−2.5794	−2.6837

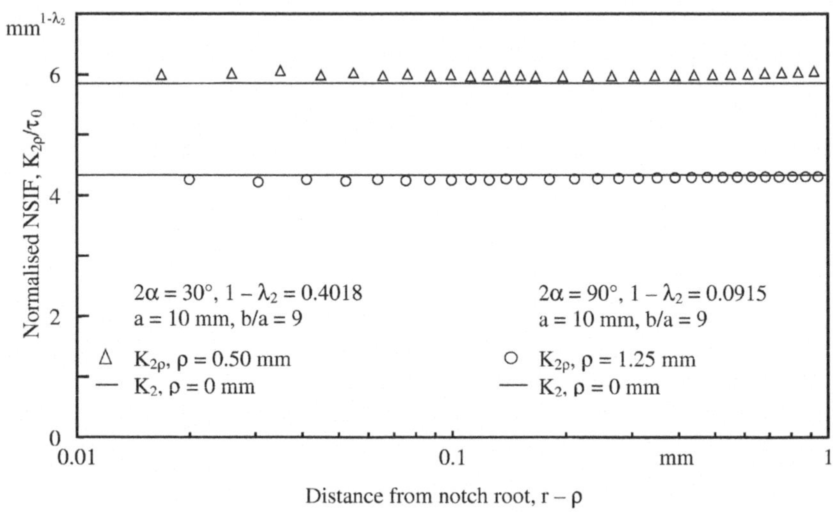

Fig. 2.51 V-notches with root hole: NSIFs $K_{2\rho}$ and K_2 dependent on distance from notch root; evaluation based on Eq. (2.144); (Lazzarin et al. 2011)

In the case of mode 2 loading, the stresses $\tau_{r\theta}(r, 0)$ in the bisector plane are given as follows:

$$\tau_{r\theta}(r, 0) = \frac{K_{2\rho}}{\sqrt{2\pi}} r^{\lambda_2 - 1} \left[1 + h_1 \left(\frac{\rho}{r} \right)^{2\lambda_2} + h_2 \left(\frac{\rho}{r} \right)^{2\lambda_2 + 1} + h_3 \left(\frac{\rho}{r} \right)^{2\lambda_2 + 2} \right] \tag{2.144}$$

with the coefficients h_1 to h_3 depending on the notch opening angle 2α according to Table 2.7. The NSIF $K_{2\rho}$ may be determined by evaluation of $\tau_{r\theta}(r, 0)$ from a FE analysis of the considered V-notch with root hole, solving Eq. (2.144) for $K_{2\rho}$ dependent on the distance from the notch root. Such an evaluation is shown in Fig. 2.51. Identical values of $K_{2\rho}$ and K_2 are found, evaluating Eq. (2.144) for $\rho > 0$ in comparison to $\rho = 0$. The values are constant over a distance from the notch root, where, with the evaluations for blunt V-notches hitherto considered, substantial variations and oscillations occur.

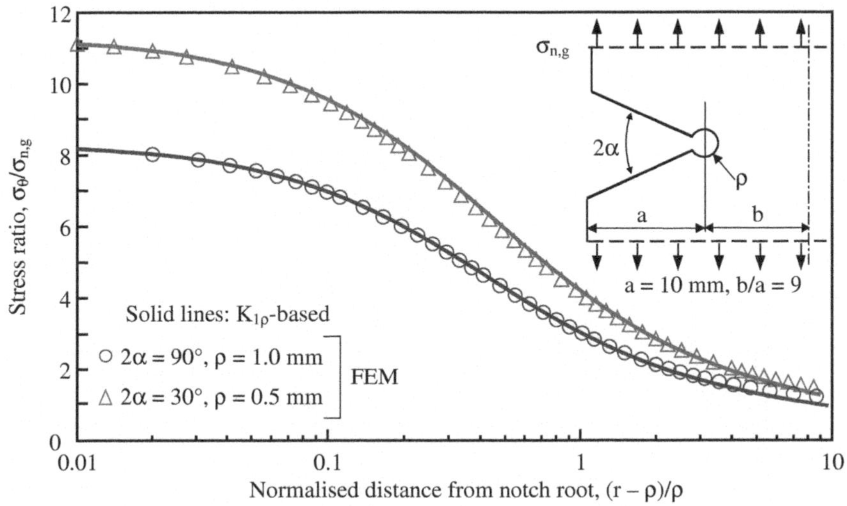

Fig. 2.52 Tensile loaded V-notch with root hole: stress σ_θ in notch bisector ahead of the notch root; two notch opening angles 2α combined with different notch radii ρ; direct compared with $K_{1\rho}$-based FE analysis results; (Zappalorto and Lazzarin 2011[(1)])

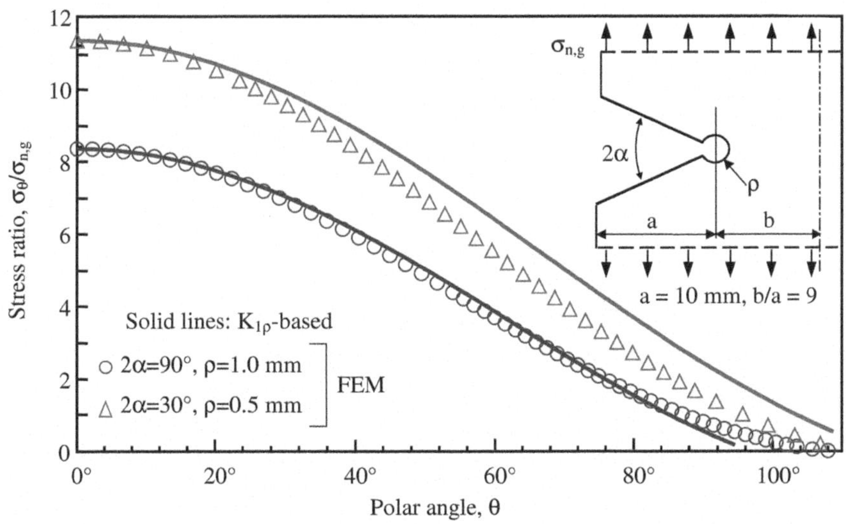

Fig. 2.53 Tensile loaded V-notch with root hole: stress σ_θ at hole edge; two notch opening angles 2α combined with different notch radii ρ; direct compared with $K_{1\rho}$-based FE analysis results; (Zappalorto and Lazzarin 2011[(1)])

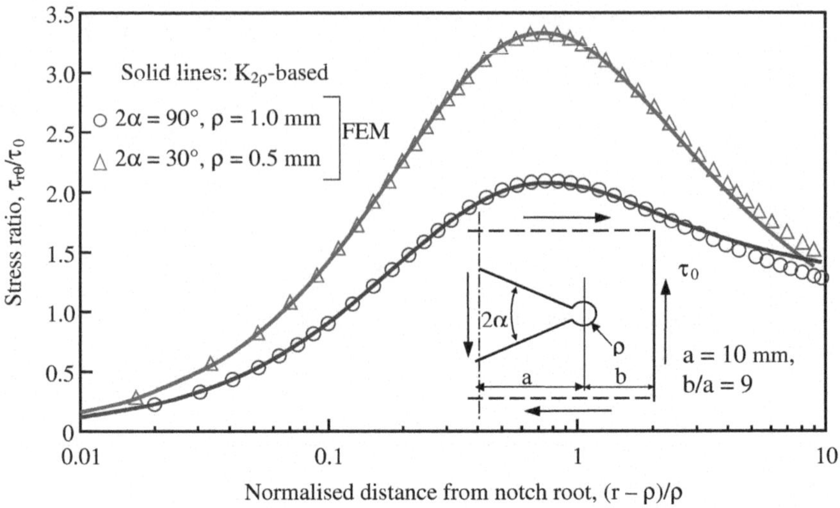

Fig. 2.54 In-plane shear-loaded V-notch with root hole: stress $\tau_{r\theta}$ in notch bisector ahead of the notch root; two notch opening angles 2α combined with different notch radii ρ; direct compared with $K_{2\rho}$-based FE analysis results; (Zappalorto and Lazzarin 2011$^{(1)}$)

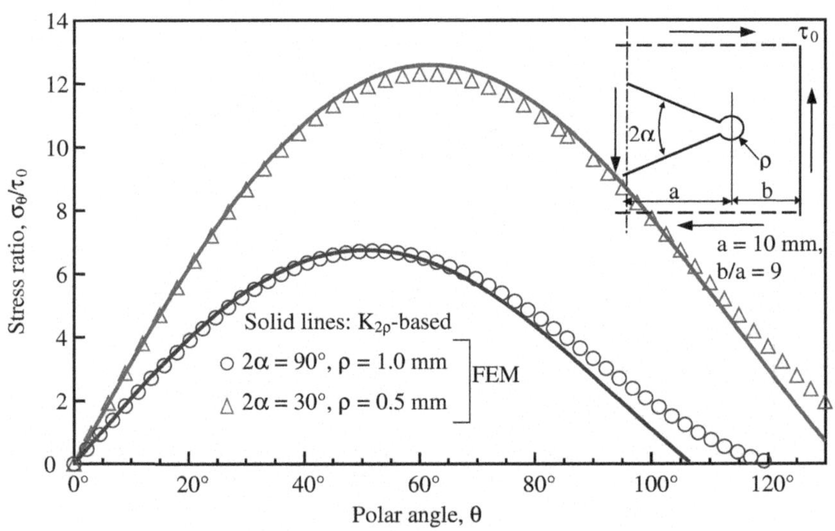

Fig. 2.55 In-plane shear-loaded V-notch with root hole: stress σ_θ at hole edge; two notch opening angles 2α combined with different notch radii ρ; direct compared with $K_{2\rho}$-based FE analysis results; (Zappalorto and Lazzarin 2011$^{(1)}$)

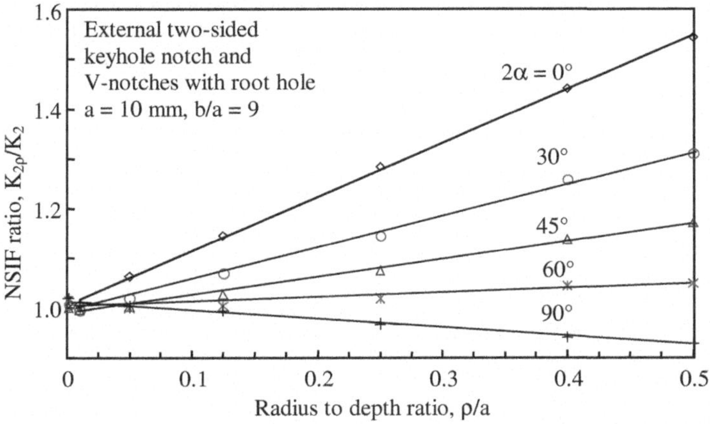

Fig. 2.56 V-notches with root hole: NSIF ratio $K_{2\rho}/K_2$ dependent on the ratio ρ/a of notch radius to notch depth, based on FE analysis results; various notch opening angles 2α; (Lazzarin et al. 2011[(1)])

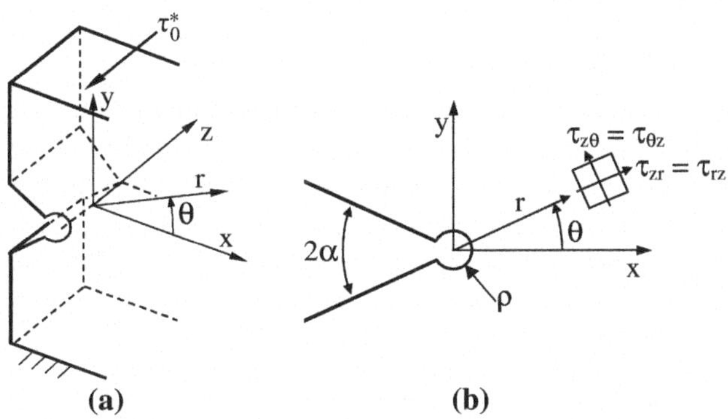

Fig. 2.57 V-notch with root hole subjected to out-of-plane shear loading (only the antiplane shear load component τ_0^* is shown); 3d view (**a**) and cross-sectional view (**b**); polar coordinate system with its origin in the centre of the hole; out-of-plane shear stress components; (Zappalorto and Lazzarin 2011[(1)])

For $\rho = 0$ (pointed notches), $K_{1\rho} \to K_1$ and $K_{2\rho} \to K_2$, as demanded by Eq. (1.138). The Eqs. (2.143) and (2.144) in the case of the keyhole ($2\alpha = 0$) are identical with those given in the literature (Kullmer 1992; Radaj et al. 2001).

The stresses σ_θ in mode 1 loading evaluated from the FE model mentioned above (converted to external notches) in the bisector plane ($\theta = 0$) and at the hole edge ($r = \rho$) are shown in Figs. 2.52 and 2.53. Direct FEM evaluation results are compared with the analytical stress distributions based on identical values of σ_{max} or $K_{1\rho}$. Some deviations are visible in the edge stress diagram, indicating the approximative character of the analytical solution.

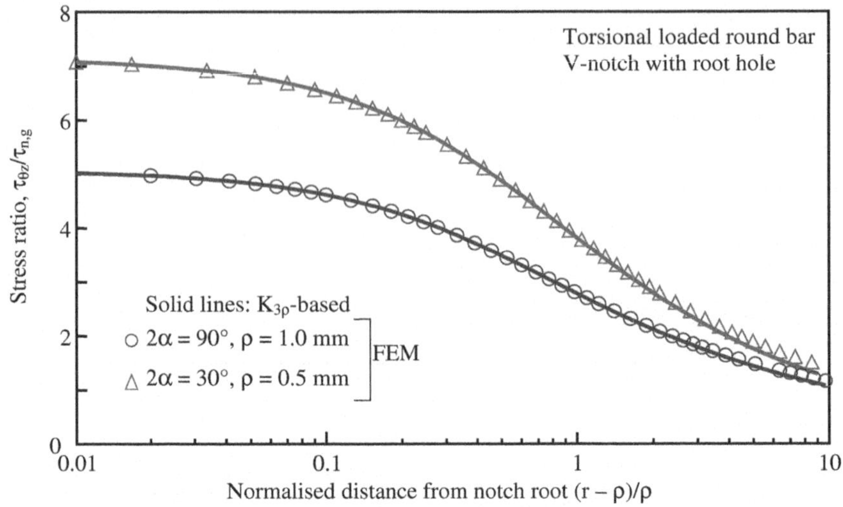

Fig. 2.58 Torsional-loaded round bar (mode 3), circumferential V-notch with root hole: shear stress $\tau_{\theta z}$ in notch bisector ahead of the notch root; two notch opening angles 2α combined with different notch radii ρ; original compared with $K_{3\rho}$-based FE analysis results; (Zappalorto and Lazzarin 2011[(1)])

The stresses $\tau_{r\theta}$ in mode 2 loading from the FE model mentioned above (with internal notches) in the bisector plane ($\theta = 0$) are plotted in Fig. 2.54. The stresses σ_θ at the hole edge ($r = \rho$) are compared in Fig. 2.55. Once more, some minor deviations occur for large values of θ.

The rise of $K_{2\rho}/K_2$ with the ratio of notch radius to notch depth, ρ/a, is shown in Fig. 2.56 for V-notches with root hole characterised by different values of the notch opening angle, 2α. All curves are linear and converge to $K_{2\rho}/K_2 = 1.0$ for $\rho/a = 0$, as demanded by Eq. (2.138). The rise is strongest for $2\alpha = 0°$ (keyhole). It is inverted to a decrease for $2\alpha \geq 70°$.

For analysing the stress field under out-of-plane shear loading conditions for the V-notch with root hole, Fig. 2.57, the following analytic function has been used (Zappalorto and Lazzarin 2011[(1)]; Zappalorto et al. 2010):

$$H(z) = \mathrm{i}az^{\lambda} + \mathrm{i}bz^{-\lambda} \tag{2.145}$$

where λ, a and b are real coefficients. Only the antimetric stress field associated with mode 3 loading is considered, compare Eq. (2.122).

The following shear stress in the bisector plane ($\theta = 0$) is derived:

$$\tau_{\theta z} = \frac{K_{3\rho}}{\sqrt{2\pi}} r^{\lambda_3 - 1} \left[1 + \left(\frac{\rho}{r} \right)^{2\lambda_3} \right] \tag{2.146}$$

In the case of torsional loading of a round bar with circumferential V-notch with root hole, the linear increase of the nominal shear stress over the radial

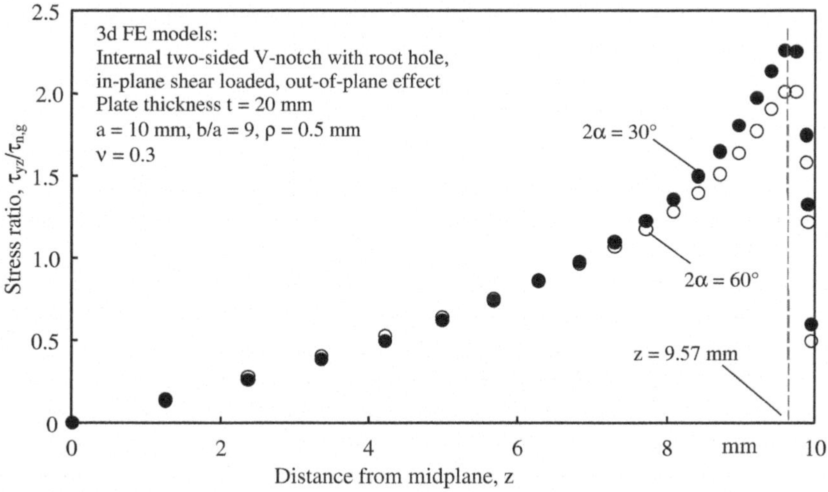

Fig. 2.59 Transverse (out-of-plane) loading effect produced by in-plane shear loading of V-notch with root hole in plate of finite thickness; out-of-plane shear stress plotted over thickness direction; FE analysis results; (Lazzarin et al. 2011)

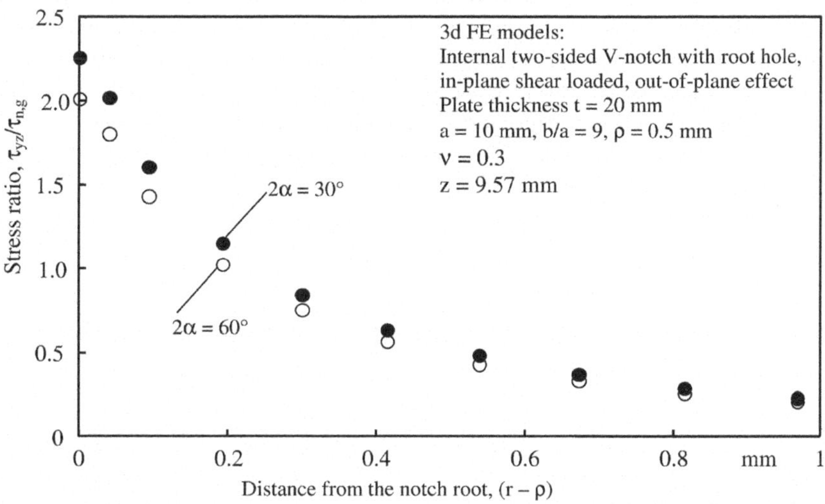

Fig. 2.60 Transverse (out-of-plane) loading effect produced by in-plane shear loading of V-notch with root hole in plate of finite thickness; decrease of maximum shear stress at $z = 9.57$ mm plotted over bisector direction; FE analysis results; (Lazzarin et al. 2011)

distance from the centre of the bar must be taken into account. This can be done by the following approximation (Zappalorto et al. 2008):

$$\tau_{\theta z} = \frac{K_{3\rho}}{\sqrt{2\pi}} r^{\lambda_3 - 1} \left[1 + \left(\frac{\rho}{r}\right)^{2\lambda_3}\right] \left(1 - \frac{r-\rho}{R_n}\right) \tag{2.147}$$

where the radius R_n denotes the net cross-section of the round bar at the notch root. An evaluation example based on FE modelling with axisymmetric elements is shown in Fig. 2.58.

Finally, the transverse (out-of-plane) shear loading effect produced by in-plane shear loading at load-free surfaces is considered. The transverse singular effects in mode 2 loaded pointed V-notches have been described in Sect. 2.3.3 with the relevant literature being quoted. Here, the corresponding transverse non-singular effect associated with rounded V-notches is considered.

The effect can only be detected using three-dimensional FE models. It produces an out-of-plane shear stress which has its maximum in the close vicinity of the plate surfaces. The effect is documented by Figs. 2.59 and 2.60 (Lazzarin et al. 2011). The rectangular plate specimen shown in Fig. 2.40 is analysed, with a plate thickness of 20 mm, containing a two-sided V-notch with root hole ($\rho = 0.5$ mm). The semiwidth of the plate is 100 mm, the notch semidepth $a = 10$ mm and the notch opening angle $2\alpha = 30°$ or $60°$. The plane model hitherto used is simply extended into the thickness direction z. The out-of-plane shear stress τ_{yz}, plotted over the distance z from the midplane of the plate specimen has its maximum at short distance from the plate surface and is zero in the midplane ($z = 0$) and at the two plate surfaces ($z = \pm 10$ mm), Fig. 2.59. The maximum out-of-plane shear stress is in the range of the original maximum in-plane shear stress (Fig. 2.54), but it is substantially smaller than the original maximum tangential stress at the hole edge (Fig. 2.55). The decrease of the out-of-plane stresses (maximum at $z = 9.57$ mm) with the distance $(r - \rho)$ from the notch root is shown in Fig. 2.60.

2.4.7 *Generalised NSIFs of Parabolic and Keyhole Notches in Comparison*

The stress field levels in the vicinity of pointed slit tips ('near field'), where the stresses are singular, are described by the SIFs K_I, K_{II}, K_{III}, which are related to the loading modes I, II, III:

$$K_I = \sigma_n \sqrt{\pi a}\, Y_I \tag{2.148}$$

$$K_{II} = \tau_n \sqrt{\pi a}\, Y_{II} \tag{2.149}$$

$$K_{III} = \tau_n^* \sqrt{\pi a}\, Y_{III} \tag{2.150}$$

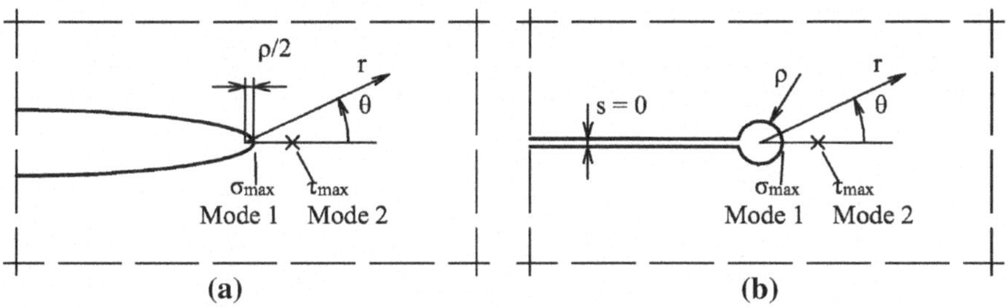

Fig. 2.61 Slender parabolic notch (**a**) and keyhole (**b**) with polar coordinate systems and indication of the position of σ_{max} in mode 1 loading and τ_{max} in mode 2 loading; (Radaj 2010)

with σ_n, τ_n, τ_n^* being the nominal or basic normal and shear stresses at the slit tip, with a being the slit length or semilength and Y_I, Y_{II}, Y_{III} being geometry factors describing the influence of finite boundaries by dimensional ratios of geometry and load position.

The SIFs quantify the first (singular) terms of a series expansion of the stress field for radial distances $r/a \ll 1.0$. The second order terms, T-stress and S-stress, may be superimposed. Any mixed mode loading state at the slit tip can be evaluated in respect of K_I, K_{II}, K_{III}, T and S, primarily by numerical methods (finite element or boundary element). Compilations of K_I, K_{II}, K_{III} formulae for basic geometric and loading cases gained by function-analytical and other methods are available (Murakami 1987). A compendium of T-stress solutions has also been published (Sherry et al. 1995).

In the following, the NSIFs $K_{1\rho}$ and $K_{2\rho}$ of slender parabolic and keyhole notches in comparison are considered, Fig. 2.61, supplemented by $K_{3\rho}$. These NSIFs describe the stress field level in the vicinity of the rounded slit ends. The maximum notch stresses σ_{max} (mode 1) and τ_{max} (mode 2) occur in the bisector plane, the former at the notch root, the latter ahead of the notch root. The maximum stress τ_{max}^* (mode 3) occurs also at the notch root. The three loading modes and their characteristic maximum stresses appear as decoupled under mixed mode loading conditions. The influence of the slit-parallel basic stress $\sigma_{\parallel}$ (the T-stress effect) is neglected below.

The slit, considered as the limit case $\rho \to 0$ of a slender parabolic notch or keyhole notch, provides the means to establish the relationships between the SIFs K_I, K_{II}, K_{III} and the maximum stresses σ_{max}, τ_{max}, τ_{max}^* for $\rho \to 0$ first, and then the relationships between the NSIFs $K_{1\rho}$, $K_{2\rho}$, $K_{3\rho}$ and σ_{max}, τ_{max}, τ_{max}^* for finite values of ρ. These relationships depend on the notch shape (elliptic, parabolic, keyhole or U-shaped). Once more, the relationships for mode 1, mode 2 and mode 3 are decoupled by considering σ_{max}, τ_{max} and τ_{max}^* in the bisector plane. The resulting 'defining equations' for $K_{1\rho}$, $K_{2\rho}$ and $K_{3\rho}$ are given below. The bisector direction is associated with $\theta = 0°$, and the radial distance r is measured from the origin of the polar coordinate system, whose distance from the notch root is $r_0 = \rho/2$ for the parabolic notch and $r_0 = \rho$ for the keyhole.

Converting the limit value equations for K_{I}, K_{II}, K_{III}, Eqs. (2.127), (2.128), (2.130), to the equations for $K_{1\rho}$, $K_{2\rho}$, $K_{3\rho}$ (Lazzarin et al. 2009), the following relationships are derived for slender parabolic notches:

$$K_{1\rho} = \frac{\sqrt{\pi\rho}}{2}\sigma_{\max} \quad (\theta = 0,\ r = 0.5\rho,\ T = 0) \tag{2.151}$$

$$K_{2\rho} = \frac{\sqrt{\pi\rho}}{0.3849}\tau_{\max} \quad (\theta = 0,\ r = 1.5\rho) \tag{2.152}$$

$$K_{3\rho} = \sqrt{\pi\rho}\,\tau^{*}_{\max} \quad (\theta = 0,\ r = 0.5\rho) \tag{2.153}$$

The corresponding relationships for the keyhole notch in Lazzarin's terminology are the following (Kullmer 1992; Radaj et al. 2001; Smith 2006):

$$K_{1\rho} = \frac{\sqrt{\pi\rho}}{2.1213}\sigma_{\max} \quad (\theta = 0,\ r = \rho,\ T = 0) \tag{2.154}$$

$$K_{2\rho} = \frac{\sqrt{\pi\rho}}{0.7132}\tau_{\max} \quad (\theta = 0,\ r = 1.723\rho) \tag{2.155}$$

$$K_{3\rho} = \frac{\sqrt{\pi\rho}}{1.4142}\tau^{*}_{\max} \quad (\theta = 0,\ r = \rho) \tag{2.156}$$

Obviously, $K_{1\rho}$, $K_{2\rho}$ and $K_{3\rho}$ are smaller for the keyhole notch than for the parabolic notch when $\sigma_{\max}$, $\tau_{\max}$ or $\tau^{*}_{\max}$ have the same value but using the same remote loading stresses results in identical NSIF values for the two notch shapes.

Some general questions arise in respect of a potential use of the NSIFs in engineering applications where structural boundaries may occur close to the considered notches, e.g. when applying the reference notch concept to welded joint models. Analytical expressions for the NSIFs are difficult to derive in such cases. Also, compiled data are not available. Therefore, the NSIF must be determined individually, using the FE or boundary element method and evaluating $\sigma_{\max}$, $\tau_{\max}$ or $\tau^{*}_{\max}$ for the preset notch radius ρ. Provided the notch shape in the model and in the derivation of the NSIF formula is identical (the normal choice), the evaluated NSIF value will be unambiguous and accurate. This means that the notch shape should be chosen in correspondence to the actual modelling requirements. Mostly, a keyhole or a U-shaped notch will be appropriate. In any case, the applied notch shape should be specified when presenting NSIF values. Only then, these can be used to reconstruct the notch stress field from given $K_{1\rho}$, $K_{2\rho}$, $K_{3\rho}$ and ρ values on the basis of the resulting $\sigma_{\max}$, $\tau_{\max}$ and $\tau^{*}_{\max}$ values. In the case of superimposed mode 1 and mode 2 stress fields, the resulting $\sigma_{\theta\ \max}$ value at the notch edge outside the bisector plane can easily be determined. In the case of a non-zero slit-parallel basic stress $\sigma_{\parallel}$, the appertaining notch stress field can additionally be superimposed, so that the stress field near the rounded slit tip is completely and accurately determined.

The necessary theoretical limit conditions already mentioned, Eq. (2.138), are the following:

$$K_{\mathrm{I}} = \lim_{\rho \to 0} K_{1\rho} \quad (\sigma_{\|} = 0) \tag{2.157}$$

$$K_{\mathrm{II}} = \lim_{\rho \to 0} K_{2\rho} \tag{2.158}$$

$$K_{\mathrm{III}} = \lim_{\rho \to 0} K_{3\rho} \tag{2.159}$$

The above considerations show that the values of $K_{1\rho}$, $K_{2\rho}$, $K_{3\rho}$ can be approximated by the values of K_{I}, K_{II}, K_{III} in the case of a microkeyhole or a slit-like U-shaped notch. The designation 'micro' means that the keyhole radius ρ is substantially smaller than the structural dimensions which control the basic stresses in the slit tip area so that major cross-sectional weakening effects are avoided. Inversely, the K_{I} and K_{II} values may be determined by the limit value procedure expressed by Eqs. (2.157–2.159). This was the original usage in the historical development. In strength assessments by engineers, the micronotch is applied as a means to quantify the near-field stress level. Because cross-sectional weakening is not at all occurring, the SIFs K_{I}, K_{II}, K_{III} are better suited for strength assessments than the NSIFs $K_{1\rho}$, $K_{2\rho}$, $K_{3\rho}$.

2.4.8 The T-Stress Effect in Rounded Notch Configurations

As already stated, the SIFs, NSIFs and generalised NSIFs are comprehensive parameters characterising not only the maximum stress at a point, but the whole stress field in the vicinity of the crack tip, notch tip or notch root. They are supplemented by the non-singular stress field components, described by the *T*-stress and *S*-stress where appropriate, Eqs. (2.13), (2.15), (2.73). A necessary condition on all these parameters is that they are approximately constant in the considered near field of the notch. This constancy has been proven by the diagrams showing the simple or generalised NSIF values over the distance from the notch tip or notch root. Higher order stress field terms may be added in unconventional cases such as application to thin-sheet lap joints.

In summary, there is no problem in using these parameters for defining the stress fields near pointed or sharply rounded notches as far as these can be approximated and superimposed by the considered singular and non-singular stress fields. Substantially three-dimensional configurations must be excluded.

The situation may become complicated when determining the SIF parameters from FE analysis models in cases of application to specimens and structural members. Here, the correct separation of the parameters is important: the separation of the mode 1 and mode 2 related constituents and the separation of the equivalent to the *T*-stress in rounded notch configurations.

The separation of mode 1 and mode 2 related constituents is not possible when evaluating the out-of-bisector maximum notch stresses, whereas the separation is no problem when evaluating the in-bisector stresses σ_θ and $\tau_{r\theta}$, as shown in the preceding subsections. The separation of the *T*-stress equivalent in rounded notch configurations is dealt with below, followed by a glance at the corresponding *S*-stress problem.

The *T*-stress is defined for pointed crack or slit tips ($2\alpha = 0$) subjected to in-plane loading. The term 'slit' (comprising the crack) is preferred in the following paragraphs of this subsection. The eigenvalue-based series expansion of the in-plane stress field at the slit tip, Eq. (2.9), is truncated in general after the second term. The resulting stress field equation can be split into an even and an odd part in respect of the polar angle θ, the even part connected with the parameters K_{I} and T, the odd part with the parameter K_{II} (Sneddon 1946; Irwin 1957). The first order term defines the stress singularity at the pointed slit tip ($r = 0$), resulting in the SIFs K_{I} and K_{II}. The second order term defines a uniform slit-parallel tensile or compressive stress named '*T*-stress' (Larsson and Carlsson 1973). These parameters control the stress field near to the slit tip, the 'near field', now written with the usual nomenclature, compare Eqs. (2.10) and (2.11):

$$\sigma_{\mathrm{I,ij}}^{(1,2)} = \frac{K_I}{\sqrt{2\pi r}} f_{\mathrm{I,ij}}^{(1)}(\theta) + T f_{\mathrm{T,ij}}^{(2)}(\theta) \quad (\mathrm{i,j} = r,\, \theta) \tag{2.160}$$

$$\sigma_{\mathrm{II,ij}}^{(1,2)} = \frac{K_{\mathrm{II}}}{\sqrt{2\pi r}} f_{\mathrm{II,ij}}^{(1)}(\theta) \quad (\mathrm{i,j} = r,\, \theta) \tag{2.161}$$

The *T*-stress may be determined from analytical solutions after series expansion of the stress field. In the case of FE modelling, it is usually determined from the medium stress in the two flank sides of the slit, evaluating the midsection values in case of internal slits (Williams 1957). Also, it may result from the difference in the stresses σ_r and σ_θ acting in the ligament close to the slit tip, Eq. (2.19); (Lazzarin et al. 2009).

When the pointed slit tip with occurrence of a *T*-stress is substituted by a rounded slit notch (e.g. a keyhole), the *T*-stress may deteriorate the usual evaluation procedures for $K_{1\rho}$. The reason is, that the $K_{1\rho}$ related notch stresses are superimposed by the *T*-stress related notch stresses. Evaluations of K_{I} and $K_{1\rho}$ based on the total notch stresses may thus be affected, Eqs. (2.127) and (2.129). If the *T*-stress effect is included in $K_{1\rho}$, the limit value relationship, Eq. (2.138), is not applicable, because non-singular stress field components are now wrongly allotted to the singular stress parameter. It is desirable, to separate the *T*-stress effect in order to ensure the correct application of the limit value relationship (Radaj 2010).

A problem of language is first addressed. Just as K_{I} and K_{II} do not occur at rounded slit tips, the same is true for the *T*-stress. The in-bisector slit-parallel stress $\sigma_r(r, 0)$ representing the *T*-stress at the pointed slit tip, is decreasing to zero at the rounded slit tip. So, the *T*-stress in a rounded slit tip model is always that slit-parallel stress which would occur without slit tip rounding in the midsection of the relevant internal slit model. This *T*-stress can be used to approximate the *T*-stress related notch stresses. The accuracy of the solution will be all the better, the smaller the notch radius in relation to the characteristic dimensions of the problem is.

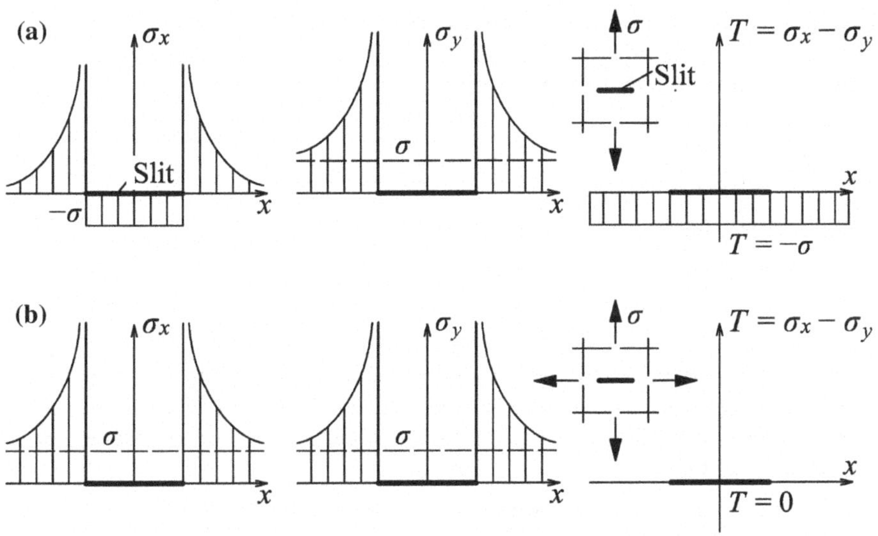

Fig. 2.62 Infinite plate with internal slit subjected to uniaxial (**a**) and biaxial (**b**) tension load; remote tensile stress σ; stresses σ_x, σ_y and T in ligament and slit line; (Radaj 2010)

The reference model of the slit is the already considered keyhole or slender elliptical notch subjected to tensile loading in slit-parallel direction. The notch stresses at the keyhole under a uniform slit-parallel tensile or compressive stress $\sigma_{\parallel}$ are given by (Radaj et al. 2001):

$$\sigma_\theta = \sigma_{\parallel}(1 - 2\cos 2\theta) \tag{2.162}$$

The solution is identical with Kirsch's solution for a single hole, with $\sigma_{\theta\ \max} = 3\sigma_{\parallel}$ at $\theta = 90°$ and $\sigma_{\theta\ \min} = -\sigma_{\parallel}$ at $\theta = 0°$ (Kirsch 1898). A more accurate solution for the keyhole gives $\sigma_{\theta\ \max} = 3.06\sigma_{\parallel}$ and $\sigma_{\theta\ \min} = -\sigma_{\parallel}$. The U-shaped notch results in $\sigma_{\theta\ \max} = 2\sigma_{\parallel}$ at $\theta = 69°$ (Radaj and Schilberth 1977, ibid. Fig. 5, $r/b = 1$, $r/a = 0.01$), whereas $\sigma_{\theta\ \min}$ remains unchanged. Considering the elliptical (or parabolic) slender notch, no pronounced stress maximum occurs at the notch edge, whereas $\sigma_{\theta\ \min} = -\sigma_{\parallel}$ at $\theta = 0°$. The latter value is confirmed by Neuber's formulae for uniaxial and biaxial tensile loading in comparison (Neuber 1985). One application-relevant conclusion from the above is that the value of the original T-stress is expressed by the notch stress at $\theta = 0°$ with opposite sign. This has to be taken into account when evaluating the $K_{1\rho}$ parameter from FE models using the notch stress at $\theta = 0°$.

Before doing so, the difference between the T-stresses at mode 1 loaded internal and external slits is demonstrated. At external slits (with an unloaded external boundary), no T-stress occurs whereas at internal slits it generally does. In order to clarify the situation in the latter case, the slit-parallel and slit-normal stresses σ_x and σ_y in the slit and ligament line $y = 0$ is discussed based on Westergaard's closed-form solution, Fig. 2.62 (Westergaard 1939; Hahn 1976). The infinite plate, subjected to uniaxial or biaxial remote tensile stress σ is considered.

The solution for uniaxial tension gives:

$$\sigma_x = \begin{cases} -\sigma \\ \sigma x/\sqrt{x^2 - a^2} - \sigma \end{cases} \qquad \sigma_y = \begin{cases} 0 & (x<a) \\ \sigma x/\sqrt{x^2 - a^2} & (x>a) \end{cases} \tag{2.163}$$

The solution for biaxial tension gives:

$$\sigma_x = \sigma_y = \begin{cases} 0 & (x<a) \\ \sigma x/\sqrt{x^2 - a^2} & (x>a) \end{cases} \tag{2.164}$$

The T-stress can be determined from the difference in the stresses σ_x and σ_y on the ligament close to the slit tip (Lazzarin et al. 2009). The expression for the T-stress is reformulated here as a limit value using the stress components σ_x and σ_y instead of σ_r and σ_θ:

$$T = \lim_{x \to a}(\sigma_x - \sigma_y) \quad (x > a) \tag{2.165}$$

Considering the above mentioned Westergaard solution, one gets $T = -\sigma$ in uniaxial tension and $T = 0$ in biaxial tension. Comparing this result with the plots in the figure, it is obvious that the T-stress in the two loading cases can be identified from the slit-parallel stress σ_x in the midsection of the internal slit, but not from the remote stress at the outside boundary of the plate. On the other hand, any superimposed loading $\sigma_{x\infty}$ at infinity will directly change the T-stress by $T = \sigma_{x\infty}$. Then, the T-stress consists of two parts, the one component connected with $\sigma = \sigma_{y\infty}$, the other with $\sigma_{x\infty}$. When considering external notches, these correspond to the condition $\sigma_{x\infty} = 0$, in general, for example the keyhole notch solution (Kullmer 1992; Radaj et al. 2001). Therefore, the solution for uniform slit-parallel tension loading is separately given by the latter authors. Altogether, it is concluded that, under uniaxial tension loading of internal rounded slit tips, the slit-parallel basic stress $\sigma_\parallel$ is equal in size and opposite in sign to the remote slit-transverse tensile stress, provided the notch radius is small in relation to the outer dimensions of the structural member.

Evaluating $K_{1\rho}$ in cases of $\sigma_\parallel \neq 0$ from FE analysis results, it is necessary to determine $K_{1\rho}^*$ without the T-stress effect and to register the T-stress or the corresponding slit-parallel basic stress $\sigma_\parallel$ separately, because different stress fields are connected with the two parameters. The equations corresponding to Eqs. (2.151) and (2.154) for the parabolic and keyhole notch types, respectively, then read as follows:

$$K_{1\rho}^* = \frac{\sqrt{\pi\rho}}{2}(\sigma_{max} + \sigma_\parallel) \quad (\theta = 0,\ r = 0.5\rho) \tag{2.166}$$

$$K_{1\rho}^* = \frac{\sqrt{\pi\rho}}{2.1213}(\sigma_{max} + \sigma_\parallel) \quad (\theta = 0,\ r = \rho) \tag{2.167}$$

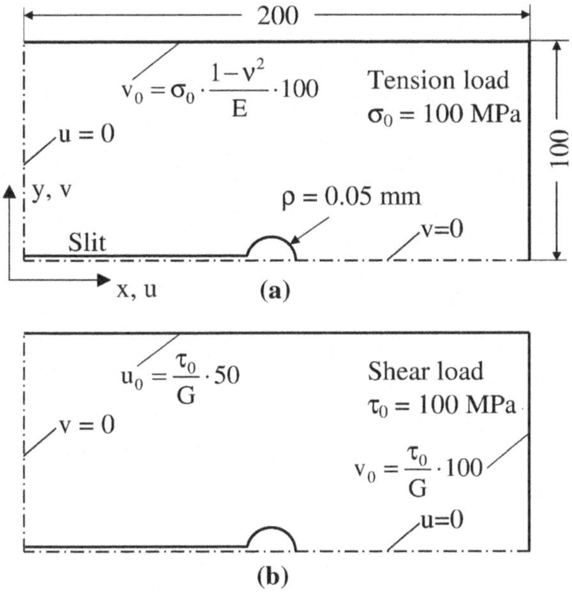

Fig. 2.63 Symmetry quarter of rectangular plate with two-sided keyhole slit, subjected to tensile loading by prescribed displacement v_0 (**a**) and to in-plane shear loading by prescribed displacements u_0 and v_0 (**b**) of the external boundary; (Radaj et al. 2009)

The necessary limit value of Eq. (2.157) is substituted as follows:

$$K_{\mathrm{I}} = \lim_{\rho \to 0} K^*_{1\rho} \tag{2.168}$$

Considering $K^*_{1\rho}$ with $\rho \to 0$ as indicated by Eq. (2.168), the slit-parallel stress $\sigma_{\parallel}$ in Eqs. (2.166) and (2.167) can be substituted by T. Therefore, the designation 'T-stress-corrected NSIF $K^*_{1\rho}$' is introduced. In cases of rounded slit tips, this correction is an approximation.

As an example for the T-stress correction, the plots of the K_{I}, $K_{1\rho}$ and K_{II}, $K_{2\rho}$ values for the two-sided keyhole slit in a rectangular plate with outer dimensions substantially larger than the notch radius, Fig. 2.63, has been reconsidered (Radaj 2010). The plate is subjected to uniaxial tensile and pure shear loading by prescribed remote boundary displacements with a reference stress of $\sigma_0 = 100$ MPa and $\tau_0 = 100$ MPa, respectively (Lazzarin et al. 2009). The $K_{1\rho}$ and $K_{2\rho}$ values were evaluated from FE models using Eqs. (2.154) and (2.155) not only at $r = \rho$ or $r = 1.723\rho$, but also at other distances $(r - \rho)$ close to the slit tip without considering the T-stress effect. The K_{I} and K_{II} values were also evaluated from FE models. Actually, the T-stress was found to be $T = -60$ MPa. Therefore, the $K_{1\rho}$ values are exactly 1.09 % too large at $(r - \rho) = 0.001$ mm, upper graph of Fig. 2.64. The difference decreases for $(r - \rho) > 0.001$ mm. The well known maximum notch stress gradient $-2/\rho$ perpendicular to the hole edge results in a full cancellation of the enlargement at $(r - \rho) = \rho/2 = 0.025$ mm. Obviously, the T-stress corrected $K^*_{1\rho}$ curve is then identical with the K_{I} curve within the plotting accuracy. The limit condition of Eq. (2.168) is precisely met, which means that the hole is sufficiently small to possess the same geometry factor Y_{I} as the pointed slit tip, or, in other words, that the cross-sectional weakening by the hole is negligibly small.

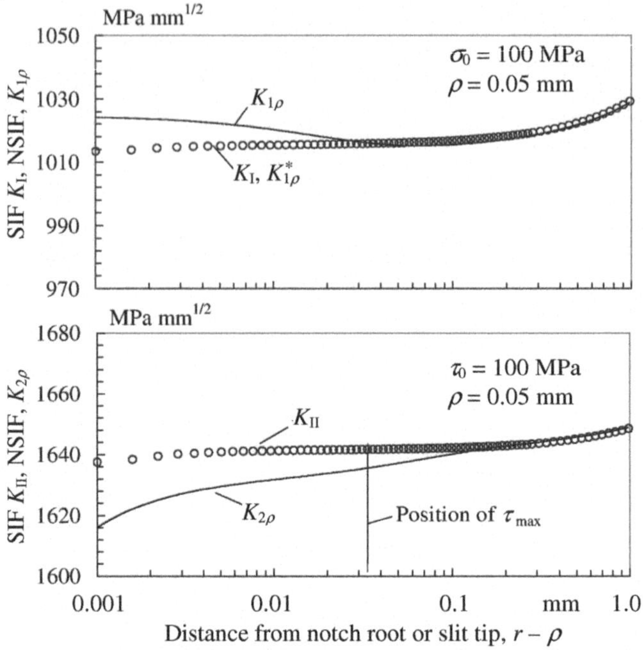

Fig. 2.64 Generalised NSIFs $K_{1\rho}$, $K^*_{1\rho}$, $K_{2\rho}$ in comparison to SIFs K_I, K_{II}, evaluated along bisector or ligament dependent on distance from notch root or slit tip, FE model of rectangular plate with two-sided keyhole or slit subjected to tensile and in-plane shear loading; (Lazzarin et al. 2009)

The deviations between the $K_{2\rho}$ and K_{II} curves in the lower graph of Fig. 2.64 have another reason. They occur for distances smaller than the τ_{max} position at $(r - \rho) = 0.723\rho = 0.0362$ mm. They mirror the fact that the correct $K_{2\rho}$ value is determined from τ_{max} at the above distance from the notch edge, with no convergence for evaluations at distances closer to the notch edge. The necessary convergence condition is $K_{2\rho} \rightarrow K_{II}$ for $\rho \rightarrow 0$, but not for $(r - \rho) \rightarrow 0$. The deviations are not a T-stress effect, because the evaluated shear stresses are independent of any T-stress.

Another, more convincing proof of the necessity to apply the T-stress correction has been given by re-analysing Lazzarin's tensile-shear loaded lap joint model (Radaj 2010).

The second order term in mode 1 loading which constitutes the T-stress at slit tips may also occur with V-notches (Ayatollahi and Dehghany 2010; Ayatollahi and Nejati 2011). An effect on $K_{1\rho}$ similar to the T-stress effect above can be expected, but has not yet been investigated. The second order term in mode 3 loading which has been named S-stress, does not affect the limit value $K_{3\rho} \rightarrow K_{III}$ because it does not contribute to the shear stresses $\tau_{\theta z}(r, 0)$ in the bisector plane which are related to $K_{3\rho}$.

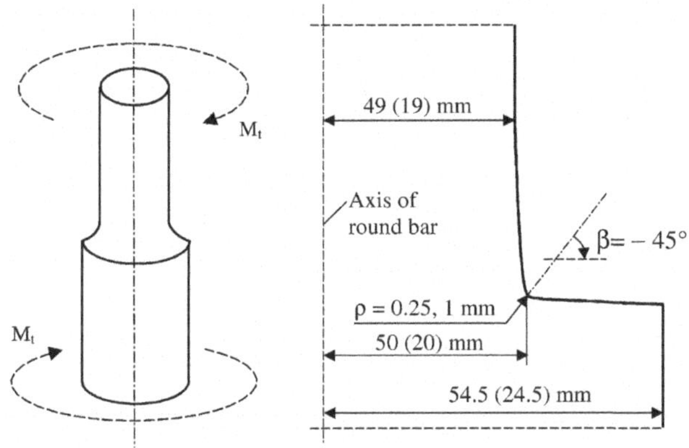

Fig. 2.65 Round bar with hyperbolic shoulder fillet subjected to torsional loading; (Zappalorto and Lazzarin 2011[(3)])

2.4.9 Application of the Generalised NSIF Concept to Round Bars with Shoulder Fillet

The generalised NSIF concept allows to describe the stress field at rounded notches, especially at sharply rounded notches, by a single parameter per loading mode. The intensity of the stress field, for which the NSIF is determined, must be available from other sources, in engineering applications from a FE analysis of the considered notch configuration.

The NSIFs thus found do not allow to perform a strength assessment directly. On the other hand, incorporation of the FE analysis results into the analytical frame of the NSIF concept allows fundamental insights in the setting of the model and its behaviour under modified conditions. This may be illustrated by the following example (Zappalorto and Lazzarin 2011[(3)]).

A round bar with shoulder fillet subjected to torsional loading is considered, Fig. 2.65. The shoulder fillet is approximated by a hyperbolic notch with notch opening angle $2\alpha = 90°$, inclined by $\beta = 45°$ against the bar axis. The notch radius is $\rho = 0.25$ or 1.0 mm. The radius up to the gross cross-section amounts to $R_g = 54.5$ or 25.5 mm, the radius up to the notch root thus being $R_n = 50$ or 20 mm.

The notch stress field (out-of-plane shear) is separated into two parts, one part antimetric, the other part symmetric, Eq. (2.122). Engineers are not always aware of the symmetric part, which corresponds to the *S*-stress originally defined at slit tips ($2\alpha = 0°$). The two parts are controlled by the antimetric and symmetric NSIFs, $K_{3\rho}^{(a)}$ and $K_{3\rho}^{(s)}$. They are determined according to Eqs. (2.124) and (2.125) and are related to the nominal torsional stress $\tau_{n,g}$ in the gross cross-section,

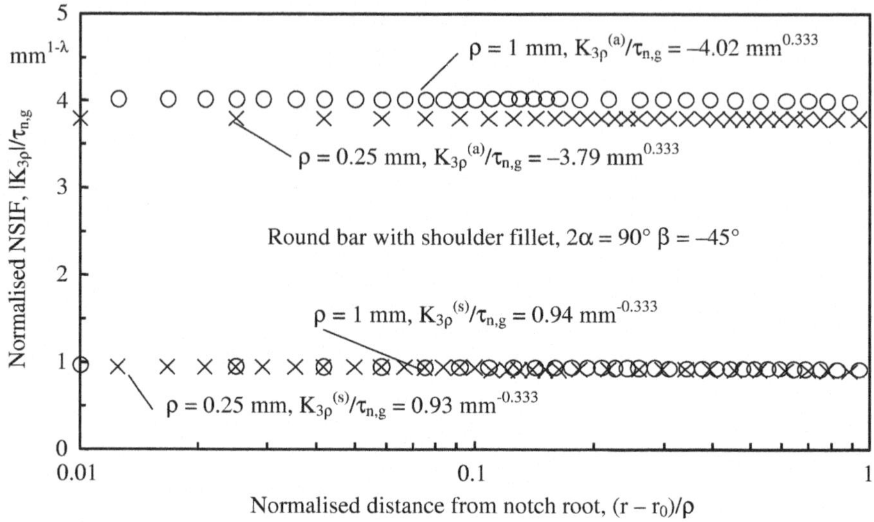

Fig. 2.66 Generalised NSIFs $K_{3\rho}^{(a)}$ and $K_{3\rho}^{(s)}$ of torsional-loaded round bar with hyperbolic shoulder fillet ($2\alpha = 90°$, two notch radii ρ) dependent on distance from notch root; evaluation of FE analysis results; nominal stress $\tau_{n,g}$ in gross cross-section; (Zappalorto and Lazzarin 2011[(3)])

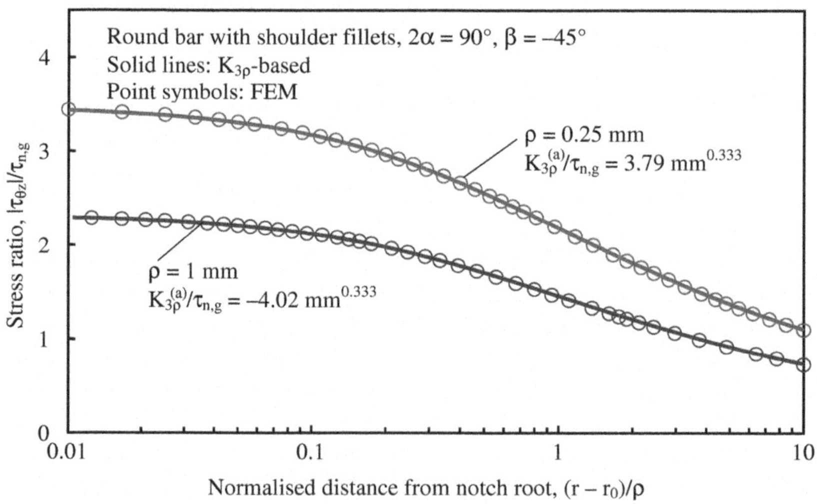

Fig. 2.67 Stress $\tau_{\theta z}$ in notch bisection over distance form notch root; torsional-loaded round bar with hyperbolic shoulder fillet; original compared with $K_{3\rho}$-based FE analysis results; nominal stress $\tau_{n,g}$ in gross cross-section; (Zappalorto and Lazzarin 2011[(3)])

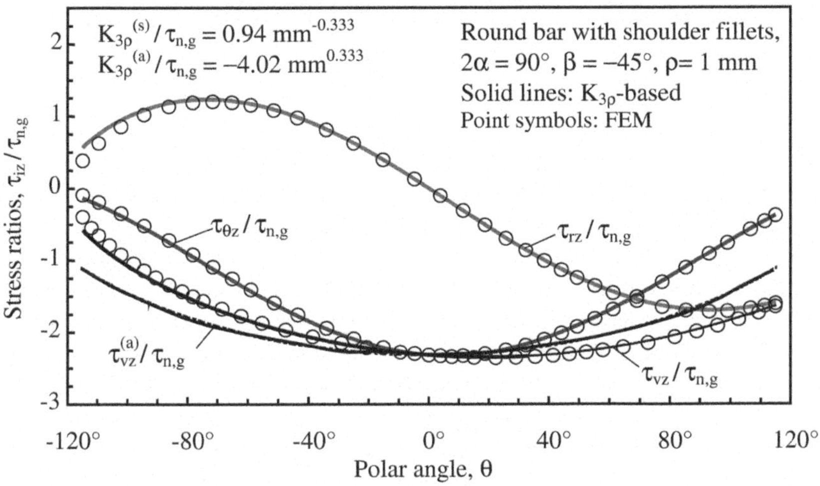

Fig. 2.68 Stresses $\tau_{\theta z}$, τ_{rz}, τ_{vz} and $\tau_{vz}^{(a)}$ at notch edge over polar angle θ; torsional-loaded round bar with hyperbolic shoulder fillet; original compared with $K_{3\rho}$-based FE analysis results; nominal stress $\tau_{n,g}$ in gross cross-section; (Zappalorto and Lazzarin 2011[(3)])

Fig. 2.66. Obviously, the notch radius $\rho \leq 1$ mm has no major influence on the NSIFs. The values of $K_{3\rho}^{(a)}$ and $K_{3\rho}^{(s)}$ should not directly be set into comparison, because they are different in their dimension. The NSIF $K_{3\rho}^{(a)}$ defines a stress field which is singular for $\rho \rightarrow 0$, whereas $K_{3\rho}^{(s)}$ is associated with non-singular stresses under the limit condition.

The out-of-plane shear stresses $\tau_{\theta z}/\tau_{n,g}$ in the bisector plane are plotted over the distance from the notch root for $\rho = 0.25$ and 1.0 mm, Fig. 2.67. FE analysis results and $K_{3\rho}$-based analytical curves are compared. There is no conceivable difference up to 10 mm ahead of the notch root with $R_g = 54.5$ mm. For $R_g = 24.5$ mm, this range is reduced to 1.0 mm. The stresses $\tau_{\theta z}$, τ_{rz} and τ_{vz} at the (hyperbolic) notch edge are plotted over the polar angle θ, Fig. 2.68. The shear stress τ_{vz} acts in planes normal to the notch edge, whereas the shear stresses $\tau_{\theta z}$ and τ_{rz} are related to the planes in the r–θ system before being converted to the u–v system. The curve of τ_{vz} deviates from the antimetric part $\tau_{vz}^{(a)}$ of the analytic solution outside the bisector plane. Rather confusing but correctly, the antimetric shear stresses in the r–θ system appear as symmetric in the u–v system following the notch edge in the positive direction of v.

In the same investigation, also the torsional loaded round bar with an inclined parabolic notch is considered. Further solutions are available for U-notches and blunt V-notches with an accurate mapping of the circular notch edge (Lazzarin et al. 2007). In both cases, the NSIFs are not defined.

2.4.10 Fatigue Limit Expressed by Notch Stress Intensity Factors

The generalised NSIF approach is closely related to the simple NSIF approach in respect of fatigue assessment. Both approaches can be used to assess the crack initiation life of notched members inclusive of fillet-welded joints (Boukharouba et al. 1995; Verreman and Nie 1996, 1997). The fatigue assessment of notched members comprises first the determination of the high-cycle fatigue limit (e.g. the technical endurance limit), second the extension of this limit value to the medium-cycle and low-cycle fatigue range (e.g. via the notch stress or strain *S–N* curve) and third the formulation of the multiaxial failure criterion (e.g. the von Mises equivalent stress in ductile materials). The fatigue process is understood as crack nucleation, short-crack growth and long-crack propagation. Crack nucleation is induced by repeated plastic notch tip deformation. Short cracks initiated at sharp notches may be arrested, provided the notch stress gradient is sufficiently steep and the basic stress level sufficiently low (non-propagating short cracks). The high-cycle fatigue limit may be described on the basis of elastic material behaviour.

The high-cycle fatigue limit for notched members has been mainly considered in relation to the fatigue limit of the unnotched material. This is expressed by the fatigue notch factor K_f which turns out to be smaller than the elastic notch stress concentration factor K_t. The relation between the fatigue strengths of rounded compared with pointed notches (inclusive of cracks) has only exceptionally been investigated. Stress intensity factors are basic for such investigations. They allow an assessment without a detailed short-crack propagation analysis, which may be too complicated for engineering purposes. Instead, the stress field in the damaging zone near the notch tip or notch root characterised by the simple or generalised NSIFs is evaluated. This 'near field' produces the driving force for crack initiation and propagation. Widely used are average stress criteria (notch stresses averaged over a material-characteristic microstructural length, area or volume at the notch tip or notch root) or, alternatively, critical distance criteria (notch stress at a material-characteristic microstructural distance from the notch tip).

The basic idea behind the approach presented in the following is the consideration of two different fatigue limits in the case of sharp notches (the notch tip radius is zero or very small, the notch stress concentration factor is correspondingly high). One fatigue limit is related to the nucleation of cracks that do not propagate, governed by the stress concentration factor K_t. The other fatigue limit is related to cracks that nucleate and then propagate, the threshold stress intensity factor ΔK_th applied to crack length plus notch depth being relevant.

It is straightforward with regard to the crack propagation fatigue limit of notches to introduce a crack-propagation-relevant microstructural length derived from the threshold stress intensity factor ΔK_th of long cracks and from the relevant crack propagation fatigue limit $\Delta\sigma_\mathrm{th}$. Such a length parameter a_0 has been defined in the following form (El Haddad et al. 1979):

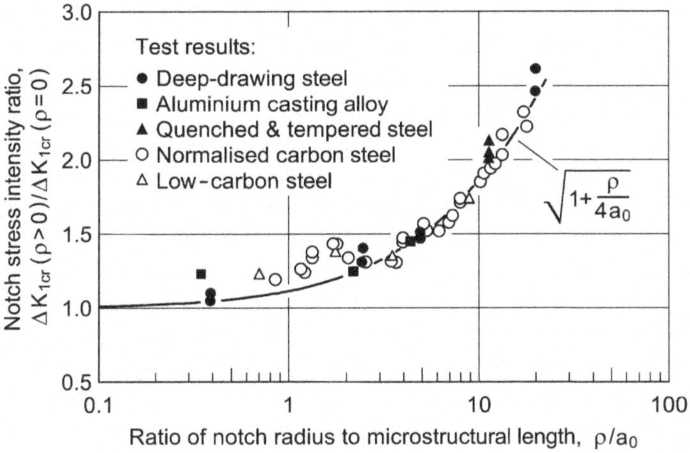

Fig. 2.69 Ratio of critical ranges of $K_{1\rho}$ and K_1, for rounded compared with pointed V-notches in different materials dependent on ratio of notch radius ρ to microstructural length a_0; analytical curve in comparison to test results; (Atzori et al. 1999[(1)])

$$a_0 = \frac{1}{\pi}\left(\frac{\Delta K_{\text{th}}}{\Delta\sigma_{\text{th}}}\right)^2 \tag{2.169}$$

It has been shown that the proper integration length for fatigue-effective stresses at crack and notch tips is $2a_0$, independent of the notch tip radius and the notch opening angle (Lazzarin et al. 1997).

The following relationships for mode 1 loading conditions have been derived based on the concept just mentioned (Atzori et al. 1999[(1)], 2001):

$$K_{\text{f}} = \frac{K_{\text{t}}}{\sqrt{1+4a_0/\rho}} = \frac{2\Delta K_1}{\Delta\sigma_{\text{n,g}}}\frac{1}{\sqrt{\pi(\rho+4a_0)}} \tag{2.170}$$

$$\Delta K_{1\text{cr}}(\rho>0) = \Delta K_{1\text{cr}}(\rho=0)\sqrt{1+\rho/4a_0} \tag{2.171}$$

with the gross area nominal stress range $\Delta\sigma_{\text{n,g}}$, the critical NSIF range $\Delta K_{1\rho\ \text{cr}} = \Delta K_{1\text{cr}}(\rho>0)$ and the critical NSIF range $\Delta K_{1\text{cr}} = \Delta K_{1\text{cr}}(\rho=0)$ beyond which initiated cracks will propagate. The derivation above uses the approximation $(1-\lambda_1)\approx 0.5$ which is justified for $2\alpha \le 60°$. Obviously, the expression on the right-hand side of Eq. (2.170) is thus only approximately dimensionless. The prediction of fatigue strength according to Eq. (2.171) for different notch radii and materials is plotted in Fig. 2.69 and compared with various experimental results taken from the literature (Lazzarin et al. 1997; Noda et al. 1995; Yao et al. 1995). The degree of correspondence seems to be satisfactory.

The simple or generalised NSIF may directly characterise the fatigue limit as shown above or as shown for welded joints in Sect. 2.3.7. These parameters can also indirectly be relevant for fatigue as can be seen from the *J*-integral criterion (see Sect. 2.3.8) or from the local strain energy density criterion (see Sect. 3.2.3).

Proceeding from the high-cycle fatigue limit, especially at the branching point between the crack nucleation and crack propagation curves in Frost's endurance limit over SCF diagram, $\sigma_{nE} = f(K_t)$, the medium-cycle *S–N* curve can be empirically approximated using an inverse slope exponent *k* which depends on the amount of elastic-plastic notch support during crack initiation and subsequent crack propagation. A more rational approach consists in using crack initiation criteria based on elastic or plastic NSIFs in the medium-cycle and low-cycle fatigue range, respectively, allowing extensions into the crack propagation range.

Multiaxial failure criteria are needed for the fatigue assessment under mixed mode loading conditions. These can be based on superposition of the effects of the individual loading modes as far as elastic NSIFs are applicable. An elliptical curve relationship has been used for fillet-welded joints in analogy to the Gough ellipse (see Sect. 2.3.6).

2.4.11 Conclusions

The stress distribution at sharply rounded notches can be described by the loading mode related generalised NSIFs $K_{1\rho}$, $K_{2\rho}$, $K_{3\rho}$ as the governing field parameters. These NSIFs are linked not only to the maximum notch stresses σ_{max}, τ_{max}, τ^*_{max}, which constitute the conventional SCFs in engineering science, but to the whole notch stress field. The field information is needed for assessing failure processes such as crack initiation and propagation or microstructural damaging phenomena. The size effect on the endurable nominal stresses is already included in the endurable NSIFs.

The generalised NSIFs can also be applied to notches with a substantially increased notch radius, provided a proper evaluation over some distance from the notch root in the bisector plane is performed, proving a sufficiently constant value of the NSIF. This value may depend on the notch radius. A substantial increase with the notch radius may be observed, especially under mode 2 loading conditions where the maximum notch stress occurs outside the bisector plane. The generalised NSIFs and the appertaining stress equations depend on the notch shape (parabolic, hyperbolic, elliptical, blunt V-notch, V-notch with circular root arc or V-notch with root hole) and on the applied analytical frame. The internal or external setting of the notch may include a minor influence of loading in the bisector direction. Structural boundary, loading and support conditions have an influence, especially if they act close to the notch. Only the effect of a finite notch depth relative to the notch radius has been systematically investigated to some extent.

The numerical basis of generalised NSIF evaluations in configurations with finite boundaries is the FE analysis method, performed with sufficiently fine meshes. The theoretical frame of the NSIFs allows interpretations and generalisations of the FE analysis results.

Fatigue assessments are possible on the basis of the generalised NSIFs, but this access has not yet been elaborated. One reason may be, that the NSIFs depend on the overall notch shape, on the geometrical details at the notch root and on the applied analytical frame. So, their generality is restricted. As far as the stress distribution of the sharply rounded notch expressed by the generalised NSIF remains close to that of the pointed notch, expressed by the simple NSIF, failure criteria and data are transferable to some extent.

2.5 Plastic Notch Stress Intensity Factor Concept

2.5.1 Survey of Section Contents

The preceding sections refer to the stress fields and NSIFs of pointed and rounded notches under the condition of linear-elastic material behaviour. Now, the presentations are extended to nonlinear-elastic and thus approximated elastic-plastic material behaviour. The stress singularity at pointed notches continues to exist, provided strain hardening is taken into account. Plastic NSIFs are defined as distinguished from the elastic NSIFs. They include plastic strain intensity factors, which exist also for a non-hardening material behaviour.

Plastic stress intensity factors were first proposed for cracked plates under nonlinear conditions of material behaviour (Hilton and Hutchinson 1971; Hilton 1973) and later on extended to pointed V-notches with arbitrary notch opening angle (Lazzarin et al. 2001). Plastic NSIFs can be stated in terms of the elastic NSIFs, provided the local elastic and elastic-plastic strain energy densities averaged over a circle sector volume around the pointed V-notch are set equal, which is appropriate under small scale yielding conditions (Lazzarin and Zambardi 2002).

Basic information is given on the deformation theory of plasticity which leads to the HRR elastic-plastic stress and strain fields at the notch tip, named after Hutchinson, Rice and Rosengren. The elastic-plastic fields at V-notches subjected to tensile loading inclusive of the definition of the plastic NSIF are presented. The corresponding fields in out-of-plane shear loading are easier to describe analytically. The parabolic notch is treated first, followed by V-notches in general. A uniform analytical frame based on the Neuber nonlinear material law is presented. The Neuber rule linking the maximum elastic-plastic stresses and strains at rounded notch tips and Glinka's alternative concept are contrasted with findings by Zappalorto and Lazzarin, who identified an influence of the notch opening angle and of the hardening exponent on the Neuber rule. A final subsection is devoted to the analytical description of the elastic-plastic material laws.

When reviewing the available publications on the plastic NSIF concept, a lack of conceptual clarity in respect of the yield limit parameters has to be overcome in some cases. As explained in more detail in Sect. 2.5.10, the real (engineering) yield limit parameters ε_Y, σ_Y or γ_Y, τ_Y (marked or offset yield limit) are not identical with the substitute (analytical) yield limit parameters ε_0, σ_0 or γ_0, τ_0 resulting from the intersection point of the linear-elastic and the asymptotic plastic limit curves. On the other hand, there is no general relationship available for relating the two types of yield parameters. The real yield limit is sometimes introduced in comparative investigations based on the FE method, when using the Ramberg–Osgood material law. It was not possible to remove any possible confusion in the following. The choice of the original authors had to be respected.

The expositions are mainly based on the analytical development of the plastic NSIF concept by Lazzarin and Zappalorto in a set of excellent publications which can be found in the list of references at the end of Chap. 2. A list of symbols is also available there.

2.5.2 Deformation Theory of Plasticity Founding the HRR Fields

When loading of the material exceeds a certain level, plastic deformation (yielding) occurs immediately (i.e. independently of time). The usual assumption is, that the elastic and plastic strains or strain increments add up to their total values:

$$\varepsilon_{ij} = \varepsilon_{ij,e} + \varepsilon_{ij,p}, \quad d\varepsilon_{ij} = d\varepsilon_{ij,e} + d\varepsilon_{ij,p} \tag{2.172}$$

The elastic strains are connected with the stresses according to Hooke's law. The plastic strains are conforming to the multiaxial yield condition, mostly according to von Mises. Two different formulations of the relationships between plastic strains and (elastic) stresses are in use. The 'incremental theory' refers to plastic strain increments in relation to the analytical yield surface. It is not used in the following. The 'deformation theory' founding the HRR fields considers the total strains. Both theories assume in general that no plastic volume change occurs, $\varepsilon_{p,kk} = 0$.

In the deformation theory leading to the HRR fields described below, a hardening material is considered whose uniaxial stress–strain curve is represented by the Ramberg–Osgood material law in the following normalised form related to ε_0, σ_0 which deviates from the original form related to E and H, Eq. (2.264) (Gross and Seelig 2001), Fig. 2.70a:

$$\frac{\varepsilon}{\varepsilon_0} = \frac{\sigma}{\sigma_0} + \alpha \left(\frac{\sigma}{\sigma_0}\right)^n \quad (1 < n < \infty) \tag{2.173}$$

where the terms on the right-hand side of the equation denote the elastic and plastic strain ratios. The quantities ε_0, σ_0 refer to the initiation of yielding, $\varepsilon_0 \approx \varepsilon_Y$, $\sigma_0 \approx \sigma_Y$, provided the factor α is sufficiently small. Under this special condition, the factor α results approximately from $\sigma_Y^{n-1} E/H^n$ with the yield limit $\sigma_Y = E\varepsilon_Y$, the modulus of elasticity E and the hardening (or strength) coefficient

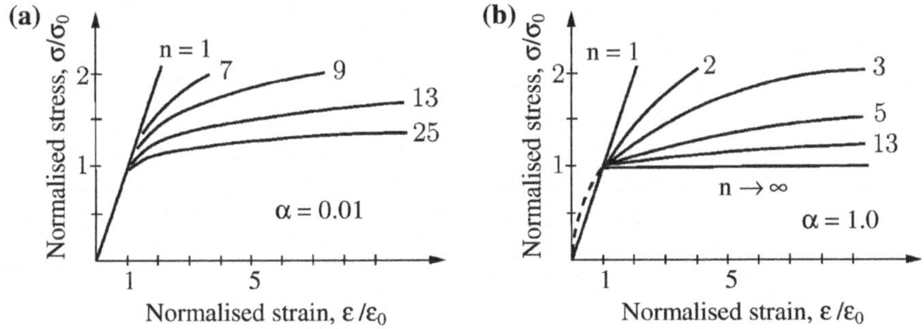

Fig. 2.70 Uniaxial stress–strain curves according to the Ramberg–Osgood material law (**a**) and according to a combination of linear-elastic and total strain power law (**b**); different factors α in the two graphs; (Gross and Seelig 2001)

H (where H is equal to σ at $\varepsilon_p = 1$). The 'strain hardening exponent' n is limited by $n = 1$ for linear-elastic and $n \to \infty$ for elastic-perfectly-plastic behaviour. It has to be noted that many authors prefer the reciprocal value $1/n$ instead of n.

At pointed notch tips, the plastic strains are found to be singular, so that the non-singular elastic strains are negligible in relation to the plastic strains, and Eq. (2.173) can be reduced to the total strain power law, in which the originally plastic strains are now interpreted as total strains:

$$\frac{\varepsilon}{\varepsilon_0} = \alpha \left(\frac{\sigma}{\sigma_0} \right)^n \tag{2.174}$$

The total strain power law, Eq. (2.174), when generalised to multiaxial stress states based on the invariant J_2 of the deviatoric stress state, can be written as follows:

$$\frac{\varepsilon_{ij}}{\varepsilon_0} = \frac{3\alpha}{2} \left(\frac{\sigma_e}{\sigma_0} \right)^{n-1} \frac{s_{ij}}{\sigma_0}, \quad \sigma_e = \sqrt{\frac{3}{2} s_{ij} s_{ij}} \tag{2.175}$$

with the effective stress σ_e according to the von Mises criterion and the deviatoric stress component s_{ij}.

The material law, Eq. (2.175), is also considered valid for a stress–strain relationship consisting of a linear-elastic rise up to the intersection point ε_0, σ_0 followed by the total strain power law in the plastic range, Fig. 2.70b:

$$\begin{aligned} \frac{\varepsilon}{\varepsilon_0} &= \frac{\sigma}{\sigma_0} \quad (\varepsilon \leq \varepsilon_0) \\ \frac{\varepsilon}{\varepsilon_0} &= \left(\frac{\sigma}{\sigma_0} \right)^n \quad (\varepsilon \geq \varepsilon_0) \end{aligned} \tag{2.176}$$

Obviously, $\alpha = 1.0$ when comparing Eq. (2.173), i.e. α is not small so that the approximations $\varepsilon_0 \approx \varepsilon_Y$ and $\sigma_0 \approx \sigma_Y$ are not anymore valid. In the following, the quantities ε_0, σ_0 are termed ‘substitute yield limit’ in order to distinguish them from the real yield limit ε_Y, σ_Y.

In the deformation theory, the plastic material behaviour is specified as non-linear-elastic. For nonlinear-elastic solid bodies with a crack undergoing monotonic, small-strain deformation, Hutchinson, Rice and Rosengren have shown that the intensity of the near-tip field is expressed by the J-integral and that the stresses, strains and displacements exhibit a singularity described by $r^{-1/(n+1)}$, $r^{-n/(n+1)}$ and $r^{-1/(n+1)}$, respectively (Hutchinson 1968; Rice and Rosengren 1968). These ‘HRR fields’ close to the crack tip are written as follows:

$$\sigma_{ij} = \sigma_0 \left(\frac{J}{\alpha \sigma_0 \varepsilon_0 I r} \right)^{1/(n+1)} \tilde{\sigma}_{ij}(\theta, n) \tag{2.177}$$

$$\varepsilon_{ij} = \alpha \varepsilon_0 \left(\frac{J}{\alpha \sigma_0 \varepsilon_0 I r} \right)^{n/(n+1)} \tilde{\varepsilon}_{ij}(\theta, n) \tag{2.178}$$

$$u_i = \alpha \varepsilon_0 r \left(\frac{J}{\alpha \sigma_0 \varepsilon_0 I r} \right)^{n/(n+1)} \tilde{u}_i(\theta, n) \tag{2.179}$$

with the ‘universal functions’ $\tilde{\sigma}_{ij}(\theta, n)$, $\tilde{\varepsilon}_{ij}(\theta, n)$ and $\tilde{u}_i(\theta, n)$ dependent on the polar angle θ, the strain hardening exponent n and the state of stress (plane stress or plane strain), with the J-integral and with the Integral I depending mildly on n. Deviations occur with large-scale yielding under plane strain conditions, which may be expressed by a ‘constraint factor’ (Shlyannikov et al. 2011).

In the equations above, the J-integral does not separate the mode 1 and mode 2 loading components. Evaluated HRR-fields refer to pure mode 1 loading in general. Such evaluations show, that the HRR-dominated region at the crack tip is substantially reduced with increasing values of n. Simultaneously, the deformed crack tip gets more and more blunted.

Even in the considered case of crack tips ($2\alpha = 0$), it has not been possible to find a complete numerical solution. Only modelling the material as perfectly-plastic has allowed substantial progress (Unger 2001, 2005, 2007). Moving to the more general case of pointed V-notches in strain-hardening materials, the problem becomes much more demanding. Not only the ‘universal functions’ require numerical solution techniques, but also the exponent s characterising the degree of the singularity. Nonlinear differential equations and nonlinear eigenvalue problems have to be solved, and the superposition principle, facilitating solutions in the linear range, is not anymore applicable.

2.5.3 *Elastic-Plastic Fields at Tensile Loaded V-Notches*

A partial solution of the problem of elastic-plastic stress and strain fields at tensile loaded V-notches is available, originally referring to large-angle V-notches simulating weld-like geometries (Lazzarin et al. 2001) and later on to pointed V-notches in general (Filippi et al. 2002[1]). Lazzarin proposed plastic notch stress intensity factors as parameters which combine the effects of member size, nominal stress level and constitutive material law on the notch stress and strain fields at pointed V-notches. Considering weld-like pointed corner notches subjected to a combination of mode 1 and mode 2 loading in the bisector plane, the notch stresses in the HRR field follow a series expansion:

$$\sigma_{ij}(r,\theta) = K_{1p} r^{-s} \tilde{\sigma}_{ij}^{(1)}(\theta) + K_{2p} r^{-z} \tilde{\sigma}_{ij}^{(2)}(\theta) + \ldots \qquad (2.180)$$

where K_{1p} and K_{2p} are the plastic notch stress intensity factors in mode 1 and mode 2 loading, representing the symmetric singular stress field (first term) and the antimetric non-singular stress field (second term) for $2\alpha > 102°$. This equation does not express the superposition principle, because K_{1p} and K_{2p} are assumed as interconnected. Their unambiguous definition requires $\tilde{\sigma}_{e,\,max}^{(1)} = \tilde{\sigma}_{e,\,max}^{(2)}$ (absolute values), appertaining to the effective (or equivalent) stress σ_e according to von Mises.

The plastic NSIF K_{1p} is defined in analogy to the elastic NSIF K_1 as follows:

$$K_{1p} = \lim_{r \to 0} \sqrt{2\pi} r^s \sigma_\theta(r, 0) \qquad (2.181)$$

with the exponent s depending on the hardening exponent and the notch opening angle, Eq. (2.187).

The plastic NSIFs K_{1p} and K_{2p} depend on the notch opening angle 2α, the nominal stress σ_n, the plate thickness t, the reference stress σ_0 and the hardening exponent n. The third material parameter in the Ramberg–Osgood law (besides E and n), the hardening (or strength) coefficient H, is dependent on σ_0, ε_0.

The FE analysis for K_{1p} of a weld-like corner notch model, $2\alpha = 120°$, $t = 20$ mm and $n = 4.0$, using the Ramberg–Osgood law, is presented in Fig. 2.71. The linearised approximation of K_{1p} in the range $\sigma_n = 100$–500 MPa has the following general form:

$$K_{1p} = (k_1 \sigma_n + k_{10} \sigma_0) t^s \qquad (2.182)$$

Actually, the K_{1p} curves are nonlinear and start from the origin of the diagram. This relationship is found tentatively based on the analytical assumption for K_{1p}. Deviations of the analytical HRR solution (power law) from the FE solution (Ramberg–Osgood law) are not explicitly visible.

The result of the nonlinear eigenvalue analysis in respect of the exponent s is plotted in Fig. 2.72. The dependency on 2α is weak for not too large angles 2α, but the dependency on n is generally strong. The exponent z in Eq. (2.180) is negative in the considered range of large values of 2α, which means, that no singularity occurs. An example for the universal function $\tilde{\sigma}_{ij}(\theta)$ is given in Fig. 2.73, with exact values and approximations in comparison.

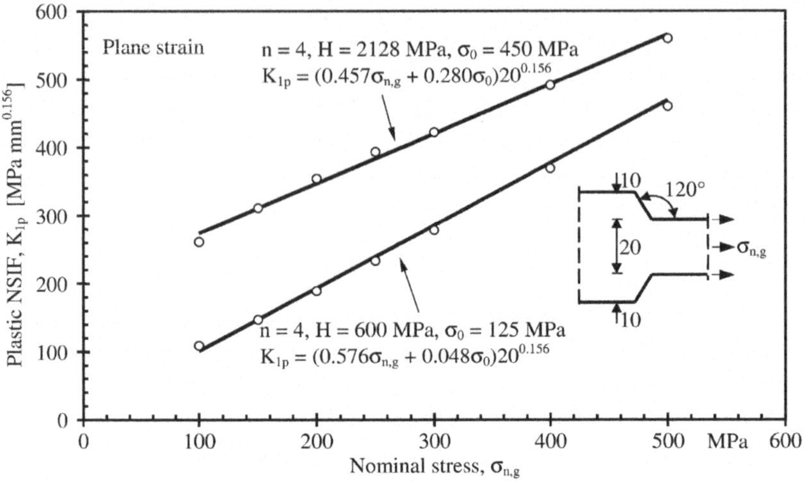

Fig. 2.71 Weld-like corner notch: plastic stress intensity factor K_{1p} dependent on nominal stress $\sigma_{n,g}$ with material parameters n, H, σ_0; linearised approximation in the range $\sigma_{n,g} = 100$ to 500 MPa; plate thickness $t = 20$ mm (Lazzarin et al. 2001)

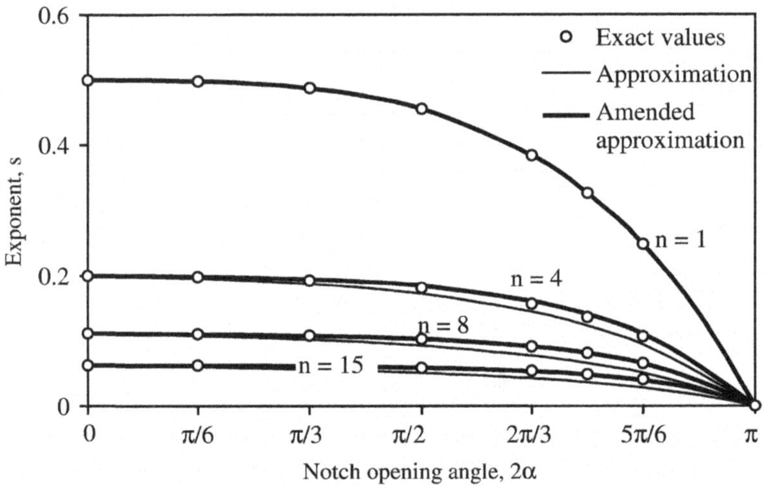

Fig. 2.72 Exponent s of the nonlinear-elastic V-notch subjected to mode 1 (symmetric) loading; exact values (Runge-Kutta) and approximations; (Filippi et al. 2002(1))

A more concise derivation and definition of the plastic NSIF under mode 1 loading conditions is possible based on the equivalent strain energy density approach (Lazzarin and Zambardi 2002). In this approach, the elastic-plastic strain energy density at the notch tip is found coincident with that determined under linear-elastic conditions (Glinka and Molski 1981). The approach is not directly applicable to pointed V-notches, since the strain energy density at the notch tip

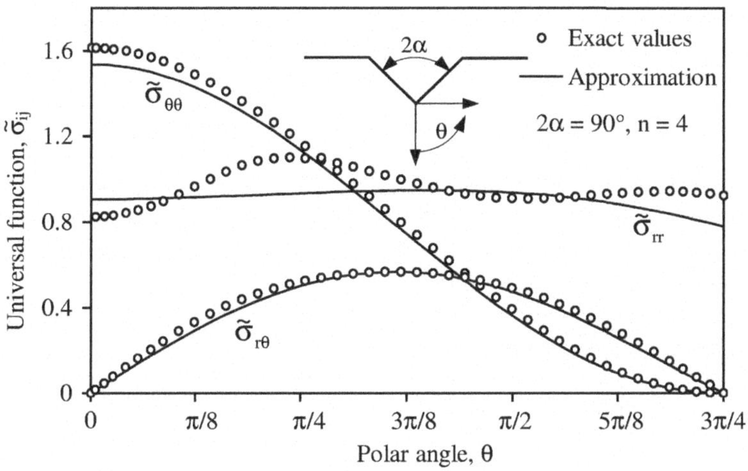

Fig. 2.73 Universal functions $\tilde{\sigma}_{ij}(\theta)$ for $2\alpha = 90°$ and $n = 4$, mode 1 (symmetric) loading; exact values (Runge-Kutta) and approximations; (Filippi et al. 2002[(1)])

tends toward infinity both for the linear-elastic and the power law hardening material. But a reformulation is possible, when applying elastic and plastic NSIFs to a finite size circle sector around the notch tip. The hypothesis is introduced that, under plane strain conditions, the value of the energy concentration due to the notch is equal for the two material laws. Under small-scale yielding conditions, this results in an identical value of the strain energy averaged over the considered circle sector. As a consequence, the plastic NSIFs of pointed V-notches can be predicted on the basis of the linear-elastic NSIFs. The essential steps of the derivation by Lazzarin and Zambardi are reviewed below.

The Ramberg–Osgood material law is the basis in the following form, generalised for multiaxial stress conditions according to the von Mises criterion:

$$\varepsilon_{ij} = \frac{1+\nu}{E} s_{ij} + \frac{1-2\nu}{E} \sigma_h \delta_{ij} + \frac{3}{2} \frac{\sigma_e^{n-1}}{C^n} s_{ij} \tag{2.183}$$

where s_{ij} is the deviatoric stress tensor, σ_h is the hydrostatic stress, δ_{ij} the Kronecker delta and σ_e the von Mises equivalent stress. Actually, the asymptotic approximation by the total strain power law is used in the analytic derivations. The material characteristic constants are the hardening exponent n and the hardening coefficient C (better known as K or H). The multiaxiality condition (plane stress or plane strain) comes in via the equivalent stress σ_e.

Based on Eq. (2.183), the plastic component W_p of the strain energy density can be calculated for monotonic loading:

$$W_p = \frac{n}{(n+1)C^n} \sigma_e^{n+1} \tag{2.184}$$

The first term of the elastic-plastic stress series expansion in the close neighbourhood of a pointed V-notch under mode 1 loading conditions, compare Eq. (2.180), turns out to be:

$$\sigma_{ij}(r,\,\theta) = K_{1p}\, r^{-s}\tilde{\sigma}_{ij}(\theta), \quad \sigma_e(r,\,\theta) = K_{1p} r^{-s}\tilde{\sigma}_e(\theta) \tag{2.185}$$

where K_{1p} is the plastic NSIF quantifying the intensity of the local stress field. The determination of the universal functions $\tilde{\sigma}_{ij}$ and $\tilde{\sigma}_e$ requires a numerical solution. They are normalised in the same manner by setting the maximum value of $\tilde{\sigma}_e(\theta)$ over the considered domain to unity.

Inserting Eq. (2.185) into Eq. (2.184), the elastic-plastic strain energy density close to the notch tip is given by:

$$W_p(r,\,\theta) = \frac{n}{(n+1)C^n}(K_{1p})^{n+1} r^{-s(n+1)}\tilde{\sigma}_e(\theta)^{n+1} \tag{2.186}$$

In the plastic zone very close to the notch tip (small-scale yielding), the elastic contribution to the total strain energy is very small. Therefore, the total energy density W can be set equal to the elastic-plastic energy density W_p which, on the other hand, is equal to the original linear-elastic energy density W_e. This should apply to any value of r within the considered notch tip region. Therefore, comparing the linear-elastic and elastic-plastic singularities, an expression for the exponent s can be gained:

$$W_e(r,\theta) \propto r^{2(\lambda_1 - 1)}, \quad W_p(r,\theta) \propto r^{-s(n+1)}, \quad s = \frac{2(1-\lambda_1)}{n+1} \tag{2.187}$$

The expression for $s(\lambda_1,\, n)$ has also been derived by other authors (Kuang and Xu 1987; Chen and Ushijima 2000). It is independent of the applied load value. For the crack, $\lambda_1 = 0.5$ and therefore $s = 1/(n+1)$ are valid.

The equivalent strain energy density approach (Glinka and Molski 1981) is based on the well-founded hypothesis that different constitutive models (elastic-plastic compared with linear-elastic) affect the relationship between stresses and strains but leave the value of the averaged energy density at the notch tip (sector radius R) compared with the same in the nominal stress region (energy concentration factors $K_{W,p}$ and $K_{W,e}$) unchanged:

$$K_{W,p}(R) = K_{W,e}(R) \tag{2.188}$$

This allows to express the plastic by the elastic stress intensity factor which can easily be determined by FE analysis:

$$K_{1p} = \left[\frac{n+1}{n}\frac{I_e(\gamma)}{I_p(\gamma,\,n)}K_{1e}^2\left(\frac{C^n}{2E}+\frac{n}{n+1}\sigma_n^{n-1}\right)\right]^{1/(n+1)} \quad (\gamma = \pi - \alpha) \tag{2.189}$$

where $I_e(\gamma)$ and $I_p(\gamma)$ are integrals of the elastic and plastic universal functions, the first evaluated analytically, the second numerically. The sector radius R should be within the range of validity of the asymptotic one-term expansion of the notch tip stress field.

Under conditions of local yielding, the second term in the parenthesis is substantially smaller than the first term, so that the following simplified formula is applicable:

$$K_{1p} = AK_{1e}^{2/(n+1)} \tag{2.190}$$

$$A = \left(\frac{n+1}{n} \frac{C^n}{2E} \frac{I_e(\gamma)}{I_p(\gamma, n)} \right)^{1/(n+1)} \tag{2.191}$$

These relationships are dependent on the notch opening angle and the material parameters but do not depend on the applied load value.

The introduced hypothesis of identical energy concentration factors, Eq. (2.188), is only a well-conceived approximation. It neglects any redistribution of the stresses from the higher stressed towards the lower stressed areas. Such redistributions vary with the considered special case. Several numerical tests using the FE method have therefore been performed by Lazzarin and Zambardi with the aim to mark the range of applicability and the degree of accuracy of the approximate solutions.

The FE analyses were carried out on a flat bar specimen with two-sided V-notch subjected to a tensile load or, alternatively, on a flat bar specimen with one-sided V-notch subjected to a bending load. Plane strain conditions are assumed. The nominal stress refers to the gross cross-section. The material parameters correspond to the following two low-carbon and medium-carbon steels:

- AISI 1008: $E = 2.06 \times 10^5$ MPa, $C = 600$ MPa, $n = 4.00$, $\sigma_Y = 125$ MPa, $\nu = 0.3$
- AISI 1045: $E = 2.06 \times 10^5$ MPa, $C = 950$ MPa, $n = 8.33$, $\sigma_Y = 450$ MPa, $\nu = 0.3$

Actually, the total strain power law is applied with the substitute yield limit according to Eq. (2.283), $\sigma_0 = (H^n/E)^{1/(n-1)}$. Thus, $\sigma_0 = 85.7$ MPa for the AISI 1008 steel, and $\sigma_0 = 456.1$ MPa for the AISI 1045 steel. Only tensile loading, the notch opening angle $2\alpha = 60°$ and the AISI 1008 steel are referenced in the two figures below. The FE analysis is carried out with the Ramberg–Osgood material law which is identical with the total strain power law for the very large strain values which define the plastic NSIF.

The stresses σ_θ normal to the bisector plane are plotted over the distance from the notch tip for various (increasing) nominal stress values in Fig. 2.74. Direct FEM results are compared with the K_{1p} analysis results. Deviations from the analytical curves (linear in logarithmic scales) occur for larger distances from the notch tip.

The plastic NSIF K_{1p} is plotted over the nominal stress $\sigma_{n,g}$ in Fig. 2.75. Analysis results according to the unreduced Eq. (2.189) and the simplified Eq. (2.190) are compared with the direct FEM results. The simplified formula may not

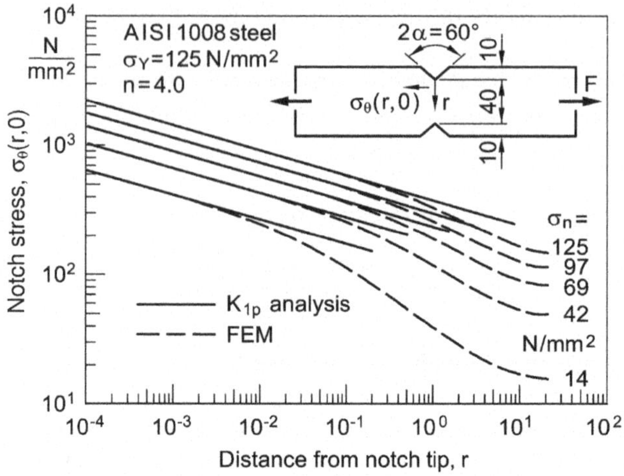

Fig. 2.74 Notch stress $\sigma_\theta(r,0)$ in the bisector plane over distance r from the V-notch tip for various nominal stresses σ_n in the gross cross-sectional area; direct FEM results compared with K_{1p} analysis results; (Lazzarin and Zambardi 2002)

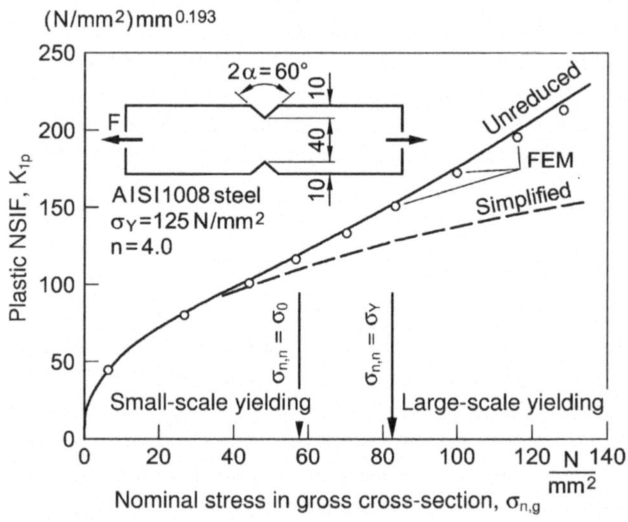

Fig. 2.75 Plastic NSIF K_{1p} over nominal stress $\sigma_{n,g}$ in the gross cross-sectional area; unreduced and simplified analytical solution compared with FEM results; ranges of small-scale and large-scale yielding; divergence point at $\sigma_{n,n} = 1.5\sigma_{n,g} = \sigma_0$; (Lazzarin and Zambardi 2002)

be used in the range of large-scale yielding, defined by $\sigma_{n,n} > \sigma_0$. The divergence point of the two curves is indicated by σ_0 and not by σ_Y, because the total strain power law is relevant. Under bending load conditions, the local yielding range, where the simplified formula is applicable, is extended to substantially larger

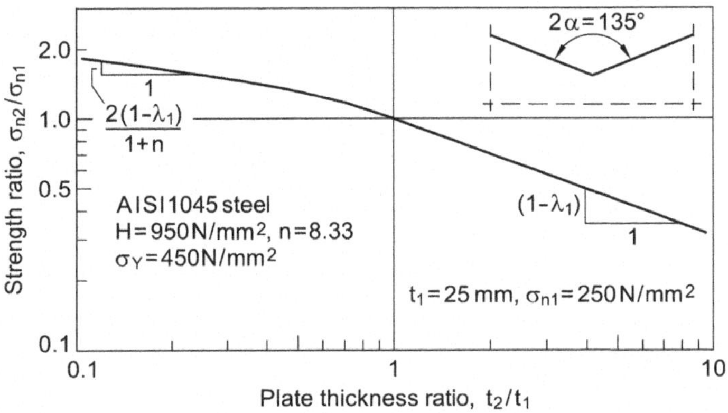

Fig. 2.76 Strength ratio σ_{n2}/σ_{n1} over gross plate thickness ratio t_2/t_1 for tensile loaded weld-like V-notched members ($2\alpha = 135°$) in medium-carbon steel AISI 1045; the fully plastic condition on the left hand side in contrast to the purely elastic condition on the right-hand side; reference values $t_1 = 25$ mm, $\sigma_{n1} = 250$ N/mm^2 in fatigue assessments according to Eurocode 3; (Lazzarin and Zambardi 2002)

nominal (bending) stress values because of the elastic bending support, characterised by the factor 1.5.

Finally, the size effect on strength is considered when varying all geometrical dimensions of the notched member in a self-similar manner, member 2 compared with member 1. The gross cross-sectional width, interpreted as plate thickness t in the weld-like models, is taken as the characteristic parameter. Based on the elastic NSIF, $K_1 = k_1 \sigma_n t^{1-\lambda_1}$, Eq. (2.87), the following relationship for the endurable nominal stresses σ_{n2} and σ_{n1} is derived:

$$\frac{\sigma_{n2}}{\sigma_{n1}} = \left(\frac{t_1}{t_2}\right)^{1-\lambda_1} \tag{2.192}$$

The corresponding formula for the plastic NSIF, $K_{1p} = AK_1^{2/(n+1)}$, Eq. (2.190), results in:

$$\frac{\sigma_{n2}}{\sigma_{n1}} = \left(\frac{t_1}{t_2}\right)^{2(1-\lambda_1)/(n+1)} \tag{2.193}$$

The ratio σ_{n2}/σ_{n1} may be interpreted as a strength ratio (fatigue or brittle fracture). The size effect on strength is thus reduced by $2/(1 + n)$ in the exponent of the plate thickness ratio t_1/t_2. The strength ratio σ_{n2}/σ_{n1} over plate thickness ratio t_2/t_1 for weld-like corner-notched members ($2\alpha = 135°$) is plotted in Fig. 2.76, using the material parameters of the AISI 1045 high-strength steel. The fully plastic condition approximated in thin-sheet material on the left-hand side contrasts with the purely elastic condition approximated in very thick plates on the right-hand side.

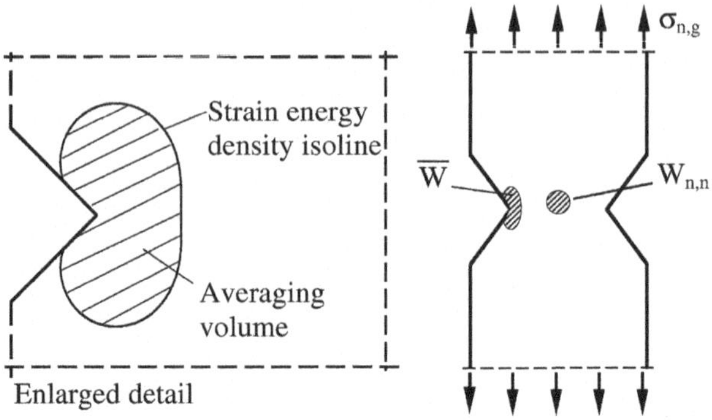

Fig. 2.77 Local volume for averaging the strain energy density W at the notch tip (resulting in $\overline{W}$) in comparison to its nominal value $W_{n,n}$ in the net cross-section; (Zappalorto and Lazzarin 2011$^{(2)}$)

2.5.4 Elastic-Plastic Fields at Tensile Loaded V-Notches, SED-Based Approach

It can be concluded from the expositions on tensile loaded V-notches above in comparison to the analytical work on out-of-plane shear loaded V-notches reviewed in the following subsections, that the governing equations in the former loading case (mode 1) are more complex so that they cannot be solved in a closed form as it is possible within the mode 3 approach (Unger 2001). A strain energy based approach to the mode 1 loading case is reviewed below (Zappalorto and Lazzarin 2011$^{(2)}$). The basic idea of the approach is, to determine the strain energy density isolines around the notch tip and to compare the elastic and plastic NSIF strain energy densities averaged within a volume or area bordered by the relevant isolines, Fig. 2.77.

The averaging volume or area has to be fully contained in the near-field, where the elastic and plastic singularity, respectively, is fully valid. The isolines of the linear-elastic strain energy density for two notch opening angles under plane stress and plane strain conditions are plotted in Fig. 2.78. Normalised coordinates $\overline{x}$, $\overline{y}$ are introduced which refer to the linear-elastic condition (complex analytical expressions). The isolines of the elastic-plastic strain energy density for different hardening exponents are also given in the quoted article.

Under plane stress conditions for in-plane loading (mode 1), the positioning of the strain energy density isolines, ranging from the linear-elastic to the perfectly plastic behaviour, is similar to that under out-of-plane shear loading (mode 3). Therefore, the following relationship between $\overline{W}_p$ and $\overline{W}_e$ is transferred from the mode 3 loading case, where it is strictly valid, to the mode 1 loading case as a well founded hypothesis:

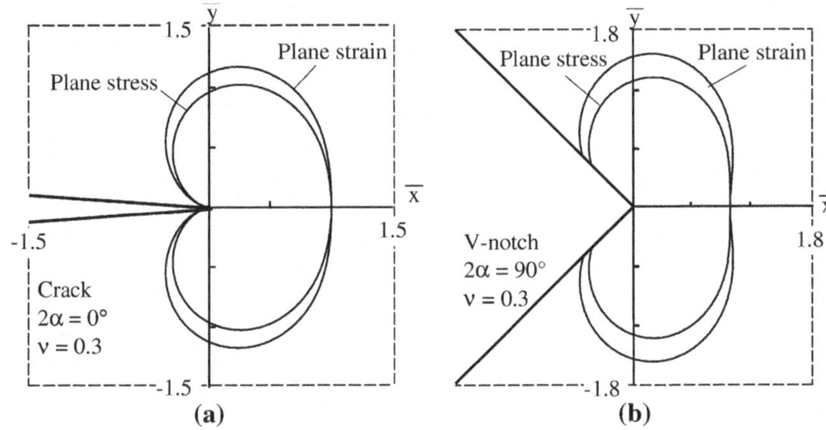

Fig. 2.78 Linear-elastic strain energy density isolines for V-notch with two different notch opening angles, $2\alpha = 0°$ (**a**) and $2\alpha = 90°$ (**b**) under plane stress and plane strain conditions in normalised coordinates $\bar{x}$, $\bar{y}$; Poisson's ratio $\nu = 0.3$; (Zappalorto and Lazzarin 2011[(2)])

$$\frac{\overline{W}_{\mathrm{p}}}{\overline{W}_{\mathrm{e}}} = \frac{2n}{n+1} \tag{2.194}$$

The following relationship is therefrom derived under the condition of small-scale yielding:

$$K_{1\mathrm{p}} = \left[K_{1\mathrm{e}}^2 \frac{H^n}{E}\left(\frac{I_{\mathrm{e}}}{I_{\mathrm{p}}}\right)^{1-\lambda_1}\right]^{1/n+1} \tag{2.195}$$

where I_{e} and I_{p} are integrals for evaluating the linear-elastic and nonlinear-elastic (named 'elastic-plastic') strain energy, respectively, within the isolines.

The analytical results are extended from the small-scale to the large-scale yielding conditions by writing Eq. (2.194) in terms of the 'energy concentration factors', $K_{W,\mathrm{p}}$ and $K_{W,\mathrm{e}}$:

$$K_{W,\mathrm{p}} = \frac{2n}{n+1} K_{W,\mathrm{e}}, \quad K_{W,\mathrm{p}} = \frac{\overline{W}_{\mathrm{p}}}{W_{\mathrm{n,n}}}, \quad K_{W,\mathrm{e}} = \frac{\overline{W}_{\mathrm{e}}}{W_{\mathrm{n,n}}} \tag{2.196}$$

Based on this relationship, the extended formula for $K_{1\mathrm{p}}$ reads:

$$K_{1\mathrm{p}} = \left[K_{1\mathrm{e}}^2 \frac{H^n}{E}\left(\frac{I_{\mathrm{e}}}{I_{\mathrm{p}}}\right)^{1-\lambda_1}\right]^{1/n+1} \times \left(1 + \frac{2n}{n+1}\frac{E}{H^n}\sigma_{\mathrm{n,n}}^{n-1}\right)^{1/n+1} \tag{2.197}$$

Under plane strain conditions for in-plane loading (mode 1), the above similarity to the out-of-plane shear behaviour (mode 3) does not exist because of the constraint effects ahead of the crack or notch tip. Therefore, the assumption

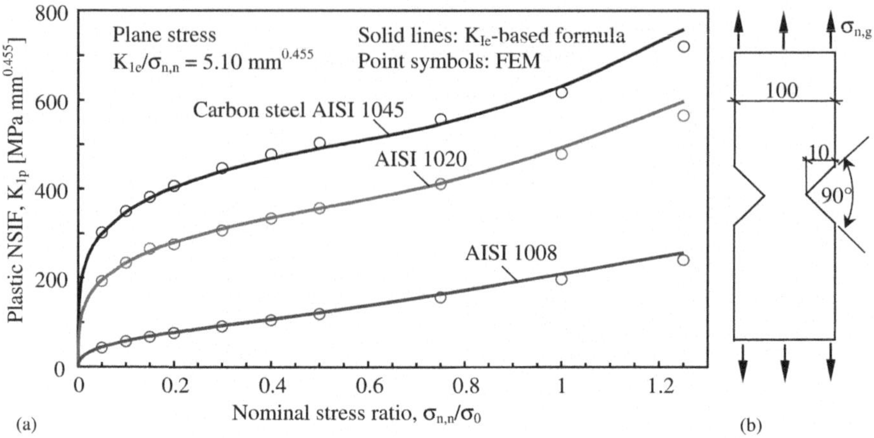

Fig. 2.79 Plastic NSIF (**a**) of V-notch ($2\alpha = 90°$) in tensile loaded flat bar (**b**) plotted over nominal stress $\sigma_{n,n}$ in net cross-section referred to substitute yield limit σ_0; plane stress conditions; (Zappalorto and Lazzarin 2011[(2)])

already used in an earlier investigation (Lazzarin and Zambardi 2002) remains valid, which referred to a circle sector of constant radius around the notch tip for averaging the strain energy density:

$$\frac{\overline{W}_p}{\overline{W}_e} = \frac{K_{W,p}}{K_{W,e}} = 1 \tag{2.198}$$

Then, applying the formal structure of Eq. (2.197), the result is:

$$K_{1p} = \left[\frac{n+1}{2n} K_{1e}^2 \frac{H^n}{E} \left(\frac{I_e}{I_p}\right)^{1-\lambda_1}\right]^{1/n+1} \times \left(1 + \frac{2n}{n+1} \frac{E}{H^n} \sigma_{n,g}^{n-1}\right)^{1/n+1} \tag{2.199}$$

One result from an investigation based on FE analysis verifying the above equations for K_{1p} as a function of K_{1e} under plane stress and plane strain conditions, respectively, is shown in Fig. 2.79. A tensile loaded flat bar with two-sided V-notch is considered. The material parameters correspond to the following three low-carbon and medium-carbon steels:

- AISI 1008: $E = 2.06 \times 10^5$ MPa, $H = 600$ MPa, $n = 4.00$, $\sigma_Y = 125$ MPa, $\nu = 0.3$
- AISI 1020: $E = 2.06 \times 10^5$ MPa, $H = 850$ MPa, $n = 6.66$, $\sigma_Y = 334$ MPa, $\nu = 0.3$
- AISI 1045: $E = 2.06 \times 10^5$ MPa, $H = 950$ MPa, $n = 8.33$, $\sigma_Y = 450$ MPa, $\nu = 0.3$

The K_{1p} curves determined according to Eq. (2.197) based on K_{1e} from a linear FE analysis are well in correspondence with the point symbols, which denote the K_{1p} values gained from nonlinear FE analysis using the total strain power law.

Based on the hypothesis for plane stress conditions, Eq. (2.194), the Neuber rule appears in terms of averaged stresses and strains corresponding to $\overline{W}_{\mathrm{p}}$ and $\overline{W}_{\mathrm{e}}$. The hypothesis for plane strain conditions, Eq. (2.198), on the other hand corresponds to the Glinka–Molski criterion in terms of averaged strain energy densities.

2.5.5 *Elastic-Plastic Fields at Out-of-Plane Shear-Loaded Parabolic Notches*

As shown in the preceding subsection, it has not been possible to solve the in-plane problem of elastic-plastic stress and strain fields at sharp notches (pointed or rounded) in a closed analytical form. The feasibility of a closed theory is substantially enhanced when considering the elastic-plastic fields in out-of-plane shear loading. Here, the governing nonlinear differential equations are easier to handle. But even here, the demands on mathematical analysis are very high.

In this subsection, an analytical study carried out by Zappalorto and Lazzarin is evaluated which is directed to the elastic-plastic field ahead of parabolic notches subjected to out-of-plane shear loading (Zappalorto and Lazzarin 2007). Small-scale yielding, but with inclusion of the stress redistribution effect, is investigated. Linear-elastic combined with perfectly plastic or power law elastic-plastic material behaviour is presumed. The analysis presents a first order ('leading term') field approximation and is therefore restricted to a small zone close to the notch tip or root. Torsional loaded round bars with a circumferential notch being the current application of out-of-plane shear field solutions, it is necessary to mention that the first order approximation must be combined with a correction term in this case, compare Eq. (2.147). The out-of-plane stress and strain symbols are used without the asterisk which was originally introduced in contrast to the in-plane shear symbols.

First, the linear-elastic followed by perfectly-plastic material behaviour is considered. The shear stresses τ_{xz} and τ_{yz} are proportional to the elastic generalised NSIF $K_{3\rho,\mathrm{e}}$. Referring to Eq. (2.153), $K_{3\rho,\mathrm{e}} = \tau_{\max}(\pi\rho)^{1/2}$ with the notch radius ρ. The shear strains γ_{xz} and γ_{yz} are proportional to $(K_{3\rho,\mathrm{e}})^2$.

Second, the linear-elastic followed by a power law elastic-plastic material behaviour, Eq. (2.281), is used as the basis of analysis. The nonlinear differential equations are transformed to a linear form by the approach described with some details in Sect. 2.5.6.

The power law of nonlinear-elastic or elastic-plastic material behaviour is written in the form of Eq. (2.286):

$$\frac{\gamma}{\gamma_0} = \left(\frac{\tau}{\tau_0}\right)^n \tag{2.200}$$

The considered resultant and simultaneously equivalent stresses and strains are:

$$\tau = \sqrt{\tau_{zx}^2 + \tau_{zy}^2} \tag{2.201}$$

$$\gamma = \sqrt{\gamma_{zx}^2 + \gamma_{zy}^2} \tag{2.202}$$

The stress and strain components follow from:

$$\begin{Bmatrix} \gamma_{zy} \\ \gamma_{zx} \end{Bmatrix} = \frac{\gamma_0}{\tau_0}\left(\frac{\tau}{\tau_0}\right)^{n-1}\begin{Bmatrix} \tau_{zy} \\ \tau_{zx} \end{Bmatrix} \tag{2.203}$$

The stress field in the near-tip plastic zone is controlled by the generalised elastic NSIF $K_{3\rho,\mathrm{e}}$:

$$\begin{Bmatrix} \tau_{zy} \\ \tau_{zx} \end{Bmatrix} = \left[\frac{nK_{3\rho,\mathrm{e}}^2}{\pi(n+1)}\tau_0^{n-1}\right]^{1/(n+1)}\left(\frac{\tilde{F}}{r}\right)^{1/(n+1)}\begin{Bmatrix} \cos\overline{\theta} \\ -\sin\overline{\theta} \end{Bmatrix} \tag{2.204}$$

where $\tilde{F} = \tilde{F}(n, \overline{\theta})$ and $\overline{\theta} = \overline{\theta}(n, \theta)$ are auxiliary functions. The corresponding strain field follows from Eq. (2.203).

The stress field equation above can be rewritten in analogy to the elastic case introducing the plastic NSIF $K_{3\rho,\mathrm{p}}$:

$$\begin{Bmatrix} \tau_{zy} \\ \tau_{zx} \end{Bmatrix} = \frac{K_{3\rho,\mathrm{p}}}{\sqrt{2\pi}}\left(\frac{\tilde{F}}{r}\right)^{1/(n+1)}\begin{Bmatrix} \cos\overline{\theta} \\ -\sin\overline{\theta} \end{Bmatrix} \tag{2.205}$$

With reference to the HRR field solution, the plastic NSIF is defined by the following limit value:

$$K_{3\rho,\mathrm{p}} = \lim_{r\to\rho/2} \sqrt{2\pi}\, r^{1/(n+1)}\tau_{zy}(r, 0) \tag{2.206}$$

Equating Eqs. (2.204) and (2.205), a relationship between plastic and elastic NSIFs under the condition of small-scale yielding (but with stress redistribution in the plastic zone) is established:

$$K_{3\rho,\mathrm{p}} = \sqrt{2\pi}\left[\frac{n}{\pi(n+1)}\tau_0^{n-1}K_{3\rho,\mathrm{e}}^2\right]^{1/(n+1)} \tag{2.207}$$

For the crack tip ($\rho = 0$), the following special form results (Rice 1967):

$$K_{\mathrm{III,p}} = \lim_{\rho\to 0} K_{3\rho,\mathrm{p}} = \sqrt{2\pi}\left[\frac{n}{\pi(n+1)}\tau_0^{n-1}K_{\mathrm{III,e}}^2\right]^{1/(n+1)} \tag{2.208}$$

The elastic and plastic NSIFs in terms of the maximum notch stresses are given as follows:

$$K_{3\rho,\mathrm{e}} = \tau_{\max,\mathrm{e}}\sqrt{\pi\rho} \tag{2.209}$$

$$K_{3\rho,\mathrm{p}} = \tau_{\max,\mathrm{p}}\sqrt{2\pi}\left(\frac{\rho}{2}\right)^{1/(n+1)} \tag{2.210}$$

When considering the maximum notch strains in parallel, a deviation from the original Neuber rule has been identified (compare Sect. 2.5.9).

In conclusion, an easy to survey closed form analytical solution for the elastic-plastic fields at parabolic notches (with inclusion of crack tips) subjected to out-of-plane shear loading has been derived by Zappalorto and Lazzarin. Uncertainties arise from the definition of the material law, because the Ramberg–Osgood relationship of metallic engineering materials has to be converted into the linear-elastic followed by power law elastic-plastic relationship of the analytical solution.

2.5.6 *Elastic-Plastic Fields at Out-of-Plane Shear-Loaded Pointed V-Notches*

Pointed V-notches subjected to out-of-plane shear loading are considered following the available analytical presentation (Lazzarin and Zappalorto 2008). The presumed material behaviour is linear-elastic followed by power-law strain hardening according to Eq. (2.200). Plastic and elastic NSIFs can be related under the condition of small-scale yielding. The near-tip stress and strain fields are then used to give closed-form expressions for the strain energy density averaged over a circle sector around the notch tip, as well as for the J-integral, both parameters as a function of the relevant plastic NSIF, these expressions being valid both for small-scale and large-scale yielding.

The basic nonlinear field equations resulting from the adoption of the nonlinear material law are converted to linear equations by introducing 'physical' coordinates as functions of stresses or of strains, similar to the hodograph transformation of fluid mechanics (Hult and McClintock 1956; Rice 1967). The resulting linear partial differential equation of second order in two parameters is solved without a series expansion (Rice 1967; Wang and Kuang 1999), based only on the 'leading order term' expressed by the nonlinear NSIF. The analysis details are passed over here, presenting the analysis results immediately.

In the near-tip plastic zone, the plastic NSIF $K_{3\mathrm{p}}$ is defined in analogy to the elastic NSIF K_3 as follows:

$$K_{3\mathrm{p}} = \lim_{r\to 0}\sqrt{2\pi}r^{1/(1-m)}\tau_{zy}(r,0) \tag{2.211}$$

where the parameter m in the exponent is dependent on n and 2α:

$$m = \frac{1}{2}\left(1 - n - \sqrt{(n-1)^2 + 4\omega^2 n}\right), \quad \omega = \frac{\pi}{\pi - 2\alpha} \tag{2.212}$$

Then, the elastic-plastic stress and strain fields are given dependent on $K_{3\mathrm{p}}$:

$$\left\{ \begin{array}{c} \tau_{zy} \\ \tau_{zx} \end{array} \right\} = \frac{K_{3p}}{\sqrt{2\pi}} \left(\frac{\tilde{F}}{r} \right)^{1/(1-m)} \left\{ \begin{array}{c} \cos \overline{\theta} \\ -\sin \overline{\theta} \end{array} \right\} \tag{2.213}$$

$$\left\{ \begin{array}{c} \gamma_{zy} \\ \gamma_{zx} \end{array} \right\} = \left(\frac{K_{3p}}{\sqrt{2\pi}} \right)^n \frac{\gamma_0}{\tau_0^n} \left(\frac{\tilde{F}}{r} \right)^{n/(1-m)} \left\{ \begin{array}{c} \cos \overline{\theta} \\ -\sin \overline{\theta} \end{array} \right\} \tag{2.214}$$

where τ_0 and γ_0 refer to the resultant or equivalent stresses and strains, Eq. (2.201) and (2.202), $\tilde{F} = \tilde{F}(m, \omega, \overline{\theta})$ and $\overline{\theta} = \overline{\theta}(m, \omega, \theta)$.

The relationship between plastic and elastic NSIFs under the condition of small-scale yielding (but with stress redistribution in the plastic zone) has the following form:

$$K_{3p} = \sqrt{2\pi} \left[-\frac{m}{\lambda_3(1-m)} \left(\frac{K_{3e}}{\sqrt{2\pi}} \right)^{1/(1-\lambda_3)} \frac{1}{\tau_0^{m+\omega}} \right]^{1/(1-m)} \tag{2.215}$$

For crack tips ($2\alpha = 0$, $\lambda_3 = 0.5$, $\omega = 1$, $m = -n$), Eq. (2.208) is confirmed.

The plastic zone considerations below refer to the reference stress τ_0 as the substitute yield limit. It is not identical with the real yield limit τ_Y. Therefore, the parameters r_p and x_p below are also substitute parameters.

The shape and size of the plastic zone is found by considering the shear stress τ_0 isolines (Tresca yield criterion). Under the condition of small-scale yielding (but with stress redistribution), the well known circular shapes at crack tips ($2\alpha = 0$) occur, but for V-notches ($2\alpha > 0$) oval shapes are typical. The oval shape gets more slender when strain hardening is weak (or the hardening exponent large).

In applications, the plastic zone size is often estimated by applying the yield criterion on the elastic stress field, resulting in the radius r_p of a circular zone around the crack or notch tip:

$$r_p = \left(\frac{K_{3e}}{\sqrt{2\pi}\tau_0} \right)^{1/(1-\lambda_3)} \tag{2.216}$$

According to the analysis above, the extension x_p of the plastic zone along the notch bisector line is enlarged by the stress redistribution:

$$x_p = -\frac{m}{\lambda_3(1-m)} r_p \tag{2.217}$$

For the crack ($2\alpha = 0$, $\lambda_3 = 0.5$, $m = -n$), this equation reproduces a result from the literature (Rice 1967):

$$x_p = \frac{2n}{n+1} r_p \tag{2.218}$$

The strain energy density, averaged over a small circle sector zone around the notch tip is the basis of fatigue assessment for pointed notches in engineering.

Therefore, it is appropriate to express the strain energy density by the plastic NSIF and vice versa.

Considering the power law in terms of total strains, Eq. (2.200), and assuming that the plastic strains are substantially larger than the elastic strains, the total and plastic strain energy densities can be set equal, approximated by:

$$W_{\mathrm{p}} = \int_0^{\gamma} \tau \, d\gamma = \int_0^{\gamma} \tau_0 \left(\frac{\gamma}{\gamma_0}\right)^{1/n} d\gamma = \frac{n}{n+1} \frac{\tau^{n+1}}{G\tau_0^{n-1}} \tag{2.219}$$

where τ and γ are the resultant or equivalent stresses and strains, Eqs. (2.201) and (2.102). Inserting the stress components τ_{zy} and τ_{zx}, Eqs. (2.213) and (2.214), the result is:

$$W_{\mathrm{p}} = \frac{n}{n+1} \frac{1}{G\tau_0^{n-1}} \left[\frac{K_{3\mathrm{p}}}{\sqrt{2\pi}} \left(\frac{\tilde{F}}{r}\right)^{1/(1-m)}\right]^{n+1} \tag{2.220}$$

The strain energy density averaged over the circle sector with radius R at the notch tip results by integration of Eq. (2.220), stating the parameter B_W:

$$\overline{W}_{\mathrm{p}} = B_W(2\alpha, n) \frac{K_{3\mathrm{p}}^{n+1}}{G\tau_0^{n-1}} R^{(n+1)/(m-1)} \tag{2.221}$$

For the J-integral, a similar expression is found, stating the parameter B_J:

$$J_{3\mathrm{p}} = B_J(2\alpha, n) \frac{K_{3\mathrm{p}}^{n+1}}{G\tau_0^{n-1}} R^{(n+m)/(m-1)} \tag{2.222}$$

Several FE analyses have been performed, in order to scale the plastic NSIFs of the investigated notched member and then to demonstrate their potential to describe the elastic-plastic field at the pointed notch tips inclusive of the average strain energy density and the J-integral. Out-of-plane shear loading is generated by a torsional moment applied to round bars with a circumferential slit ($2\alpha = 0$) or V-notch ($2\alpha = 120°$). The outer diameter is 60 mm, the inner diameter 40 mm. The two steels AISI 1008 and AISI 1045 with the material parameters specified in Sect. 2.5.3 were simulated. An important point was, to what extent the total strain power law in the theoretical analysis comes up to the well-proven Ramberg–Osgood law in the FE analysis.

The material parameters in the total strain power law in terms of the parameters in the Ramberg–Osgood law are defined as follows. It is generally accepted, not to modify the exponent n which characterises the degree of the singularity. Actually, there is no difference in the two laws when considering very large strains, which constitute the singularity. The reference values σ_0, ε_0 or τ_0, γ_0 result from the intersection point of the linear-elastic with the nonlinear power law curve (for the conversion from σ, ε to τ, γ see Sect. 2.5.10). Lazzarin and Zappalorto propose to equate the strain energy densities W_{p} resulting with the two material laws.

Integration of the plastic component in the Ramberg–Osgood law gives:

$$W_{\rm p} = \frac{n}{n+1}\frac{\sigma_{\rm eq}^{n+1}}{H^n} = \frac{n\,3^{(n+1)/2}}{n+1}\frac{\tau^{n+1}}{H^n} \quad (\sigma_{\rm eq} = \sqrt{3}\tau) \tag{2.223}$$

where $\sigma_{\rm eq}$ is the von Mises equivalent stress.

Equating this expression for $W_{\rm p}$ with that of the total strain power law, Eq. (2.220), the result is:

$$H = \left(G\tau_0^{n-1}3^{(n+1)/2}\right)^{1/n}, \quad \tau_0 = \left(\frac{H^n}{3^{(n+1)/2}G}\right)^{1/(n-1)} \tag{2.224}$$

The combination of Eqs. (2.221) and (2.224) gives a further expression for the plastic NSIF:

$$K_{3\rm p} = \left(\frac{\overline{W}_{\rm p}H^n}{3^{(n+1)/2}B_W(2\alpha,\, n)}\right)^{1/(n+1)} R^{1/(1-m)} \tag{2.225}$$

Considering two of the carbon steels in the comparative investigation of Sect. 2.5.4, the following conversions were carried out:

- Steel AISI 1008: $\sigma_{\rm Y} = 125$ MPa, $\tau_{\rm Y} = \sigma_Y/\sqrt{3} = 72$ MPa, $\tau_0 = 47$ MPa
- Steel AISI 1045: $\sigma_{\rm Y} = 450$ MPa, $\tau_{\rm Y} = \sigma_Y/\sqrt{3} = 259$ MPa, $\tau_0 = 258$ MPa

The numerical results in the article of Lazzarin and Zappalorto are demonstrated in the following by a set of typical plots referring to the V-notch ($2\alpha = 120°$) in a round bar specimen subjected to torsional loading. All the plotted data are based on FE analysis results, because a fully analytical solution for the finite size configuration is impossible. The analytically derived relationships have been scaled by FE analysis results. What is checked by the plots is, whether or to what extent the theoretically predicted correlations are confirmed in actual configurations.

In Fig. 2.80, the determination of the plastic NSIFs $K_{3\rm p}$ from the shear stresses in the notch bisector plane (FE analysis results) according to Eq. (2.211) is shown for different nominal shear stresses $\tau_{\rm n,n}$ in the net cross-section. Horizontal curves over the distance from the notch tip are found. Their length increases with the extension of the plastic zone, i.e. with the nominal stress level. In Fig. 2.81, the shear stresses in the bisector plane are plotted, determined from the FE analysis results directly and, in comparison, from the FE-based $K_{3\rm p}$ values. The gradients in the theoretical analysis are fully confirmed by the FE analysis results.

In Fig. 2.82, the plastic NSIF $K_{3\rm p}$ is plotted over the nominal stress $\tau_{\rm n,n}$ normalised by the yield limit $\tau_{\rm Y}$. The $K_{3\rm p}$ values determined from the FE stresses τ_{zy} and from the average strain energy density $\overline{W}_{\rm p}$ are well in agreement, whereas the $K_{3\rm p}$ values based on the elastic $K_{3,\rm e}$ value are too low for nominal stresses above the yield limit. In Fig. 2.83, it is shown that $K_{3\rm p}$ can be approximated by a linear function of $\tau_{\rm n,n}/\tau_{\rm Y}$.

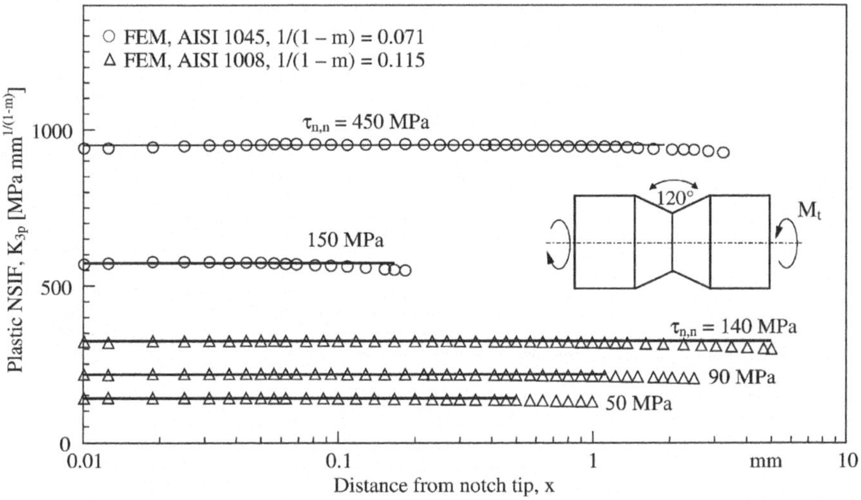

Fig. 2.80 Plastic NSIF of V-notch ($2\alpha = 120°$) in round bar subjected to torsional loading; evaluation of FE analysis results according to Eq. (2.211) (point symbols with medium lines); steels AISI 1045 and AISI 1008; nominal stress $\tau_{n,n}$ in net cross-section; (Lazzarin and Zappalorto 2008)

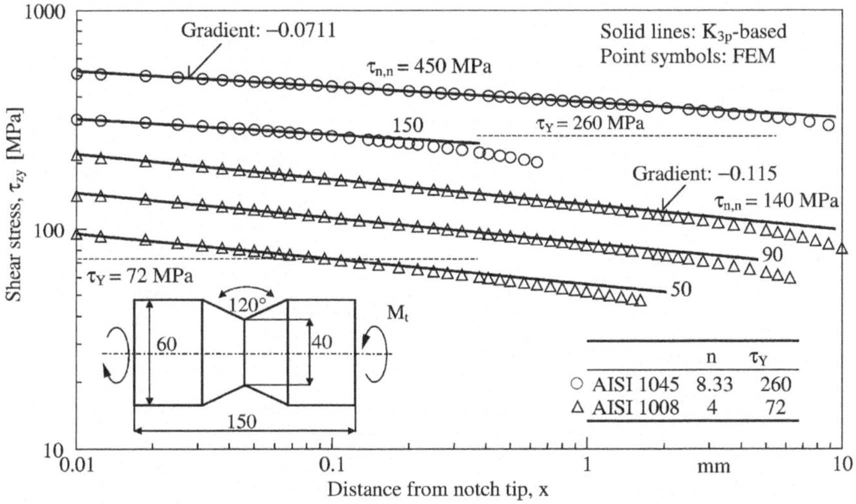

Fig. 2.81 Shear stress in bisector plane of V-notch ($2\alpha = 120°$) in round bar subjected to torsional loading; FE analysis results (point symbols) compared with K_{3p}-based values (solid lines); steels AISI 1045 and AISI 1008; nominal stress $\tau_{n,n}$ in net cross-section; (Lazzarin and Zappalorto 2008)

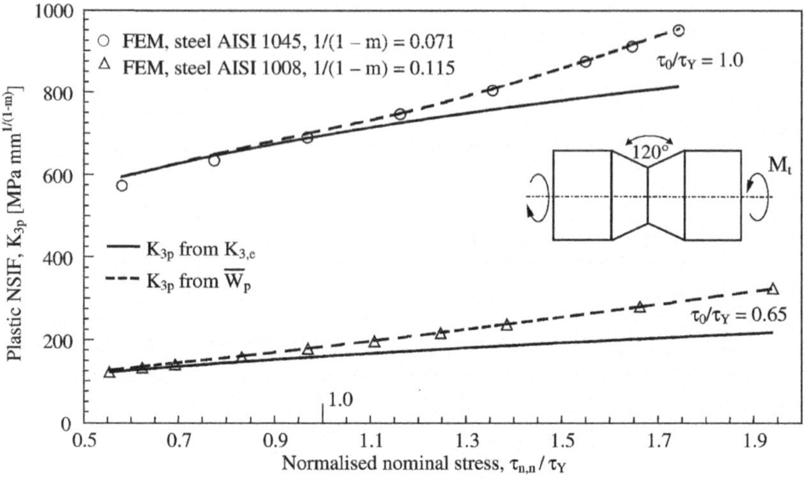

Fig. 2.82 Plastic NSIF of V-notch ($2\alpha = 120°$) in round bar subjected to torsional loading; derived from elastic NSIF (solid lines), from averaged strain energy density in circular sector of the FE model (broken lines) and from FE stresses (point symbols); steels AISI 1045 and AISI 1008; nominal stress $\tau_{n,n}$ in net cross-section; (Lazzarin and Zappalorto 2008)

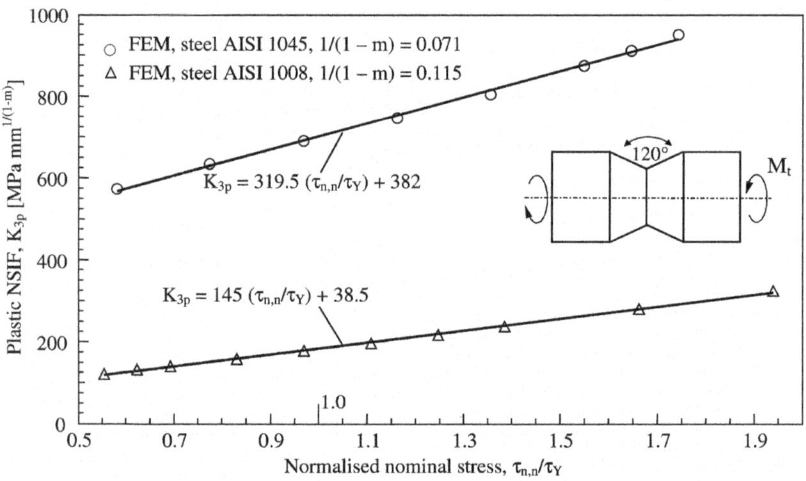

Fig. 2.83 Plastic NSIF at V-notch ($2\alpha = 120°$) in round bar subjected to torsional loading; evaluation of FE analysis results with linear medium line; steels AISI 1045 and AISI 1008; nominal stress $\tau_{n,n}$ in net cross-section; (Lazzarin and Zappalorto 2008)

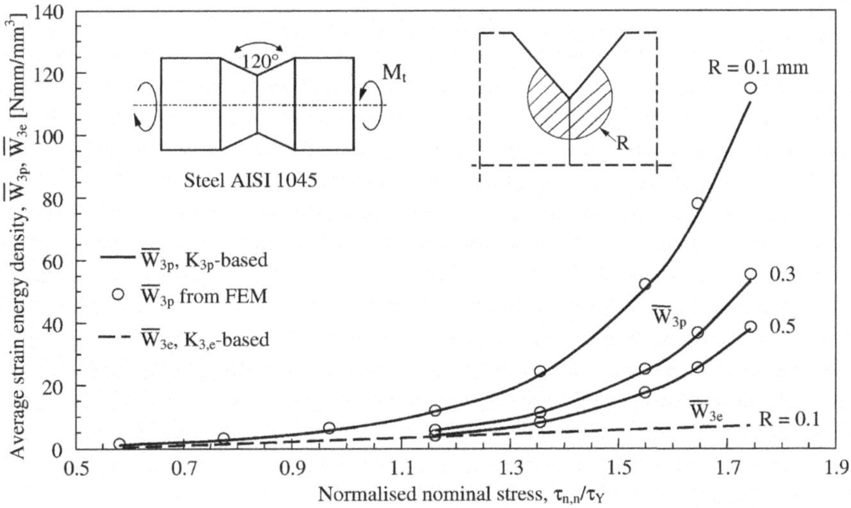

Fig. 2.84 Average plastic and elastic strain energy density at V-notch ($2\alpha = 120°$) in round bar subjected to torsional loading; direct evaluation of FE analysis results (point symbols) versus FE-based evaluation of plastic NSIFs K_{3p}, Eq. (2.221); linear-elastic behaviour in comparison; steel AISI 1045; nominal stress $\tau_{n,n}$ in net cross-section; (Lazzarin and Zappalorto 2008)

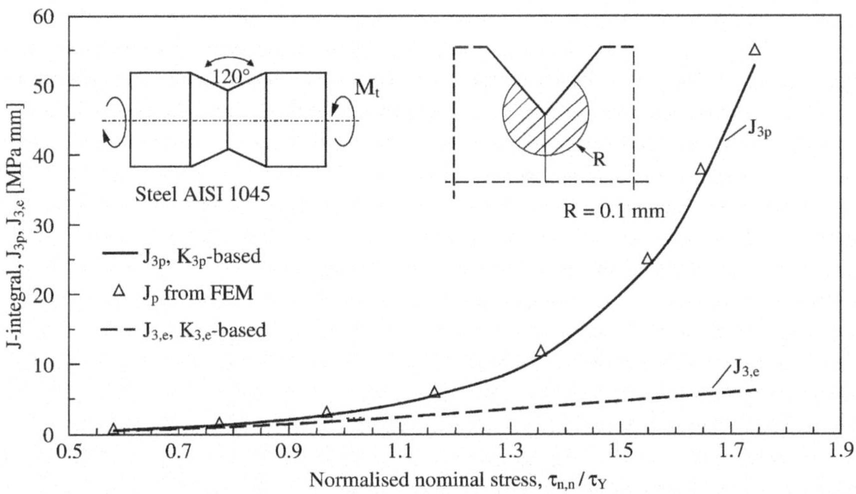

Fig. 2.85 Plastic and elastic *J*-integral at V-notch ($2\alpha = 120°$) in round bar subjected to torsional loading; direct evaluation of FE analysis results (point symbols) compared with FE-based evaluation of plastic NSIFs K_{3p}, Eq. (2.222); linear-elastic behaviour in comparison; steel AISI 1045; nominal stress $\tau_{n,n}$ in net cross-section; (Lazzarin and Zappalorto 2008)

In Fig. 2.84, the average plastic strain energy density $\overline{W}_{\rm p}$ is plotted over $\tau_{\rm n,n}/\tau_{\rm Y}$, direct evaluation of FE analysis results compared with FE-based evaluation of plastic NSIFs $K_{\rm 3p}$, Eq. (2.215). The average elastic strain energy density $\overline{W}_{\rm e}$ is shown in comparison. In Fig. 2.85, the plastic J-integral $J_{\rm 3p}$ is drawn over $\tau_{\rm n,n}/\tau_{\rm Y}$, direct evaluation of FE analysis results contrasted with FE-based evaluation of $J_{\rm 3p}$, Eq. (2.222). The elastic J-integral $J_{\rm 3,e}$ is shown in comparison.

The following conclusions are drawn from the investigation reported by Lazzarin and Zappalorto: The elastic-plastic boundary assumes an oval shape for V-notches ($2\alpha > 0$) under small-scale yielding conditions. The entire stress and strain field in the plastic zone around the notch tip is unambiguously expressed by the plastic NSIF. A simple relationship exists between elastic and plastic NSIFs under the condition of small-scale yielding. The plastic NSIF can be approximated as linearly dependent on the ratio $\tau_{\rm n}/\tau_{\rm Y}$ in the full range of small-scale and large-scale yielding. The plastic strain energy density averaged over a circle sector centred at the notch tip can be given in closed form as a function of the plastic NSIF. The same applies to the J-integral.

2.5.7 Uniform Analysis of Nonlinear Fields at Out-of-Plane Shear-Loaded V-Notches

In this subsection, an overview is given of the uniform approach to the analysis of nonlinear stress and strain fields at out-of-plane shear loaded V-notches (pointed or rounded) and cracks, which was presented by Zappalorto and Lazzarin in a pioneering article (Zappalorto and Lazzarin 2010). The mathematical tools were already developed in previous works (Hult and McClintock 1956; Neuber 1958; Rice 1967). Most important in respect of the uniform approach is the conversion of the governing nonlinear field equations resulting from the nonlinear material law into linear equations by introducing ‘physical coordinates’ as functions of stresses or strains (similar to the hodograph transformation in fluid mechanics). The appertaining series expansion can be confined to the leading order term, when the analysis is limited to the near-field at the notch or crack tip. Higher order fields have been investigated by other authors (Yuan and Yang 1994/95, Yang et al. 1996). The transformation above makes it possible to find a closed form solution of the anti-plane problem as soon as the nonlinear material law is defined. The reviewed publication is related primarily to the Neuber nonlinear material law, but comprises also the nonlinear power law and an attempt to include the Ramberg–Osgood law.

Considering the Neuber nonlinear material law according to Eqs. (2.287) and (2.288), Fig. 2.96, and taking advantage of a new polynomial form (after Legendre) of the above transformation, the closed form solution is gained under small-scale yielding conditions (Zappalorto and Lazzarin 2010). As the inversion to the r–θ plane is far from easy, the authors propose to combine two limit solutions, one for large strains close to the notch tip, the other for small strains at a certain distance from it.

For a clearer formal presentation of the results, notch strain intensity factors are introduced in analogy to the notch stress intensity factors, extending definitions used in the linear-elastic into the nonlinear-elastic range. Considering the notch bisector plane ($\theta = 0$) these elastic and plastic notch strain intensity factors are defined as follows:

$$K^{(\gamma)}_{3\rho,\mathrm{e}} = \lim_{r \to r_0} \sqrt{2\pi} r^{1-\lambda_3} \gamma(r, 0), \quad K^{(\gamma)}_{3\rho,\mathrm{p}} = \lim_{r \to r_0} \sqrt{2\pi} r^{-s_{\gamma,\mathrm{p}}} \gamma(r, 0) \tag{2.226}$$

where the subscript ρ indicates that the notch root radius may be different from zero and the exponent $s_{\gamma,\mathrm{p}}$ depends on the material law.

Considering the Neuber nonlinear material law ($s_{\gamma,\mathrm{p}} = -1$), the following relationships are derived for the linear-elastic and nonlinear-elastic (i.e. plastic) generalised notch strain and notch stress intensity factors:

$$K^{(\gamma)}_{3\rho,\mathrm{e}} = \sqrt{2\pi}(-\omega_1 C_1 \gamma_0^{\omega_1})^{1-\lambda_3}, \quad K_{3\rho,\mathrm{e}} = G\sqrt{2\pi}(-\omega_1 C_1 \gamma_0^{\omega_1})^{1-\lambda_3} \tag{2.227}$$

$$K^{(\gamma)}_{3\rho,\mathrm{p}} = -\sqrt{2\pi}\omega_1 2^{1-\omega_1}(1+\omega_1)C_1, \quad K_{3\rho,\mathrm{p}} = \sqrt{2\pi}\tau_0 \tag{2.228}$$

where $\omega_1 = \pi/(\pi - 2\alpha)$ and where C_1 is a parameter which defines the level of the strains both in the elastic and plastic zone.

The following relationship between the plastic notch strain intensity factor and the elastic NSIF has been derived:

$$K^{(\gamma)}_{3\rho,\mathrm{p}} = \sqrt{2\pi} 2^{1-\omega_1}(1+\omega_1)\left(\frac{K_{3\rho,\mathrm{e}}}{\sqrt{2\pi}\tau_0}\right)^{\omega_1+1} \gamma_0 \tag{2.229}$$

The nonlinear (elastic-plastic) strains close to the notch tip are dependent on the plastic notch strain intensity factor:

$$\begin{Bmatrix} \gamma_{zx} \\ \gamma_{zy} \end{Bmatrix} = \frac{K^{(\gamma)}_{3\rho,\mathrm{p}}}{\sqrt{2\pi}} \frac{1}{r} \begin{Bmatrix} -\cos\omega_1\theta \sin\theta \\ \cos\omega_1\theta \cos\theta \end{Bmatrix} \tag{2.230}$$

The corresponding stresses are given by:

$$\begin{Bmatrix} \tau_{zx} \\ \tau_{zy} \end{Bmatrix} = \tau_0 \begin{Bmatrix} -\sin\theta \\ \cos\theta \end{Bmatrix} \tag{2.231}$$

The linear-elastic strains and stresses at a sufficient distance from the notch tip are expressed by the elastic NSIF:

$$\begin{Bmatrix} \gamma_{zx} \\ \gamma_{zy} \end{Bmatrix} = \frac{K_{3\rho,\mathrm{e}} r^{\lambda_3 - 1}}{\sqrt{2\pi} G} \begin{Bmatrix} -\sin(1-\lambda_3)\theta \\ \cos(1-\lambda_3)\theta \end{Bmatrix} \tag{2.232}$$

$$\begin{Bmatrix} \tau_{zx} \\ \tau_{zy} \end{Bmatrix} = \frac{K_{3\rho,\mathrm{e}} r^{\lambda_3 - 1}}{\sqrt{2\pi}} \begin{Bmatrix} -\sin(1-\lambda_3)\theta \\ \cos(1-\lambda_3)\theta \end{Bmatrix} \tag{2.233}$$

When the notch root radius is different from zero, the solution can also be expressed by the maximum notch stress or strain, based on the following relationships:

$$K_{3\rho,\mathrm{p}}^{(\gamma)} = \sqrt{2\pi}\gamma_{\max,\mathrm{p}} r_0, \quad \tau_{\max,\mathrm{p}} = \tau_0, \quad K_{3\rho,\mathrm{e}} = \sqrt{2\pi}\tau_{\max,\mathrm{e}} r_0^{1-\lambda_3} \tag{2.234}$$

where r_0 is the distance of the notch root from the origin of the polar coordinate system.

The stress and strain distributions along the notch bisector plane of a pointed V-notch ($\rho = 0$, $2\alpha = 120°$) subjected to out-of-plane shear loading are plotted in Fig. 2.86. The complete solution based on Legendre polynomials and its two limit lines, both according to Zappalorto and Lazzarin, are compared with Neuber's different analytical approach (Neuber 1961). The numerical results are fully corresponding. The plot refers to the pointed notch ($\rho = 0$), but is transferable 'in principle' to cases of a small finite notch radius ($\rho > 0$), provided the coordinate x is truncated on the left hand side of the notch root.

The above theory has also been used to determine the substitute elastic-plastic boundary under small-scale yielding conditions by imposing the condition $\gamma = \gamma_0$. The complete solution is compared with the purely elastic and purely plastic solutions. The plastic zone according to the complete solution presented by Zappalorto and Lazzarin is substantially larger than according to the elastic and plastic limit solutions.

Considering the total strain power law according to Eq. (2.286), Fig. 2.95, and applying the same analytical tools as before, the following expressions for the stresses and strains in the close vicinity of the notch tip are derived from a stress-based (as opposed to strain-based) formulation of the nonlinear problem:

$$\begin{Bmatrix} \tau_{zx} \\ \tau_{zy} \end{Bmatrix} = \frac{K_{3\rho}}{\sqrt{2\pi}} \left(\frac{\tilde{F}}{r}\right)^{1/(1-m)} \begin{Bmatrix} -\sin\overline{\theta} \\ \cos\overline{\theta} \end{Bmatrix} \tag{2.235}$$

$$\begin{Bmatrix} \gamma_{zx} \\ \gamma_{zy} \end{Bmatrix} = \left(\frac{K_{3\rho}}{\sqrt{2\pi}}\right)^{n} \frac{\gamma_0}{\tau_0^n} \left(\frac{\tilde{F}}{r}\right)^{n/(1-m)} \begin{Bmatrix} -\sin\overline{\theta} \\ \cos\overline{\theta} \end{Bmatrix} \tag{2.236}$$

where $K_{3\rho}$ is the nonlinear NSIF, generalised for any value of ρ; compare Eqs. (2.213) and (2.214), where $\rho = 0$ and where 'nonlinear' is interpreted as 'plastic'. The exponent m corresponds to Eqs. (2.211) and (2.212). $\tilde{F} = \tilde{F}(m, \omega_1, \overline{\theta})$ and $\overline{\theta} = \overline{\theta}(m, \omega_1, \theta)$ are auxiliary parameters.

Finally, an attempt has been made by Zappalorto and Lazzarin to achieve a solution for the stress and strain field at the notch tip when considering the Ramberg–Osgood material law in the following special form:

$$G\gamma = \tau + \delta\tau^n \quad (1 < n < \infty) \tag{2.237}$$

where G is the shear modulus, n the hardening exponent, δ a material constant ($\delta = \tau_0^{1-n} = G/H_\tau^n$) and τ_0, γ_0 marking the intersection point of the linear-elastic

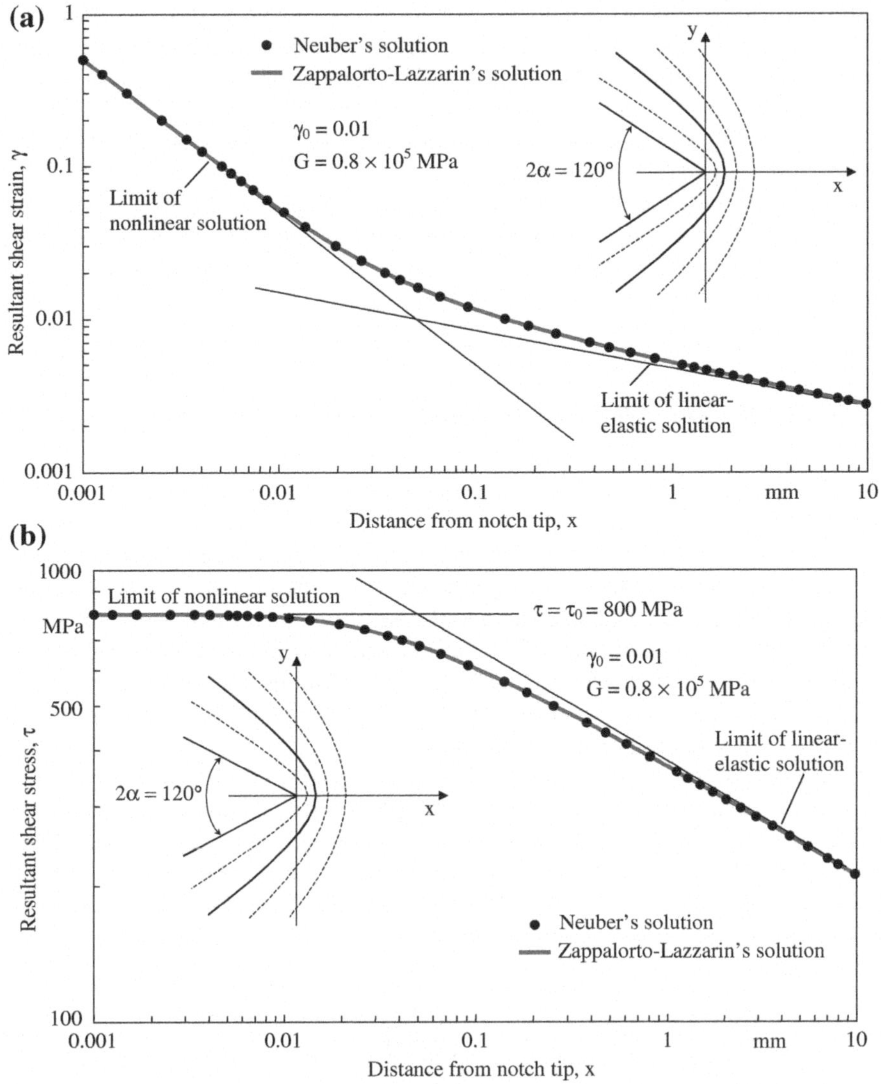

Fig. 2.86 Elastic-plastic strain distribution (**a**) and stress distribution (**b**) along the notch bisector plane of pointed V-notch ($\rho = 0$, $2\alpha = 120°$) subjected to out-of-plane shear loading; solutions based on the Neuber nonlinear material law; Neuber's compared with Zappalorto–Lazzarin's analytical approach; (Zappalorto and Lazzarin 2010)

curve and the asymptotic plastic total strain power law curve. Equation (2.237) can be derived from the normalised form of the Ramberg–Osgood relationship, Eq. (2.173), by setting $\alpha = 1.0$ there. This corresponds to the total strain power law combined with the initially linear-elastic behaviour. The Eq. (2.237) may be termed 'the asymptotic form' of the Ramberg–Osgood relationship.

A complete solution for V-notches with notch opening angle different from zero by application of the above analytical tools is not possible, but the two limit conditions for τ allow special solutions. The limit condition $\tau \rightarrow \infty$ of purely plastic behaviour at the notch tip results in the already derived Eqs. (2.235) and (2.236) with $K_{3\rho} = K_{3\rho,\mathrm{p}}$. The limit condition $\tau \rightarrow 0$ of linear-elastic behaviour results in the Eqs. (2.232) and (2.233).

For the special case of a notch opening angle equal to zero corresponding to pointed or blunt crack tips, a complete solution based on the asymptotic form of the Ramberg–Osgood relationship has been presented (Zappalorto and Lazzarin 2010).

2.5.8 *Plastic Zone Related to Elastic Higher Order Singularities*

The higher order singularities at crack tips and V-notch tips are generally omitted in the linear-elastic analysis with the argument that the strain energy and the displacements in the near-tip region should remain bounded. But it has been shown that the higher singular terms are significant in the elastic field outside the non-linear zone around the crack tip (Hui and Ruina 1995). Later on, it has been proven that the higher order singular terms can be included in the elastic part of the complete elastic-plastic stress and strain field solutions for out-of-plane shear loading (Zappalorto and Lazzarin 2011[(4)]). The findings of the two authors just mentioned are summarised in the following.

At crack tips, higher order singularities of the order (1 − 2j)/2 with j = 2, 3, 4, 5,... are found which are governed by the order 1 stress intensity factor K_{III}. The Neuber material law, Eq. (2.287), is applied under small-scale yielding conditions. The higher order terms are found to be dependent, besides on K_{III}, on the substitute radius r_{p} of the plastic zone around of the crack tip. The strains in the ligament under out-of-plane shear loading conditions read as follows, whereas the stresses result from $\tau = G\gamma$:

$$\gamma = \frac{K_{\mathrm{III}}}{G\sqrt{2\pi x}}\left[1 + \frac{r_{\mathrm{p}}}{4x} + \frac{1}{32}\left(\frac{r_{\mathrm{p}}}{x}\right)^2 - \frac{1}{128}\left(\frac{r_{\mathrm{p}}}{x}\right)^3 - \frac{5}{2048}\left(\frac{r_{\mathrm{p}}}{x}\right)^4 + \ldots\right] \tag{2.238}$$

$$r_{\mathrm{p}} = \frac{K_{\mathrm{III}}^2}{2\pi\tau_0^2} \tag{2.239}$$

For crack tips, now using the Ramberg–Osgood material law in the form of Eq. (2.237), $G\gamma = \tau + \delta\tau^n$ with $\delta = \tau_0^{1-n} = G/H_\tau^n$, the following relationship is derived, with the hardening exponent $n = 4.0$ as an example, once more in the ligament under out-of-plane shear loading conditions:

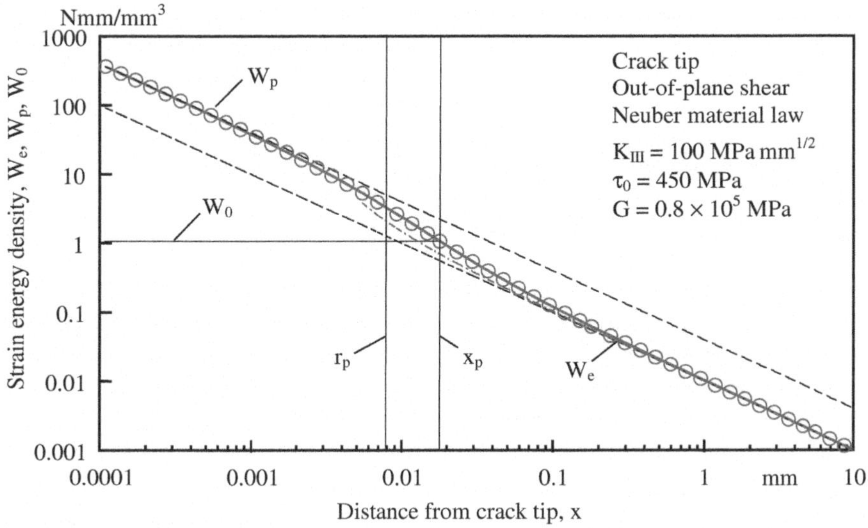

Fig. 2.87 Plastic (W_p) and elastic (W_e) strain energy density over distance x from the crack tip; substitute yield limit γ_0, γ_0 defines r_p from linear-elastic analysis and x_p from elastic-plastic analysis (Zappalorto and Lazzarin 2011(4))

$$\tau = \frac{K_{\mathrm{III}}}{\sqrt{2\pi x}}\left[1 + \frac{2\pi}{3\sqrt{6}}\left(\frac{r_{\mathrm{p}}}{x}\right) - \sqrt{2}\left(\frac{r_{\mathrm{p}}}{x}\right)^{3/2} + \frac{\sqrt{2}\pi^2}{9}\left(\frac{r_{\mathrm{p}}}{x}\right)^2 - \frac{8\sqrt{2}\pi}{3\sqrt{3}}\left(\frac{r_{\mathrm{p}}}{x}\right)^{5/2} + \ldots\right] \tag{2.240}$$

$$\gamma = \frac{1}{G}(\tau + \delta\tau^4) = \frac{1}{G}\left[1 + \left(\frac{\tau}{\tau_0}\right)^3\right]\tau \quad (n = 4) \tag{2.241}$$

The conclusion from Eqs. (2.238) and (2.240) is, that the intensities of the higher order terms depend on the plastic zone radius r_{p}, which is defined by the substitute yield limit τ_0 (not identical with the real yield limit τ_{Y}).

The physical necessity of the higher order singular terms is proven on the basis of the strain energy density distribution. The asymptotic densities along the ligament in the elastic zone, W_{e}, and in the plastic zone, W_{p}, considering the Neuber material law are given by:

$$W_{\mathrm{e}} = \frac{1}{2}G\gamma^2 = \frac{K_{\mathrm{III}}^2}{4\pi G}\frac{1}{x} \tag{2.242}$$

$$W_{\mathrm{p}} = \tau_0\gamma = \frac{K_{\mathrm{III}}}{\pi G}\frac{1}{x} \quad (\gamma \gg \gamma_0) \tag{2.243}$$

It is obvious from the above asymptotic behaviour, that the strain energy density level in the plastic zone is four times greater than in the elastic zone with a smooth transition according to the complete solution, Fig. 2.87. The strain energy density corresponding to the substitute yield limit $\gamma = \gamma_0$ is given by:

$$W_0 = \tau_0 \gamma_0 \left(\sqrt{2} - 1\right) \quad (\gamma = \gamma_0) \tag{2.244}$$

This strain energy density occurs on the ligament at the distance x_p:

$$x_p = \left[\sqrt{2} + \ln\left(1 + \sqrt{2}\right)\right] r_p = 2.3\, r_p \tag{2.245}$$

For V-notch tips, similar relationships for the stress and strain field have been derived by the same authors based on the Neuber material law under small-scale yielding conditions. Here, it was found that the higher order terms are governed by the first linear-elastic eigenvalues and not by higher order solutions. Once more, the higher order terms can be presented within a relationship governed by K_3 and dependent on the radius r_p of the substitute plastic zone around the V-notch tip.

The notch opening angle $2\alpha = 135°$, $\lambda_3^{(1)} = 0.8$, is considered as an example. The strains in the bisector plane under out-of-plane shear loading conditions read as follows, whereas the stresses result from $\tau = G\gamma$:

$$\gamma = \frac{K_3}{G\sqrt{2\pi}x^{1/5}} \left[1 + \frac{1}{3}\left(\frac{r_p}{x}\right)^{2/5} + \frac{1}{8}\left(\frac{r_p}{x}\right)^{4/5} + \frac{5}{54}\left(\frac{r_p}{x}\right)^{6/5} + \frac{41}{2592}\left(\frac{r_p}{x}\right)^{8/5} + \ldots\right] \tag{2.246}$$

$$r_p = \left(\frac{K_3}{\tau_0\sqrt{2\pi}}\right)^{1/(1-\lambda_3)} = \left(\frac{K_3}{\tau_0\sqrt{2\pi}}\right)^5 \tag{2.247}$$

The conclusion is that the higher order singular terms can fully be included in the elastic part of the complete elastic-plastic stress and strain field solutions. The intensity of these additional terms depends on the elastic stress intensity factor and the substitute plastic zone radius r_p.

2.5.9 Neuber Rule Including the Influence of the Notch Opening Angle

The reviewed analytical developments related to elastic-plastic fields at sharp notches make it possible to include the influence of the notch opening angle in the Neuber rule where an unexpected effect in total is observed. Before presenting these results in detail, the Neuber rule is explained in general. It refers to rounded notches with exclusion of pointed notches, in contrast to the NSIF approach, which refers to pointed notches with inclusion of rounded notches.

The elastic-plastic notch stresses and strains can be approximated by predominantly uniaxial models and correspondingly simple formulae. Widely used is Neuber's 'macrosupport formula' besides an approximation by Glinka based on

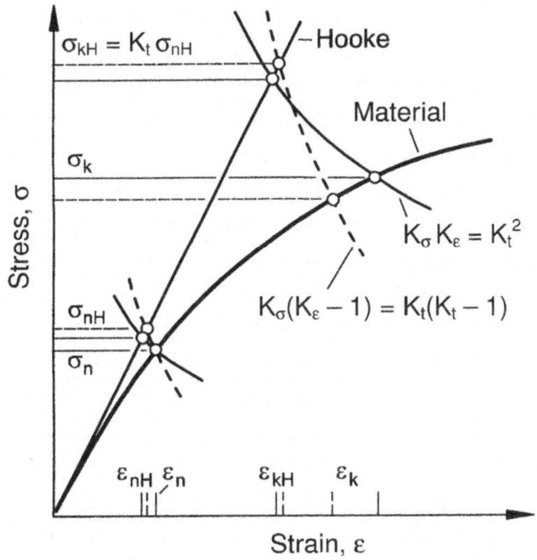

Fig. 2.88 Illustration of the analysis of elastic-plastic notch stresses and strains for sharp and mild notches; maximum notch stress σ_k, maximum notch strains ε_k, nominal stress σ_n, nominal strain ε_n, subscript H for 'Hooke' (after Neuber 1961, 1968(1), 1968(2), 1985)

the strain energy density. As soon as the yield limit is locally exceeded, the notch stresses rise underproportional and the notch strains overproportional. The relation between the elastic-plastic stress and strain concentration factors, K_σ and K_ε, on the one hand and the elastic stress concentration factor K_t on the other hand, is given by Neuber for sharp and mild notches, respectively, by the following equations (Neuber 1961, 1968(1), 1961(2), 1985):

$$K_\sigma K_\varepsilon = K_t^2 \quad (K_t \gg 1.0) \tag{2.248}$$

$$K_\sigma (K_\varepsilon - 1) = K_t (K_t - 1) \tag{2.249}$$

These equations are expressed by the maximum notch stresses $\sigma_k = \sigma_{max}$ and maximum notch strains $\varepsilon_k = \varepsilon_{max}$ (in the elastic-plastic range) as well as by the nominal stresses σ_n and nominal strains ε_n (in the elastic range) as follows:

$$\sigma_k \varepsilon_k = \sigma_n \varepsilon_n K_t^2 = \frac{\sigma_n^2}{E} K_t^2 \tag{2.250}$$

$$\sigma_k (\varepsilon_k - \varepsilon_n) = \sigma_n \varepsilon_n K_t (K_t - 1) = \frac{\sigma_n^2}{E} K_t (K_t - 1) \tag{2.251}$$

The content of Eqs. (2.248–2.251) is illustrated by Fig. 2.88, where the condition of elastic nominal stresses and strains referring to the net cross-section is deleted in favour of possibly occurring elastic-plastic conditions. Conforming to Neuber's analysis, the maximum actual notch stresses and strains on the stress–strain curve of the material are determined starting from the maximum linear-elastic notch stress $\sigma_{kH} = \sigma_{max,e}$ (with appertaining maximum notch strain ε_{kH}) on the Hooke straight line curve. The connection between appertaining points on the two curves is established by hyperbola-like curve sections, which run flatter for sharp notches,

Eq. (2.250) and steeper for mild notches, Eq. (2.251). Thus, slightly higher notch root stresses and strains are found with the former than with the latter equation.

The maximum linear-elastic notch stress σ_{kH} according to Hooke result from the Hooke nominal stress σ_{nH} (with appertaining nominal strain $\varepsilon_{\mathrm{nH}}$) by multiplication with the elastic stress concentration factor K_{t}. The Hooke nominal stress is connected with the actual nominal stress σ_{n} (with appertaining nominal strain ε_{n}) by the hyperbola-like curve sections mentioned above. They are slightly higher than the actual nominal stress and differently high for the two curve sections. Thus, different values of σ_{kH} appear with the same value of K_{t}. The increase and the difference in σ_{kH} disappear if the actual nominal stress σ_{n} remains linear-elastic. As can be seen from Fig. 2.88, too high notch stresses and notch strains are determined, if the original Neuber rule, Eq. (2.248), is applied to mild notches.

The Neuber rule was originally proposed based on general mathematical considerations on an appropriate 'leading function' (Neuber 1961). The Neuber formulae, Eqs. (2.248) and (2.250), were subsequently analytically derived for parabola-like notches subjected to out-of-plane shear loading with an arbitrary nonlinear material law (Neuber 1985, ibid. pp. 86–87). Actually, a parabolic notch was the basis in combination with the Neuber material law in the unconventional form $G\gamma = \tau/(1 - \tau/\tau_0)$. The law does not allow for hardening under large strains. The resulting formula was then rewritten in terms of tensile stresses and strains with reference to the von Mises criterion.

As an alternative to the (original) Neuber rule, Glinka has proposed an approach based on the strain energy density in order to determine the elastic-plastic stresses and strains (Glinka 1985). The assumption is introduced that the strain energy density in the elastically supported plastic area at the notch tip (small-scale yielding) remains equal to the value of the linear-elastic solution (equivalent strain energy density concept). This assumption suggests itself for uniaxial and proportionally biaxial stress states, because the stresses and strains exceeding the yield limit deviate in opposite directions from the linear-elastic behaviour, whereas the principal directions of the stress and strain state remain unchanged. The Glinka approach offers a higher solution accuracy than the original Neuber formula. This has been proven on an experimental basis and was further substantiated by the following comparison of the two approaches.

The comparison is based on the Ramberg–Osgood material law. It is simplified by the assumption that the nominal stress remains in the linear-elastic range. The elastic strain energy density rising with the squared product of nominal stress times stress concentration factor is equated with the correspondingly rising elastic and plastic components of the elastic-plastic strain energy density at the notch tip (Glinka 1985):

$$\frac{\sigma_{\mathrm{n}}^2}{2E} K_{\mathrm{t}}^2 = \frac{\sigma_{\mathrm{k}}^2}{2E} + \frac{\sigma_{\mathrm{k}}}{n' + 1} \left(\frac{\sigma_{\mathrm{k}}}{H}\right)^{1/n'} \tag{2.252}$$

where $1/n'$ is used instead on n as the hardening exponent ($0 < n' < 1.0$).

The original Neuber equation in terms of the Ramberg–Osgood material law, considered as a strain energy density equation, reads:

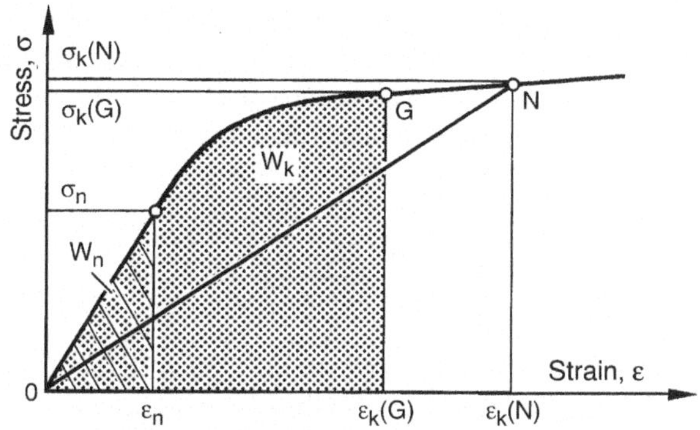

Fig. 2.89 Illustration of the elastic-plastic notch stress and strain analytical solutions by Glinka (point G) and by Neuber (point N); Glinka's solution on the basis of a constant ratio of strain energy densities W_k und W_n; (Glinka 1985)

$$\frac{\sigma_n^2}{2E}K_t^2 = \frac{\sigma_k}{2}\left[\frac{\sigma_k}{E} + \left(\frac{\sigma_k}{H}\right)^{1/n'}\right] \tag{2.253}$$

The solutions according to Glinka (curve point G) and Neuber (curve point N) are contrasted in Fig. 2.89. The strain energy density W_n of the nominal stress appears as a hatched triangular area. The corresponding strain energy density W_k of the maximum notch stress and strain is represented by the halftone area below the curve. The strain energy density according to the Neuber solution is characterised by the area below the straight line between the points 0 and N. The notch strain according to Neuber is substantially larger than according to Glinka. Experimental data confirm Glinka's result. Also, Neuber's modified solution, Eq. (2.251), reduces the notch strain.

The plastic NSIF approach by Zappalorto and Lazzarin allows further comparisons with the Neuber and Glinka results. The characteristic feature of this approach in respect of the Neuber rule is that the influence of the hardening exponent n and of the notch opening angle 2α is explicitly taken into account.

A first relationship is gained from the elastic-plastic stress and strain field analysis for parabolic notches ($2\alpha = 0$) subjected to out-of-plane shear loading (Zappalorto and Lazzarin 2007). The material law consists of linear-elastic behaviour up to the substitute yield limit τ_0 and power law behaviour in terms of the total strains above this limit. Continuing the exposition in Sect. 2.5.5, the following equation for the elastic-plastic notch stresses and strains is derived:

$$\tau_{max,p}\gamma_{max,p} = \tau_0\gamma_0\frac{2n}{n+1}\frac{1}{\pi\rho}\left(\frac{K_{3\rho,e}}{\tau_0}\right)^2 = \frac{2n}{n+1}\frac{(\tau_{max,e})^2}{G} \tag{2.254}$$

By dividing both sides of the equation by $\tau_n\gamma_n = \tau_n^2 G$, one obtains the Neuber rule in the following form:

$$K_\tau K_\gamma = \frac{2n}{n+1}K_t^2 \tag{2.255}$$

This relationship is different from Neuber's original expression $K_\tau K_\gamma = K_t^2$ by a factor 1.0–2.0, identical only for the linear-elastic limit case ($n = 1$) and with the largest deviation for the elastic perfectly-plastic material ($n \to \infty$).

Based on the above analytical frame, an explicit formula for the maximum stress $\tau_{\max,p}$ is derived:

$$\tau_{\max,p} = \tau_0 \left[\frac{1}{2\pi} \left(\frac{K_{3\rho,e}}{\tau_0} \right)^2 \frac{2}{\rho} \right]^{1/(n+1)} = \left(\tau^2_{\max,e} \tau_0^{n-1} \right)^{1/(n+1)} \tag{2.256}$$

Here, larger stresses and strains than with the original Neuber rule are found, in contrast to the tendency in Glinka's experimental results, after conversion of the tensile stresses there to shear stresses here.

A second investigation is devoted exclusively to the Neuber rule, using the Neuber nonlinear material law (Zappalorto and Lazzarin 2009). Pointed and rounded V-notches are considered, subjected to out-of-plane shear loading. The basic plastic and elastic NSIF relationships are stated in Sect. 2.5.7: $K_{3\rho,p}$, $K^{(\gamma)}_{3\rho,p}$, $K_{3\rho,e}$, $K^{(\gamma)}_{3\rho,e}$ and also $K_{3\rho,p} = f(K_{3\rho,e})$ according to Eqs. (2.227–2.229).

With the aim to check the Neuber rule, the following product is formulated:

$$K_{3\rho,p} K^{(\gamma)}_{3\rho,p} = 2\pi\gamma_0\tau_0\beta \left(\frac{K_{3\rho,e}}{\sqrt{2\pi}\tau_0} \right)^{\omega_1+1} \tag{2.257}$$

$$\beta = 2^{1-\omega_1}(1+\omega_1), \quad \omega_1 = \frac{\pi}{\pi - 2\alpha} \tag{2.258}$$

The plastic NSIFs in terms of the maximum stresses and strains read as follows:

$$K_{3\rho,p} = \sqrt{2\pi}\tau_{\max,p}, \quad K^{(\gamma)}_{3\rho,p} = \sqrt{2\pi}\gamma_{\max,p} r_0 \tag{2.259}$$

where $r_0 = \rho\,(q-1)/q$ with $(1 < q < 2)$ for hyperbolic notches and $q = 2$ for parabolic notches.

The elastic NSIF is given by the following expression (Zappalorto et al. 2008):

$$K_{3\rho,e} = \sqrt{2\pi}\tau_{\max,e} r_0^{1-\lambda_3} \tag{2.260}$$

Equation (2.257) written in the maximum stresses and strains according to Eqs. (2.25) and (2.260) reads as follows:

$$\tau_{\max,p}\,\gamma_{\max,p} = \tau_0\gamma_0\beta \left(\frac{\tau_{\max,e}}{\tau_0} \right)^{\omega_1+1} \tag{2.261}$$

where β and ω_1 are notch angle dependent parameters. Thus, the product of the stress and strain concentrations depends on the notch opening angle, contrary to the original Neuber rule.

In the special case of the parabolic notch ($2\alpha = 0$, $\omega_1 = 1$, $\beta = 2$), Eq. (2.261) results in:

$$\tau_{\max,\,p}\,\gamma_{\max,\,p} = \frac{2\tau^2_{\max,\,e}}{G} \tag{2.262}$$

$$K_\tau\,K_\gamma = 2K_t^2 \tag{2.263}$$

This result confirms Eq. (2.255) for $n \to \infty$, the condition, which brings the two different material laws more or less in correspondence (elastic perfectly-plastic behaviour).

The difference between Neuber's and Zappalorto–Lazzarin's final formulae, both derived for the same nonlinear material law, is discussed in detail by the second-mentioned authors. They uncover an inconsistency in Neuber's derivation, which expresses itself in the factor 1 instead of 2 on the nonlinear strains. But the removal of this deficiency does not solve the problem. Obviously, Neuber's original formula is applicable when approaching the limit case of purely elastic material behaviour. This is an indication that Neuber's original formula is appropriate in the range of small-scale yielding. Experimental and numerical (FE analysis) investigations have repeatedly demonstrated the applicability of the original formula with the reservation of a modification for mild notches. The modification has the effect that the resulting notch strains are slightly reduced. On the other hand, Zappalorto and Lazzarin's relationships, Eqs. (2.255) and (2.263), are appropriate for the extremely large plastic strains which occur close to pointed notch tips.

2.5.10 Analytical Description of Elastic-Plastic Material Behaviour

The analytical HRR and Neuber solutions for the stress and strain field at sharp notches under the condition of elastic-plastic material behaviour are actually nonlinear-elastic solutions. The solutions are bound to the condition of monotonic loading in every point of the field. Any unloading, globally or locally, is not allowed. The plastic strains are simply the nonlinear strain components, total strain minus linear-elastic strain. This condition applies to the analytical solutions and must be ascertained in comparative models using the FE method. The demanding mathematical formulations are easier to understand, if one considers them as referring to nonlinear-elastic material behaviour. Different types and variants of nonlinear material laws are in use.

Uniaxial Ramberg–Osgood Material Law

The relationship between stresses and strains in uniaxial monotonic loading, which is widely used by engineers, is the Ramberg–Osgood material law describing a single smooth curve without a distinct yield limit, Fig. 2.90:

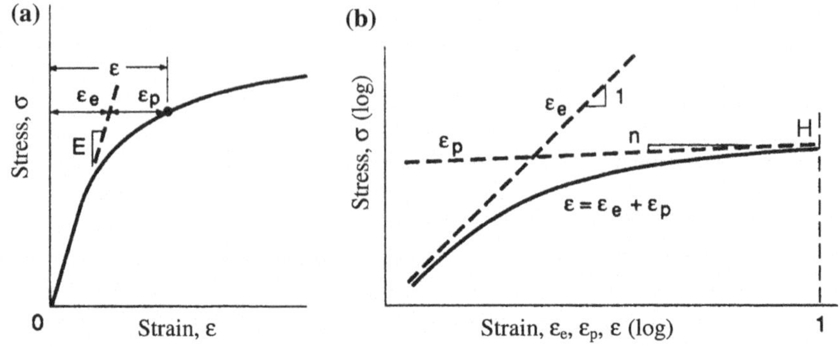

Fig. 2.90 Ramberg–Osgood material law: stress–strain curves drawn in linear (**a**) and logarithmic (**b**) coordinates; (Dowling 1993)

$$\varepsilon = \varepsilon_{\mathrm{e}} + \varepsilon_{\mathrm{p}} = \frac{\sigma}{E} + \left(\frac{\sigma}{H}\right)^{n} \tag{2.264}$$

with the elastic strain ε_{e}, the plastic strain ε_{p}, the modulus of elasticity E, the hardening (or strength) coefficient H and the hardening exponent n. The exponent n is determined from log σ over log ε plots. For the metallic engineering materials, $n = 2.5$–20.0 is found. In low-cycle fatigue assessments, a corresponding cyclic Ramberg–Osgood relationship is used. The cyclic material parameters may deviate substantially from the monotonic ones.

In engineering science, the offset yield limit $\sigma_{\mathrm{Y0.2}}$ is introduced, which denotes the stress where a plastic strain $\varepsilon_{\mathrm{Y0.2}} = 0.2\ \%$ occurs. Therefore, the following formula applies:

$$\frac{\varepsilon_{\mathrm{p}}}{\varepsilon_{\mathrm{Y0.2}}} = \left(\frac{\sigma}{\sigma_{\mathrm{Y0.2}}}\right)^{n} \quad (\varepsilon_{\mathrm{Y0.2}} = 0.002) \tag{2.265}$$

Equating ε_{p} from Eq. (2.264) and (2.265) gives:

$$\sigma_{\mathrm{Y0.2}} = \varepsilon_{\mathrm{Y0.2}}^{1/n} H \tag{2.266}$$

which means that $\sigma_{\mathrm{Y0.2}}$ depends on n and H for a defined $\varepsilon_{\mathrm{Y0.2}}$ and is not a third independent material parameter.

Multiaxial Ramberg–Osgood Material Law

The Ramberg–Osgood material law, generalised for multiaxial stress conditions, is usually presented in tensor form related to the stress and strain components, Eq. (2.183). Rather lengthy expressions for stresses, strains and strain energies result from it. A simplified procedure, preferred by engineers, is possible. The necessary conditions are monotonic loading, constant ratios of the stress components and validity of the octahedral shear stress criterion (corresponding to the von Mises criterion). Then, according to Hencky's approach (Hencky 1924; Chakrabarty 2006), the components of the plastic strain tensor are set proportional to the components of the coaxial deviatoric stress tensor:

$$\varepsilon_{ij,p} = \lambda s_{ij} \tag{2.267}$$

For work-hardening materials, the coefficient λ depends on the von Mises equivalent stress σ_{eq}, which is considered to be a function of the equivalent plastic strain $\varepsilon_{p,eq}$:

$$\sigma_{eq} = \sqrt{\frac{3}{2} s_{ij} s_{ij}} \tag{2.268}$$

$$\varepsilon_{eq,p} = \sqrt{\frac{2}{3} \varepsilon_{ij,p} \varepsilon_{ij,p}} \tag{2.269}$$

The relationship between σ_{eq} and $\varepsilon_{eq,p}$ is given by the uniaxial stress versus plastic strain curve. The coefficient λ can be expressed by σ_{eq} and $\varepsilon_{eq,p}$ or by the secant modulus E_p according to Eq. (2.270):

$$\lambda = \frac{3\varepsilon_{eq,p}}{2\sigma_{eq}} = \frac{3}{2E_p} \tag{2.270}$$

It is a peculiarity of Hencky's approach that the coefficient λ is related to the stress–strain curve. In its original, more general definition, it is a constant factor which denotes the occurrence of plastic strain rates or strain increments as correlated with the deviatoric stress state.

The von Mises equivalent stress σ_{eq} expressed by the principal stresses σ_1, σ_2, σ_3 is related to the octahedral shear stress τ_o:

$$\sigma_{eq} = \frac{1}{2}\sqrt{(\sigma_1 - \sigma_2)^2 + (\sigma_2 - \sigma_3)^2 + (\sigma_3 - \sigma_1)^2} = \frac{3}{\sqrt{2}}\tau_o \tag{2.271}$$

The corresponding equivalent plastic strain $\varepsilon_{eq,p}$ expressed by the principal plastic strains $\varepsilon_{1,p}$, $\varepsilon_{2,p}$, $\varepsilon_{3,p}$ is related to the octahedral plastic shear strain $\gamma_{o,p}$:

$$\varepsilon_{eq,p} = \frac{\sqrt{2}}{3}\sqrt{(\varepsilon_{1,p} - \varepsilon_{2,p})^2 + (\varepsilon_{2,p} - \varepsilon_{3,p})^2 + (\varepsilon_{3,p} - \varepsilon_{1,p})^2} = \frac{1}{\sqrt{2}}\gamma_{o,p} \tag{2.272}$$

The equivalent total strain ε_{eq} is the sum of its elastic and plastic components:

$$\varepsilon_{eq} = \frac{\sigma_{eq}}{E} + \varepsilon_{eq,p} \tag{2.273}$$

The equivalent stress and strain, σ_{eq} and $\varepsilon_{eq,p}$ or ε_{eq}, is identical with the stress and strain, σ_1 and $\varepsilon_{1,p}$ or ε_1, in the uniaxial tensile test case ($\sigma_{eq} = \sigma_1$, $\sigma_2 = \sigma_3 = 0$, $\varepsilon_{eq,p} = \varepsilon_{1,p}$, $\varepsilon_{2,p} = \varepsilon_{3,p} = -0.5\varepsilon_{1,p}$) so that any monotonic stress–strain curve is a sufficient basis for describing arbitrary stress and strain states. The

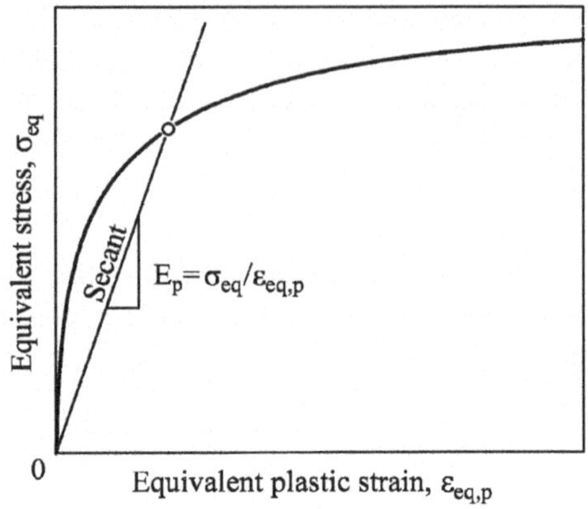

Fig. 2.91 Plastic modulus E_p defined as secant modulus of the equivalent stress over equivalent strain curve, identical with uniaxial stress σ_1 over strain ε_1 curve; (Dowling 1993)

plastic strain components are related to the stress components by equations which are structured analogous to the Hooke law equations, but with the elastic modulus E replaced by the plastic modulus E_p and with Poisson's ratio $v = 0.5$ related to the incompressibility condition for the plastic strains:

$$\begin{aligned}
\varepsilon_{x,p} &= \frac{1}{E_p}\left[\sigma_x - 0.5(\sigma_y + \sigma_z)\right], \quad \gamma_{xy,p} = \frac{3}{E_p}\tau_{xy} \\
\varepsilon_{y,p} &= \frac{1}{E_p}\left[\sigma_y - 0.5(\sigma_x + \sigma_z)\right], \quad \gamma_{yz,p} = \frac{3}{E_p}\tau_{yz} \\
\varepsilon_{z,p} &= \frac{1}{E_p}\left[\sigma_z - 0.5(\sigma_x + \sigma_y)\right], \quad \gamma_{zx,p} = \frac{3}{E_p}\tau_{zx}
\end{aligned} \tag{2.274}$$

The plastic tensile modulus E_p or alternatively the plastic shear modulus $G_p = E_p/3$ (with $v = 0.5$) is dependent on σ_{eq} and $\varepsilon_{eq,p}$ or on τ_o and $\gamma_{o,p}$:

$$E_p = \frac{\sigma_{eq}}{\varepsilon_{eq,p}}, \quad G_p = \frac{\tau_o}{\gamma_{o,p}}, \quad G_p = \frac{E_p}{3} \tag{2.275}$$

These plastic moduli are also named 'secant moduli' as illustrated by Fig. 2.91.

The multiaxial Ramberg–Osgood law has the general form:

$$\varepsilon_{eq} = \varepsilon_{eq,e} + \varepsilon_{eq,p} = \frac{\sigma_{eq}}{E} + \left(\frac{\sigma_{eq}}{H}\right)^{1/n} \tag{2.276}$$

The factor λ in Eq. (2.270) and the secant modulus E_p in Eq. (2.275), applied to the Ramberg–Osgood relationship, result in:

$$\lambda = \frac{3}{2}\frac{\sigma_{eq}^{n-1}}{H^n}, \quad E_p = \frac{H^n}{\sigma_{eq}^{n-1}} \tag{2.277}$$

Based on the equivalent stress and strain relationships, Eqs. (2.270–2.272), any arbitrary multiaxial stress state can be provided with the appertaining stress–strain relationship in terms of its characteristic stress and strain components.

Pure Shear Ramberg–Osgood Material Law

Most important is the application of the multiaxial Ramberg–Osgood material law to pure shear loading conditions, especially out-of-plane shear loading. The following relationship is thus gained (Dowling 1993):

$$\gamma = \frac{\tau}{G} + \sqrt{3}\left(\frac{\sqrt{3}\tau}{H}\right)^{1/n} \tag{2.278}$$

or alternatively:

$$\gamma = \frac{\tau}{G} + \left(\frac{\tau}{H_\tau}\right)^{1/n}, \quad H_\tau = \frac{H}{3^{(n+1)/2}} \tag{2.279}$$

This relationship is confirmed by the power law conversion from σ, ε to τ, γ components based on equated strain energy densities, Eq. (2.284).

In the case of out-of-plane shear loading, the resultant or equivalent stresses and plastic strains in the cross-sectional plane are:

$$\tau = \sqrt{\tau_{zx}^2 + \tau_{zy}^2}, \quad \gamma_{\mathrm{p}} = \sqrt{\gamma_{zx,\mathrm{p}}^2 + \gamma_{zy,\mathrm{p}}^2} \tag{2.280}$$

No simple relationship exists between the offset yield limit $\sigma_{\mathrm{Y}0.2}$ in the tensile test defined by $\varepsilon_{\mathrm{Y}0.2} = 0.2\ \%$, and the corresponding offset yield limit in the shear test, were γ_{Y} is initially undefined. The two shear test parameters may be determined by transfer from the equivalent stress–strain curve to the pure shear stress–strain curve. Alternatively, the parameters may be gained from equating the strain energy densities at the two yield limits. This has not yet been done.

Total Strain Power Law

Another way to describe elastic-plastic material behaviour in the considered theoretical solutions is to assume linear-elastic behaviour up to the reference values γ_0, τ_0 (or ε_0, σ_0) and a power law behaviour in terms of the total strains above the reference values, Figs. 2.92 and 2.93. This may be considered as the asymptotic form of the Ramberg–Osgood material law. The equations are given for out-of-plane shear loading in terms of γ, τ, but are transferable to tensile loading by introducing ε, σ instead of γ, τ [compare Eq. (2.176)]:

$$\begin{aligned} \frac{\gamma}{\gamma_0} &= \frac{\tau}{\tau_0} \quad (\gamma \leq \gamma_0) \\ \frac{\gamma}{\gamma_0} &= \left(\frac{\tau}{\tau_0}\right)^n \quad (\gamma \geq \gamma_0) \end{aligned} \tag{2.281}$$

The reference values γ_0, τ_0 (or ε_0, σ_0) represent the linear-elastic stress and strain on the nonlinear $\tau{-}\gamma$ (or $\sigma{-}\varepsilon$) curve:

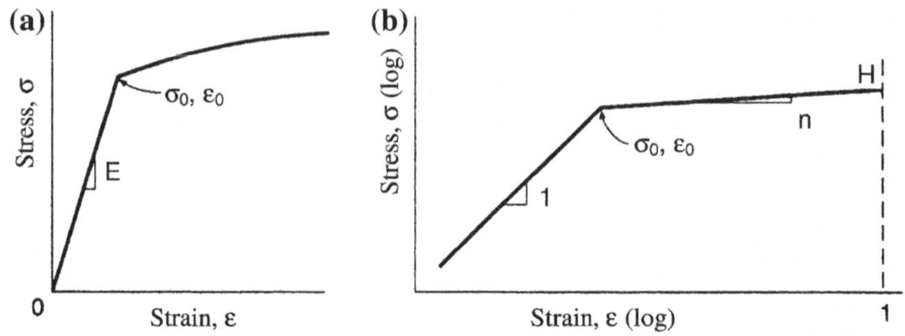

Fig. 2.92 Linear–elastic followed by total strain power law material behaviour: stress–strain curves drawn in linear (**a**) and logarithmic (**b**) coordinates; (Dowling 1993)

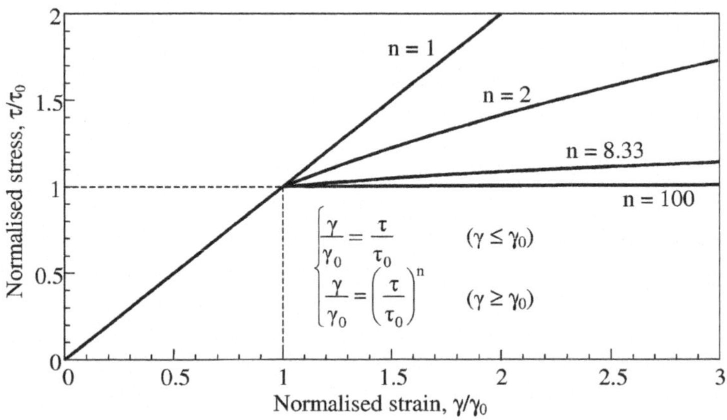

Fig. 2.93 Linear-elastic followed by total strain power law material behaviour: normalised stress–strain curves drawn in linear coordinates; (Lazzarin and Zappalorto 2008)

$$\gamma_0 = \frac{\tau_0}{G} = \left(\frac{\tau_0}{H_\tau}\right)^n, \quad \tau_0 = \left(\frac{H_\tau^n}{G}\right)^{1/(n-1)} \tag{2.282}$$

$$\varepsilon_0 = \frac{\sigma_0}{E} = \left(\frac{\sigma_0}{H}\right)^n, \quad \sigma_0 = \left(\frac{H^n}{E}\right)^{1/(n-1)} \tag{2.283}$$

There is no difference between the total strain power law and the Ramberg–Osgood relationship for very large strains, as they occur in the close vicinity of the notch tip, so that n and H_τ or H can be set equal, but there are major differences in the values of γ_0, τ_0 (or ε_0, σ_0) compared with those of γ_Y, τ_Y (or ε_Y, σ_Y), Fig. 2.94. The mentioned differences are relevant when considering the plastic zone size. One has to differentiate between the real size defined by ε_Y, σ_Y and the substitute size given by ε_0, σ_0.

In order to avoid confusion when comparing NSIF values from analytical expressions based on the total strain power law with those from FE analysis based

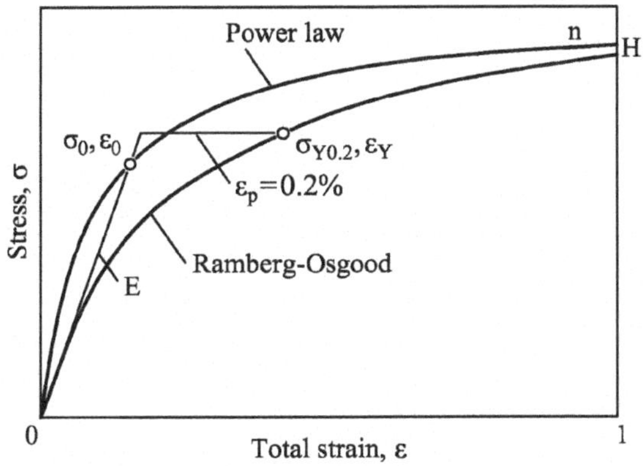

Fig. 2.94 Stress–strain curves, total strain power law in comparison to the Ramberg–Osgood law with characteristic material parameters; (Radaj, unpublished)

on the Ramberg–Osgood relationship, it should be kept in mind that identical values of H and n in the two producers are sufficient and that the reference values ε_0, σ_0 in the power law are a substitute for the real yield limit in the Ramberg–Osgood relationship, the substitute based on the elastic modulus E.

The power law in terms of ε, σ may be converted to the power law in terms of γ, τ based on the expressions for the equivalent stresses and strains, Eqs. (2.271–2.273), combined with the assumption of identical strain energy densities (Zappalorto and Lazzarin 2009). These are determined by integration of the power law curve starting at the origin, i.e. without the linear-elastic initial curve rise. They are termed W_{nl} ('nonlinear') instead of W_{p} ('plastic') as in the original publication above. Also, the derivation is formally modified; compare Eqs. (2.223) and (2.224):

$$W_{\mathrm{nl}} = \frac{n}{n+1}\frac{\sigma_{\mathrm{eq}}^{n+1}}{H^n} = \frac{n}{n+1}\frac{\tau^{n+1}}{H_\tau^n}, \quad H_\tau = \frac{H}{3^{(n+1)/2}} \tag{2.284}$$

$$W_{\mathrm{nl}} = \frac{n}{n+1}\frac{\sigma_{\mathrm{eq}}^{n+1}}{E\sigma_{\mathrm{eq},0}^{n-1}} = \frac{n}{n+1}\frac{\tau^{n+1}}{G\tau_0^{n-1}}, \quad \tau_0 = \sigma_{\mathrm{eq},0}\left(\frac{2(1+\nu)}{3^{(n+1)/2}}\right)^{1/(n-1)} \tag{2.285}$$

with $\sigma_{\mathrm{eq}} = \sqrt{3}\tau$ and $G = E/2(1+\nu)$. The hardening coefficient H_τ of the power law, Eq. (2.284), is identical with H_τ of the Ramberg–Osgood relationship, Eq. (2.279).

The total strain power law in Eq. (2.281) may alternatively be considered valid in the whole range of γ without the restriction $\gamma \geq \gamma_0$. Using γ_0, τ_0 as reference quantities for arbitrary values of γ, τ, a nonlinear-elastic material law is described, Fig. 2.95:

$$\frac{\gamma}{\gamma_0} = \left(\frac{\tau}{\tau_0}\right)^n \quad (1 < n < \infty) \tag{2.286}$$

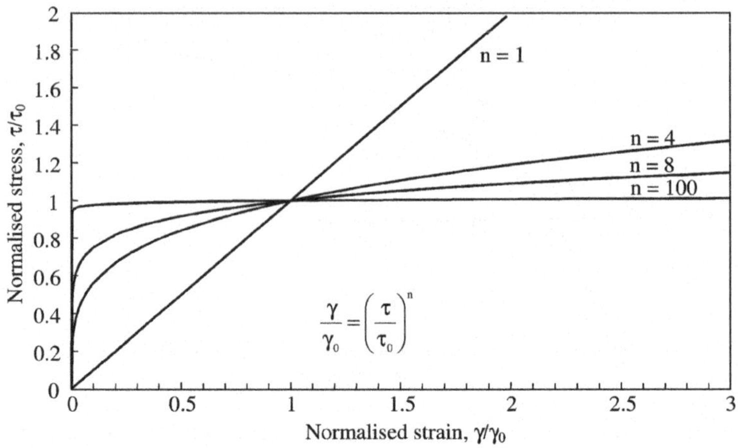

Fig. 2.95 Nonlinear-elastic total strain power law of material behaviour: normalised stress–strain curves drawn in linear coordinates; (Zappalorto and Lazzarin 2010)

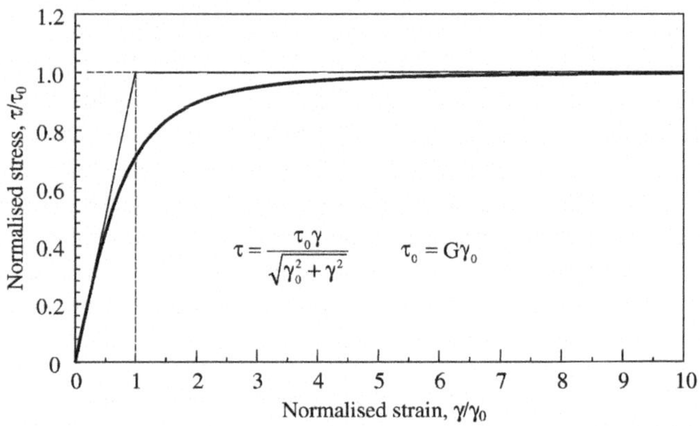

Fig. 2.96 Neuber nonlinear material law: normalised stress–strain curves drawn in linear coordinates; (Zappalorto and Lazzarin 2010)

Neuber's Nonlinear Material Law

Finally, the nonlinear stress–strain law used by Neuber is presented (Neuber 1958, 1961), Fig. 2.96:

$$\tau = \frac{\tau_0 \gamma}{\sqrt{\gamma_0^2 + \gamma^2}}, \quad \tau_0 = G\gamma_0 \tag{2.287}$$

where one of the two equations can also be substituted by the original Neuber version:

$$G\gamma = \frac{\tau}{\sqrt{1 - (\tau/\tau_0)^2}} \tag{2.288}$$

This material law has the following properties: $\tau \to \tau_0$ for $\gamma \to \infty$, $\tau \to G\gamma$ for $\gamma \to 0$ and a continuous strain hardening behaviour in the γ range between. Virtually no hardening occurs for values $\gamma/\gamma_0 \geq 5.0$. A stress singularity at pointed notches cannot be formed under these conditions, but a strain singularity very well.

The different material laws reviewed above will have an influence on the stress and strain fields at pointed or rounded notches. This influence manifests itself in different plastic NSIFs. Therefore, the plastic NSIFs depend on the material law. An exception is non-hardening material behaviour, where the plastic NSIFs are identical.

2.5.11 Conclusions

The elastic-plastic stress and strain fields at pointed or rounded notch tips can be uniformly described by plastic NSIFs as governing parameters. The plastic NSIFs are combined for that purpose with definite 'universal functions'. Both the stress and strain fields are handled within this analytical frame, differentiating between stress intensity factors and strain intensity factors. Definite relationships exist between elastic and plastic NSIFs, or, following the loading curve, between initial linear and final nonlinear NSIFs.

The available analytical solutions are derived for nonlinear-elastic material laws which may be interpreted as elastic-plastic behaviour under conditions of monotonic loading. Two different material laws are the basis of the analytical theory: the Ramberg–Osgood relationship asymptotically approximated by the total strain power law for strain-hardening materials (the HRR solutions) and Neuber's material law without hardening at large strains. These material laws allow closed form solutions for the antiplane fields and approximate solutions for the in-plane fields.

There remains the problem, how to interpret the derived analytical solutions in terms of the Ramberg–Osgood material law which is preferred by engineers and actually fits the real-world material behaviour better than the power material law or the Neuber material law does. There is no problem with the limit states of small elastic strains and large plastic strains at the notch tip, but there is a problem in the intermediate range, where the elastic-plastic boundary is established.

Valuable basic insights have been gained in respect of the following engineering issues, mainly under small-scale yielding conditions: plastic NSIFs for V-notches and parabolic notches, the size effect for V-notched specimens, expressions for the strain energy density and the *J*-integral, elastic-plastic stress and strain distributions in the bisector plane for various notch opening angles. The solutions for configurations with finite geometries are scaled by FE-analysis results. The Neuber rule for extremely large strains is specified in terms of hardening exponent and notch opening angle for the extremely large strains at pointed notch tips.

2.6 Stress Intensity Factor Concept for Rigid Inclusions

2.6.1 Survey of Section Contents

Any geometrical or material discontinuity in an, apart from this, homogeneous elastic continuum produces local stress concentrations when the continuum or the discontinuity is subjected to any loadings. At the pointed edges of such discontinuities stress singularities occur. The stress singularities at crack tips, a geometrical discontinuity, are well known. When the crack is substituted by a rigid thin inclusion (also termed 'layer', 'lamella' or simply 'line'), the stresses at the pointed tip of the inclusion remain singular.

This is also the case if the thin inclusion is not rigid, but only stiffer or less stiff than the surrounding material. The thin inclusion can have membrane stiffness, bending stiffness or both. Rigid or elastic thin inclusions are of some application relevance in respect of coated or bonded members, lamella-reinforced composites and materials containing flat inclusions.

Despite a limited practical applicability of the thin inclusion models compared with crack models, a remarkable number of analytical solutions has been published (e.g. Sih 1965; Brussat and Westermann 1975; Wang et al. 1985, 1986; Chen 1986, 1991, 1992; Chen and Cheung 1987; Cheung and Chen 1989). Evaluation of these contributions shows that a sufficiently complete systematics of the basic loading modes (singular and non-singular) is not available, that the stress intensity factors are differently defined, and that the relevant limit value formulae are only partially given. The exposition below is based on a uniform redefinition of the basic models and their analytical treatment by the Kolosov–Muskhelishvili complex stress function theory (Radaj and Zhang 1993[(2)]).

The basic loading modes at rigid thin inclusion tips are defined in analogy to the basic loading modes at crack tips. The stresses and displacements in the close vicinity of these tips are described in a uniform format by stress intensity factors, angular functions and non-singular terms. The limit value formulae for the stress intensity factors are given. The flat and curved rigid thin inclusion is considered as an example of practical application.

The concept is open for extensions to wedge-shaped rigid inclusions without and with wedge tip rounding (Hasebe et al. 1990).

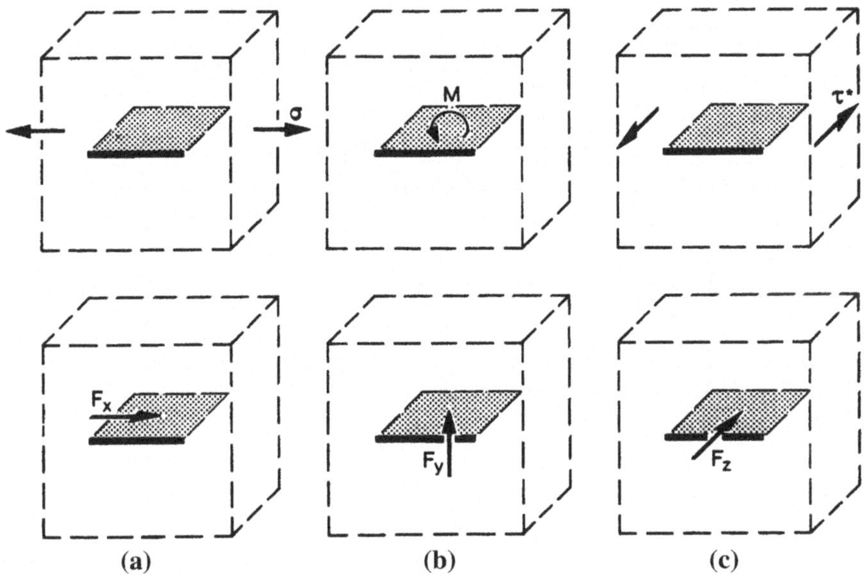

Fig. 2.97 Basic singular loading modes I*(**a**), II*(**b**) and III*(**c**) of rigid thin inclusion in infinite plate; plane and antiplane strain fields; arrows designate remote boundary stresses or resultant forces acting on the inclusion; (Radaj and Zhang 1993(2))

2.6.2 Basic Loading Modes at Rigid Thin Inclusion Tips

The basic loading modes at the tips of a rigid thin inclusion are defined in analogy to the basic loading modes at crack tips. The basic singular loading modes I*, II* and III* are shown in Fig. 2.97. A rigid thin inclusion strip permeates an infinite elastic body (or continuum). This configuration is substituted by a plate of finite thickness containing the inclusion. The multiaxiality condition with mode I* and mode II* loading is plane strain. Otherwise, out-of-plane singular effects will occur where the inclusion front line butts on the plate surface. The solutions below are also given for plane stress conditions, but this is only of academic interest. In mode III* loading, the out-of-plane deflections of the plate surfaces correspond to the condition in the out-of-plane shear loaded infinite body. Mode I* loading is characterised by longitudinal in-plane tensile (or compressive) basic stresses, mode II* loading by transverse in-plane basic shear stresses and mode III* loading by out-of-plane basic shear stresses. The basic stress state may be produced by a resultant force or moment (F_x, F_y, F_z, M), or by remote boundary stresses (σ, τ^*).

The basic non-singular loading modes 0I*, 0II* and 0III* are shown in Fig. 2.98. Mode 0I* loading is defined by triaxial tensile (or compressive) basic stresses (σ_{0x}, σ_{0y}, σ_{0z}) which suppress any displacement in the plane of the inclusion, $\varepsilon_{0x} = \varepsilon_{0z} = 0$, resulting in $\sigma_{0x} = \sigma_{0z} = \sigma_{0y}\nu/(1 - \nu)$. Mode 0II* loading is defined by in-plane basic shear stresses τ_0, which have no stress raising effect at the

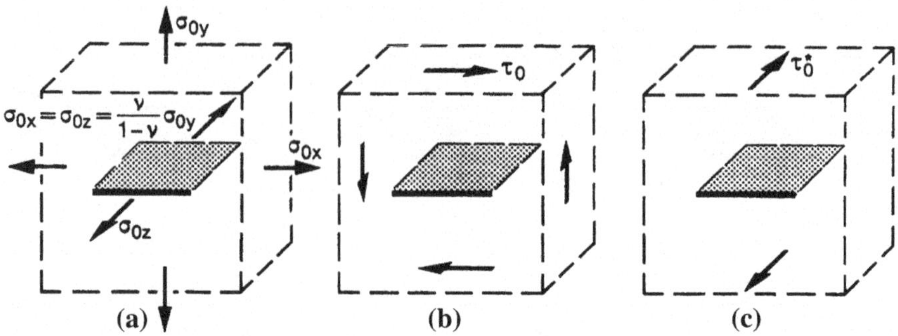

Fig. 2.98 Basic non-singular loading modes 0I*(**a**), 0II*(**b**) and 0III*(**c**) of rigid thin inclusion in infinite plate; plane and antiplane strain fields; arrows designate remote boundary stresses; (Radaj and Zhang 1993(2))

inclusion. The same applies to mode 0III* loading by out-of-plane basic shear stresses τ_0^*.

The analogy with the relevant crack models is obvious in respect of their number (three singular and three non-singular loading modes) and in respect of some basic characteristics, but the external or internal loading conditions are not the same.

2.6.3 Asymptotic Stress Field Close to Rigid Thin Inclusion Tips

The three basic loading modes with singular stresses at the rigid thin inclusion tip produce the following asymptotic stress distribution close to these tips. Actually, a crack with the flank sides rigidly supported is considered, Fig. 2.99:

$$\sigma_{\mathrm{ij}} = \frac{1}{\sqrt{2\pi r}}\left[K_{\mathrm{I}}^{*}f_{\mathrm{I,ij}}^{*}(\theta) + K_{\mathrm{II}}^{*}f_{\mathrm{II,ij}}^{*}(\theta) + K_{\mathrm{III}}^{*}f_{\mathrm{III,kz}}^{*}(\theta)\right] \qquad (2.289)$$
$$(\mathrm{i,j} = x, y \text{ and } \mathrm{k} = x, y \text{ or } \mathrm{i,j} = r, \theta \text{ and } \mathrm{k} = r, \theta)$$

The mode-related stress intensity factors K_{I}^*, K_{II}^* and K_{III}^* depend on the magnitude of the load, the length of the rigid thin inclusion and further geometrical parameters of the considered configuration. The mode-related functions $f_{\mathrm{I,ij}}^*(\theta)$, $f_{\mathrm{II,ij}}^*(\theta)$ and $f_{\mathrm{I,ij}}^*(\theta)$, describe the angular distribution of the stresses at the inclusion tip. The relationship above is strictly valid for $r \to 0$ and approximately valid for values of r which are small in relation to the inclusion length and other geometrical parameters of the configuration.

The three basic loading modes with non-singular stresses at the rigid thin inclusion tip are simply identical with the basic tensile and shear stresses in Cartesian format shown in Fig. 2.98. Their conversion into the polar coordinate system is elementary.

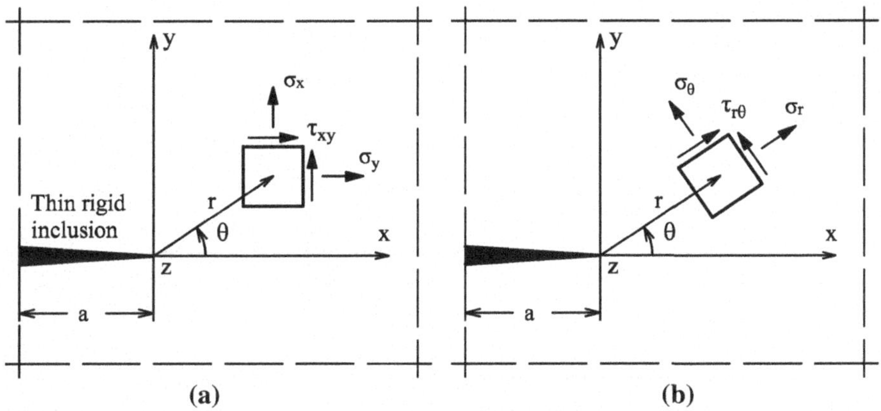

Fig. 2.99 Cartesian (**a**) and polar (**b**) coordinate system with correspondingly defined stresses at the tip of rigid thin inclusion; (Lazzarin and Tovo 1998)

The expressions for the stresses and displacements at the rigid thin inclusion tips are gained based on the Kolosov–Muskhelishvili complex stress function theory. The equations for in-plane loading (mode I* and II*) are derived from the stress functions $\varphi(\zeta)$ and $\psi(\zeta)$ for the infinite elastic plate containing a rigid elliptical inclusion (Muskhelishvili 1963, ibid. pp. 355–358). The equations for out-of-plane shear loading (mode III*) are derived from the Westergaard–type stress function $\varphi^*(z)$ for the infinite elastic plate containing a rigid thin inclusion (method according to Tada et al. 1999). Reference is also made to the author's relevant work (Radaj and Zhang 1993(2)).

The rigid thin inclusion of length $2a$ is considered as the geometrical limit case of a rigid elliptical inclusion with the semiaxis b reduced to zero. The function which maps the line ($-a \leq x \leq a$) in the z-plane ($z = x + \mathrm{i}y$) onto the unit circle in the ζ-plane ($\zeta = \rho e^{\mathrm{i}\theta}$) reads:

$$z(\zeta) = \frac{a}{2}\left(\zeta + \frac{1}{\zeta}\right) \tag{2.290}$$

The inverse function $\zeta(z)$ is used to retransform the stress and displacement functions from the ζ-plane into the z-plane, thereby establishing bipolar coordinates in the z-plane represented by the radii r_0, r_1, r_2 and the angles θ_0, θ_1, θ_2, Fig. 2.100:

$$\zeta(z) = \frac{1}{a}\left(z \pm \sqrt{z^2 - a^2}\right) \tag{2.291}$$

Such bipolar coordinates have originally been used when solving the corresponding crack problem (Hahn 1976, ibid. pp. 77–78).

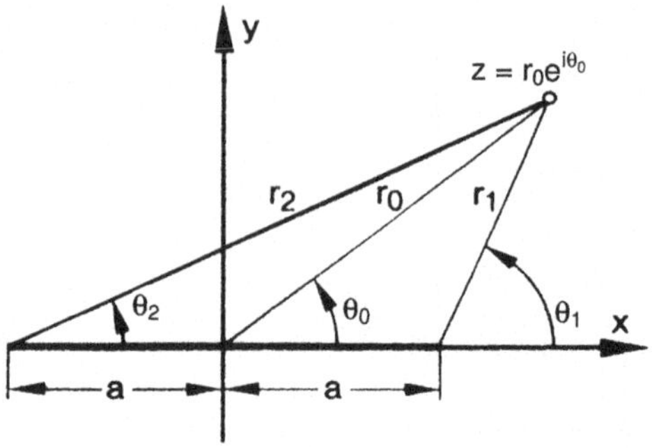

Fig. 2.100 Bipolar coordinates at rigid thin inclusion of length $2a$ in infinite plate (Radaj and Zhang 1993[(2)])

The stress and displacement equations for in-plane loading conditions (mode I* and II*) are valid both for plane strain ($\varepsilon_z = 0$) and plane stress ($\sigma_z = 0$) conditions, using different expressions for the parameter κ:

$$\kappa = 3 - 4\nu \quad (\varepsilon_z = 0) \tag{2.292}$$

$$\kappa = \frac{3-\nu}{1+\nu} \quad (\sigma_z = 0) \tag{2.293}$$

where $0 \leq \nu \leq 0.5$ is the Poisson's ratio, resulting in $1 \leq \kappa \leq 3$ for plane strain or $5/3 \leq \kappa \leq 3$ for plane stress. For structural steels with $\nu = 0.28$–0.30, the plane strain condition results in $\kappa = 1.8$–1.9.

The mode I* loading condition (longitudinal tension or compression) at the rigid thin inclusion is considered by the examples of remote longitudinal tension σ and of a longitudinal force F_x applied to the inclusion, Fig. 2.97a. The complete equations of the stress functions, the stresses (σ_x, σ_y, τ_{xy}) and the displacements (u, v) are reduced to the following relationships in the close vicinity of the inclusion tip ($r = r_1$, $\theta = \theta_1$, $r \ll a$), neglecting terms which are small of higher order (Radaj and Zhang 1993[(2)]):

$$\begin{Bmatrix} \sigma_x \\ \sigma_y \\ \tau_{xy} \end{Bmatrix} = \frac{K_\mathrm{I}^*}{\sqrt{2\pi r}} \begin{Bmatrix} \cos\theta/2\ (3+\kappa-2\sin\theta/2\ \sin 3\theta/2) \\ \cos\theta/2\ (1-\kappa+2\sin\theta/2\ \sin 3\theta/2) \\ \sin\theta/2\ (1+\kappa+2\cos\theta/2\ \cos 3\theta/2) \end{Bmatrix} \tag{2.294}$$

$$\begin{Bmatrix} u \\ v \end{Bmatrix} = \frac{K_\mathrm{I}^*}{\pi G}\sqrt{2\pi r} \begin{Bmatrix} \cos\theta/2\ (\kappa+\sin^2\theta/2) \\ -\sin\theta/2\ \cos^2\theta/2 \end{Bmatrix} \tag{2.295}$$

where G is the shear modulus which may be expressed in terms of the elastic modulus E by $G = E/2(1+\nu)$.

The stress intensity factors in the two considered loading cases are:

$$K_\mathrm{I}^* = \frac{1+\kappa}{8\kappa}\sigma\sqrt{\pi a} \tag{2.296}$$

$$K_{\mathrm{I}}^{*} = \frac{1}{2(1+\kappa)} \frac{F_x}{\sqrt{\pi a}} \tag{2.297}$$

These SIFs at the inclusion tip are smaller than the SIFs in the comparable crack loading cases ($K_{\mathrm{I}} = \sigma(\pi a)^{1/2}$, $K_{\mathrm{I}} = F_x/(\pi a)^{1/2}$). Similar definitions of the SIFs are found in the literature (Sih 1965; Wang et al. 1985).

Some further phenomena are worth to be mentioned. In the loading case of longitudinal tension σ, a constant transverse tensile stress $\sigma_{\mathrm{y}} = \sigma(\kappa - 1/\kappa)/8$ is acting on both sides of the rigid inclusion (non-reduced equations, $-a \leq x \leq a$), equilibrated by transverse compressive stress singularities ahead of the inclusion tips. The tensile force T_{x} in the rigid inclusion (per unit of plate thickness) is determined from τ_{xy} at $\theta = \pi$ and $\theta = -\pi$, adding up the two values and integrating them starting at the inclusion tip ($r = 0$):

$$T_x = 4(1+\kappa) K_{\mathrm{I}}^{*} \sqrt{\frac{r}{2\pi}} \quad (r \ll a) \tag{2.298}$$

The mode II* loading condition (transverse shear loading) at the rigid thin inclusion is considered by the examples of a turning moment M and of a transverse force F_y applied to the inclusion, Fig. 2.97b. The complete equations of the stress functions, the stresses (σ_x, σ_y, τ_{xy}) and the displacements (u, v) are reduced to the following relationships in the close vicinity of the inclusion tip ($r = r_1$, $\theta = \theta_1$, $r \ll a$), neglecting terms which are small of higher order (Radaj and Zhang 1993(2)):

$$\begin{Bmatrix} \sigma_x \\ \sigma_y \\ \tau_{xy} \end{Bmatrix} = \frac{K_{\mathrm{II}}^{*}}{\sqrt{2\pi r}} \begin{Bmatrix} \sin\theta/2\,(-3+\kappa-2\cos\theta/2\,\cos 3\theta/2) \\ \sin\theta/2\,(-1-\kappa+2\cos\theta/2\,\cos 3\theta/2) \\ \cos\theta/2\,(1-\kappa-2\sin\theta/2\,\sin 3\theta/2) \end{Bmatrix} \tag{2.299}$$

$$\begin{Bmatrix} u \\ v \end{Bmatrix} = \frac{K_{\mathrm{II}}^{*}}{\pi G} \sqrt{2\pi r} \begin{Bmatrix} \sin\theta/2\,\cos^2\theta/2) \\ -\cos\theta/2\,(\kappa - \sin^2\theta/2 \end{Bmatrix} \tag{2.300}$$

The stress intensity factors in the two considered loading cases are:

$$K_{\mathrm{II}}^{*} = \frac{1}{(1+\kappa)} \frac{M}{a\sqrt{\pi a}} \tag{2.301}$$

$$K_{\mathrm{II}}^{*} = \frac{1}{2(1+\kappa)} \frac{F_y}{\sqrt{\pi a}} \tag{2.302}$$

The transverse force T_y in the rigid inclusion (per unit of plate thickness) is determined from σ_y at $\theta = \pi$ and $\theta = -\pi$, adding up the two values and integrating them proceeding from the inclusion tip ($r = 0$):

$$T_y = -4(1+\kappa)K_{\mathrm{II}}^* \sqrt{\frac{r}{2\pi}} \quad (r \ll a) \tag{2.303}$$

The mode III* loading condition (out-of-plane shear loading) at the rigid thin inclusion is considered by the examples of remote out-of-plane shear stress $\tau^*$and of an out-of-plane force F_z applied to the inclusion, Fig. 2.97c. The Westergaard-type stress function $\varphi^*(z)$ for the two loading cases has been defined based on more general equations (Tada et al. 1973, p. 1.23). The stresses (τ_{xz}, τ_{yz}) and displacements w are derived therefrom in bipolar coordinates. The reduction to the stresses and displacements in the close vicinity of the inclusion tip is then performed (Radaj and Zhang 1993$^{(2)}$). A solution available from the literature is based on Poisson's potential equation for the displacement function (Wang et al. 1986). The reduced equations for $r \ll a$ read:

$$\left\{ \begin{matrix} \tau_{xz} \\ \tau_{yz} \end{matrix} \right\} = \frac{K_{\mathrm{III}}^*}{\sqrt{2\pi r}} \left\{ \begin{matrix} \cos\theta/2 \\ \sin\theta/2 \end{matrix} \right\} \tag{2.304}$$

$$w = \frac{K_{\mathrm{III}}^*}{\pi G} \sqrt{2\pi r} \cos\theta/2 \tag{2.305}$$

The stress intensity factors in the two considered loading cases are:

$$K_{\mathrm{III}}^* = \tau^* \sqrt{\pi a} \tag{2.306}$$

$$K_{\mathrm{III}}^* = \frac{F_z}{2\sqrt{\pi a}} \tag{2.307}$$

The out-of-plane shear force T_z in the rigid inclusion (per unit of plate thickness) is determined from τ_{yz} at $\theta = \pi$ and $\theta = -\pi$, adding up the two values and integrating them proceeding from the inclusion tip ($r = 0$):

$$T_z = 4K_{\mathrm{III}}^* \sqrt{\frac{r}{2\pi}} \quad (r \ll a) \tag{2.308}$$

2.6.4 *Limit Value Formulae for Stress Intensity Factors at Rigid Thin Inclusion Tips*

The stress intensity factors K_{I}^*, K_{II}^*, K_{III}^* at rigid thin inclusion tips can be determined from any of the stress and displacement equations for mode I*, II*, III* loading in any polar plane, but the ligament plane ($\theta = 0$) is best suited for that purpose because of the characteristic stresses in that plane and because the stresses from different modes acting simultaneously can be separated here. The following limit value formulae are applicable evaluating the stresses in the ligament plane ($\theta = 0$), Eqs. (2.294), (2.299) and (2.304):

$$K_{\mathrm{I}}^{*} = \lim_{r \to 0} \frac{\sigma_x \sqrt{2\pi r}}{3 + \kappa} \tag{2.309}$$

$$K_{\mathrm{II}}^{*} = \lim_{r \to 0} \frac{\tau_{xy} \sqrt{2\pi r}}{1 - \kappa} \tag{2.310}$$

$$K_{\mathrm{III}}^{*} = \lim_{r \to 0} \tau_{xz} \sqrt{2\pi r} \tag{2.311}$$

Evaluation of the displacements in the ligament plane, Eqs. (2.295), (2.300) and (2.305) gives:

$$K_{\mathrm{I}}^{*} = \lim_{r \to 0} \frac{\pi G}{\kappa} \frac{u}{\sqrt{2\pi r}} \tag{2.312}$$

$$K_{\mathrm{II}}^{*} = \lim_{r \to 0} \frac{-\pi G}{\kappa} \frac{v}{\sqrt{2\pi r}} \tag{2.313}$$

$$K_{\mathrm{III}}^{*} = \lim_{r \to 0} \pi G \frac{w}{\sqrt{2\pi r}} \tag{2.314}$$

Evaluation of the forces in the inclusion Eqs. (2.298), (2.303) and (2.308), results in:

$$K_{\mathrm{I}}^{*} = \lim_{r \to 0} \frac{\pi}{2(1 + \kappa)} \frac{T_x}{\sqrt{2\pi r}} \tag{2.315}$$

$$K_{\mathrm{II}}^{*} = \lim_{r \to 0} \frac{-\pi}{2(1 + \kappa)} \frac{T_y}{\sqrt{2\pi r}} \tag{2.316}$$

$$K_{\mathrm{III}}^{*} = \lim_{r \to 0} \frac{\pi}{2} \frac{T_z}{\sqrt{2\pi r}} \tag{2.317}$$

Another limit value procedure for determining the stress intensity factor at the rigid inclusion tip is based on the notch stress concentration in the principal plane ($\theta = 0$) of the comparable rigid elliptical inclusion. Evaluating formulae for the loading cases of remote tension loading (mode I*), transverse shear loading (mode II*) and out-of-plane shear loading (mode III*), the following limit values for the notch radius $\rho \to 0$ are found (Radaj 1971; Radaj and Schilberth 1977; Mura 1987; Radaj and Zhang 1993[(2)]):

$$K_{\mathrm{I}}^{*} = \lim_{\rho \to 0} \frac{1}{1 + \kappa} \sigma_{x\,\max} \sqrt{\pi \rho} \tag{2.318}$$

$$K_{\mathrm{II}}^{*} = \lim_{\rho \to 0} \frac{-1}{1 + \kappa} \tau_{xy\,\max} \sqrt{\pi \rho} \tag{2.319}$$

$$K_{\mathrm{III}}^{*} = \lim_{\rho \to 0} \tau_{xz\,\max} \sqrt{\pi \rho} \tag{2.320}$$

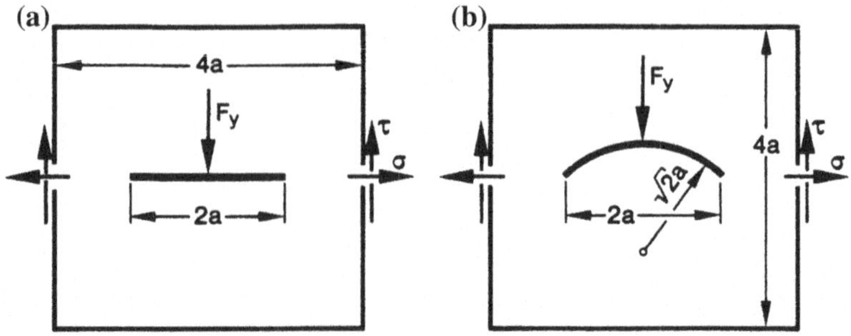

Fig. 2.101 Flat rigid thin inclusion (**a**) and curved rigid thin inclusion (**b**) in quadratic plate subjected to inclusion-parallel tensile loading (σ) or transverse shear loading (F_y, τ); (Radaj and Zhang 1993[(2)])

2.6.5 Example of Stress Intensity Factor Evaluation

The limit value formulae, Eqs. (2.309–2.317) allow to determine unknown stress intensity factors at rigid thin inclusion tips based on a FE or boundary element analysis of the stresses and displacements in the ligament plane ($\theta = 0$). As an example of application, a boundary element analysis has been performed for a flat and a curved (quarter-circular) rigid thin inclusion in a quadratic plate subjected to tensile or transverse shear loading, Fig. 2.101. The plate width is two times the crack length. The tensile stress σ has a constant value, the shear stress τ is variable according to a triangular shape, with a zero value at the corners of the quadratic plate and a maximum value in the midpoint of the loaded plate edge.

The stresses σ_x or τ_{xy} and the displacements $u = u_x$ or $v = u_y$ (with the coordinate x brought into line with the local direction of the inclusion at its tip) are determined at small distances from the inclusion tip ($r/a < 1.0$), placing the substructure boundary into the ligament plane ($\theta = 0$). These stress and displacement values are inserted into Eqs. (2.309), (2.310), (2.312) and (2.313) without the limit, resulting in $K_{\mathrm{I}}^*(r)$ and $K_{\mathrm{II}}^*(r)$, the medium values of the stress and displacement evaluation then being extrapolated to the inclusion tip applying a least square fit, Fig. 2.102. The values for the quadratic plate, K_{I}^* and K_{II}^*, are normalised by the values for the infinite plate, K_{I0}^* and K_{II0}^*. In the tensile loading case, K_{I}^* related to K_{I0}^* is slightly reduced. In the transverse shear loading case, K_{II}^* related to K_{II0}^* meets exactly the value 1.5 which results from the parabolic shear stress distribution in a comparable transverse-shear-loaded beam.

An evaluation has also been performed for the stress intensity factor ratios at flat and curved rigid thin inclusions in comparison, both in the tensile and transverse shear loading case, Table 2.8. The resultant stress intensity factor is determined according to $K^{*2} = K_{\mathrm{I}}^{*2} + K_{\mathrm{II}}^{*2}$. In the case of the curved rigid inclusion, non-singular stresses may occur at the inclusion tip which have not been evaluated.

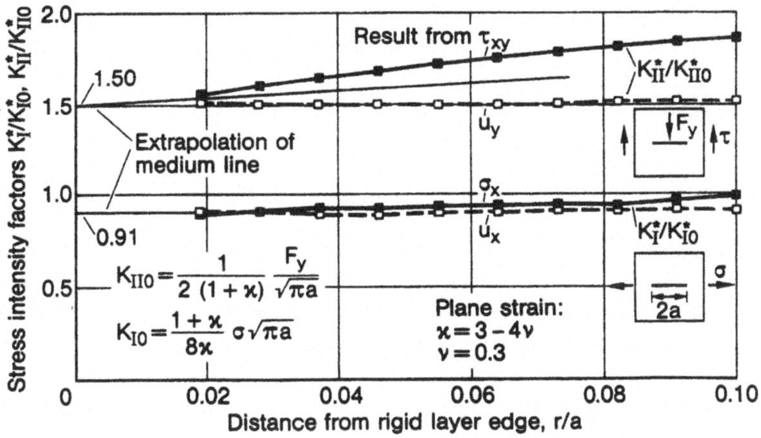

Fig. 2.102 Stress intensity factors for flat rigid thin inclusion subjected to tensile loading and transverse shear loading determined from ligament stresses (σ_x or τ_{xy}) and displacements ($u = u_x$, $v = u_y$), quadratic plate values K_I^* and K_{II}^* normalised by infinite plate values K_{I0}^* and K_{II0}^*; (Radaj and Zhang 1993[(2)]))

Table 2.8 Stress intensity factor ratios for flat and curved rigid thin inclusion in quadratic plate subjected to tensile or transverse shear loading; (Radaj and Zhang 1993[(2)])

	Tensile loading			Transverse shear loading		
Inclusion type	K_I^*/K_{I0}^*	K_{II}^*/K_{I0}^*	K_{res}^*/K_{I0}^*	K_I^*/K_{II0}^*	K_{II}^*/K_{II0}^*	K_{res}^*/K_{II0}^*
Flat rigid	0.91	0.00	0.91	0.00	1.50	1.50
Curved rigid	0.58	0.77	0.96	0.75	0.93	1.19

2.6.6 Outlook to Wedge-Shaped Rigid Inclusion

Just as the SIF concept at crack tips has been extended to the NSIF concept at V-shaped notches, the SIF concept at rigid thin inclusion tips can be extended to an NSIF concept of rigid wedge-shaped inclusion tips. The crack or inclusion tips can either be pointed or rounded. A rigid wedge-shaped inclusion is identical with a V-shaped notch with its flanks rigidly fixed. This problem has early been solved for tensile and in-plane shear loading (modes 1* and 2*); (Hasebe et al. 1990).

The eigenvalue equations under the considered rigidly fixed notch flank condition read as follows:

$$\kappa \sin \lambda_1^* q\pi - \lambda_1^* \sin q\pi = 0 \tag{2.321}$$

$$\kappa \sin \lambda_2^* q\pi + \lambda_2^* \sin q\pi = 0 \tag{2.322}$$

Compared with the corresponding equations representing the free notch flank condition, Eqs. (2.66) and (2.67), the factor κ modifies the first term and the sign between the two terms is inverted.

The corresponding eigenvalue equation for out-of-plane loading is identical with the V-notch equation, Eq. (2.68):

$$\sin \lambda_3^*(2\pi - 2\alpha) = 0 \tag{2.323}$$

Comparing the published values of $\lambda_1^* = m_1 + 1$ and $\lambda_2^* = m_2 + 1$ for $\kappa = 2$ (Hasebe et al. 1990) with the values of λ_1 and λ_2, considering values for $2\alpha = 90°$, the degree of the singularity is diminished in tensile loading ($\lambda_1^* = 0.748$ compared with $\lambda_1 = 0.545$), but enlarged in transverse shear loading ($\lambda_2^* = 0.602$ compared with $\lambda_2 = 0.909$).

The maximum stress at the notch root in the bisector plane in the case of tensile loading and transverse shear loading, respectively, are given as a power series expansion (Hasebe et al. 1990):

$$\sigma_{r\,\max} = h_r \rho^{\lambda_1 - 1} + h_{r,2} \rho^{\lambda_{1,2} - 1} + h_{r,3} \rho^{\lambda_{1,3} - 1} + \ldots \tag{2.324}$$

$$\tau_{r\theta\,\max} = h_{r\theta} \rho^{\lambda_2 - 1} + h_{r\theta,2} \rho^{\lambda_{2,2} - 1} + h_{r\theta,3} \rho^{\lambda_{2,3} - 1} + \ldots \tag{2.325}$$

Since the series expansions in Eqs. (2.324) and (2.325) converge fast, two or three terms are sufficient in general to provide acceptable accuracy. When two or three SCFs for different values of ρ are known, e.g. by a FE analysis, then the full functional dependency is defined which is valid in the close vicinity of the inclusion tip.

Just as reference is made to the rhombic cutouts in Sect. 2.4.5 dealing with the SCFs of rounded V-notches, the corresponding rhombic inclusion problem may be addressed (Radaj and Schilberth 1977).

2.6.7 Conclusions

The SIF and NSIF concepts describing the stress, strain and displacement fields in the close vicinity of crack or notch tips can be transferred to the corresponding rigid inclusion tips. The basic singular and non-singular loading modes are defined in an analogous manner.

The asymptotic stress, strain and displacement fields close to the rigid inclusion tip are characterised by stress intensity factors linked with trigonometric function expressions. Based on these equations, limit value ($r \to 0$) formulae are derived which allow to determine the SIFs or NSIFs (not yet accomplished) from FE or boundary element analysis models. Other limit value ($\rho \to 0$) formulae relate the maximum notch stresses of comparable elliptical inclusions to the SIFs of flat thin inclusions.

A boundary element analysis of a flat and a curved rigid thin inclusion in a quadratic plate subjected to tensile or transverse shear loading is performed, demonstrating the application of the SIF concept. Finally, a wedge-shaped rigid inclusion without and with notch rounding is considered, providing an outlook to the more general NSIF concept.

In conclusion, it is shown that the rigid inclusion problems can be solved within the same analytical frame as has been developed for the open notch problems.

List of Symbols

A	Parameter relating $K_{1\mathrm{p}}$ to K_1
A_{11}, A_{22}, A_{33}	Coefficients in K_{eq} formula
A_{k}	Coefficient in K–N relationship
a	Elliptic crack depth, notch depth, crack length
a, b	Axes of ellipse
a, b, c	Coefficients in complex stress function
$a_{\mathrm{i}}, a_{\mathrm{cr}}$	Initiated and critical crack length
a_0	Microstructural length parameter
a_1, a_2	Parameters substituting K_1, K_2
$a_{2\tau}, a_{2\sigma}$	Factors in K_2 formulae
B_J, B_W	Parameters of J and $\overline{W}_{\mathrm{p}}$, mode 3
b	Notch depth, semiwidth of net cross-section
C	Hardening coefficient ($C = H$)
C_1	Strain level parameter in $K_{3\rho}$
c	Elliptic crack semilength
d	Mesh width, element size
d, e, f	Coefficients in complex stress function
E, E'	Modulus of elasticity, original and substitute
$e_{\mathrm{d1}}, e_{\mathrm{d2}}$	Distortional strain energy density coefficients
F	Tensile force
$\tilde{F}$	Angle-dependent function of antiplane shear stress
F_x, F_y, F_z	Resultant forces acting on inclusion
f_T	Angle-dependent function of T-stress
f_1, f_2, f_3	Angle-dependent function of mode 1, 2, 3 stress
$f_{\mathrm{I}}, f_{\mathrm{II}}, f_{\mathrm{III}}$	Angle-dependent function of mode I, II, III stress
$G_{\mathrm{res}}, G_{\mathrm{max}}$	Energy release rate, resultant and maximum value
G	Shear modulus
$g_1, g_2, g_3, g_{\mathrm{n}}$	Coefficients in expression for σ_θ in bisector
H	Complex stress function
H, H_τ	Hardening coefficient, tensile and shear loading

h	Plate semithickness
h_1, h_2, h_3	Coefficients in expression for $\tau_{r\theta}$ in bisector
h_r, $h_{r\theta}$	Leading term factors of $\sigma_{r\ \max}$ and $\tau_{r\theta\ \max}$ at rigid wedge
I, I_e, I_p	Integral in HRR field equation, total, elastic, plastic
J	J-integral
J_2, J_K, J_T	J-integrals for kinking crack
J_{3p}	J-integral related to K_{3p}
J_I, J_{II}, J_{III}	J-integral, mode I, II, III
J_V, J_L	J-integrals of V-notches
K_t	Theoretical stress concentration factor
K_σ, K_ε	Stress and strain concentration factor, nonlinear
K_τ, K_γ	Shear stress and strain concentration factor, nonlinear
$K_{W,e}$, $K_{W,p}$	Strain energy concentration factor, elastic, plastic
K_I, K_{II}, K_{III}	Stress intensity factor, mode I, II, III
K_{Ic}, $K_{I,c}$	Fracture toughness, critical stress intensity factor
K_I^*, K_{II}^*, K_{III}^*	Stress intensity factor, mode I*, II*, III*, rigid inclusion
K_{I0}^*, K_{II0}^*	Reference values of K_I^*, K_{II}^* in infinite plate
K_{res}, K_{eq}	Resultant and equivalent stress intensity factor
K_{th}	Threshold stress intensity factor
K_0, K_0^*	Threshold stress intensity factor, coplanar and kinking crack
K_{1cr}	Critical notch stress intensity factor
K_1, K_2, K_3	Notch stress intensity factor, mode 1, 2, 3
$K_{1\rho}$, $K_{2\rho}$, $K_{3\rho}$	Generalised notch stress intensity factor, mode 1, 2, 3
$K_{1\rho}^*$	Generalised NSIF, T-stress-corrected
$K_{3\rho,e}^{(\gamma)}$, $K_{3\rho,p}^{(\gamma)}$	Notch strain intensity factor, mode 3, elastic, plastic
$\overline{K}_{i\rho}$	Average NSIF, mode i = 1, 2, 3
K_O	Stress intensity factor, transverse singular mode O
$K_{2,n}$	Nominal NSIF, mode 2
k	Inverse slope exponent, S–N curve
k_1, k_2, k_3	Geometry coefficients on K_1, K_2, K_3
k_1, k_{10}	Factors in $K_{1\rho}$ formula
M	Turning moment acting on the inclusion
M_t	Torsional moment
M_{km}, M_{kb}	Magnification factor on K_I, membrane, bending
m	Elastic-plastic antiplane shear exponent
N, N_E, N_C	Number of cycles, endurable values
n	Hardening exponent
P_f	Failure probability
q	Factor on $\pi/2$ for notch internal angle
R	Ratio of lower to upper load
R, R_0	Radius of integration path, of control volume

R_g, R_n	Radius of gross and net cross-section
R_p	Radius of plastic zone
r	Polar coordinate, radial distance
r^*	Radius of core region
r_0	Notch root distance
r_0, r_1, r_2	Radial distances in bipolar coordinate system
r_2	V-notch radius at rhombic cut-out
r_p	Radius of plastic zone
S	S-stress, non-singular
S_{min}, S_{max}	Strain energy density factors
s	Elastic-plastic eigenvalue exponent
s_{ij}	Deviatoric stress tensor
$s_{\gamma,p}$	Elastic-plastic strain eigenvalue exponent
T	T-stress, non-singular
T_x, T_y, T_z	Forces acting on rigid inclusion
T_σ, T_N	Scatter range indices related to σ and N
t, t_u, t_l	Plate thickness, upper and lower
u, v, w	Displacements in Cartesian coordinates
u_0, v_0	Remote boundary displacements
u_x, u_y, u_z	Displacements in Cartesian coordinates
u_i, $\tilde{u}_i$	Displacement vector, its angle-dependent term
W, W_e, W_p	Strain energy density, total, elastic, plastic
$\overline{W}$, $\overline{W}_e$, $\overline{W}_p$	Locally averaged strain energy densities
$W_{n,n}$	Strain energy density of nominal stress in net cross-section
w	Complex coordinate
x, y, z	Cartesian coordinates
x_p	Extension of plastic zone
Y_1, Y_2, Y_3	Geometry factors on K_I, K_{II}, K_{III}
Y_m, Y_b	Geometry factors on K_I, membrane, bending
z	Complex coordinate
z	Elastic-plastic eigenvalue exponent
α	Factor in normalised Ramberg–Osgood material law
α, α_{02}	Notch opening semi-angle
β	Angle of inclination against bar axis
β	Notch angle dependent parameter in Neuber rule for V-notch
γ	Internal semi-angle of V-notch
γ, γ_e, γ_p	Total, elastic and plastic shear strain
γ_0	Substitute or reference yield shear strain
$\gamma_{max,p}$	Maximal notch shear strain, nonlinear
$\gamma_{p,eq}$	Equivalent plastic shear strain
γ_Y	Yield limit shear strain, marked and offset
Δ	Relative deviation
δ	Thickness ratio

δ	Factor in reduced Ramberg–Osgood material law
δ_{ij}	Kronecker delta
ε, ε_e, ε_p	Total, elastic and plastic strain
ε_{ij}, $\tilde{\varepsilon}_{ij}$	Strain tensor, its angle-dependent term
ε_k, ε_{kH}	Maximum notch strain, nonlinear and Hooke
ε_n, ε_{nH}	Nominal strain, nonlinear and Hooke
$\varepsilon_{p,\ eq}$	Equivalent plastic strain
ε_0	Substitute or reference yield strain
ε_Y, $\varepsilon_{Y0.2}$	Yield limit strain, marked and 0.2 % offset
ζ	Complex coordinate
η	Factor on ρ for averaging $K_{i\rho}$
θ, $\bar{\theta}$	Polar angle, crack propagation angle
θ_0, θ_1, θ_2	Polar angles in bipolar coordinate system
κ	Stiffness parameter in plate theory
κ	In-plane multiaxiality parameter
λ_1, λ_2, λ_3	Eigenvalues at V-notch, mode 1, 2, 3
λ_1^*, λ_2^*, λ_3^*	Eigenvalues at rigid wedge, mode 1^*, 2^*, 3^*
λ_a, λ_s	Eigenvalues, antimetric and symmetric, antiplane shear
λ_O	Eigenvalue of transverse singular mode
μ	Exponent in complex stress function
ν	Poisson's ratio
ρ	Radius of notch curvature, notch radius
ρ^*	Microstructural support length
σ, σ_0, $\sigma_{x\infty}$	Remote tensile stress
σ_0	Substitute or reference yield stress
σ_0^*	Normal stress parallel to crack front
σ_{0x}, σ_{0y}, σ_{0z}	Remote boundary stresses, mode $0I^*$ loading
σ_1	First principal stress
σ_{cb}, σ_{tb}	Counter-bending and tensile-bending stress
σ_E, $\bar{\sigma}_E$	Endurance limit stress, original and averaged
σ_e, σ_{eq}	Effective and equivalent stress, von Mises
$\tilde{\sigma}_e$	Angle-dependent term of σ_e
σ_h	Hydrostatic stress
σ_{ij}, $\tilde{\sigma}_{ij}$	Stress tensor, its angle-dependent term
σ_k, σ_{kH}	Maximum notch stress, nonlinear and Hooke
σ_m, σ_b	Membrane and bending stress
σ_{max}, $\sigma_{\theta\ max}$	Maximum notch stress
σ_n, $\sigma_{n,g}$, $\sigma_{n,n}$	Nominal stress, in gross and net cross-section
σ_{n1}, σ_{n2}	Endurable nominal stress
σ_{nE}	Endurance limit nominal stress
σ_p	Peak stress
σ_r, σ_θ	Normal stresses in polar coordinates
σ_t, $\sigma_{t\ max}$	Tangential stress, its maximum value

σ_{th}	Threshold stress of crack propagation
σ_u, σ_l	Stress in upper and lower plate
σ_x, σ_y, σ_z	Normal stresses in Cartesian coordinates
σ_Y, $\sigma_{Y0.2}$	Yield limit, marked and 0.2 % offset
σ_I, σ_{II}, τ_{III}	Stresses related to mode I, II, III
$\sigma_{\|}$	Slit-parallel stress
τ	Resultant shear stress, antiplane
τ_0	Substitute or reference yield shear stress
τ_0, τ_0^*	Remote shear stress, in-plane and out-of-plane
τ_1	Principal shear stress
τ_E, $\bar{\tau}_E$	Endurance limit shear stress, original and averaged
τ_{eq}	Equivalent or resultant shear stress
τ_{max}, τ_{max}^*	Maximum notch shear stress, in-plane and out-of-plane
$\tau_{max,p}$	Maximum notch shear stress, nonlinear or plastic
τ_n, τ_n^*	Nominal shear stress, in-plane and out-of-plane
$\tau_{n,g}$, $\tau_{n,n}$	Nominal shear stress, gross and net cross-section
$\tau_{n,g}^*$	Nominal torsional stress, gross cross-section
$\tau_{r\theta}$	Shear stress in polar coordinates
τ_{rz}, $\tau_{\theta z}$	Out-of-plane shear stresses
τ_Y	Shear yield limit, marked or offset
$\tau_{\perp}$, $\tau_{\|}$	Shear stresses, transverse and longitudinal to edge
Φ	Elliptical integral, stress resultant function
φ, φ^*	Complex stress function
χ	Trigonometric eigenvalue coefficient
ψ	Complex stress or harmonic displacement function
ω, ω_1	Notch angle dependent parameter in $K_{3\rho}$
$\tilde{\omega}_1$, $\tilde{\omega}_2$	Auxiliary parameter in $K_{1\rho}$, $K_{2\rho}$
FE	Finite element
FEM	Finite element method
HRR	Hutchinson–Rice–Rosengren
NSIF	Notch stress intensity factor
SCF	Stress concentration factor
SIF	Stress intensity factor

References

Akin JE (1976) The generation of elements with singularities. Int J Numer Meth Engng 10:1249–1259

Anderson TL ([2]1995) Fracture mechanics – Fundamentals and applications. CRC-Press, Boca Raton Fla

Atzori B (1985) Fracture mechanics or notch effect for fatigue design (in Italian). In: Proc 13th AIAS Conf, Bergamo, pp 294–307

Atzori B, Meneghetti G (2001) Fatigue strength of fillet welded structural steels – finite elements, strain gauges and reality. Int J Fatigue 23:713–721

Atzori B, Lazzarin P, Tovo R (1997) Stress distribution for V-shaped notches under tensile and bending loads. Fatigue Fract Engng Mater Struct 20:1083–1092

Atzori B, Lazzarin P, Tovo R ($1999^{(1)}$) From a local stress approach to fracture mechanics: a comprehensive evaluation of the fatigue strength of welded joints. Fatigue Fract Engng Mater Struct 22:369–381

Atzori B, Lazzarin P, Tovo R ($1999^{(2)}$) Stress field parameters to predict the fatigue strength of notched components. J Strain Analysis 34:437–453

Atzori B, Lazzarin P, Filippi S (2001) Crack and notches: analogies and differences of the relevant stress distributions and practical consequences in fatigue limit predictions. Int J Fatigue 23:355–362

Ayatollahi MR, Dehghany M (2010) On *T*-stresses and V-notches. Int J Fract 165:1–126

Ayatollahi MR, Nejati M (2011) Determination of NSIFs and coefficients of higher order terms for sharp notches using finite element method. Int J Mech Sciences 53:64–177

Benthem JP (1977) State of stress at the vertex of a quarter-infinite plate in a half space. Int J Solids Struct 13:479–492

Benthem JP (1987) Stresses in the region of rounded corners. Int J Solids Struct 23:239–252

Berto F, Lazzarin P (2010) On higher order terms in the crack tip stress field. Int J Fract 161:221–226

Berto F, Lazzarin P, Kotousov A, Harding S (2011) Out-of-plane singular stress fields in V-notched plates and welded lap joints induced by in-plane shear load conditions. Fatigue Fract Engng Mater Struct 34:291–304

Bogy DB (1971) Two edge-bonded elastic wedges of different materials and wedge angles under surface tractions. J Appl Mech 38:377–386

Boukharouba T, Tamine T, Nui L, Chehimi C, Pluvinage G (1995) The use of notch intensity factor as a fatigue crack initiation parameter. Engng Fract Mech 52:503–512

Brussat TB, Westermann RA (1975) A Westergaard-type stress function for line inclusion problems. Int J Solids Struct 11:665–667

Carpenter WC (1984) Calculation of fracture mechanics parameters for a general corner. Int J Fract 24:45–58

Chakrabarty J (^{3}2006) Theory of plasticity. Oxford, Elsevier

Chen DH (1995) Stress intensity factors for V-notched strip under tension or in-plane bending. Int J Fract 70:81–97

Chen DH, Ushijima K (2000) Elastic-plastic stress singularity near the tip of V-notch. Int J Fract 106:117–134

Chen YZ (1986) Singular behaviour at fixed line tip in plane elasticity. Engng Fract Mech 25:11–16

Chen YZ (1991) Investigation of stress singularity coefficient for a finite plate containing rigid line. Engng Fract Mech 40:17–24

Chen YZ (1992) Evaluation of stress singularity coefficient from the solution of the notch problem. Comput Struct 42:563–567

Chen YZ, Cheung YK (1987) Stress singularity coefficients in an infinite plate containing rigid line and applied by concentrated forces. Engng Fract Mech 26:729–739

Cheng YZ (1988) Evaluation of K_2 values from the solution of notch problem. Int J Fract 38:R61–R64

Cheung YK, Chen YZ (1989) Multiple rigid line problems in an infinite plate. Engng Fract Mech 34:379–391

Cottrell B, Rice JR (1980) Slightly curved or kinked cracks. Int J Fract 16:155–169

Creager M, Paris PC (1967) Elastic field equations for blunt cracks with reference to stress corrosion cracking. Int J Fract Mech 3:247–252

Dini D, Hills DA (2004) Asymptotic characterisation of nearly sharp notch root stress fields. Int J Fatigue 130:651–666.

Dowling NE (1993) Mechanical behaviour of materials. Prentice Hall, New Jersey
Dunn ML, Suwito W, Cunningham S (1997[1]) Stress intensities at notch singularities. Engng Fract Mech 57:417–430
Dunn ML, Suwito W, Cunningham S (1997[2]) Fracture initiation at sharp notches: correlation using critical stress intensities. Int J Solid Struct 34:3873–3883
El Haddad MH, Topper TH, Smith KN (1979) Prediction of non-propagating cracks. Engng Fract Mech 11:573–584
Elber W (1970) Fatigue crack closure under cyclic tension. Engng Fract Mech 2:37–45
Elber W (1971) The significance of fatigue crack closure. ASTM STP 486, Philadelphia Pa, pp 230–242
Erdogan F, Sih GC (1963) On the crack extension in plates under plane loading and transverse shear. J Basic Engng 85:519–527
Eurocode 3 (2005) Design of steel structures, Part 1-9, Fatigue, EN 1993-1-9, Brussels, CEN
Filippi S, Lazzarin P (2004) Distributions of the elastic principal stress due to notches in finite size plates and rounded bars uniaxially loaded. Int J Fatigue 26:377–391
Filippi S, Ciavarella M, Lazzarin P (2002[1]) An approximate, analytical approach to the 'HRR'-solution for sharp V-notches. Int J Fract 117:269–286
Filippi S, Lazzarin P, Tovo R (2002[2]) Developments of some explicit formulas useful to describe elastic stress fields ahead of notches in plates. Int J Solids Struct 39:4543–4565
Gdoutos EE (1990) Fracture mechanics criteria and applications. Kluwer Acad Publ, Dordrecht
Glinka G (1985) Energy density approach to calculation of inelastic strain–stress near notches and cracks. Engng Fract Mech 22:485–508
Glinka G, Molski K (1981) A method of elastic-plastic stress and strain calculation at a notch root. Mater Sci Engng 50:93–100
Glinka G, Newport A (1987) Universal features of elastic notch tip stress fields. Int J Fatigue 9:143–150
Gough HJ (1950) Engineering steels under combined cyclic and static stresses. J Appl Mech 50:113–125
Gough HJ, Pollard HV (1935) The strength of metals under combined alternating stresses. In: Proc Inst Mech Eng 131:3–103
Gough HJ, Pollard HV (1937) Properties of some materials for cast crankshafts with special reference to combined stresses. In: Proc Inst Automob Eng 31:821
Gross R, Mendelson A (1972) Plane elastostatic analysis of V-notched plates. Int J Fract Mech 8:267–276
Gross D, Seelig T ([3]2001) Bruchmechanik. Springer Verlag, Berlin
Haefele PM, Lee JD (1995) The constant stress term. Engng Fract Mech 50:869–882
Hahn HG (1976) Bruchmechanik. Teubner, Stuttgart
Harding S, Kotousov A, Lazzarin P, Berto F (2010) Transverse singular effects in V-shaped notches stressed in mode II. Int J Fract 164:1–14
Hasebe N, Kutanda Y (1978) Calculation of stress intensity factors from stress concentration factors. Engng Fract Mech 10:215–221
Hasebe N, Nakamura T, Iida J (1990) Notch mechanics for plane and thin-plate bending problems. Engng Fract Mech 37:87–99
Hein VL, Erdogan F (1971) Stress singularities in a two-material wedge. Int J Fract Mech 7:317–330
Hencky H (1924) Zur Theorie plastischer Deformationen und der hierdurch im Material hervorgerufenen Nachspannungen. Z Angew Math Mech 4:323–334
Hilton PD (1973) Plastic intensity factors for cracked plates subjected to biaxial loading. Int J Fract 9:149–156
Hilton PD, Hutchinson JW (1971) Plastic intensity factors for cracked plates. Engng Fract Mech 3:435–451
Hobbacher A (ed) (2009) Fatigue design of welded joints and components. IIW-Doc XIII-2151-07/XV-1254-07. Welding Research Council, Bulletin 520, New York

Hui CY, Ruina A (1995) Why *K*? Higher order singularities and small-scale yielding. Int J Fract 72:97–120
Hult JA, McClintock FA (1956) Elastic-plastic stress and strain distribution around sharp notches under repeated shear. In: 9th Int Congress Appl Mech, 8, Brussels
Hussain MA, Pu SL, Underwood J (1974) Strain energy release rate for a crack under combined mode I and mode II. In: Fracture analysis, ASTM STP 560, Philadelphia Pa, pp 2–28
Hutchinson JW (1968) Singular behaviour at the end of a tensile crack tip in a hardening material. J Mech Phys Solids 16:13–31
Irwin GR (1957) Analysis of stresses and strains near the end of a crack traversing a plate. J Appl Mech 24:361–364
Irwin GR (1958) Fracture. In: Handbuch der Physik, Vol 4, pp 551–590. Springer, Berlin
Kane TR, Mindlin RD (1956) High frequency extensional vibrations of plates. J Appl Mech 23:277–283
Kirsch G (1898) Die Theorie der Elasitzität und die Bedürfnisse der Festigkeitslehre. Z VDI 42:797–807
Koibuchi K, To K, Iida M, Hosomi T (1999) Fatigue strength assessment of welded toes and defects of structures based on cyclic plastic zone size. In: Engng Against Fatigue. Rotterdam, Balkema, pp 373–380
Kotousov A (2005) On stress singularities at angular corners of plates of arbitrary thickness under tension. Int J Fract 132:L29–L36
Kotousov A (2007) Fracture in plates of finite thickness. Int J Solids Struct 44:8259–8273
Kotousov A, Lew YT (2006) Stress singularities resulting from various boundary conditions in angular corners of plates of arbitrary thickness in extension. Int J Solids Struct 43:5100–5109
Kotousov A, Wang CH (2002) Three-dimensional solutions for transversally isotropic composite plates with a notch. Compos Struct 57:443–450
Kuang ZB, Xu XP (1987) Stress and strain fields at the tip of a V-notch in a power law hardening material. Int J Fract 35:39–53
Kullmer G (1992) Elastic stress fields in the vicinity of a narrow notch with circular root. In: Proc Eur Conf on Reliability and Structural Integrity of Advanced Materials, ECF9, pp 905–910
Kurath P (1992) Multiaxiality fatigue criteria for spot welds. SAE Techn Paper 920668, SAE. Warrendale Pa
Larsson SG, Carlsson AJ (1973) Influence of non-singular stress terms and specimen geometry on small-scale yielding at crack tips in elastic-plastic materials. J Mech Phys Solids 21:263–277
Lazzarin P (1986) Influence of welded joint geometry on stress fields in the highly stressed zones (in Italian). Riv Ital Saldatura 2:107–115
Lazzarin P, Filippi S (2006) A generalized stress intensity factor to be applied to rounded V-shaped notches. Int J Solids Struct 43:2461–2478
Lazzarin P, Livieri P (2000) Plastic zone size at the weld toe in welded joints made of steel and light metals (in Italian). Rev Ital Della Saldatura 52:421–437
Lazzarin P, Livieri P (2001) Notch stress intensity factors and fatigue strength of aluminium and steel welded joints. Int J Fatigue 23:225–232
Lazzarin P, Tovo R (1996) A unified approach to the evaluation of linear-elastic stress fields in the neighbourhood of cracks and notches. Int J Fract 78:3–19
Lazzarin P, Tovo R (1998) A notch stress intensity factor approach to the stress analysis of welds. Fatigue Fract Engng Mater Struct 21:1089–1103
Lazzarin P, Zambardi R (2002) The equivalent strain energy density approach reformulated and applied to sharp V-shaped notches under localized and generalized plasticity. Fatigue Fract Engng Mater Struct 25:917–928
Lazzarin P, Zappalorto M (2008) Plastic notch stress intensity factors for pointed V-notches under antiplane shear loading. Int J Fract 152:1–25
Lazzarin P, Tovo R, Meneghetti G (1997) Fatigue crack initiation and propagation phases near notches in metals with low notch sensitivity. Int J Fatigue 19:647–657
Lazzarin P, Tovo R, Filippi S (1998) Elastic stress distributions in finite size plates with edge notches. Int J Fract 91:269–282

Lazzarin P, Zambardi R, Livieri P (2001) Plastic notch stress intensity factors for large V-shaped notches under mixed mode load conditions. Int J Fract 107:361–377

Lazzarin P, Livieri P, Zambardi R (2002) A *J*-integral based approach to predict the fatigue strength of components weakened by sharp V-shaped notches. Int J Comp Appl Techn 15:202–210

Lazzarin P, Sonsino CM, Zambardi R (2004) A notch stress intensity approach to assess the multiaxial fatigue strength of welded tube-to-flange joints subjected to combined loadings. Fatigue Fract Engng Mater Struct 27:127–140

Lazzarin P, Berto F, Radaj D (2006) Uniform fatigue strength of butt and fillet welded joints in terms of local strain energy density. In: Proc 9th Int Fatigue Congress. Oxford, Elsevier Science, pp 1–10

Lazzarin P, Zappalorto M, Yates JR (2007) Analytical study of stress distributions due to semi-elliptic notches in shafts under torsion loading. Int J Engng Sci 45:308–328

Lazzarin P, Berto F, Radaj D (2009) Fatigue-relevant stress field parameters of welded lap joints – Pointed slit tip compared with keyhole notch. Fatigue Fract Engng Mat Struct 32:713–735

Lazzarin P, Zappalorto M, Berto F (2011) Generalized stress intensity factors for rounded notches in plates under in-plane shear loading. Int J Fract 170:123–144

Lee H, Kim N (2004) Fatigue life prediction of multi-spot welded panel structures using an equivalent stress intensity factor. Int J Fatigue 26:403–412

Lin KY, Tong P (1980) Singular finite elements for the fracture analysis of V-notched plate. Int J Numer Meth Engng 15:1343–1354

Linder J, Melander A, Larsson M, Bergengren Y (1998) Fatigue design of spot-welded austenitic and duplex stainless sheet steels. Fatigue Fract Engng Mater Struct 21:673–686

Livieri P, Tovo R (2009) The use of the J_V parameter in welded joints – Stress analysis and fatigue assessment. Int J Fatigue 31:153–163

Meneghetti G, Lazzarin P (2007) Significance of the elastic peak stress evaluated by FE analyses at the point of singularity of sharp V-notched components. Fatigue Fract Engng Mater Struct 30:95–106

Mizui M, Sekine T, Tsujimura A, Takishima T, Shimazaki Y (1988) An evaluation of fatigue strength for various kinds of spot-welded specimens. SAE Techn Paper 880375, SAE, Warrendale Pa

Mura T (1987) Micromechanics of defects in solids. Nijhoff, Dordrecht

Murakami Y (ed) (1987) Stress intensity factors handbook. Pergamon Press, Oxford

Mushkelishvili NI (1963) Some basic problems of the mathematical theory of elasticity. Noordhoff, Groningen.

Nakamura T, Parks DM (1989) Antisymmetrical 3d stress field near the crack front of a thin elastic plate. Int J Solids Struct 25:1411–1426

Neuber H (^{2}1958, ^{3}1985) Kerbspannungslehre. Springer Verlag, Berlin

Neuber H (1961) Theory of stress concentration for shear-strained prismatical bodies with arbitrary nonlinear stress–strain law. J Appl Mech 28:544–550

Neuber H (1968$^{(1)}$) A physically nonlinear notch and crack model. J Mech Phys Solids 16:289–294

Neuber H (1968$^{(2)}$) Über die Berücksichtigung der Spannungskonzentration bei Festigkeitsberechnungen. Konstruktion 20:245–251

Noda NA, Nisitani H, Harada S, Takase Y (1995) Fatigue strength of notched specimens having nearly equal sizes of ferrite. Int J Fatigue 17:237–244

Noda NA, Oda K, Inoue T (1996) Analysis of newly defined stress intensity factors for angular corners using singular integral equations of the body force method. Int J Fract 76:243–261

Noda NA, Takase Y (2003) Generalised stress intensity factors for V-shaped notch in a round bar under torsion, tension and bending. Engng Fract Mech 70:1447–1466

Nui LS, Chehimi C, Pluvinage G (1994) Stress field near a large blunted tip V-notch and application of the concept of the critical notch stress intensity factor (NSIF) to the fracture toughness of very brittle materials. Engng Fract Mech 49: 235–335

Pook LP (1989) The significance of mode I branch cracks for mixed mode fatigue crack growth threshold behaviour. In: Brown MW, Miller KJ (eds) Biaxial and multiaxial fatigue. Mech Engng Publ, London, pp 247–263

Quian J, Hasebe N (1997) Property of eigenvalues and eigenfunctions for an interface V-notch in antiplane elasticity. Engng Fract Mech 56:729–734

Radaj D (1971) Kerbspannungen an Öffnungen und starren Kernen. Habilitation thesis, TU Braunschweig, self-publishing

Radaj D (1989) Stress singularity, notch stress and structural stress at spot-welded joints. Engng Fract Mech 34:495–506

Radaj D (2010) *T*-stress corrected notch stress intensity factors with application to welded lap joints. Fatigue Fract Engng Mater Struct 33:378–389

Radaj D, Giering A (1994) Spot-welded H-specimen – The local stress parameters. Materialprüf 36:275–281

Radaj D, Giering A (1995) Spot-welded double-cup specimen – The local stress parameters. Materialprüf 37:458–462; emendation 39:430

Radaj D, Heib M (1978) Energy density fracture criteria for cracks under mixed-mode loading. Materialprüf 20:256–262

Radaj D, Schilberth G (1977) Kerbspannungen an Ausschnitten und Einschlüssen. DVS-Verlag, Düsseldorf

Radaj D, Vormwald M (^{3}2007) Ermüdungsfestigkeit – Grundlagen für Ingenieure. Springer Verlag, Berlin

Radaj D, Zhang S (1991$^{(1)}$) Stress intensity factors for spot welds between plates of unequal thickness. Engng Fract Mech 39:391–413

Radaj D, Zhang S (1991$^{(2)}$) Simplified formulae for stress intensity factors of spot welds. Engng Fract Mech 40:233–236

Radaj D, Zhang S (1992) Stress intensity factors for spot welds between plates of dissimilar materials. Engng Fract Mech 42:407–426

Radaj D, Zhang S (1993$^{(1)}$) Analogies between crack tip and rigid line tip stresses and displacements. Engng Fract Mech 44:913–919

Radaj D, Zhang S (1993$^{(2)}$) Loading modes and stress intensity factors at rigid layer edges. Engng Fract Mech 46:199–208

Radaj D, Zhang S (1993$^{(3)}$) On the relations between notch stress and crack stress intensity in plane shear and mixed-mode loading. Engng Fract Mech 44:691–704

Radaj D, Zhang S (1994). Stress distribution at bimaterial crack tips. Engng Fract Mech 47:613–617

Radaj D, Zhang S (1995$^{(1)}$). Process zone fracture criteria for crack tips. Engng Fract Mech 50:111–120

Radaj D, Zhang S (1995$^{(2)}$) Punktgeschweißte Doppelhutprofilprobe unter Torsionsbelastung – lokale Beanspruchungsparameter und Schwingfestigkeit. Materialprüf 37:458–462; emendation 39:426

Radaj D, Zheng Z, Möhrmann W (1990) Local stress parameters at the weld spot of various specimens. Engng Fract Mech 37:933–951

Radaj D, Lehrke HP, Greuling S (2001) Theoretical fatigue-effective notch stresses at spot welds. Fatigue Fract Engng Mater Struct 24:293–308

Radaj D, Sonsino CM, Fricke W (^{2}2006) Fatigue assessment of welded joints by local approaches. Woodhead Publ, Cambridge and CRC Press, Boca Raton Fla

Radaj D, Berto F, Lazzarin P (2009) Local fatigue strength parameters for welded joints based on strain energy density with inclusion of small-size notches. Engng Fract Mech 76:1109–1130

Rice JR (1967) Stresses due to a sharp notch in a work-hardening elastic-plastic material loaded by longitudinal shear. J Appl Mech 34:287–298

Rice JR (1968) A path-independent integral and the approximate analysis of strain concentration by notches and cracks. J Appl Mech 35:379–386

Rice JR, Rosengren GF (1968) Plane strain deformation near a crack tip in a power-law hardening material. J Mech Phys Solids 16:1–12

Rooke DP, Cartwright DJ (1976) Compendium of stress intensity factors. Her Majesty's Stationary Office, London
Sanford RJ (2003) Principles of fracture mechanics. Pearson Education, Upper Saddle River, NJ
Sator C (2010) Asymptotische Nahfeldanalyse ebener Multi-Materialverbindungsstellen mit der Methode der komplexen Potentiale. TU Darmstadt, Forschungsbericht Mechanik, Bd. 21
Savruk MP, Kazberuk A (2006) Relationship between the stress intensity and stress concentration factors for sharp and rounded notches. Mater Sci 42:725–738
Savruk MP, Kazberuk A (2007) A unified approach to problems of stress concentration near V-shaped notches with sharp and rounded tip. Int J Appl Mech 43:182–197
Savruk MP, Kazberuk A (2010) Two-dimenstional fracture mechanics problems for solids with sharp and rounded V-notches. Int J Fract 161:79–95
Seeger T, Greuling S, Brüning J, Leis P, Radaj D, Sonsino CM (2005) Bewertung lokaler Berechnungskonzepte zur Ermüdungsfestigkeit von Punktschweißverbindungen. FAT-Schriftenreihe 196, Frankfurt
Seweryn A (2002) Modeling of singular stress fields using the finite element method. Int J Solids Struct 39:4787–4804
Seweryn A, Molski K (1996) Elastic stress singularities and corresponding generalised stress intensity factors for angular corners under various boundary conditions. Engng Fract Mech 55:529–556
Seweryn A, Mróz Z (1995) A non-local stress failure condition for structural elements under multiaxial loading. Engng Fract Mech 51:855–973
Seweryn A, Zwolinski J (1993) Solution for the stress and displacement fields in the vicinity of a V-notch of negative wedge angles in plane problems of elasticity. Engng Fract Mech 44:275–281
Sherry AH, France CC, Goldthorpe MR (1995) Compendium of *T*-stress solutions for two- and three-dimensional cracked geometries. Fatigue Fract Engng Mater Struct 18:141–155
Shlyannikov VN, Boychenko NV, Tartygasheva AM (2011) In-plane and out-of-plane crack tip constraint effects under biaxial nonlinear deformation. Engng Fract Mech 78:1771–1783
Sih GC (1965) Plane extension of rigidly embedded line inclusions. In: Huang TC, Johnson MW (eds) Developments in mechanics, vol 3. John Wiley, London, pp 61–79
Sih GC (1973) Handbook of stress intensity factors. Lehigh Univ, Bethlehem Pa
Sih GC (1974) Strain energy density factor applied to mixed-mode crack problems. Int J Fract 10:305–321
Sih GC (1975) A three-dimensional strain energy density factor theory of crack propagation. In: Three-dimensional crack problems. Leyden, Noordhoff, pp 15–80
Sih GC, Liebowitz H (1968) Mathematical theories of brittle fracture. In: Fracture, vol. 2, Mathematical fundamentals. Academic Press
Smith E (2004) The elastic stress distribution near the root of an elliptically cylindrical notch subjected to mode III loadings. Int J Engng Sci 42:1831–1839
Smith E (2006) The mode III elastic stress distribution near the root of an intrusion-type notch and of a keyhole notch. Int J Engng Sci 44:340–344
Sneddon IN (1946) The distribution of stress in the neighbourhood of a crack in an elastic solid. Proc Roy Soc, Ser A 187:229–260
Susmel L, Tovo R (2004) Multiaxial fatigue life prediction of two welded details subjected to local complex stress states. In: Proc 7th Int Conf on Biaxial/Multiaxial Fatigue and Fracture. Berlin, DVM, pp 447–452
Tada HP, Paris C, Irwin GR (1973) The stress analysis of cracks handbook. Del Res Corp, Hellertown Pa
Tada HP, Paris C, Irwin GR (^{3}1999) The stress analysis of cracks handbook. ASME Press, New York
Tanaka K, Okajima H, Koibuchi K (2002) Fatigue strength CAE system for three-dimensional welded structures. Fatigue Fract Engng Mater Struct 25:275–282
Theocaris PS, Ioakimidis NT (1979) The V-notched elastic half plane problem. Acta Mech 32:125–140

Tovo R, Lazzarin P (1999) Relationships between local and structural stress in the evaluation of the weld toe stress distribution. Int J Fatigue 21:1063–1078

Unger DJ (2001) Analytical fracture mechanics. Dover Publ, Mineola, NY

Unger DJ (2005) A plane stress perfectly plastic mode I crack solution with continuous stress field. J Appl Mech 72:62–67

Unger DJ (2007) A complete perfectly plastic solution for the mode I crack problem under plane stress loading conditions. J Appl Mech 74:586–589

Verreman Y, Nie B (1996) Early development of fatigue cracking at manual fillet welds. Fatigue Fract Engng Mater Struct 19:669–681

Verreman Y, Nie B (1997) Short crack fatigue propagation at fillet welds. In: Proc Int Conf on Performance of Dynamically Loaded Welded Structures. New York, WRC, pp 240–253

Wang TJ, Kuang ZB (1999) Higher order aymptotic solutions of V-notch tip fields for damaged nonlinear materials under antiplane shear loading. Int J Fract 96:303–329

Wang ZY, Zhang HT, Chou YT (1985) Characteristics of the elastic field of a rigid line inhomogeneity. J Appl Mech 52:818–822

Wang ZY, Zhang HT, Chou YT (1986) Stress singularity at the tip of a rigid line inhomogeneity under antiplane shear loading. J Appl Mech 53:459–461

Westergaard HM (1939) Bearing pressures and cracks. J Appl Mech 6:A49–A53

Williams ML (1952) Stress singularities resulting from various boundary conditions in angular corners of plates in tension. J Appl Mech 19:526–528

Williams ML (1957) On the stress distribution at the base of a stationary crack. J Appl Mech 24:109–114

Williams JG, Ewing PD (1972) Fracture under complex stress – The angled crack problem. Int J Fract Mech 8:441–446

Yao W, Xia K, Gu Y (1995) On the fatigue notch factor K_f. Int J Fatigue 17:245–251

Yang W, Freund LB (1985) Transverse shear effects for through-cracks in an elastic plate. Int J Solids Struct 21:977–994

Yang S, Yuan FG, Cai X (1996) Higher order asymptotic elastic-plastic crack tip fields under antiplane shear. Engng Fract Mech 54:405–422

Yuan FG, Yang S (1994/95) Analytical solutions of fully plastic crack tip higher order fields under antiplane shear. Int J Fract 69:1–26

Yuuki R, Ohira T (1986) Fracture mechanics analysis of the fatigue strength of various spot-welded joints. Seisan Kenyu 38:475–478

Yuuki R, Ohira T (1989) Development of the method to evaluate the fatigue life of spot-welded structures by fracture mechanics. IIW Doc III-928-89/XIII-1358-89

Yuuki R, Ohira T, Kishi N, Mori N (1985$^{(1)}$) Fracture mechanics evaluating of the fatigue strength of spot-welded joints and structures. JSAE Paper 852096, pp 501–506

Yuuki R, Ohira T, Nakatsukasa H, Li W (1985$^{(2)}$) Fracture mechanics analysis and evaluation of the fatigue strength of spot-welded joints (in Japanese). Trans JSME 51:1772–1779

Zappalorto M, Lazzarin P (2007) Analytical study of the elastic-plastic stress fields ahead of parabolic notches under antiplane shear loading. Int J Fract 148:139–154

Zappalorto M, Lazzarin P (2009) A new version of the Neuber rule accounting for the influence of the notch opening angle for out-of-plane shear loads. Int J Solids Struct 46:1901–1910

Zappalorto M, Lazzarin P (2010) A unified approach to the analysis of nonlinear stress and strain fields ahead of mode III loaded notches and cracks. Int J Solids Struct 47:851–864

Zappalorto M, Lazzarin P (2011$^{(1)}$) In-plane and out-of-plane stress field solutions for V-notches with end holes. Int J Fract 168:167–180

Zappalorto M, Lazzarin P (2011$^{(2)}$) Strain energy based evaluations of plastic notch stress intensity factors at pointed V-notches under tension. Engng Fract Mech 78:2691–2706

Zappalorto M, Lazzarin P (2011$^{(3)}$) Stress fields due to inclined notches and shoulder fillets in shafts under torsion. J Strain Analysis 46:187–199

Zappalorto M, Lazzarin P (2011$^{(4)}$) On the intensity of linear-elastic high order singularities ahead of cracks and re-entrant corners. Int J Solids Struct 48:953–961

Zappalorto M, Lazzarin P, Yates JR (2008) Elastic stress distributions resulting from hyperbolic and parabolic notches in round shafts under torsion and uniform antiplane shear loadings. Int J Solid Struct 45:4879–4901
Zappalorto M, Lazzarin P, Berto F (2009) Elastic notch stress intensity factors for sharply V-notched rounded bars under torsion. Engng Fract Mech 76:439–453
Zappalorto M, Lazzarin P, Filippi S (2010) Stress field equations for U and blunt V-shaped notches in axisymmetric shafts under torsion. Int J Fract 164:253–269
Zhang S (1999) Recovery of notch stress and stress intensity factors. In: NAFEMS World Congress '99, Effective engineering analysis, vol. 2:1103–1114

Chapter 3
Local Strain Energy Density Concept

D. Radaj

Abstract The local strain energy density (SED) approach is elaborated for strength assessments in respect of brittle fracture and high-cycle fatigue. Pointed and rounded (blunt) V-notches subjected to tensile loading (mode 1) are primarily considered, occasionally extended to multiaxial conditions (mode 3, mixed mode 1 and 2). The application to brittle fracture is related to PMMA flat bar specimens with U-notches. The application to high-cycle fatigue comprises fillet-welded joints, weld-like shaped and V-notched base material specimens as well as round bar specimens with V-notch. The relation of the local SED concept to comparable other concepts is investigated, among them the Kitagawa and Atzori diagrams, the Neuber concept of fictitious notch rounding applied to welded joints and the J-integral approach. Alternative details of the local SED concept such as a semicircular control volume, microrounded notches and slit-parallel loading are also investigated. Coarse FE meshes at pointed or rounded notch tips are proven to be acceptable for accurate local SED evaluations. The peak stress method, which is based on a special coarse FE mesh for the assessment of the fatigue strength of welded joints, is also suitable.

Contents

D. Radaj and M. Vormwald, *Advanced Methods of Fatigue Assessment*,
DOI: 10.1007/978-3-642-30740-9_3,

3.1 Survey of Chapter Contents

The basic idea behind the local strain energy density (SED) approach corresponds to Neuber's concept of an 'elementary material volume' or 'microstructural support length' used for stress averaging (Neuber 1958). The average SED in a defined control volume around the pointed or rounded notch (or crack) tip is considered to be the material parameter which describes the initiation of brittle fracture or high-cycle fatigue failure.

The approach was originally named 'finite volume energy-based approach' or 'equivalent strain energy density approach' (Lazzarin and Zambardi 2001, 2002), and later on 'local strain energy density approach' Lazzarin et al. 2008[(2)]).

The approach is elaborated for pointed and rounded V-notches (inclusive of U-notches) subjected to tensile loading (mode 1) with extensions to torsional (mode 3) and multiaxial (mixed mode) loading. The static strength properties of V-notched specimens made of brittle material are thus described in a uniform manner, both for pointed notches (Lazzarin and Zambardi 2001; Yosibash et al. 2004) and for blunt notches (Lazzarin and Berto 2005[(1)]). Also, the high-cycle fatigue strength of welded joints made of steel or aluminium alloy is given in terms of the cyclic average SED of the pointed weld notch (Lazzarin et al. 2003; Livieri and Lazzarin 2005).

In Sect. 3.2, the local SED concept for pointed V-notches is presented. The basic relationships both for the total and for the distortional local SED are derived. Application to brittle fracture and to high-cycle fatigue failure is demonstrated. Fatigue *W–N* curves for welded joints and weld-like specimens are considered. The linear-elastic conditions, on which the analysis is normally based, are also extended to nonlinear elastic-plastic conditions establishing the link between the two.

In Sect. 3.3, the local SED concept for pointed V-notches is applied to multiaxial loading conditions (mode 1, 2, 3 superimposed). Axisymmetric tube-to-flange fillet-welded joints are considered under bending and torsional loads. Also, V-notched round bar specimens under tensile and torsional superimposed loading are analysed.

In Sect. 3.4, the local SED concept is developed for blunt notches subjected to tensile loading (mode 1) and extended to multiaxial conditions (mixed mode 1 and 2). This concept is applied to describe the brittle fracture strength of PMMA specimens with U-notches.

In Sect. 3.5, the relation of the local SED concept to comparable other concepts is investigated. First the relation to the Kitagawa and Atzori diagrams is demonstrated inclusive of Taylor's comparison with notch-mechanical concepts. Then, the relation to the Neuber concept of fictitious notch rounding applied to welded joints is presented. Finally, the relation to the *J*-integral is derived with focus on the ratio of elastic strain energy (within the control volume) to the *J*-integral dependent on the control volume radius.

In Sect. 3.6, special issues of the local SED concept are investigated, such as semicircular control volumes, microrounded notches and slit-parallel loading. A great number of evaluation examples are given.

In Sect. 3.7, coarse FE meshes at pointed notch tips are proven to be acceptable for accurate local SED evaluations applied to the fatigue assessment of welded joints. A special alternative is the 'peak stress method' which can be linked to the SED-based procedure.

In Sect. 3.8, the conclusions from the preceding sections of the chapter are drawn.

3.2 Pointed Notch SED Concept

3.2.1 Survey of Section Contents

The local SED concept for pointed V-notches is presented. The basic relationships both for the total and for the distortional local SED are derived. Application to the brittle fracture strength and to the high-cycle fatigue strength is demonstrated. Fatigue *W–N* curves for welded joints and weld-like specimens are considered. The linear-elastic conditions, on which the analysis is normally based, are also extended to nonlinear elastic-plastic conditions establishing the link between the two.

The expositions are primarily based on the analytical developments and their practical applications carried out by Lazzarin and Zambardi. A list of symbols and a list of references are available at the end of Chap. 3.

3.2.2 Basic Relationships of the Pointed Notch SED Concept

Not the elastic maximum notch stress controls static or cyclic crack initiation at notches of structural members, but rather the average notch stresses over a definite finite volume of the material at the notch root. This idea goes back to Neuber who formulated the hypothesis of an elementary material volume for this purpose (Neuber 1958).

Determining the average of the partly multiaxial and singular notch stresses in a finite volume around the notch tip needs further sophistication. On this track, the local strain energy density approach has been proposed and analytically developed (Lazzarin and Zambardi 2001).

In the case of pointed notch tips, not only the stresses tend towards infinity, but also the strain energy density (SED). Contrary to this, the average SED in a local finite volume around the notch tip has a finite value. This value is considered to control failure. The question is, what shape and size the finite volume around the notch tip should have. It is conceived as a sector-shaped cylinder of radius *R* along the notch tip line, termed 'control volume'. The analytical derivations below refer to the stress conditions in the sector plane, Fig. 3.1. The notch opening angle, $2\alpha = 135°$, shown in the graph corresponds to an idealised fillet weld shape. Other, less practicable choices of the control volume shape are discussed in Sect. 3.6.2.

The analytical frame of the local SED approach (Lazzarin and Zambardi 2001, 2002) refers to the in-plane stress and strain field (plane stress or plane strain) close to the notch tip, specified in a polar coordinate system brought into line with the notch bisector ($\theta = 0$), Fig. 3.2. The material is assumed to be isotropic and linear-elastic.

The stress field close to the V-notch tip is expressed by superimposing the symmetrical and antimetrical components (modes 1 and 2) as follows:

$$\sigma_{\mathrm{ij}}(r,\theta) = r^{\lambda_1 - 1} K_1 \begin{pmatrix} \tilde{\sigma}_{\theta\theta}^{(1)} & \tilde{\sigma}_{r\theta}^{(1)} & 0 \\ \tilde{\sigma}_{r\theta}^{(1)} & \tilde{\sigma}_{rr}^{(1)} & 0 \\ 0 & 0 & \tilde{\sigma}_{zz}^{(1)} \end{pmatrix} + r^{\lambda_2 - 1} K_2 \begin{pmatrix} \tilde{\sigma}_{\theta\theta}^{(2)} & \tilde{\sigma}_{r\theta}^{(2)} & 0 \\ \tilde{\sigma}_{r\theta}^{(2)} & \tilde{\sigma}_{rr}^{(2)} & 0 \\ 0 & 0 & \tilde{\sigma}_{zz}^{(2)} \end{pmatrix} \quad (3.1)$$

with the mode-related NSIFs K_1, K_2, the relevant eigenvalues λ_1, λ_2 and the functions $\tilde{\sigma}_{\mathrm{ij}}$ representing the angular dependency of the stresses, Eqs. (2.70) and (2.71). The NSIFs are defined in the form of Eqs. (2.77) and (2.78):

$$K_1 = \lim_{r\to 0} \sqrt{2\pi} r^{1-\lambda_1} \sigma_{\theta\theta}(r,0), \quad K_2 = \lim_{r\to 0} \sqrt{2\pi} r^{1-\lambda_2} \sigma_{r\theta}(r,0) \quad (3.2)$$

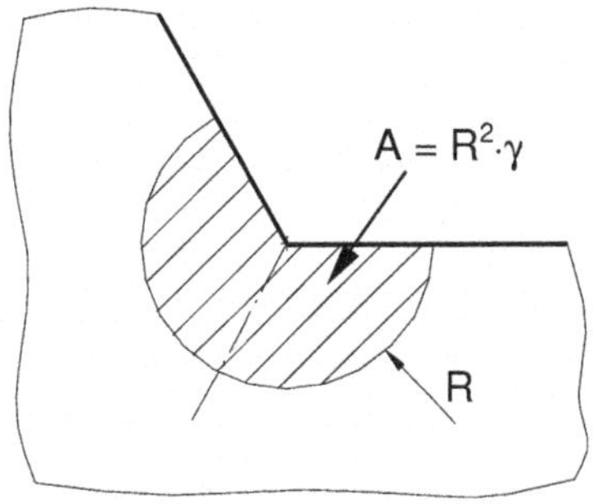

Fig. 3.1 Cylindrical control volume or circular control area around pointed V-notch tip; (Lazzarin and Zambardi 2001)

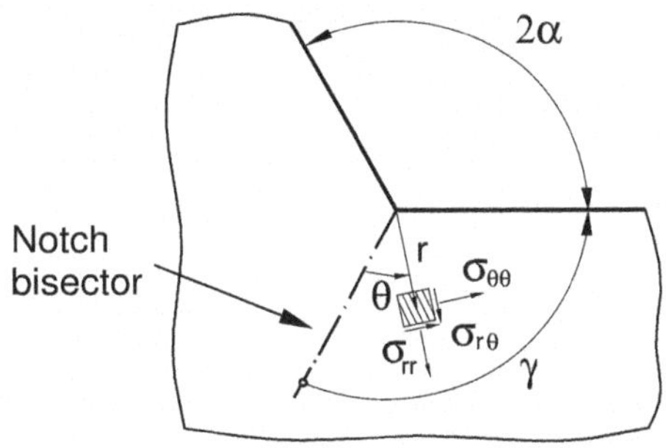

Fig. 3.2 Polar coordinate system centred at the notch tip and in line with the notch bisector ($\theta = 0$), appertaining stress components; (Lazzarin and Zambardi 2001)

The relationship between the stresses and strains in an isotropic linear-elastic material is given by:

$$\varepsilon_{\mathrm{ij}} = \frac{1+\nu}{E}\sigma_{\mathrm{ij}} - \frac{\nu}{E}\sigma_{\mathrm{kk}}\delta_{\mathrm{ij}} \tag{3.3}$$

with the modulus of elasticity E, Poisson's ratio ν, the sum of principal stresses σ_{kk} and the Kronecker delta δ_{ij}.

By combining Eqs. (3.1) and (3.3), the strain field close to the V-notch tip results:

$$\begin{aligned}
\varepsilon_{\mathrm{ij}}(r,\theta) = {} & \frac{r^{\lambda_1 - 1}K_1}{E}\begin{pmatrix} \tilde{\sigma}_{\theta\theta}^{(1)} - \nu\left(\tilde{\sigma}_{rr}^{(1)} + \sigma_{zz}^{(1)}\right) & (1+\nu)\tilde{\sigma}_{r\theta}^{(1)} & 0 \\ (1+\nu)\tilde{\sigma}_{r\theta}^{(1)} & \tilde{\sigma}_{rr}^{(1)} - \nu\left(\tilde{\sigma}_{\theta\theta}^{(1)} + \sigma_{zz}^{(1)}\right) & 0 \\ 0 & 0 & \tilde{\sigma}_{zz}^{(1)} - \nu\left(\tilde{\sigma}_{\theta\theta}^{(1)} + \sigma_{rr}^{(1)}\right) \end{pmatrix} \\
& + \frac{r^{\lambda_2 - 1}K_2}{E}\begin{pmatrix} \tilde{\sigma}_{\theta\theta}^{(2)} - \nu\left(\tilde{\sigma}_{rr}^{(2)} + \sigma_{zz}^{(2)}\right) & (1+\nu)\tilde{\sigma}_{r\theta}^{(2)} & 0 \\ (1+\nu)\tilde{\sigma}_{r\theta}^{(2)} & \tilde{\sigma}_{rr}^{(1)} - \nu\left(\tilde{\sigma}_{\theta\theta}^{(2)} + \sigma_{zz}^{(2)}\right) & 0 \\ 0 & 0 & \tilde{\sigma}_{zz}^{(1)} - \nu\left(\tilde{\sigma}_{\theta\theta}^{(2)} + \sigma_{rr}^{(2)}\right) \end{pmatrix}
\end{aligned} \tag{3.4}$$

The strain energy density W of in-plane fields can be expressed as follows:

$$W = \int_0^{\varepsilon_{ij}} \sigma_{ij} d\varepsilon_{ij} = \frac{1}{2}\sigma_{ij}\varepsilon_{ij} = \frac{1}{2}\left[\sigma_{\theta\theta}\varepsilon_{\theta\theta} + \sigma_{rr}\varepsilon_{rr} + \sigma_{zz}\varepsilon_{zz} + \sigma_{r\theta}\varepsilon_{r\theta}\right] \tag{3.5}$$

Insertion of Eqs. (3.1) and (3.4) into Eq. (3.5) gives the total SED W consisting of the components W_1 and W_2 relating to the mode 1 and 2 stress and strain fields, supplemented by the mixed mode component W_{12}:

$$\begin{aligned} W =& W_1 + W_2 + W_{12} \\ =& \frac{1}{2E} r^{2(\lambda_1 - 1)} K_1^2 f_1(\theta) + \frac{1}{2E} r^{2(\lambda_2 - 1)} K_2^2 f_2(\theta) + \frac{1}{E} r^{\lambda_1 + \lambda_2 - 2} K_1 K_2 f_{12}(\theta) \end{aligned} \tag{3.6}$$

$$f_1(\theta) = \tilde{\sigma}_{\theta\theta}^{(1)2} + \tilde{\sigma}_{rr}^{(1)2} + \tilde{\sigma}_{zz}^{(1)2} - 2\nu\left(\tilde{\sigma}_{\theta\theta}^{(1)}\tilde{\sigma}_{rr}^{(1)} + \tilde{\sigma}_{\theta\theta}^{(1)}\tilde{\sigma}_{zz}^{(1)} + \tilde{\sigma}_{rr}^{(1)}\tilde{\sigma}_{zz}^{(1)}\right) + 2(1+\nu)\tilde{\sigma}_{r\theta}^{(1)2} \tag{3.7}$$

$$f_2(\theta) = \tilde{\sigma}_{\theta\theta}^{(2)2} + \tilde{\sigma}_{rr}^{(2)2} + \tilde{\sigma}_{zz}^{(2)2} - 2\nu\left(\tilde{\sigma}_{\theta\theta}^{(2)}\tilde{\sigma}_{rr}^{(2)} + \tilde{\sigma}_{\theta\theta}^{(2)}\tilde{\sigma}_{zz}^{(2)} + \tilde{\sigma}_{rr}^{(2)}\tilde{\sigma}_{zz}^{(2)}\right) + 2(1+\nu)\tilde{\sigma}_{r\theta}^{(2)2} \tag{3.8}$$

The angle-dependent function $f_{12}(\theta)$ is not specified, because it is not further needed in case of symmetric sector areas for determining the average SED ($W_{12} = 0$). The contribution W_2 due to mode 2 loading may be neglected, provided the notch opening angle is large enough ($2\alpha \geq 102.6°$) as to make the mode 2 field non-singular.

The angle-dependent functions $f_1(\theta)$, $f_2(\theta)$ and $f_{12}(\theta)$ can be determined analytically for any notch opening angle 2α . Then, Eq. (3.6) describes the local SED dependent on r and θ provided the NSIFs K_1 and K_2 are known.

The next step is to determine the average SED within a sector area of radius R around the notch tip. The radius R should be small enough, so that the sector area remains within the range of validity of the one term stress expansion resulting in the NSIF.

The sector area of radius R, over which the average is determined, is given by:

$$A(R) = \int_0^R \int_{-\gamma}^{+\gamma} r\,dr\,d\theta = R^2\gamma \tag{3.9}$$

By integration of $W(r)$ over the symmetric sector area ($W_{12} = 0$) and division by $A(R)$, the average local SED $\overline{W}(R)$ turns out to be:

Table 3.1 Values of the integrals $I_1(\gamma)$ and $I_2(\gamma)$ as well as $I_{d1}(\gamma)$ and $I_{d2}(\gamma)$ for pointed V-notches under plane strain conditions; $\nu = 0.3$; (Lazzarin and Zambardi 2001)

2α	2γ	λ_1	λ_2	$I_1(\gamma)$	$I_2(\gamma)$	$I_{d1}(\gamma)$	$I_{d2}(\gamma)$
0°	360°	0.5000	0.500	0.8450	2.1450	0.4550	1.9550
45°	315°	0.5050	0.6597	0.8247	1.7610	0.4688	1.7206
90°	270°	0.5445	0.9085	0.7504	1.4379	0.4838	1.4746
135°	225°	0.6736	1.3021	0.6201	1.1505	0.4678	1.2183

$$\begin{aligned}\overline{W}(R) &= \frac{K_1^2}{2ER^2\gamma}\int_0^R\int_{-\gamma}^{+\gamma} r^{2(\lambda_1-1)}f_1(\theta)r\,\mathrm{d}r\,\mathrm{d}\theta + \frac{K_2^2}{2ER^2\gamma}\int_0^R\int_{-\gamma}^{+\gamma} r^{2(\lambda_2-1)}f_2(\theta)r\,\mathrm{d}r\,\mathrm{d}\theta \\ &= \frac{I_1(\gamma)}{4E\lambda_1\gamma}K_1^2R^{2(\lambda_1-1)} + \frac{I_2(\gamma)}{4E\lambda_2\gamma}K_2^2R^{2(\lambda_2-1)} \\ &= \frac{1}{E}e_1(\gamma)K_1^2R^{2(\lambda_1-1)} + \frac{1}{E}e_2(\gamma)K_2^2R^{2(\lambda_2-1)}\end{aligned} \tag{3.10}$$

$$I_1(\gamma) = \int_{-\gamma}^{+\gamma} f_1(\theta)\mathrm{d}\theta, \quad I_2(\gamma) = \int_{-\gamma}^{+\gamma} f_2(\theta)\mathrm{d}\theta \tag{3.11}$$

$$e_1(\gamma) = \frac{I_1(\gamma)}{4\lambda_1\gamma}, \quad e_2(\gamma) = \frac{I_2(\gamma)}{4\lambda_2\gamma} \tag{3.12}$$

Values of the integrals $I_1(\gamma)$ and $I_2(\gamma)$ under plane strain conditions are listed in Table 3.1 for Poisson's ratio $\nu = 0.3$.

The internal notch angle 2γ being linked to the notch opening angle 2α ($2\alpha = 2\pi - 2\gamma$), the total SED coefficients e_1 and e_2 can be plotted dependent on the notch opening angle, Fig. 3.3 (for plane strain conditions). For sufficiently small values of 2α, the value of e_2 is substantially larger than that of e_1. Even for $2\alpha \geq 102.6°$, where the SED field is non-singular, e_1 and e_2 have approximately the same value. This is not an indication that the SED values are approximately the same. Actually, $\overline{W}_2 \ll \overline{W}_1$ because of the missing singularity in the stress and strain field of mode 2 loading. The ratio $\overline{W}_2/\overline{W}_1$ can be determined from:

$$\frac{\overline{W}_2}{\overline{W}_1} = \frac{e_2}{e_1}\left(\frac{K_2}{K_1}\right)^2 R^{2(\lambda_2-\lambda_1)} \tag{3.13}$$

The energy concentration factor K_W is defined by relating the average local SED $\overline{W}$ at the notch tip, Eq. (3.10), to the nominal SED $W_{n,g} = \sigma_{n,g}^2/2E$ in the gross cross-section (nominal stress $\sigma_{n,g}$):

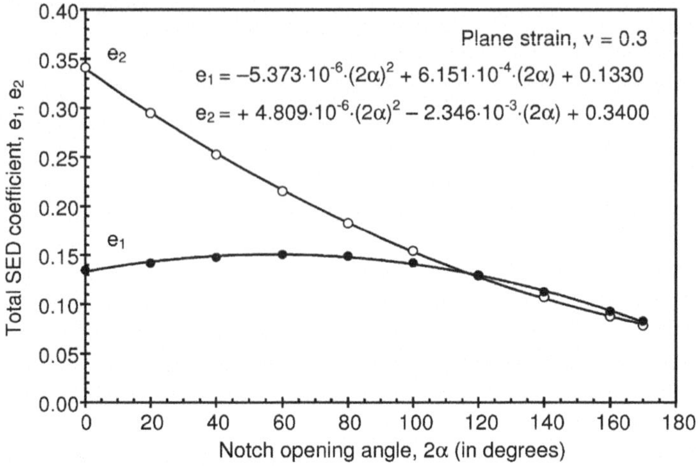

Fig. 3.3 Total SED coefficients e_1, e_2 dependent on notch opening angle 2α for plane strain conditions; (Lazzarin and Zambardi 2001)

$$K_W = \frac{\overline{W}}{W_{\mathrm{n,g}}} = \frac{I_1(\gamma)}{2\lambda_1\gamma\,\sigma^2_{\mathrm{n,g}}} K_1^2 R^{2(\lambda_1-1)} + \frac{I_2(\gamma)}{2\lambda_2\gamma\,\sigma^2_{\mathrm{n,g}}} K_2^2 R^{2(\lambda_2-1)} \tag{3.14}$$

There is some indication that the average distortional SED $\overline{W}_{\mathrm{d}}$ instead of the average total SED $\overline{W}$ is decisive for failure in components with rounded notches (von Mises criterion). Then, the following relationships should be used:

$$W_{\mathrm{d}} = W_{\mathrm{d1}} + W_{\mathrm{d2}} = \frac{1+\nu}{3E} r^{2(\lambda_1-1)} K_1^2 f_{\mathrm{d1}}(\theta) + \frac{1+\nu}{3E} r^{2(\lambda_2-1)} K_2^2 f_{\mathrm{d2}}(\theta) \tag{3.15}$$

$$f_{\mathrm{d1}}(\theta) = \tilde{\sigma}_{\theta\theta}^{(1)2} + \tilde{\sigma}_{rr}^{(1)2} + \tilde{\sigma}_{zz}^{(1)2} - \left(\tilde{\sigma}_{\theta\theta}^{(1)}\tilde{\sigma}_{rr}^{(1)} + \tilde{\sigma}_{\theta\theta}^{(1)}\tilde{\sigma}_{zz}^{(1)} + \tilde{\sigma}_{rr}^{(1)}\tilde{\sigma}_{zz}^{(1)}\right) + 3\tilde{\sigma}_{r\theta}^{(1)2} \tag{3.16}$$

$$f_{\mathrm{d2}}(\theta) = \tilde{\sigma}_{\theta\theta}^{(2)2} + \tilde{\sigma}_{rr}^{(2)2} + \tilde{\sigma}_{zz}^{(2)2} - \left(\tilde{\sigma}_{\theta\theta}^{(2)}\tilde{\sigma}_{rr}^{(2)} + \tilde{\sigma}_{\theta\theta}^{(2)}\tilde{\sigma}_{zz}^{(2)} + \tilde{\sigma}_{rr}^{(2)}\tilde{\sigma}_{zz}^{(2)}\right) + 3\tilde{\sigma}_{r\theta}^{(2)2} \tag{3.17}$$

The local average distortional SED $\overline{W}_d(R)$ turns out to be:

$$\begin{aligned}\overline{W}_d(R) &= \frac{(1+\nu)I_{\mathrm{d1}}(\gamma)}{6E\lambda_1\gamma} K_1^2 R^{2(\lambda_1-1)} + \frac{(1+\nu)I_{\mathrm{d2}}(\gamma)}{6E\lambda_2\gamma} K_2^2 R^{2(\lambda_2-1)} \\ &= \frac{e_{\mathrm{d1}}(\gamma)}{E} K_1^2 R^{2(\lambda_1-1)} + \frac{e_{\mathrm{d2}}(\gamma)}{E} K_2^2 R^{2(\lambda_2-1)}\end{aligned} \tag{3.18}$$

$$I_{\mathrm{d1}}(\gamma) = \int_{-\gamma}^{+\gamma} f_{\mathrm{d1}}(\theta)\mathrm{d}\theta, \quad I_{\mathrm{d2}}(\gamma) = \int_{-\gamma}^{+\gamma} f_{\mathrm{d2}}(\theta)\mathrm{d}\theta \tag{3.19}$$

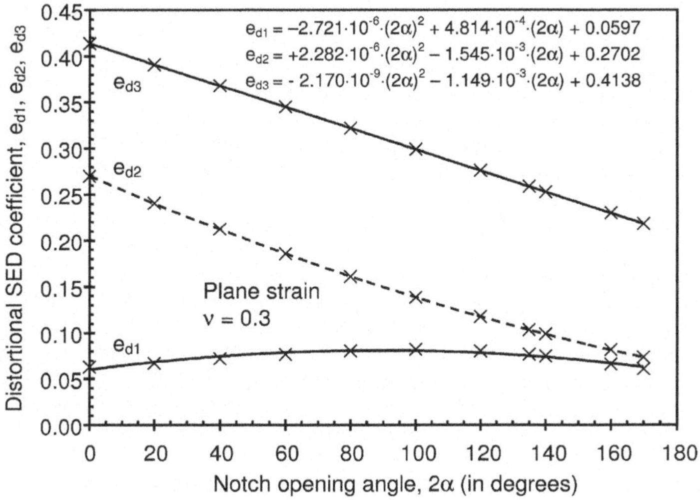

Fig. 3.4 Distortional SED coefficients e_{d1}, e_{d2}, e_{d3} dependent on notch opening angle 2α; e_{d1} and e_{d2} for plane strain conditions; (Lazzarin et al. 2004)

$$e_{d1}(\gamma) = \frac{(1+\nu)I_{d1}}{6\lambda_1\gamma}, \quad e_{d2}(\gamma) = \frac{(1+\nu)I_{d2}}{6\lambda_2\gamma} \tag{3.20}$$

Values of the integral $I_{d1}(\gamma)$ and $I_{d2}(\gamma)$ under plane strain conditions are listed in Table 3.1. They are independent of Poisson's ratio ν.

The distortional SED coefficients e_{d1} and e_{d2} are plotted dependent on the notch opening angle 2α, Fig. 3.4 (for plane strain conditions). In comparison to the e_1 and e_2 values, a marked reduction has taken place at smaller values of 2α, where appreciable hydrostatic stresses occur ahead of the notch tip.

For the sake of completeness, out-of-plane shear loading (mode 3), is now considered (Lazzarin et al. 2004). Here $W_3 = W_{d3}$, because the hydrostatic stresses are zero in the whole field:

$$W_3 = W_{d3} = \frac{1+\nu}{3E} r^{2(\lambda_3-1)} K_3^2 f_{d3}(\theta) \tag{3.21}$$

$$f_{d3}(\theta) = 3\left(\tilde{\sigma}_{\theta z}^2 + \tilde{\sigma}_{rz}^2\right) \tag{3.22}$$

The average local SED $\overline{W}_{d3}(R)$ turns out to be:

$$\overline{W}_{d3}(R) = \frac{e_{d3}(\gamma)}{E} K_3^2 R^{2(\lambda_3-1)} \tag{3.23}$$

$$e_{d3}(\gamma) = \frac{(1+\nu)I_{d3}(\gamma)}{6\lambda_3\gamma}, \quad I_{d3}(\gamma) = \int_{-\gamma}^{+\gamma} f_{d3}(\theta)\mathrm{d}\theta \tag{3.24}$$

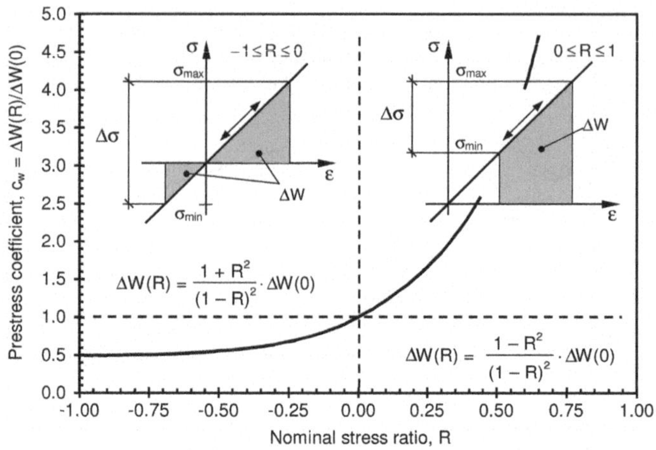

Fig. 3.5 Prestress coefficient c_W dependent on nominal stress ratio R for cyclic SED ΔW with constant $\Delta\sigma$; (Lazzarin et al. 2004)

The distortional energy coefficients e_{d3} dependent on 2α are supplemented in Fig. 3.4. The values of e_{d3} are substantially larger than those of e_{d2} and e_{d1}.

When evaluating welded joint fatigue test results in terms of the local average SED at the weld toe, the SED range ΔW (or ΔW_d) depends on the preload expressed by the nominal stress ratio $R = \sigma_l/\sigma_u$, with the lower and upper nominal stresses, σ_l and σ_u, of the load cycle (not to be confused with the radius R of the local area or volume of the average SED), Fig. 3.5:

$$\begin{aligned} \Delta W(R) &= c_W \Delta W(0) = \frac{1+R^2}{(1-R)^2}\Delta W(0) \qquad (-1 \le R \le 0) \\ \Delta W(R) &= c_W \Delta W(0) = \frac{1-R^2}{(1-R)^2}\Delta W(0) \qquad (0 \le R \le 1) \end{aligned} \tag{3.25}$$

where the SED range $\Delta W(0)$ for $R = 0$ or $\sigma_l = 0$ is the reference parameter (prestress coefficient $c_W = 1.0$ for $R = 0$ and $c_W = 0.5$ for $R = -1$). The condition $\Delta\sigma =$ const. is introduced when deriving the Eqs. (3.25).

Another influence on the average local SED $\overline{W}$ is exerted by larger values of the non-singular T-stress, e.g. in welded thin-sheet lap joints (Lazzarin et al. 2009):

$$\overline{W} = \frac{e_1}{E}\frac{K_I^2}{R} + \frac{e_2}{E}\frac{K_{II}^2}{R} + \frac{1-\nu^2}{2E}T^2 + \frac{8\sqrt{2}}{15(\pi)^{3/2}}\frac{(1+\nu)(2-5\nu)}{E}\frac{K_I T}{R^{1/2}} \tag{3.26}$$

The weld notch of such joints is characterised by $2\alpha = 0$, therefore $K_1 = K_I$ and $K_2 = K_{II}$ as well as $e_1 = 0.1345$ and $e_2 = 0.3414$ for plane strain conditions and $\nu = 0.3$.

It was found that for sheet thicknesses $t = 1$ and 5 mm combined with a control radius $R = R_0 = 0.28$ mm for welded joints of steel, the contribution of the

K_1T term is negligible in comparison to the other three terms. A numerical check based on FE analysis showed a poor accuracy of the above formula, when applied to the sheet thickness $t = 1$ mm. The radius R_0 should be small in relation to the sheet thickness in order to achieve acceptable accuracy. Otherwise, it is necessary to include higher order terms in Eq. (3.26).

3.2.3 Application of the Pointed Notch SED Concept to Brittle Fracture

The characteristic feature of brittle fracture is that it occurs under elastic stress field conditions, either without any plastic deformation (glass, ceramics) or without deformation on a macroscale (brittle technical metals).

Brittle fracture at pointed V-notches may be assumed to occur when the average local SED $\overline{W}$ reaches a critical value W_c independent of the notch opening angle and independent of the loading type (tensile or in-plane shear loading). This corresponds to the Beltrami criterion (Beltrami 1885). The material parameter W_c may be determined from the ultimate tensile strength σ_U of unnotched specimen:

$$\overline{W} \leq W_c, \quad W_c = \frac{\sigma_U^2}{2E} \tag{3.27}$$

The radius R_0 of the control volume, where the critical local SED value W_c occurs, may be determined from the plane strain fracture toughness K_{Ic} (Lazzarin and Zambardi 2001). It is considered to be a material parameter. When the mode 2 SED W_2 is zero (symmetric geometry and loading) or negligibly small (large notch opening angle with vanishing mode 2 singularity), the NSIF K_1 can be correlated with the average local SED $\overline{W}_1$, Eq. (3.10):

$$K_1 = \sqrt{\frac{4E\lambda_1\gamma}{I_1(\gamma)}\overline{W}_1\, R_0^{(1-\lambda_1)}} \tag{3.28}$$

The critical condition is expressed by:

$$K_{1c} = \sqrt{\frac{2\lambda_1\gamma}{I_1(\gamma)}}\sigma_U R_0^{(1-\lambda_1)} = f_1(2\alpha)\sigma_U R_0^{(1-\lambda_1)} \tag{3.29}$$

where the intensity coefficient $f_1(2\alpha)$ just as $f_2(2\alpha)$, not to be confused with $f_1(\theta)$ and $f_2(\theta)$ in Eq. (3.12), depends on the notch opening angle 2α, Fig. 3.6.

When the V-notch becomes a crack ($2\alpha = 0$), K_{1c} coincides with the fracture toughness K_{Ic}:

$$K_{Ic} = f_1(\theta)\sigma_U\sqrt{R_0} \tag{3.30}$$

Thus, the control volume radius R_0 turns out to be:

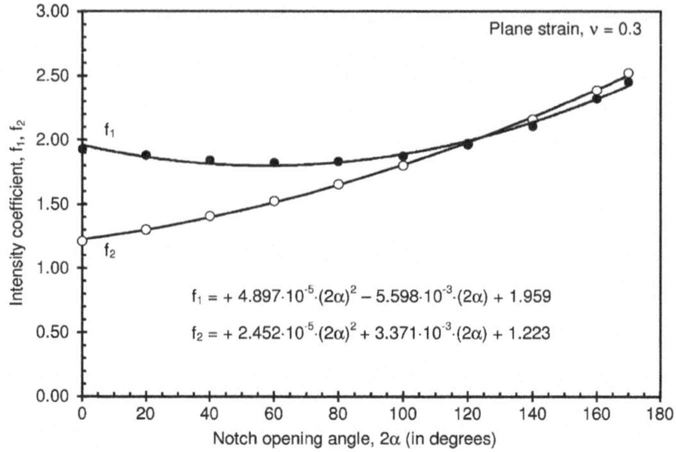

Fig. 3.6 Intensity coefficients f_1, f_2 dependent on notch opening angle 2α; plane strain conditions; (Lazzarin and Zambardi 2001)

$$R_0 = \left(\frac{K_{\mathrm{Ic}}}{f_1(0)\sigma_{\mathrm{U}}}\right)^2 = \frac{I_1(\pi)}{\pi}\left(\frac{K_{\mathrm{Ic}}}{\sigma_{\mathrm{U}}}\right)^2 \tag{3.31}$$

or in an explicit form related to plane strain conditions (Yosibash et al. 2004):

$$R_0 = \frac{(1+\nu)(5-8\nu)}{4\pi}\left(\frac{K_{\mathrm{Ic}}}{\sigma_{\mathrm{U}}}\right)^2 \tag{3.32}$$

The corresponding expression for plane stress conditions reads (Lazzarin and Berto):

$$R_0 = \frac{5-3\nu}{4\pi}\left(\frac{K_{\mathrm{c}}}{\sigma_{\mathrm{U}}}\right)^2 \tag{3.33}$$

The control volume radius R_0 was originally designated by the symbol R_{C} (subscript capital C for 'control'), but this is avoided here in order to rule out the misinterpretation 'critical', which is attributed only to $\overline{W}$. The control volume radius R_0 depends on the material. Its value decreases with rising brittleness. It does not depend on the notch opening angle per definition.

Finally, inserting R_0 from Eq. (3.31) into Eq. (3.29), the critical NSIF $K_{1\mathrm{c}}$ for pointed V-notches of brittle material is:

$$K_{1\mathrm{c}} = f_1(2\alpha)\sigma_{\mathrm{U}}\left(\frac{K_{\mathrm{Ic}}}{f_1(0)\sigma_{\mathrm{U}}}\right)^{2(1-\lambda_1)} = \sqrt{\frac{2\lambda_1\gamma}{I_1(\gamma)}}\sigma_{\mathrm{U}}\left(\sqrt{\frac{I_1(\pi)}{\pi}}\frac{K_{\mathrm{Ic}}}{\sigma_{\mathrm{U}}}\right)^{2(1-\lambda_1)} \tag{3.34}$$

The above approach has been validated on the basis of experimental data reported in the literature (Lazzarin and Zambardi 2001).

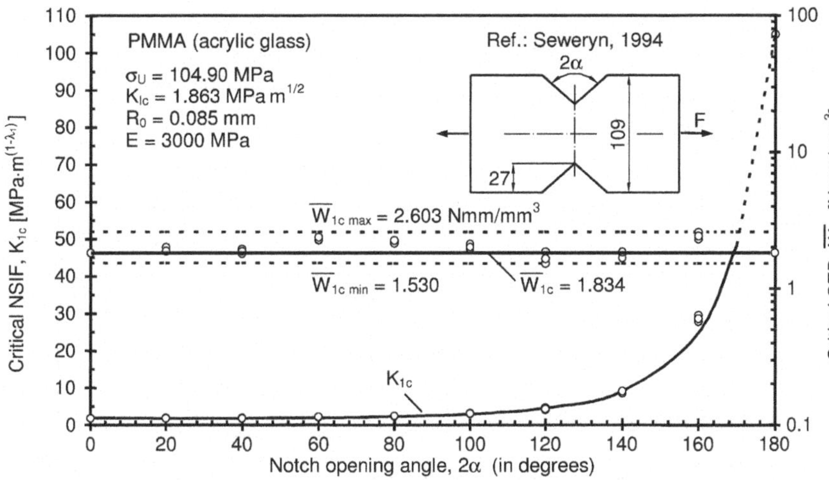

Fig. 3.7 Critical NSIF K_{1c} and critical local SED $\overline{W}_{1c}$ for tensile loaded two-sided V-notch flat bar specimens made of PMMA (acrylic glass) dependent on notch opening angle 2α; analytical expression for K_{1c} (solid curve) compared with experimental data (point symbols); local SED $\overline{W}_{1c}$ in comparison; (Lazzarin and Zambardi 2001)

The first set of evaluated experimental data refers to tensile loaded two-sided V-notch plate specimens made of PMMA (acrylic glass) and duraluminium with different notch opening angles (Seweryn 1994). In Fig. 3.7, the experimental data (point symbols) are contrasted with the analytical result (solid curve). The analytical curve of the critical NSIF K_{1c} dependent on notch opening angle results from Eq. (3.29) after insertion of the relevant K_{Ic} and σ_U values. The NSIF values K_1 from Seweryn's investigation are gained by FE analysis. The loading tests give the critical NSIF values K_{1c}. These K_{1c} values are converted to $\overline{W}_{1c}$ values by means of Eq. (3.10). The analytical value $\overline{W}_{1c} = \sigma_U^2/2E = 1.834\,\text{N mm/mm}^3$ corresponds to the rhombic point symbol at $2\alpha = 180°$. The experimental data confirm that $\overline{W}_{1c}$ is a material parameter which is independent on the notch opening angle.

Similar results are available for duraluminium ($\sigma_U = 705$ MPa, $K_{Ic} = 50.6\,\text{MPa m}^{1/2}$, $R_0 = 1.382$ mm, $E = 0.7 \times 10^5$ MPa). Whereas the fracture surface is perpendicular to the applied load direction in the PMMA specimens, it is inclined by 45° in the duraluminium specimens, indicating brittle crack initiation in the former case as against ductile crack initiation in the latter case. Lazzarin and Zambardi were encouraged by this result in their effort to extend the local SED concept into the elastic-plastic range. Reference is also made to Carpinteri's investigation of V-notched PMMA specimens with conforming results (Carpinteri 1987).

The second set of evaluated experimental data refers to the above PMMA specimen subjected to superimposed tensile and shear loading (Seweryn et al. 1997).

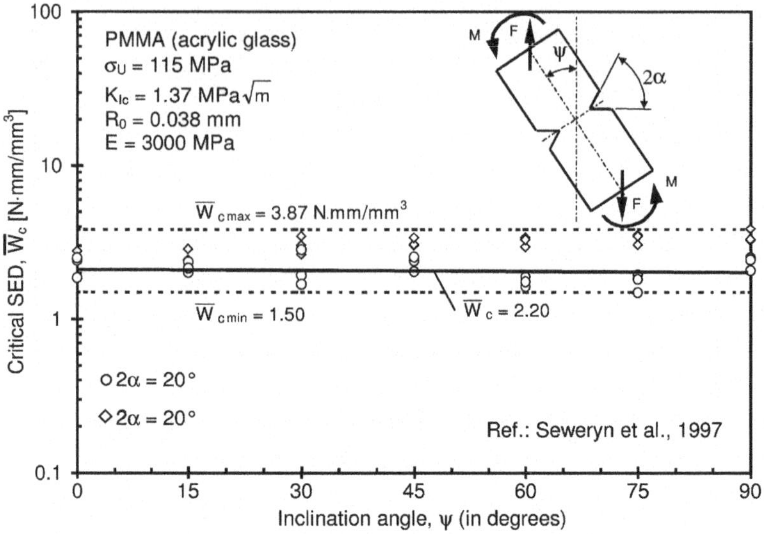

Fig. 3.8 Critical local SED $\overline{W}_c = (\overline{W}_1 + \overline{W}_2)_c$ for superimposed tensile and shear loading of two-sided V-notch flat bar specimens made of PMMA (acrylic glass) dependent on the inclination angle ψ, controlling the tensile to shear load ratio; experimental data (point symbols) compared with analytical expression (solid line); (Lazzarin and Zambardi 2001)

As can be seen from Fig. 3.8, the critical local SED $\overline{W}_c = (\overline{W}_1 + \overline{W}_2)_c$ appears to be roughly independent not only of the notch opening angle, but also of the ratio of tensile to shear load inclusive of the pure loading states. The analytical value is $\overline{W}_c = \sigma_U^2/2E = 2.20\,\text{Nmm/mm}^3$. It has to be noted that the scatter range of SED values increases quadratically when compared with the scatter range of NSIF values. Reference is also made to relevant results for V-notched PMMA specimens under mixed mode loading conditions presented by Dunn et al. (1997).

When brittle fracture is initiated mainly by plastic deformation at the notch tip (semiductile materials, plane stress conditions or rounded notch tips), the argument is well founded, that instead of the total energy its distortional component should be decisive, i.e. the von Mises criterion. The following relationships are then valid (Lazzarin and Zambardi 2001).

The limit condition for the distortional SED reads:

$$\overline{W}_d \leq W_{dc}, \quad W_{dc} = \frac{1+\nu}{3E}\sigma_U^2 \tag{3.35}$$

Because the SEDs W_d and W are linked by the uniaxial test conditions for σ_U, the following relationship applies:

$$W_{dc} = \frac{2(1+\nu)}{3} W_c \tag{3.36}$$

which results in $W_{dc} = 0.867W_c$ for $v = 0.3$.

Neglecting the K_2 component, K_1 can be given in terms of the average local distortional SED $\overline{W}_{d1}$:

$$K_1 = \sqrt{\frac{6E\lambda_1\gamma}{(1+v)I_{d1}(\gamma)}\overline{W}_{d1}\,R_{0d}^{(1-\lambda_1)}} \tag{3.37}$$

Once more, when the V-notch becomes a crack ($2\alpha = 0$), K_{1c} coincides with the fracture toughness K_{Ic}:

$$K_{Ic} = \sqrt{\frac{\pi}{I_{d1}(\pi)}}\sigma_U\sqrt{R_{0d}} \tag{3.38}$$

Thus, the control volume radius R_{0d} turns out to be:

$$R_{0d} = \frac{I_{d1}(\pi)}{\pi}\left(\frac{K_{Ic}}{\sigma_U}\right)^2 \tag{3.39}$$

or in an explicit form related to plane strain conditions:

$$R_{0d} = \frac{7-16v+16v^2}{8\pi}\left(\frac{K_{Ic}}{\sigma_U}\right)^2 \tag{3.40}$$

The corresponding expression for plane stress conditions is:

$$R_{0d} = \frac{7}{8\pi}\left(\frac{K_c}{\sigma_U}\right)^2 \tag{3.41}$$

Under plane strain conditions, $I_{d1}(\pi) = 0.455$ and $I_1(\pi) = 0.845$ for $v = 0.3$ (Lazzarin and Zambardi 2001), so that $R_{0d} = 0.54R_0$. Thus, the control volume radius is a material parameter which depends on the applied failure criterion (Beltrami or von Mises).

The resulting expression for K_{1c} reads:

$$K_{1c} = \sqrt{\frac{2\lambda_1\gamma}{I_{d1}(\gamma)}}\sigma_U\left(\sqrt{\frac{I_{d1}(\pi)}{\pi}}\frac{K_{Ic}}{\sigma_U}\right)^{2(1-\lambda_1)} \tag{3.42}$$

3.2.4 Application of the Pointed Notch SED Concept to Fatigue Failure

The total fatigue life of welded joints measured in number of cycles up to total failure, is composed of the crack initiation and crack propagation portion. The crack initiation life is terminated by a technical surface crack of about 0.25 mm depth. The relation of the crack initiation to the crack propagation portion varies considerably. The crack initiation life portion is larger in the high-cycle than in the low-cycle fatigue range, it is larger for high-strength steels and it is prevailing in unnotched or mildly notched members.

It has been shown that the NSIF is a parameter which controls the crack initiation phase in members with sharp notches, especially too in welded joints (Boukharouba et al. 1995; Verreman and Nie 1996; Pluvinage 1997). It can also be used for expressing the total fatigue life of welded joints (Lazzarin and Tovo 1998; Atzori et al. 1999). A necessary condition is, that the dimension of the relevant NSIF does not vary, which means that the notch opening angle must be the same for the evaluated specimens. The data in Figs. 2.28 and 2.29 for fillet-welded joints assume a notch opening angle $2\alpha = 135°$. The restriction to only one notch opening angle is removed by evaluating the average local SED instead of the NSIF.

The average local SED at the notch tip of the weld toe is evaluated for linear-elastic material conditions which is an appropriate approximation in the high-cycle fatigue range ($N \geq 5 \times 10^5$ cycles). It is also used in the medium-cycle fatigue range ($10^4 \leq N \leq 5 \times 10^5$ cycles) where plastic deformations occur. The plastic zone may even be larger than the control volume with radius $R_0 \approx 0.3$ mm for steels or 0.1 mm for aluminium alloys (see the data below). The justification to use the linear-elastic strain energy under small-scale yielding conditions is provided by the 'equivalent strain energy density approach' (Glinka 1985), compare Sect. 2.5.9. According to this approach, which works well under plane strain conditions, the elastic-plastic SED at the root of a rounded notch is set equal to the SED determined under purely elastic conditions. This concept is not directly applicable to pointed V-notches because the SED at the notch tip tends toward infinity both for linear-elastic and power law elastic-plastic material behaviour. If applied to the average SED in the control volume at the notch tip, this problem is removed (Lazzarin and Zambardi 2002). The elastic SED concept is thus applicable over the whole medium-cycle and high-cycle fatigue range.

The first step in the local SED approach is the determination of the control volume radius R_0 for the specimens under investigation. It depends, as already stated, on the material and on the multiaxial failure criterion. This task is more complicated for welded joints under cyclic loading producing fatigue failure than for PMMA specimens under static loading producing brittle fracture. In the former case, the material is locally inhomogeneous in the as-welded condition and the cracking phenomena change to some extent with the cyclic load level or the endured load cycles. The control volume radius R_0 is determined for fillet-welded joints ($2\alpha = 135°$) with experimental data gained in the high-cycle fatigue range. Only the mode 1 average local SED $\overline{W}_1$ is evaluated whereas the mode 2 component $\overline{W}_2$ remains negligibly small. Plane strain conditions are assumed together with the Beltrami total strain energy criterion.

The control volume radius R_0 is then given by the following expression (Lazzarin and Zambardi 2001):

$$R_0 = \left(\frac{\Delta K_{1E}}{f_1 \Delta\sigma_E}\right)^{1/(1-\lambda_1)} = \left(\frac{\sqrt{2e_1}\Delta K_{1E}}{\Delta\sigma_E}\right)^{1/(1-\lambda_1)} \tag{3.43}$$

where λ_1 and f_1 depend on the notch opening angle 2α, while ΔK_{1E} is the reference value of the endurable NSIF range for the fillet-welded joints and $\Delta\sigma_E$ is the reference value of the endurable stress range at a flush-ground butt weld, which simulates the material behaviour at the weld toe or root. The reference values ΔK_{1E} and $\Delta\sigma_E$ are mean values ($P_f = 50$ %) at $N_E = 2 \times 10^6$ cycles (IIW design recommendations) or 5×10^6 cycles (the other quotations below) with load ratio $R = 0$. With the usual inverse slope exponent $k = 3$ for welded joints of steel, the factor 1.36 occurs between the fatigue strength at 2×10^6 cycles compared with 5×10^6 cycles. With the common scatter range index, the factor 1.37 results between the data for $P_f = 50$ % compared with $P_f = 2.3$ %. For $2\alpha = 135°$, the parameters in Eq. (3.43) are $f_1 = 2.065$ and $\lambda_1 = 0.674$, whereas for $2\alpha = 0°$, $f_1 = 1.936$ and $\lambda_1 = 0.5$.

Considering welded joints of carbon steels with toe failures ($2\alpha = 135°$), the relevant data are:

- $\Delta K_{1E} = 211$ MPa mm$^{0.326}$ for fillet-welded joints (Lazzarin and Livieri 2001),
- $\Delta\sigma_E = 155$ MPa for flush-ground butt welds (Atzori and Dattoma 1983),
- $R_0 = 0.28$ mm according to Eq. (3.43).

The reference value $\Delta\sigma_E = 155$ MPa ($P_f = 50$ %) corresponding to $\Delta\sigma_E = 113$ MPa ($P_f = 2.3$ %) deviates to some extent from $\Delta\sigma_E = 100$ MPa ($P_f = 2.3$ %) in the IIW design recommendations (Hobbacher 2009).

For welded joints of carbon steels with root failures ($2\alpha = 0°$), the relvant data are:

- $\Delta K_{1E} = 180$ MPa mm$^{0.5}$ for fillet-welded joints (Livieri and Lazzarin 2005),
- $\Delta\sigma_E = 155$ MPa as above,
- $R_0 = 0.36$ mm according to Eq. (3.43).

The well-founded reference value ΔK_{1E} ($P_f = 50$ %) above is not in agreement with $\Delta K_{th} = 180$ MPa mm$^{0.5}$ ($P_f = 2.3$ %), the code-relevant lower bound threshold SIF for welded joints (Radaj et al. 2006).

Considering welded joints of aluminium alloys, the relevant data are:

- $\Delta K_{1E} = 99$ MPa mm$^{0.326}$ for fillet-welded joints (Livieri and Lazzarin 2005),
- $\Delta\sigma_E = 96$ MPa for flush-ground butt welds (Person 1971),
- $R_0 = 0.12$ mm according to Eq. (3.43).

The reference value $\Delta\sigma_E = 96$ MPa ($P_f = 50$ %) corresponding to $\Delta\sigma_E = 70$ MPa ($P_f = 2.3$ %) deviates substantially from $\Delta\sigma_E = 40$ MPa ($P_f = 2.3$ %) in the IIW design recommendations (Hobbacher 2009).

The poor correspondence of the $\Delta\sigma_E$ values mentioned above does not mean that any values are more accurate than others. The deviations designate only the possible uncertainty in the derived R_0 values. Such uncertainties are normal in generalising code regulations for engineering purposes. On the other hand, it is stated by the developers of the local SED method that certain variations $0.2 \leq R_0 \leq 0.4$ mm (for steel) have only a minor effect on the fatigue life predictions for weld toes, where the mode 2 stress field is non-singular (Lazzarin et al. 2003). The effect is more pronounced when mode 1 and 3 stress fields are superimposed, which are singular to different degrees.

A uniform material-dependent value of R_0 is used for evaluating test data not only in the high-cycle fatigue range, but also in the medium-cycle and low-cycle fatigue range (Lazzarin and Zambardi 2001; Lazzarin and Livieri 2001; Livieri and Lazzarin 2005). This is a simplification appropriate for engineers. Thinking 'more physically', the control radius is dependent on the endured number of cycles, $R_0 = R_0(N)$. Considering Eq. (3.43) for the determination of R_0, the ratio $\Delta K_{1E}/\Delta\sigma_E$ increases with decreasing values of N. This can be quantified for aluminium alloys, where the inverse slope exponent $k = 4$ of the S–N curve for welded joints differs from $k = 5.5$ for the flash-ground butt weld (Livieri and Lazzarin 2005). Using these data in Eq. (3.43) at $N = 10^5$ instead of 5×10^6 cycles, the result is $R_0 = 0.27$ mm for aluminium alloys. The endurable values $\Delta\overline{W}_E$ are correspondingly reduced. Actually, the W–N curve for the flush-ground butt weld is thus also valid for sharply notched welded joints, but only to the extent that linear-elastic conditions are an acceptable approximation. This is certainly not the case for $N \leq 10^5$ cycles.

It will be shown in Sect. 3.3.3, that the value of R_0 in out-of-plane shear loading may not be identical with the value of R_0 in tensile and bending loading, at least not for non-welded round bar specimens.

3.2.5 *Fatigue W–N Curves of Welded Joints and Weld-Like Specimens*

Accurate fatigue test data (toe failures) for fillet-welded transverse attachment joints, tensile or bending loaded, made of low and medium carbon steels (Maddox 1987; Gurney 1991; Kihl and Sarkani 1997), were re-evaluated in terms of $\Delta\overline{W}_1$. The cyclic SED values $\Delta\overline{W}$ were determined based on the NSIF values ΔK_1, Eq. (3.10), which were evaluated based on FE analysis results using fine meshes and the limit $r \to 0$ formula on the stresses in the bisector plane (Sect. 2.3.4). The resulting W–N scatter band is shown in Fig. 3.9 (Lazzarin and Zambardi 2001). A wide range of base plate thicknesses, $t = 6$–100 mm, was included. The bold solid line denotes the equivalent to the design curve ($P_f = 97.7$ %) for non-welded sheet material according to Eurocode 3. The higher position of the weld toe data may be surprising, but it has to be noted that the scatter range index of the parent material is substantially enlarged, so that the medium line ($P_s = 50$ %) will be positioned near the upper limit of the welded joint scatter band.

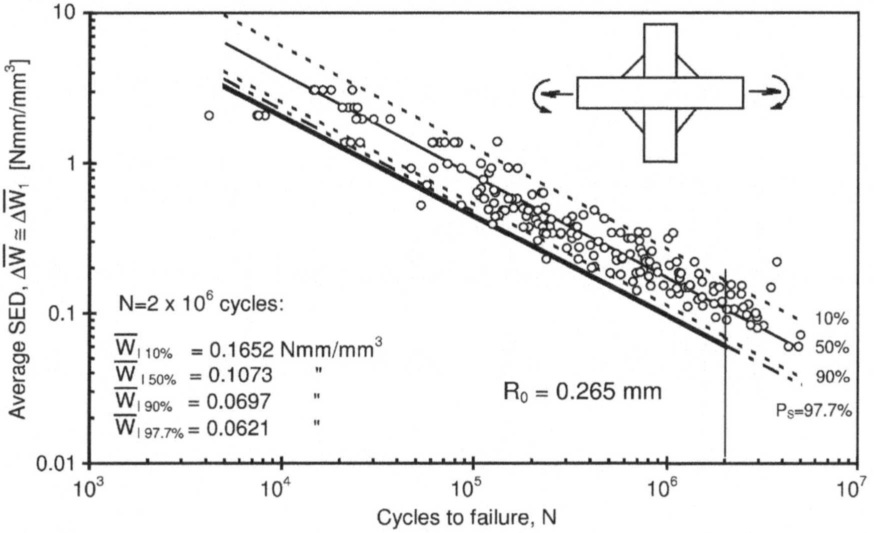

Fig. 3.9 Fatigue test data (toe failures) for fillet-welded transverse attachment joints of steel (as-welded or stress-relieved) subjected to tensile and bending loads; nominal stress ratio $R = 0$; survival probability $P_s = 97.7$ % corresponds to two standard deviations; bold solid line: non-welded sheet material according to Eurocode 3 ($\Delta\sigma_A = 160$ MPa at $N = 2 \times 10^6$ cycles, $P_s = 97.7$ %); (Lazzarin and Zambardi 2001)

The data base of test results was then enlarged by fatigue test data for fillet-welded transverse attachment joints with varying gradients of the fillet weld, 30–70° (Lazzarin et al. 2003). Also, a test series with $t = 3$ mm was included. The updated *W–N* scatter band is shown in Fig. 3.10. The inverse slope exponent $k = 1.5$ of the *W–N* curve corresponds to $k = 3.0$ of the *S–N* curve, a usual value for welded joints in design codes. The scatter range index of the *W–N* curve, $P_f = 2.3$ % compared with 97.7 %, $T_W = 1{:}3.3$, corresponds to the index of the *S–N* curve, $P_f = 10$ % compared with 90 %, $T_\sigma = 1 : (\sqrt{3.3}/1.21) = 1 : 1.50$, also a usual value for welded joints in design codes.

Finally, further fatigue test data were added to the scatter band of the *W–N* curve of Fig. 3.10: test results for cruciform fillet-welded joints, load-carrying fillet welds with root failures (evaluated with a uniform R_0 value), inclusive of two unconventional steels, Fig. 3.11 (Livieri and Lazzarin 2005). Obviously, both failure types are well represented by the original scatter band, with the exception of the test data for the two unconventional steels. Meanwhile, also test data for butt welds have been shown to correspond with the scatter band, Fig. 3.12 (Lazzarin et al. 2006; Lazzarin and Berto 2008). Later on, test results for hollow section joints were included (Berto and Lazzarin 2009), and also for three-dimensional joints. The actual version of the diagram comprises 900 test data (Lazzarin et al. 2011). A seperate diagram has been derived with test data for tensile-shear loaded spot-welded thin-sheet lap joints (Berto and Lazzarin 2009).

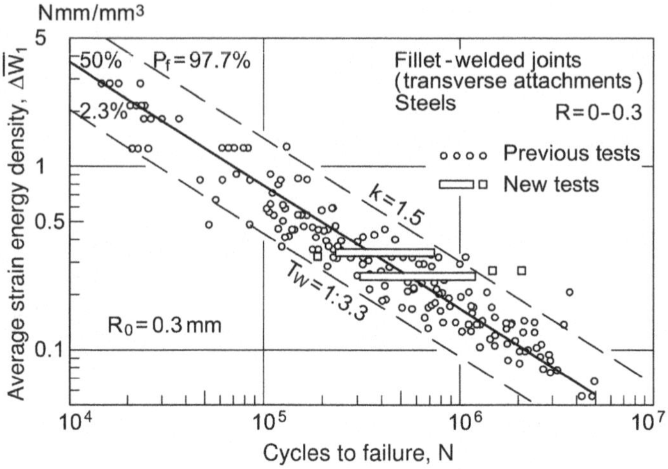

Fig. 3.10 Fatigue test data (toe failures) for fillet-welded transverse attachment joints of steel (as-welded or stress-relieved) subjected to tensile and bending loads; *W–N* curves related to average total SED under mode 1 loading; scatter band characterises mean values plus and minus two standard deviations; T_W refers to $P_f = 2.3$ compared with 97.7 %; (Lazzarin et al. 2003)

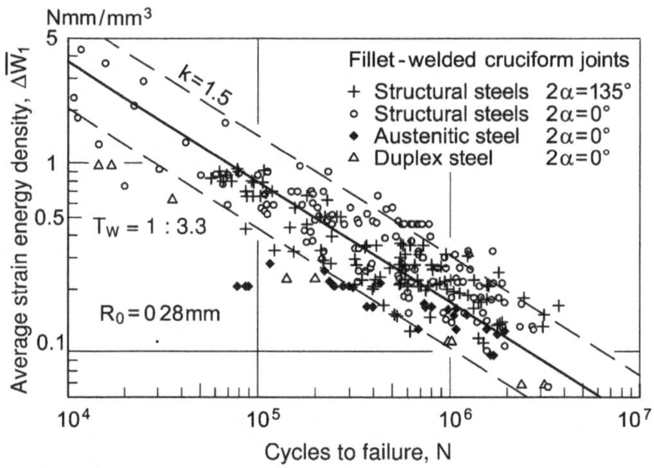

Fig. 3.11 Fatigue test data for fillet-welded cruciform joints of steel (new test series, toe and root failures) inserted in the scatter band of Fig. 3.10; load-carrying and non-load-carrying fillet welds, tensile and bending loads; (Livieri and Lazzarin 2005)

A plot of fatigue test data for fillet-welded joints made of aluminium alloys subjected to tensile loading is also available, Fig. 3.13 (Livieri and Lazzarin 2005). Cruciform and T-joints with toe failures ($2\alpha = 135°$) and lap joints with root failure ($2\alpha = 0°$) are evaluated. Surprisingly, the values of $\Delta\overline{W}$ at 2×10^6 cycles for welded joints made of aluminium alloys are more or less identical with the

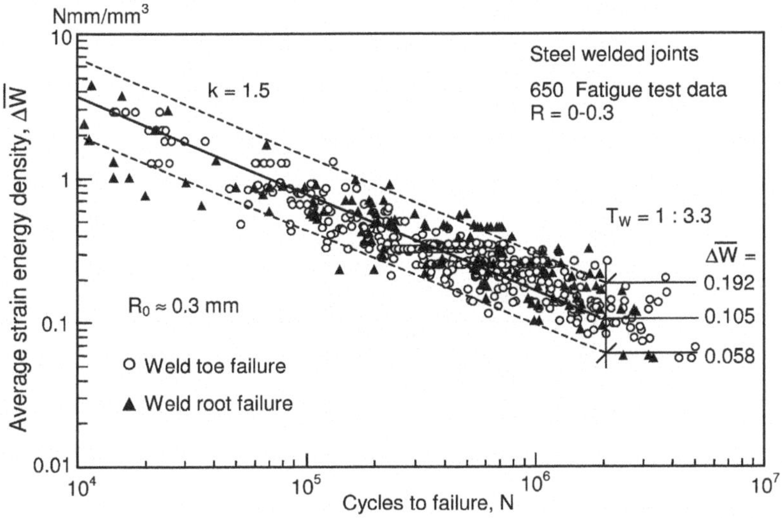

Fig. 3.12 Fatigue test data (toe and root failures) for fillet-welded cruciform joints of steel (Fig. 3.11) with inclusion of butt welds; (Lazzarin et al. 2006; Lazzarin and Berto 2008)

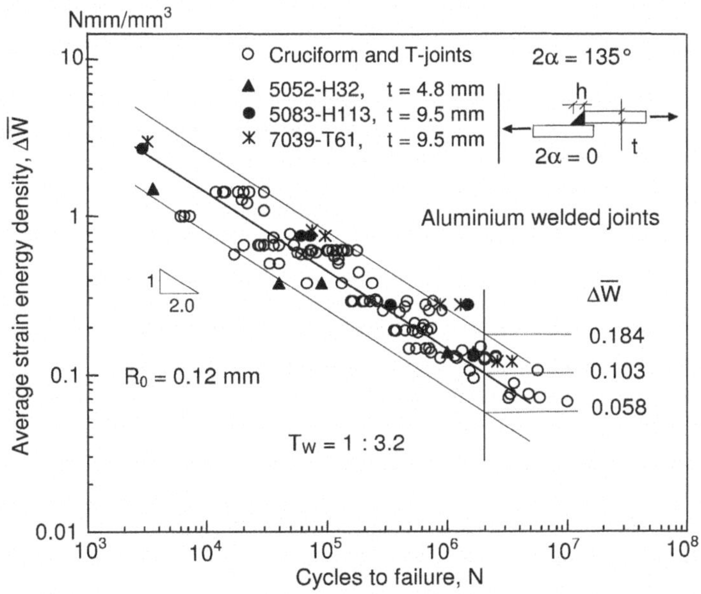

Fig. 3.13 Fatigue test data for fillet-welded joints of aluminium alloy subjected to tensile loads; cruciform and T-joints with toe failures ($2\alpha = 135°$) and lap joints with root failures ($2\alpha = 0°$); T_W refers to $P_f = 2.3$ compared with 97.7 %; (Livieri and Lazzarin 2005)

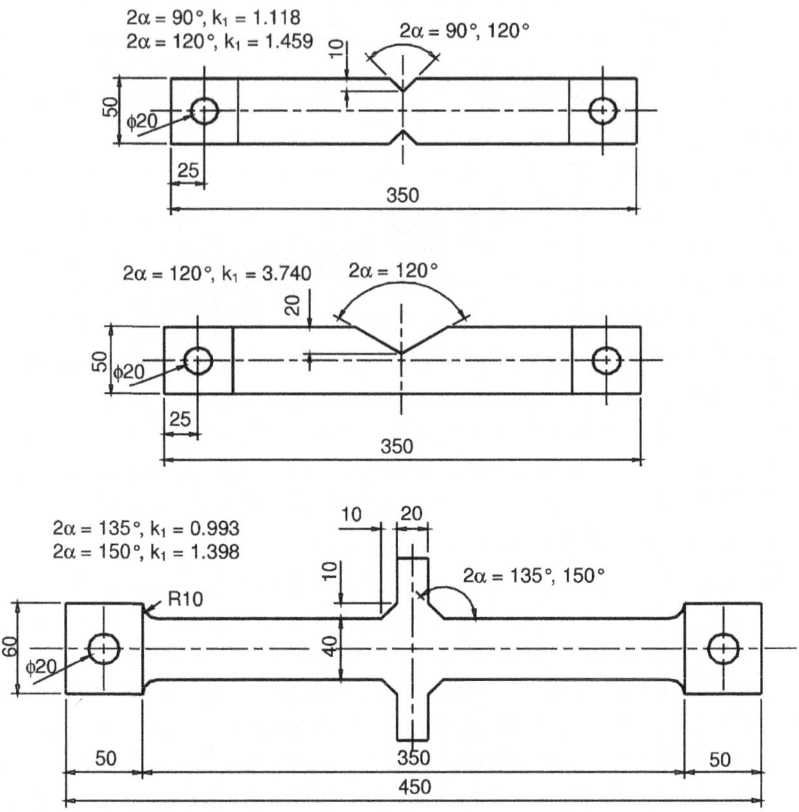

Fig. 3.14 V-notched flat bar fatigue test specimens (Kihara and Yoshii 1991), geometries and dimensions; (Lazzarin and Zambardi 2001)

corresponding values for carbon steels. The inverse slope exponent $k = 2.0$ of the *W–N* curve corresponds to $k = 4.0$ of the *S–N* curve, this being once more a usual value for welded joints of aluminium alloys in design codes.

Accurate fatigue test data for tensile loaded weld-like and V-notch flat bar specimens made of base material (Kihara and Yoshii 1991) were also evaluated in terms of $\Delta\overline{W}_1$ (Lazzarin and Zambardi 2001). The geometry and dimensions of the test specimens are shown in Fig. 3.14. The endured SED range is plotted over number of cycles N_i up to crack initiation (crack length $a_i = 0.2$ mm) for mild steel and high-strength steel, Fig. 3.15. The ratio 3.4 of the endurable $\Delta\overline{W}_1$ values at $N_i = 2 \times 10^6$ cycles corresponds to the ratio 1.8 of the $\Delta\sigma_n$ values, a realistic value for crack initiation. The inverse slope exponent $k \approx 2.3$ of the *W–N* curve corresponds to $k \approx 4.6$ of the *S–N* curve, to be compared with the usual value $k \approx 5.0$ for non-welded specimens. The scatter range indices based on $P_f = 10$ and 90 % are $T_W \approx 1 : 1.7$ corresponding to $T_\sigma = 1 : 1.3$. This is markedly less than the usual value for welded joints, $T_\sigma = 1 : 1.5$.

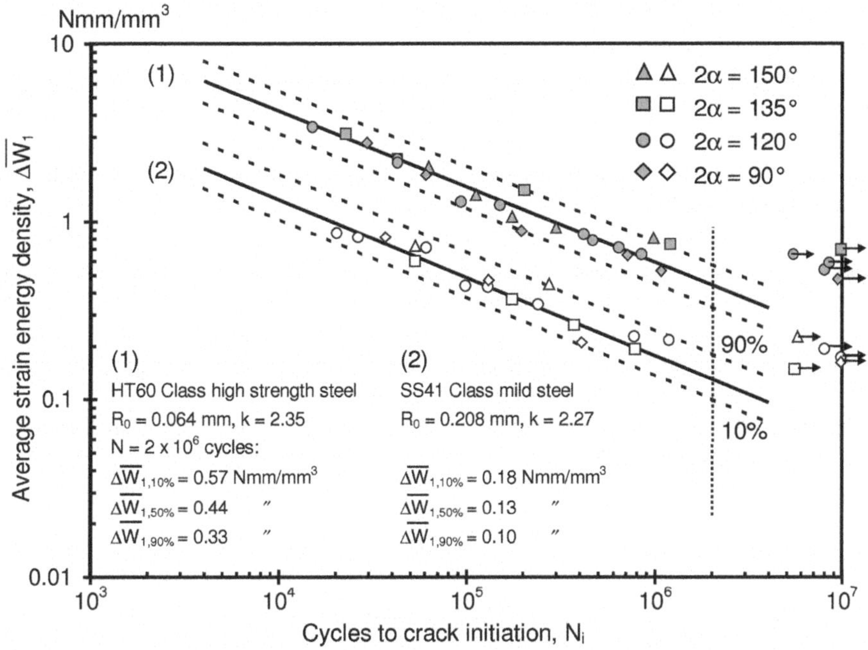

Fig. 3.15 Fatigue test results (Kihara and Yoshii 1991) in terms of $\Delta\overline{W}_1$ plotted over cycles to crack initiation N_i ($a_i = 0.2$ mm) for different V-notched flat bar specimens ($2\alpha = 90$, 120, 135, 150°) of mild steel and of high-strength steel subjected to tensile loading, nominal stress ratio $R = 0$; (Lazzarin and Zambardi 2001)

3.2.6 Extension to Elastic-Plastic Strain Energy Density

Linear-elastic conditions, exactly or approximately, are the basis of the derivations above with application to brittle fracture or high-cycle fatigue failure. Elastic-plastic conditions should be taken into account when considering ductile fracture or medium to low-cycle fatigue failure. Up to now, the linear-elastic parameters are empirically extended into the elastic plastic range for the fatigue assessment of welded joints.

As shown in the preceding Sect. 2.5, the plastic notch stress intensity factor (NSIF) concept describing the elastic-plastic fields at sharp notches is rather well developed. Use was made of the elastic-plastic strain energy density (SED) on several occasions during the developments. Most important was Glinka's equivalent SED hypothesis, which allows to equate the linear-elastic and elastic-plastic local SED values under small scale yielding conditions. The relationships between the elastic and plastic NSIFs could thus be established (Sects. 2.5.3 and 2.5.6), inclusive of the definition the elastic-plastic eigenvalues $s(\lambda_1, n)$. Also, the link between the σ–ε and τ–γ power law was based on a SED consideration (Sects.

2.5.6 and 2.5.10). The elastic-plastic fields of tensile loaded V-notches were derived with support by the SED concept (Sect. 2.5.4).

In the following, the basic formulae are summarised which express the relationships between the elastic or plastic average SED in the control volume (with radius R) and the elastic or plastic NSIF. The total strain power law is applied. The control volume radius R should remain within the range of validity of the asymptotic one-term expansion of the notch tip stress field. Under small-scale yielding conditions a link between the elastic and plastic parameters is possible based on Glinka's equivalent SED hypothesis. The formulae for large-scale yielding conditions are more complex, while the link to the elastic parameters is retained. The reverse expressions giving the NSIF in terms of the SED are also given. The presentation is restricted to the mode 1 and mode 3 parameters, whereas mode 2 is generally negligible in applications.

In-plane tensile loading (mode 1) is analysed based on the following formulae (Lazzarin and Zambardi 2002):

$$\overline{W}_{1\mathrm{e}}(R) = \frac{I_{\mathrm{e}}(\gamma)}{4\lambda_1 \gamma E} K_{1\mathrm{e}}^2 R^{2(\lambda_1 - 1)} \tag{3.44}$$

$$\overline{W}_{1\mathrm{p}}(R) = \frac{n}{(n+1)H^n} \frac{I_{\mathrm{p}}(\gamma, n)}{\gamma\,[2-(n+1)s]} K_{1\mathrm{p}}^{n+1} R^{-s(n+1)}, \quad s = \frac{2(1-\lambda_1)}{n+1} \tag{3.45}$$

where $I_{\mathrm{e}}(\gamma)$ and $I_{\mathrm{p}}(\gamma)$ are integrals of the elastic and plastic angle-dependent functions and $\gamma = \pi - \alpha$.

For small-scale yielding, Eqs. (2.190) and (2.191) result in:

$$K_{1\mathrm{p}} = \left(\frac{n+1}{n} \frac{H^n}{2E} \frac{I_{\mathrm{e}}(\gamma)}{I_{\mathrm{p}}(\gamma, n)}\right)^{1/(n+1)} K_{1\mathrm{e}}^{2/(n+1)} \tag{3.46}$$

For large-scale yielding, Eq. (2.189) has to be applied:

$$K_{1\mathrm{p}} = \left[\frac{n+1}{n} \frac{I_{\mathrm{e}}(\gamma)}{I_{\mathrm{p}}(\gamma, n)} \left(\frac{H^n}{2E} + \frac{n}{n+1} \sigma_{\mathrm{n}}^{n-1}\right)\right]^{1/(n+1)} K_{1\mathrm{e}}^{2/(n+1)} \tag{3.47}$$

The reverse relationships for determining the elastic or plastic NSIF from the elastic or plastic SED read as follows:

$$K_{1\mathrm{e}} = R^{1-\lambda_1} \sqrt{\frac{4\lambda_1 \gamma E}{I_{\mathrm{e}}(\gamma)} \overline{W}_{1\mathrm{e}}} \tag{3.48}$$

$$K_{1\mathrm{p}} = R^s \left[\frac{(n+1)H^n}{n} \frac{\gamma\,[2-(n+1)s]}{I_{\mathrm{p}}(\gamma, n)} \overline{W}_{1\mathrm{p}}\right]^{1/(n+1)}, \quad s = \frac{2(1-\lambda_1)}{n+1} \tag{3.49}$$

Out-of-plane shear loading (mode 3) is analysed based on the following formulae (Lazzarin and Zappalorto 2008):

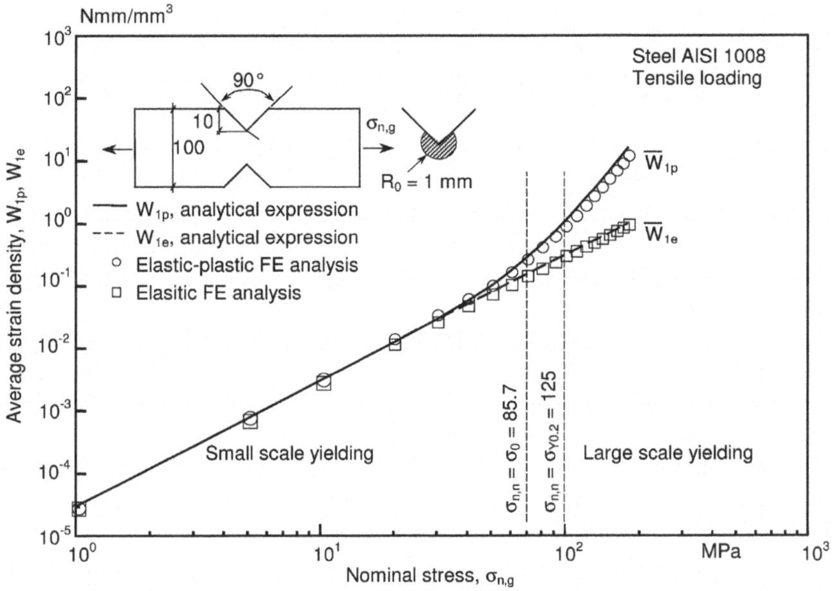

Fig. 3.16 Average SED $\overline{W}_{1p}$ and $\overline{W}_{1e}$ raising with increasing nominal stress $\sigma_{n,g}$ in gross cross-section; analytical and FE analysis results; V-notched flat bar specimen subjected to tensile loading; small-scale and large-scale yielding regime separated by $\sigma_{n,n} = \sigma_0$; (Lazzarin and Berto 2008)

$$\overline{W}_{3e}(R) = \frac{1}{4\pi\lambda_3 G} K_{3e}^2 R^{2(\lambda_3-1)} \tag{3.50}$$

$$\overline{W}_{3p}(R) = \frac{B_W(\gamma, n)}{H_\tau^n} K_{3p}^{n+1} R^{(n+1)/(m-1)}, \quad H_\tau^n = G\tau_0^{n-1} \tag{3.51}$$

For small-scale yielding, K_{3p} is expressed by K_{3e} according to Eq. (2.215):

$$K_{3p} = \sqrt{2\pi}\left(\frac{-m}{\lambda_3(1-m)}\frac{1}{\tau_0^{m+\omega}}\right)^{1/(1-m)}\left(\frac{K_{3e}}{\sqrt{2\pi}}\right)^{1/(1-\lambda_3)(1-m)} \tag{3.52}$$

The reverse relationships for determining the elastic or plastic NSIF from the elastic or plastic SED read as follows:

$$K_{3e} = R^{1-\lambda_3}\sqrt{4\pi\lambda_3 G\overline{W}_{3e}} \tag{3.53}$$

$$K_{3p} = R^{1/(1-m)}\left(\frac{H_\tau^n}{B_W(\gamma, n)}\overline{W}_{3p}\right)^{1/(n+1)} \tag{3.54}$$

The behaviour of the SED parameters $\overline{W}_{1e}$ and $\overline{W}_{1p}$ with increasing load is illustrated by Fig. 3.16. The solid line curve refers to the plastic SED $\overline{W}_{1p}$

according to Eq. (3.45) combined with the plastic NSIF K_{1p} according to Eq. (3.46). The broken line curve for the elastic SED $\overline{W}_{1e}$, Eq. (3.44), is supplemented. Confirming FE analysis results are also plotted. In the small-scale yielding regime, limited by the substitute yield limit in the net cross-section, $\sigma_{n,n} = \sigma_0$, the plastic SED is equal to the elastic SED, $\overline{W}_{1p} = \overline{W}_{1e}$, because Glinka's equivalent hypothesis was applied. Beyond the σ_0 limit of the small-scale yielding regime, $\overline{W}_{1p}$ starts to raise above the linear continuation of $\overline{W}_{1e}$. The raise accelerates with increasing nominal stress. This can be seen in Fig. 2.84, where the corresponding results for a V-notched round bar subjected to torsional loading (mode 3) are plotted.

The raise of the elastic-plastic SED in the large-scale yielding regime depends mainly on the hardening exponent n (steeper raise for lower values of n) and the control volume radius R_0 (steeper raise for lower values of R_0). A substantial influence of the multiaxiality condition (plane strain compared with plane stress or pure shear conditions) on the progress of large-scale yielding can be predicted. Some tentative results are available (Lazzarin and Berto 2008), but their reliability is doubtful. Further confirmation and an explanation of the results is required, which would be substantiate the observed increase of the inverse slope exponent k of the *S–N* or *W–N* curve for welded joints under torsional loading conditions.

The interest in reliable data on the small-scale and large-scale yielding behaviour of V-notched specimens, e.g. fillet-welded joints, results from the prospect of determining the fatigue *S–N* curve of such specimens based on a combination of strain *S–N* curve and local SED values. When asking for the inverse slope exponent k of the relevant design *S–N* curve, the additional aspects of crack initiation and propagation have to be taken into account. It has been shown that the degree of elastic support during plastic deformation and crack initiation is very influential (Gimperlein 1998; Radaj et al. 2006, ibid. pp. 231/232).

3.3 Multiaxial Loading SED Concept

3.3.1 Survey of Section Contents

The local SED concept for pointed V-notches is applied to multiaxial loading conditions (mode 1, 2, 3 superimposed). Axisymmetric tube-to-flange fillet-welded joints are considered under bending and torsional loads. Also, V-notched round bar specimens under tensile and torsional superimposed loading are analysed.

The expositions are primarily based on the analytical developments and their practical applications carried out by Lazzarin, Zambardi and Berto. A list of symbols and a list of references are available at the end of Chap. 3.

3.3.2 Multiaxial Fatigue of Welded Joints

Uniaxial loading comprises tensile and bending loads which produce two-dimensional stresses at the weld toe under plane strain conditions. Loading in the form of torsional loads produces a stress state at the weld toe which differs substantially from the above. Any loading producing deviations from the plane strain state is termed 'multiaxial loading'.

The main application of the multiaxial loading SED concept to welded joints are axisymmetric tube-to-flange fillet-welded joints subjected to pure torsional loads or to a superposition of bending moments and torsional loads. The weld toe notch, simulated by a 135 degree V-notch is then under the simultaneous action of bending and shear stresses from mode 1, 2 and 3 loading, mode 1 and 2 under plane strain conditions, mode 2 generally negligible at the weld toe.

In mode 3 or superimposed mode 1 and mode 3 loading, where the microcracks are inclined relative to the toe notch line, a notch-parallel macrocrack can only be formed over steps in the fracture surface. On the other hand, in mode 1 loading, coalescence of microcracks is easier to achieve because the crack plane does not change. Thus, the failure behaviour under mode 1 and mode 3 conditions is different. Additionally, the bending moments and torsional loads may be applied out-of-phase.

Despite the mentioned peculiarities, welded joints of the above type under multiaxial loading are assessed in respect of fatigue based on a uniform radius R_0 of the control volume. The relevant expression for the average total SED $\overline{W}$ reads as follows (Lazzarin et al. 2004):

$$c_W \Delta\overline{W} = c_W \left[\frac{e_1}{E} \left(\frac{\Delta K_1}{R_0^{1-\lambda_1}} \right)^2 + \frac{e_2}{E} \left(\frac{\Delta K_2}{R_0^{1-\lambda_2}} \right)^2 + \frac{e_3}{E} \left(\frac{\Delta K_3}{R_0^{1-\lambda_3}} \right)^2 \right] \quad (3.55)$$

Reference is made to Eqs. (3.10), (3.23) and (3.25). The coefficients e_1, e_2, e_3 (e_1 and e_2 under plane strain conditions, $e_3 = e_{d3}$) depend on the notch opening angle 2α and on Poisson's ratio ν. Simple approximations with $\nu = 0.3$ and 2α in degrees are:

$$e_1 = 0.133 + 6.151 \times 10^{-4}(2\alpha) - 5.373 \times 10^{-6}(2\alpha)^2 \quad (3.56)$$

$$e_2 = 0.340 - 2.346 \times 10^{-3}(2\alpha) + 4.809 \times 10^{-6}(2\alpha)^2 \quad (3.57)$$

$$e_3 = 0.414 - 1.149 \times 10^{-3}(2\alpha) - 2.170 \times 10^{-9}(2\alpha)^2 \quad (3.58)$$

In exceptional cases, the distortional SED $\Delta\overline{W}_d$ is used instead of the total SED $\Delta\overline{W}$. The appertaining coefficients e_{d1}, e_{d2}, e_{d3} are contained in the plot of Fig. 3.4.

The prestress coefficient c_W in Eq. (3.55) takes the influence of the stress ratio R into account, $c_W = 1.0$ for $R = 0$ and $c_W = 0.5$ for $R = -1$, Eqs. (3.25). For

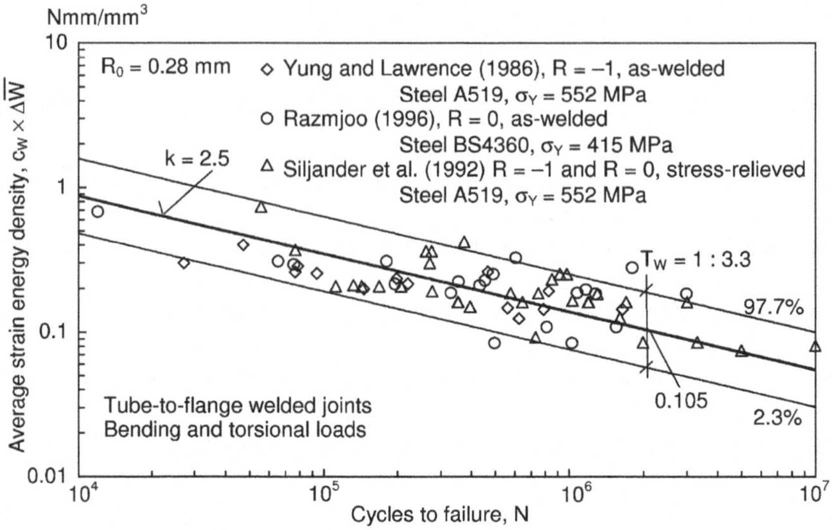

Fig. 3.17 Multiaxial fatigue test data in terms of $c_W \times \Delta\overline{W}$ for tube-to-flange welded joints made of steel; see list of references for the quoted publications; (Lazzarin et al. 2008$^{(2)}$; Berto and Lazzarin 2011)

welded joints in the as-welded condition $c_W = 1.0$ is appropriate, independently of R, because of the tensile residual stresses produced by welding. The value $c_W = 0.5$ should be used for welded joints with a stress relief treatment.

The NSIF ranges ΔK_1, ΔK_2, ΔK_3 are determined applying the FE analysis method, either with very fine meshes evaluating the limit value $r \to 0$ formulae on the stresses in the bisector plane, or with coarse meshes evaluating the average SED in the control volume (Sect. 3.7).

Fatigue test results for fillet-welded tube-to-flange joints of steel subjected to bending and torsional loads, applied separately or superimposed, from investigations by Yung and Lawrence, by Razmjoo and by Siljander et al. are plotted in Fig. 3.17 (Lazzarin et al. 2008$^{(2)}$; Berto and Lazzarin 2011). The data points are set in comparison to the scatter band defined by the $\Delta\overline{W}$ values at $N = 2 \times 10^6$ cycles of tensile or bending loaded welded joints in general, but extended in the finite life regime with the inverse slope exponent $k = 2.5$ instead of 1.5 (corresponding to $k = 5.0$ and 3.0 of the S–N curves in the Eurocode 3). Obviously, the multiaxial SED data fit well into the scatter band of the uniaxial SED. Comparable results for the tube-to-flange joint with full-penetration fillet weld (Sonsino 1995) fit the inverse slope exponent $k = 2.5$, but indicate a much higher fatigue strength (factor 3 in terms of $\Delta\overline{W}$, factor $3^{1/2} = 1.73$ in terms of $\Delta\sigma_n$), Fig. 3.18. Part of the increase can be explained by weld toe grinding. Out-of-phase loading of the unmachined weld toe gives the lowest $\Delta\overline{W}$ values. Another reason for the increase in relation of the SED data of Siljander et al. (identical S–N curve) is the fact that the conversion of Sonsino's S–N data is performed with $\rho = 0$ (actual lower bound

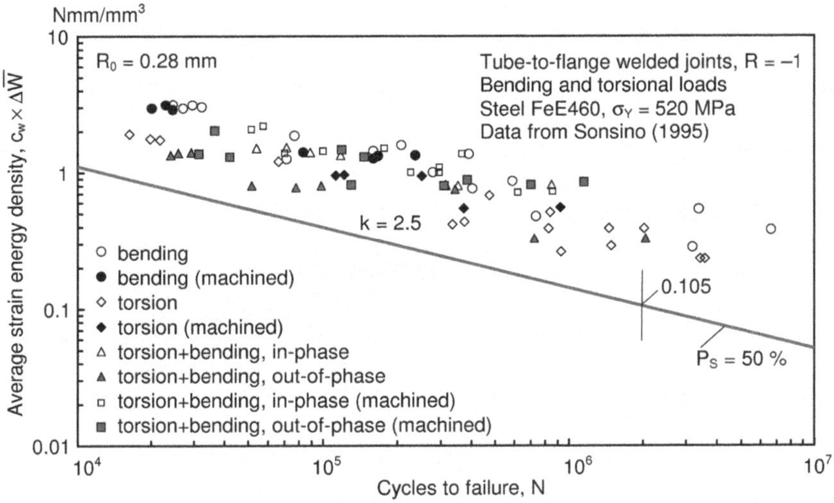

Fig. 3.18 Multiaxial fatigue test data in terms of $c_W \times \Delta\overline{W}$ for tube-to-flange welded joints made of steel; see list of references for the quoted publication; (Berto and Lazzarin 2011)

value) whereas the conversion of Siljander's data is based on $\rho = 4$ mm (actual upper bound value). An evaluation of Sonsino's results in terms of the distortional strain energy density is also available (Lazzarin et al. 2004).

Fatigue test results for fillet-welded tube-to-flange and tube-to-sleeve joints of aluminium alloys subjected to bending and torsional loads, applied separately or superimposed, from investigations of Küppers and Sonsino and from Costa et al. are plotted in Fig. 3.19 (Lazzarin et al. 2008(2); Berto and Lazzarin 2011). The SED range $\Delta\overline{W}$ is determined from the maximum notch stresses in the former case and by means of the blunt notch crescent-shaped averaging area (Sect. 3.4) in the latter case. The data points are set in comparison to the scatter band defined by the $\Delta\overline{W}$ values at $N = 2 \times 10^6$ cycles, tensile or bending loaded welded joints in general, extended in the finite life regime with the inverse slope exponent $k = 2.0$ (Livieri and Lazzarin 2005) or $k = 3.25$ (Lazzarin et al. 2008(2)). The multiaxial test results fit well into the scatter band with $k = 3.25$.

3.3.3 Multiaxial Fatigue of V-Notched Round Bar Specimens

Three major investigations have been performed with V-notched round bar specimens of medium-carbon steels subjected to multiaxial fatigue loading (Berto et al. 2011; Berto and Lazzarin 2011). Unnotched specimens and a specimen with shoulder fillet were included. Sharp notches were considered, $\rho = 0.5$ or 0.1 mm. The notch depth varied between 0.5 and 4.0 mm with a round bar diameter of

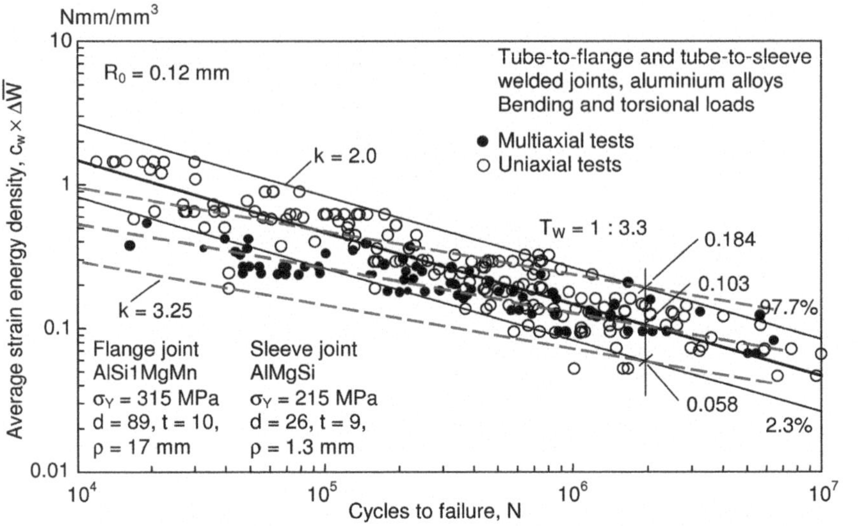

Fig. 3.19 Multiaxial fatigue test data in terms of $c_W \times \Delta\overline{W}$ for tube-to-flange and tube-to-sleeve welded joints made of aluminium alloys; (Lazzarin et al. 2008(2); Berto and Lazzarin 2011; Küppers and Sonsino 2003; Costa et al. 2005)

20 mm. Tensile and torsional cyclic loading ($R = 0$ or -1) were applied separately or superimposed (amplitude ratios $\lambda = \tau_a/\sigma_a = 0.6$–$1.6$). The phase shift angle between the two loadings was $\phi = 0$ and 90°.

In the earlier investigation referring to medium-carbon steel C40 ($\sigma_Y = 537$ MPa), one and the same control volume defined by the radius R_0 is used for tensile and torsional loading, in conformity with the procedure applied to welded joints. In the later two investigations referring to the medium-carbon steels 39NiCrMo3 ($\sigma_Y = 900$ MPa) and AISI 416 ($\sigma_Y = 570$ MPa), different control volume radii R_{01} and R_{03} in tensile and torsional loading are introduced in order to remain closer to the physical reality.

Three reasons for the difference in fatigue behaviour of the tensile and torsional loaded V-notched round bar specimens are given by the quoted authors:

- Under tensile loading resulting in a life of 5×10^6 cycles, the specimens behave linear-elastic, with the exception of a very small plastic zone ahead of the notch tip, about 0.025 mm on the bisector line, whereas under torsional loading, this zone extends over about 0.5 mm (at least according to FE analysis results).
- Under the tensile loading case above, non-propagating cracks along the bisector line are observed which are correspondingly smaller than in the torsional loading case (0.05 compared with 0.5 mm).
- The 'factory roof' topology of the fracture surface produced by torsional loading needs another 'driving energy' than the plane fracture surface occurring with tensile loading. Additionally, debris is produced by the mating roof surfaces in

the former case of loading, especially under fully reversed and out-of-phase torsional loading. Advantageous crack closure effects are the consequence.

Dealing with the round bar specimens with circumferential V-notch under superimposed tensile (mode 1) and torsional (mode 3) loading, the average total SED $\Delta\overline{W}$ under linear-elastic conditions reads as follows, compare Eq. (3.55):

$$\Delta\overline{W} = \frac{e_1}{E}\frac{\Delta K_1^2}{R_{01}^{2(1-\lambda_1)}} + \frac{e_3}{E}\frac{\Delta K_3^2}{R_{03}^{2(1-\lambda_3)}} \tag{3.59}$$

The control volume radii R_{01} and R_{03} are determined by considering the two loading modes separately (Lazzarin et al. 2008[(2)]; Berto and Lazzarin 2011):

$$R_{01} = \left(\sqrt{2e_1}\frac{\Delta K_{1E}}{\Delta\sigma_E}\right)^{1/(1-\lambda_1)} \tag{3.60}$$

$$R_{03} = \left(\sqrt{\frac{e_3}{1+\nu}}\frac{\Delta K_{3E}}{\Delta\tau_E}\right)^{1/(1-\lambda_3)} \tag{3.61}$$

with the endurable NSIF ranges ΔK_{1E}, ΔK_{3E} and the endurable nominal stress ranges $\Delta\sigma_E$, $\Delta\tau_E$, all parameters determined at 2×10^6 cycles.

For the round bar specimen ($2\alpha = 90°$, $a = 4$ mm) made of quenched and tempered steel 39NiCrMo3, the following data are relevant:

- $\Delta K_{1E} = 700$ MPa mm$^{0.455}$, $\Delta\sigma_E = 630$ MPa, $\lambda_1 = 0.5445$, $e_1 = 0.1462$, $R_{01} = 0.327$ mm
- $\Delta K_{3E} = 1{,}217$ MPa mm$^{0.333}$, $\Delta\tau_E = 530$ MPa, $\lambda_3 = 0.6667$, $e_3 = 0.3103$, $R_{03} = 1.426$ mm

For the considered round bar specimen made of quenched and tempered steel AISI 416, the following data are relevant:

- $\Delta K_{1E} = 512$ MPa mm$^{0.455}$, $\Delta\sigma_E = 697$ MPa, $\lambda_1 = 0.5445$, $e_1 = 0.1462$, $R_{01} = 0.13$ mm
- $\Delta K_{3E} = 891$ MPa mm$^{0.333}$, $\Delta\tau_E = 474$ MPa, $\lambda_3 = 0.6667$, $e_3 = 0.3103$, $R_{03} = 0.78$ mm

Based on known control volume radii R_{01} and R_{03}, the average total SED $\Delta\overline{W}$ can be calculated for any test result based on Eq. (3.55) with R_0 substituted by R_{01} and R_{03}. An accumulation of data points $\Delta\overline{W}$ over N is thus generated, which is now statistically analysed.

The statistical analysis (least square fit) assumes a linear $\log\Delta\overline{W}$ over $\log N$ relationship. Run-out data ($N > 5 \times 10^6$ cycles) were excluded from the analysis. The following parameters are gained as a result: the inverse slope exponent k (sometimes simplified to even numbers), the reference values of $\Delta\overline{W}$ at $N = 2 \times 10^6$ cycles and the scatter band range index T_W referring to 10 and 90 %.

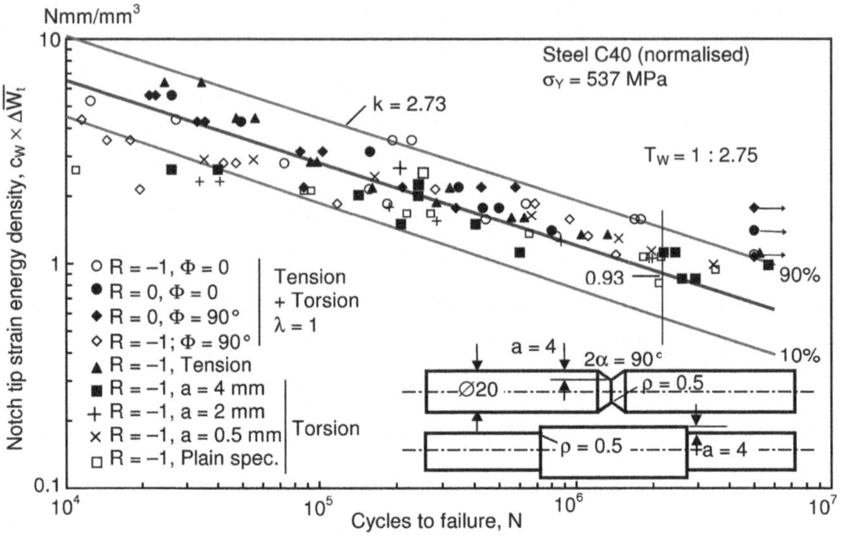

Fig. 3.20 Multiaxial fatigue test data in terms of $c_W \times \Delta \overline{W}_t$ for sharply notched and unnotched (plain) round bar specimens made of normalised steel C40; tensile and torsional loading, separate and superimposed, without and with phase shift (angle ϕ); (Atzori et al. 2006; Berto and Lazzarin 2011)

Multiaxial fatigue test data of sharply notched (but not pointed) and unnotched round bar specimens made of the normalised steel C40 ($\sigma_Y = 537$ MPa, $\sigma_U = 715$ MPa), subjected to separate and superimposed tensile and torsional loading, both in-phase and out-of-phase, are evaluated and plotted in terms of two different SED parameters (Atzori et al. 2006). In Fig. 3.20, the total SED range $c_W\, \Delta W_t$ at the point of maximum notch stress is evaluated:

$$c_W \Delta W_t = \frac{c_W}{2E}\left[\Delta\sigma_{max}^2 + 2(1+\nu)\Delta\tau_{max}^2\right] \tag{3.62}$$

where c_W is the prestress coefficient, and $\Delta\sigma_{max}$ and $\Delta\tau_{max}$ are the maximum tensile and torsional notch stress ranges. This variant of the SED approach for rounded notches under multiaxial loading has already been used earlier (Park and Nelson 2000; Moftakhar et al. 1995). In Fig. 3.21, the average total SED range $c_W \Delta \overline{W}$ is plotted, which refers to the blunt notch crescent-shaped averaging area. The control radius R_0 is defined based on the arithmetic mean of the relevant SED values in tensile and torsional loading at 2×10^6 cycles.

Another set of multiaxial fatigue test data of sharply notched and unnotched (plain) round bar specimens, now made of the quenched and tempered steel 39NiCrMo3 ($\sigma_Y = 900$ MPa, $\sigma_U = 995$ MPa), is evaluated and plotted in Fig. 3.22 (Berto and Lazzarin 2011). The variation of the loading parameters is

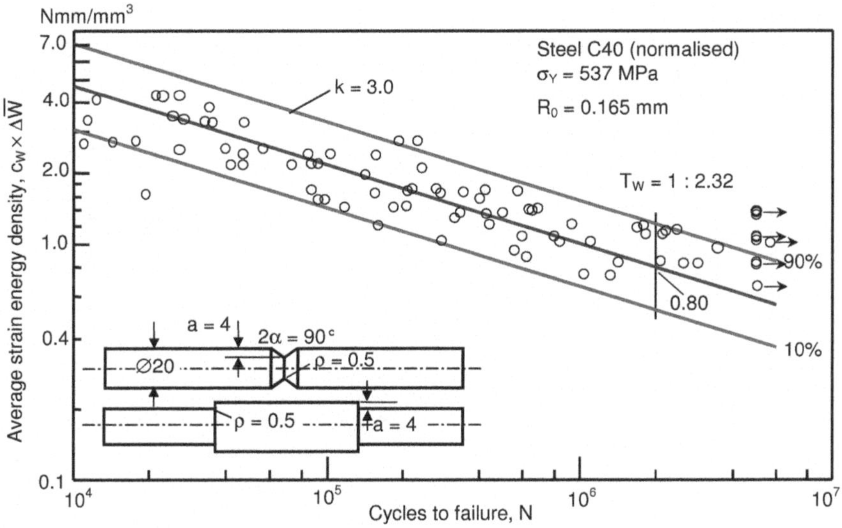

Fig. 3.21 Multiaxial fatigue test data in terms of $c_W \times \Delta\overline{W}$ for sharply notched and unnotched (plain) round bar specimens made of normalised steel C40; tensile and torsional loading, separate and superimposed (amplitude ratio $\lambda = \tau_a/\sigma_a$), without phase shift (angle $\phi = 0$); V-notch model; (Atzori et al. 2006; Berto and Lazzarin 2011)

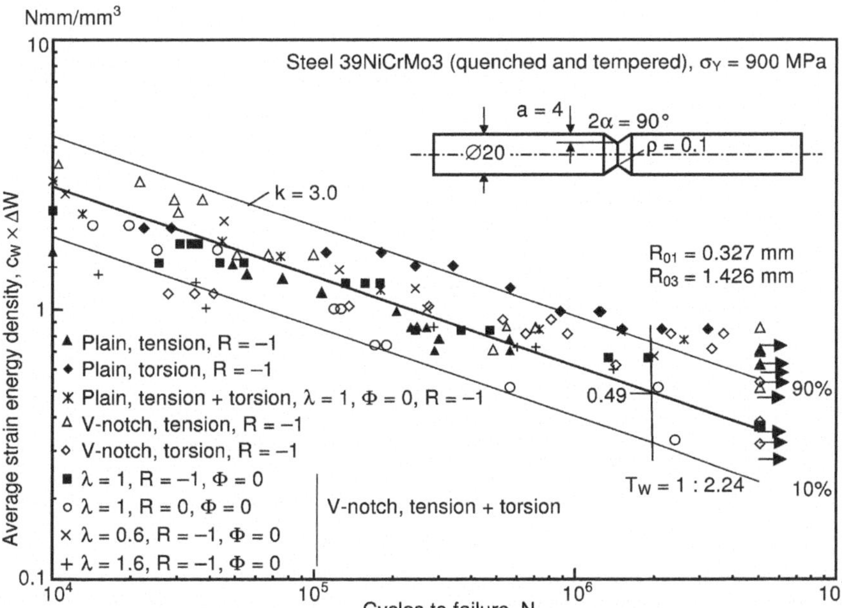

Fig. 3.22 Multiaxial fatigue test data in terms of $c_W \times \Delta\overline{W}$ for sharply notched and unnotched (plain) round bar specimens made of quenched and tempered steel 39NiCrMo3; tensile and torsional loading, separate and superimposed (amplitude ratio $\lambda = \tau_a/\sigma_a$), without and with phase shift (angle ϕ); blunt V-notch model; (Berto and Lazzarin 2011)

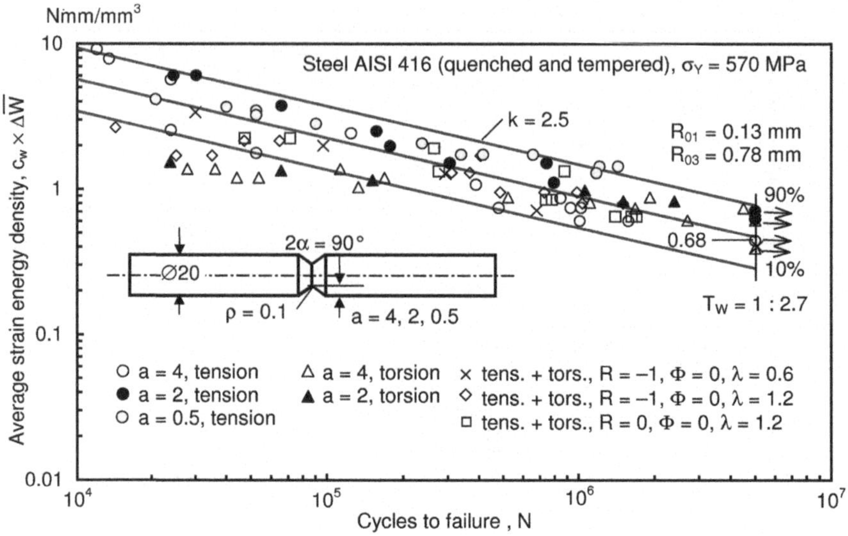

Fig. 3.23 High-slope multiaxial fatigue test data in terms of $c_W \times \Delta\overline{W}$ for sharply notched round bar specimens made of quenched and tempered steel AISI 416; tensile and torsional loading, separate and superimposed (amplitude ratio $\lambda = \tau_a/\sigma_a$), without phase shift (angle $\phi = 0$); blunt V-notch model; (Berto and Lazzarin 2011)

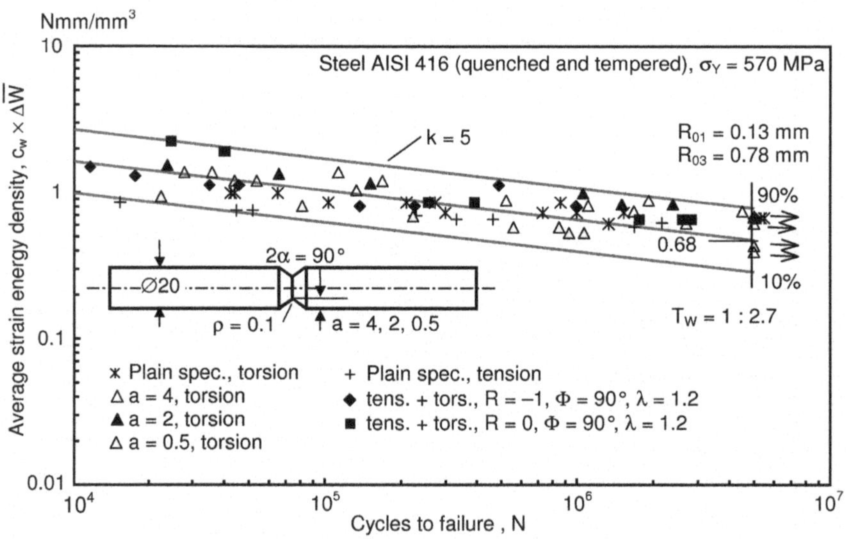

Fig. 3.24 Low-slope multiaxial fatigue test data in terms of $c_W \times \Delta\overline{W}$ for unnotched (plain) and for some notched round bar specimens made of quenched and tempered steel AISI 416; tensile and torsional loading, separate and superimposed (amplitude ratio $\lambda = \tau_a/\sigma_a$), partly with phase shift (angle $\phi = 90°$); blunt V-notch model; (Berto and Lazzarin 2011)

similar to the above. The blunt notch crescent-shaped averaging area is used. The peculiarity is that different control radii R_{01} and R_{03} are used for tensile and torsional loading.

Finally multiaxial fatigue test data for sharply notched and unnotched (plain) round bar specimens made of the quenched and tempered steel AISI 416 ($\sigma_Y = 570$ MPa, $\sigma_U = 700$ MPa) are presented on the same line as above (Berto and Lazzarin 2011). The blunt notch crescent-shaped averaging area is used. The peculiarity here is, that the data are subdivided into two sets, one resulting in a higher slope of the *W–N* curve typical for notched specimens, Fig. 3.23 (the plotted data for pure torsional loading are not included in the statistical analysis), the other resulting in a lower slope of the *W–N* curve typical for unnotched (plain) specimens, Fig. 3.24 (but the data for notched specimens under pure torsional loading or superimposed loading with phase shift are included).

3.4 Blunt Notch SED Concept

3.4.1 Survey of Section Contents

The local SED concept is developed for blunt notches subjected to tensile loading (mode 1) and extended to multiaxial conditions (mixed mode 1 and 2). This concept is applied for determining the brittle fracture strength of PMMA specimens with U-notches.

The expositions are primarily based on the analytical developments carried out by Lazzarin and Berto and applied to brittle fracture by Gómez and Elices. A list of symbols and a list of references are available at the end of Chap. 3.

3.4.2 Basic Relationships

The local SED concept has been extended from pointed to blunt V-notches inclusive of U-notches. Tensile loading (mode 1) is considered first in the analytic derivations, which are then transferred to mixed mode 1 and 2 loading conditions based on a semi-empirical, but numerically confirmed procedure (Lazzarin and Berto 2005(1)).

The geometrical parameters of the blunt V-notch and the stress field parameters in polar coordinates are shown in Fig. 3.25. The notch radius ρ is geometrically defined. The distance r_0 between the centre of the polar coordinates and the notch root is part of the analytical theory. As can be seen from the formulae for r_0 and q in the diagram, $r_0 = \rho/2$ for $2\alpha = 0$ and $r_0 \rightarrow 0$ for $2\alpha \rightarrow \pi$, i.e. the considered centre of coordinates moves closer to the notch root for larger notch opening angles.

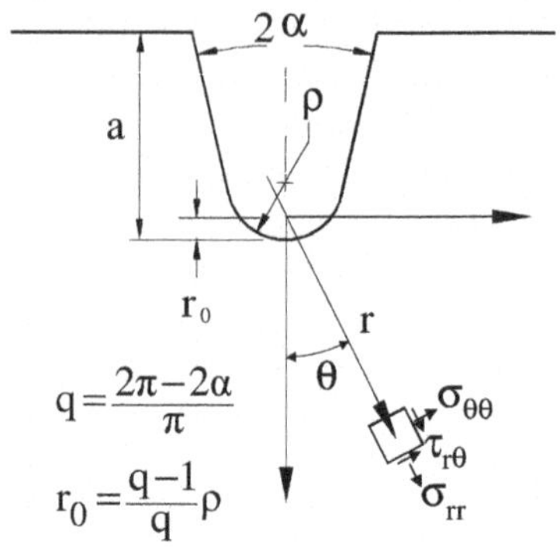

Fig. 3.25 Geometrical and stress field parameters at blunt V-notch; (Lazzarin and Berto 2005[(1)])

The determining geometrical parameters of the crescent-shaped control volume area at blunt V-notches are shown in Fig. 3.26c. The area A of the average local SED has the circular notch edge with radius ρ as the boundary on one side and a circle arc with radius $R_0 + r_0$ as the boundary on the other side. The two radii are differently centred. The centre of the radius ρ is geometrically given. The centre of the radius $R_0 + r_0$ is identical with the centre of the polar coordinate system in the analytical solution for the blunt V-notch problem. The control radius R_0 characterising the control volume area is introduced as a material parameter which is independent of the notch geometry.

More precisely, the crescent-shaped control volume area A is determined as follows, Fig. 3.27:

$$A = \int_{R_1}^{R_2} \int_{-\bar{\theta}}^{+\bar{\theta}} r \, \mathrm{d}r \, \mathrm{d}\theta \tag{3.63}$$

where the radius R_1 of the inner boundary depends on the polar angle θ, whereas the radius $R_2 = R_0 + r_0$ is independent of θ.

The (elastic) average local SED under mode 1 loading is expressed in the following form (Lazzarin and Berto 2005[(1)]):

$$\overline{W}_1 = F(2\alpha)H(2\alpha,\ R_0/\rho)\frac{\sigma_{\max}^2}{E} \tag{3.64}$$

Values of the functions $F(2\alpha)$ and $H(2\alpha,\ R_0/\rho)$, the latter for Poisson's ratio $\nu = 0.3$ (metals) and $\nu = 0.4$ (PMMA), are listed in Table 3.2.

By introducing the NSIF $K_{1\rho}$ together with the notch radius ρ instead of the maximum notch stress $\sigma_{\max}$, a more compact form of Eq. (3.64) is gained:

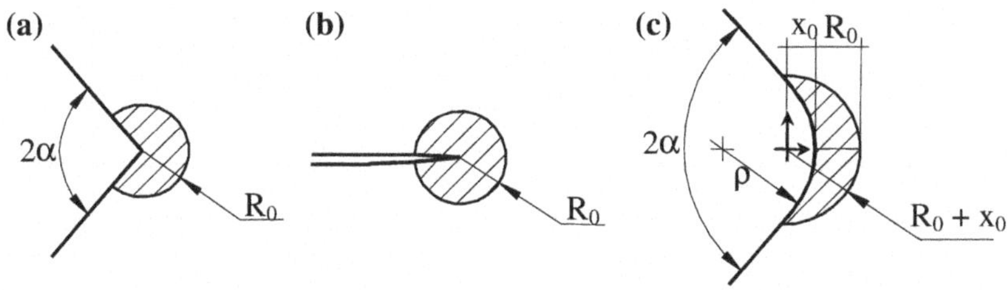

Fig. 3.26 Geometrical parameters of the control volume area used for averaging the SED in the close vicinity of the notch root; pointed V-notch (**a**), crack or slit (**b**) and blunt V-notch (**c**); (Lazzarin and Berto)

Fig. 3.27 Further geometrical parameters characterising the crescent-shaped control area *A* at blunt V-notches; circle arc of radius R_2 intersects the curvilinear notch root edge or the rectilinear part of the V-notch edge; (Lazzarin and Berto 2005[(1)])

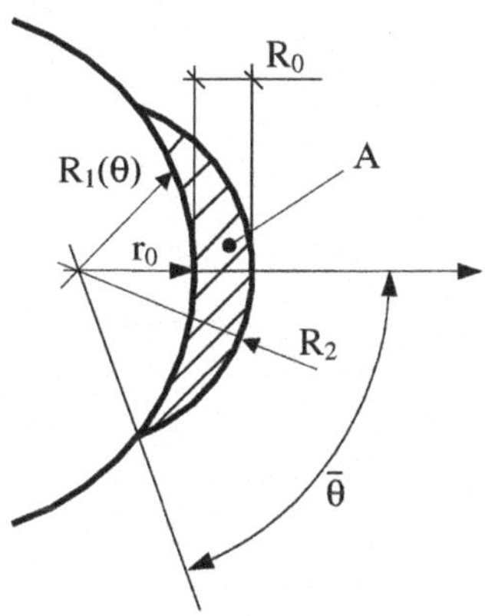

$$\overline{W}_1 = H(2\alpha,\ R_0/\rho)\frac{K_{1\rho}^2}{E}\rho^{2(\lambda_1-1)} \tag{3.65}$$

When the notch opening angle is zero, $2\alpha = 0$, corresponding to the parabolic notch, Eqs. (3.61) and (3.62) are simplified (Gómez et al. 2007):

$$\overline{W}_1 = H(0,\ R_0/\rho)\frac{\pi\sigma_{\max}^2}{4E} \tag{3.66}$$

$$\overline{W}_1 = H(0,\ R_0/\rho)\frac{K_{1\rho}^2}{E\rho} \tag{3.67}$$

In cases of superimposed tensile and shear loading (mixed mode 1 and 2), the procedure above, developed for tensile loading can be transferred on a semi-empirical, but numerically confirmed basis. Under mixed mode conditions, the maximum principal stress does not occur at the notch root in the bisector plane, but is shifted to a notch edge point out of the bisector plane. It has been proven numerically (FE analysis) that correct average SED values are gained, if the control volume and the appertaining base point of r_0 are rotated by the polar angle

Table 3.2 Values of the functions $F(2\alpha)$ and $H(2\alpha, R_0/\rho)$ for blunt V-notches under plane strain conditions; $\nu = 0.3$ (metals) and $\nu = 0.4$ (PMMA); (Lazzarin and Berto 2005[(1)])

2α	$F(2\alpha)$	$H(2\alpha, R_0/\rho)$, $\nu = 0.3$, R_0/ρ =			$H(2\alpha, R_0/\rho)$, $\nu = 0.4$, R_0/ρ =		
		0.01	0.1	1	0.01	0.1	1
0°	0.7850	0.5638	0.4518	0.1314	0.5194	0.4099	0.1110
45°	0.6692	0.6609	0.5264	0.1447	0.6093	0.4802	0.1252
90°	0.7049	0.6290	0.4955	0.1328	0.5801	0.4535	0.1174
135°	1.0717	0.4114	0.3206	0.1037	0.3795	0.2938	0.0932

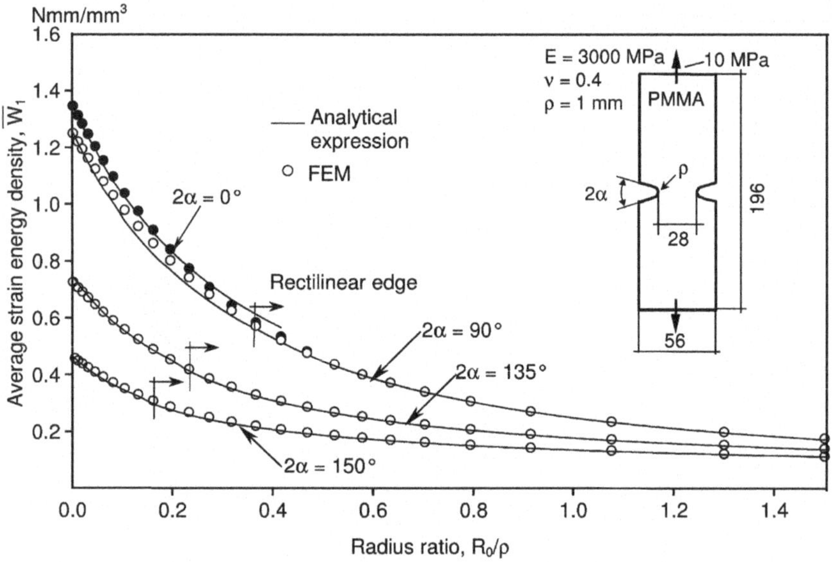

Fig. 3.28 Comparison between analytical and numerical results, average SED plotted over radius ratio R_0/ρ for different notch opening angles 2α; nominal stress in the gross cross-sectional area; (Lazzarin and Berto 2005[(1)])

θ^* around the base point of ρ into the direction of the point of maximum stress (assuming a circle arc). The angle θ^*, which denotes simultaneously the crack initiation point and the crack propagation direction, depends on the ratio of tensile to shear load.

The degree of accuracy of Eq. (3.64) has been checked by means of FE analyses, performed under linear-elastic and plane strain conditions (Lazzarin and Berto 2005[(1)]). The check was necessary, mainly because the analytical stress field solution is approximate and its degree of accuracy is found to decrease with the distance r from the notch tip and with the angle θ.

A typical geometry used for the check is shown in Fig. 3.28, where the notch root radius was kept constant ($\rho = 1$ mm) whereas the notch opening angle 2α ranged from 0 to 150°. The values of the modulus of elasticity E and of Poisson's ratio ν are typical for PMMA. Very fine FE meshes were used at the notch root.

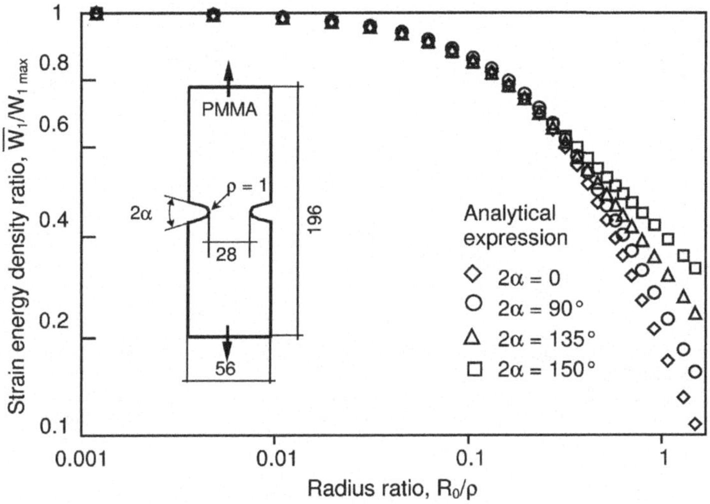

Fig. 3.29 SED ratio $\overline{W}_1/W_1$ plotted over radius ratio R_0/ρ for different notch opening angles 2α; (Lazzarin and Berto 2005[(1)])

The check comprised both notch boundary cases, the one where the notch edge of the control volume is completely defined by the notch radius ρ, the other where the notch edge includes also rectilinear parts of the V-notch flanks. The SED $\overline{W}_1$ dependent on the radius ratio R_0/ρ is considered for different notch opening angles 2α. The numerical results plotted in the figure correspond well with the analytical predictions. The plot of the SED $\overline{W}_1$, normalised by the point-related maximum value W_{1max}, is shown in Fig. 3.29. There is no influence of the notch opening angle 2α for $R_0/\rho < 0.2$.

3.4.3 Application of the Blunt Notch SED Concept to Brittle Fracture

The blunt notch SED concept has been applied by Lazzarin and Berto to accurate experimental data provided by Gómez and Elices in several publications dealing with the brittle fracture behaviour of U-notched and V-notched specimens made of acrylic glass PMMA and subjected to tensile or bending loads at room temperature or at −60 °C (Lazzarin and Berto 2005[(1)]).

The following specimens and loading conditions are analysed and tested, which are shown with their dimensions as part of the diagrams:

- Single edge notch specimen under bending load (SENB)
- Single edge notch specimen under tensile load (SENT)
- Double edge notch specimen under tensile load (DENT)

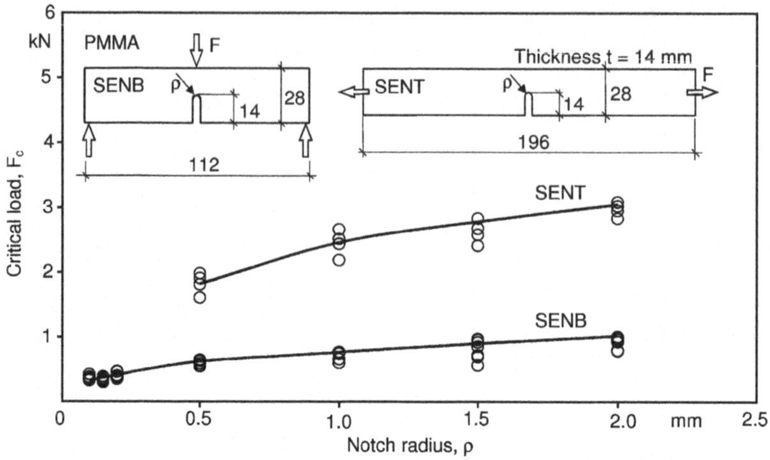

Fig. 3.30 Critical load F_c at U-notched PMMA specimens dependent on the notch radius ρ; analytical expression (solid line curves) compared with test results (point symbols); (Lazzarin and Berto 2005(1); Gómez et al. 2000; Gómez and Elices 2003(1), 2003(2))

The material parameters at room temperature (first data set) and at −60 °C (second and third data set are the following:

• Temperature T [°C]	20	−60	−60
• Modulus of elasticity E [N/mm^2]	3,000	5,290	5,050
• Poisson's ratio ν	0.4	0.4	0.4
• Ultimate tensile strength σ_U [N/mm^2]	136.0	130.0	128.4
• Fracture toughness K_{Ic} [MPa m$^{0.5}$]	$\approx$1.0	1.8	1.7
• Critical SED W_c [Nmm/mm^3], Eq. (3.27)	3.08	1.60	1.69
• Control volume radius R_0 [mm], Eq. (3.32)	0.011	0.035	0.035

A first set of experimental data refers to investigations at room temperature (Gómez et al. 2000; Gómez and Elices 2003(1), 2003(2)). The behaviour of the material in the unnotched tensile specimen was not fully brittle insofar as a nonlinear raise of the σ–ε curve was observed with an abrupt semibrittle fracture at the critical load without any necking. Fracture of the notched specimens was fully brittle. The critical load F_c is determined experimentally and analytically dependent on the notch radius ρ, Fig. 3.30, on the notch depth a, Fig. 3.31, and on the notch opening angle 2α, Fig. 3.32. The hypothesis behind the analysis is a constant value of the critical SED W_c. Correspondence between analytical and experimental results is proven with the exception of the SENT specimen with $2\alpha = 150°$, where $F_c \approx 2.9$ kN is determined analytically, whereas the experimental result is $F_c \approx 1.9$ kN.

A second set of experimental data refers to U-notched and V-notched specimens tested at −60 °C (Gómez and Elices 2004). The specimen geometry comprised the notch opening angles $2\alpha = 0$, 90, 120 and 150°, the notch depths 5 and

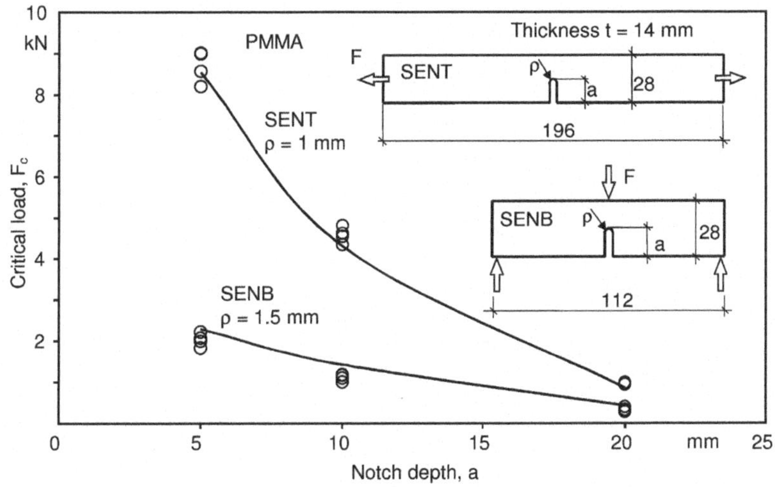

Fig. 3.31 Critical load F_c at U-notched PMMA specimens dependent on the notch depth a; analytical expression (solid line curves) compared with test results (point symbols); (Lazzarin and Berto 2005[1]; Gómez et al. 2000; Gómez and Elices 2003[1], 2003[2])

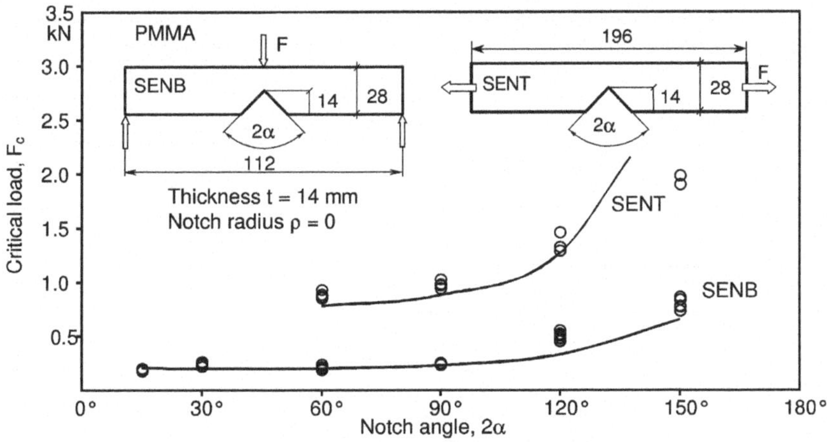

Fig. 3.32 Critical load F_c at U-notched PMMA specimens dependent on notch opening angle 2α; analytical expression (solid line curves) compared with test results (point symbols); (Lazzarin and Berto 2005[1]; Gómez et al. 2000; Gómez and Elices 2003[1], 2003[2])

14 mm and the notch radii $\rho = 0.05$–4.0 mm. Here, also the unnotched tensile specimen exhibited a fully brittle behaviour (no nonlinearity in the σ–ε curve). The critical load F_c is plotted dependent on the notch radius ρ for the SENB specimen in Fig. 3.33 and for the DENT and SENT specimens in Fig. 3.34. In both cases, different notch depths a are considered. Once more, correspondence between analytical and experimental results is proven. Finally, the critical SED $\overline{W}_{1c}$ is

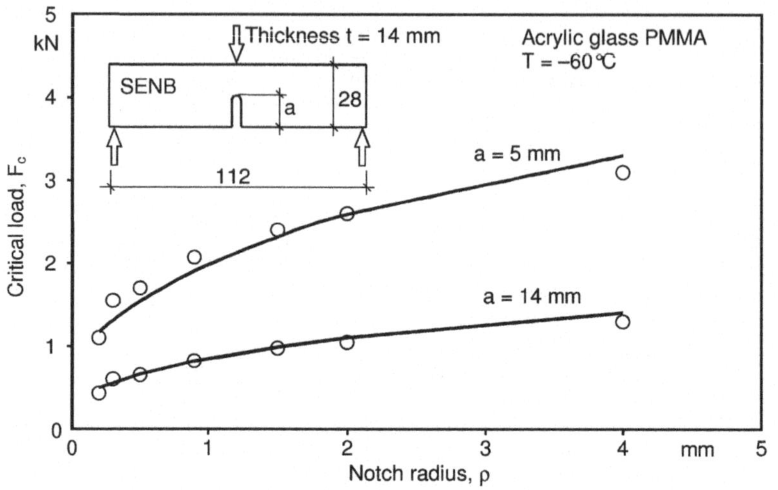

Fig. 3.33 Critical load F_c at U-notched SENB specimens (PMMA, $T = -60°$) dependent on the notch radius ρ for two notch depths a; analytical expression (solid line curves) compared with test results (point symbols); (Lazzarin and Berto 2005[(1)]; Gómez and Elices 2004)

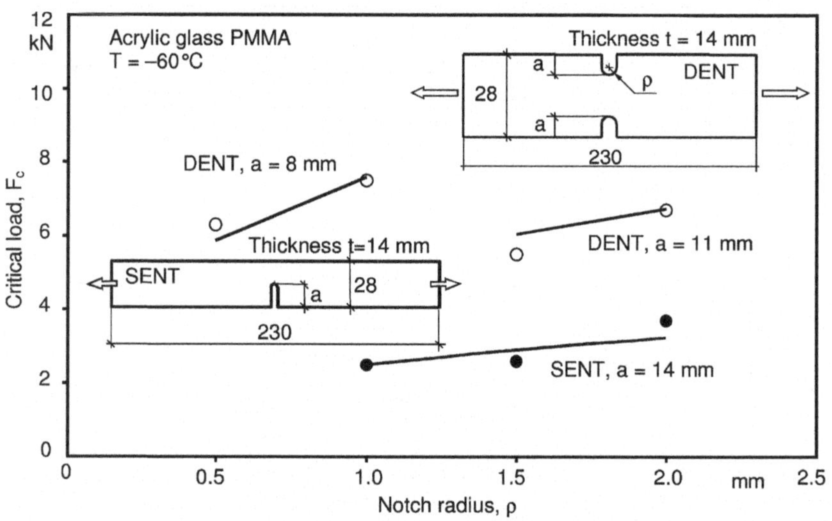

Fig. 3.34 Critical load F_c at U-notched DENT and SENT specimens (PMMA, $T = -60°$) dependent on the notch radius ρ for different notch depths a; analytical expression (solid line curves) compared with test results (point symbols); (Lazzarin and Berto 2005[(1)]; Gómez and Elices 2004)

plotted over the critical load F_c for the different specimen geometries and loading conditions, Fig. 3.35. The point symbols refer to the experimental data, the bold solid line denotes the hypothesis of a constant $\overline{W}_c$ value behind the analytical

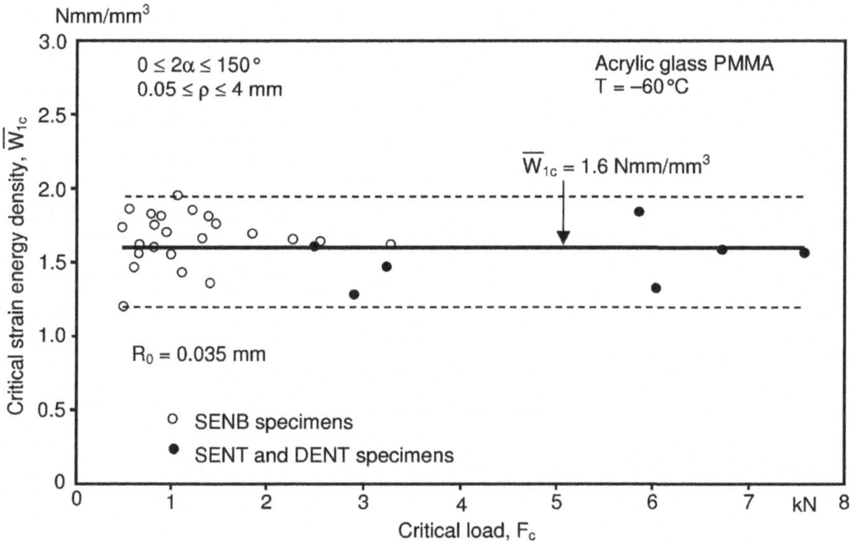

Fig. 3.35 Critical SED $\overline{W}_{1c}$ for all specimens and geometries (PMMA, $T = -60°$) dependent on the critical load F_c; analytical hypothesis (bold solid line) compared with test results (point symbols); (Lazzarin and Berto 2005[(1)]; Gómez and Elices 2004)

solution. The scatter range is relatively small, similar to that in Fig. 3.7, i.e. there is no difference between the ranges for pointed and blunted notches.

A third set of experimental data refers to U-notched SENB specimens at −60 °C subjected to three point bending loads, producing mixed mode conditions at the U-notch root by an eccentric loading and support arrangement, Fig. 3.36 (Gómez et al. 2007). The ratio of bending-tensile to in-plane shear load in the net cross-section ahead of the notch is adjusted by the varying eccentricity b of the applied load. The influence of the notch radius $\rho = 0$–4.0 mm on the critical load F_c is investigated, with the notch depth a kept constant. The experimental results are compared with the analytical results based on fine-meshed FE models.

The blunt V-notch solution for $2\alpha = 0$ does not refer to U-shaped notches, where the radius of notch curvature ρ_c remains constant, but to parabolic notches, where the radius of notch curvature ρ_c increases progressively when moving away from the bisector direction $\theta = 0$:

$$\rho_c = \frac{\rho}{(\cos\theta/2)^3} \tag{3.68}$$

According to this relationship, $\rho_c = 1.11\rho$, 1.54ρ and 2.83ρ for $\theta = 30$, 60 and 90°. Therefore, the blunt notch solution cannot be directly applied to U-shaped notches in cases of mixed mode loading, whereas mode 1 tensile loading ($\theta = 0$, $\rho_c = \rho$) is well represented. The procedure described in the paragraph after Eq. (3.67) is applied for calculating the critical load F_c under mixed mode loading

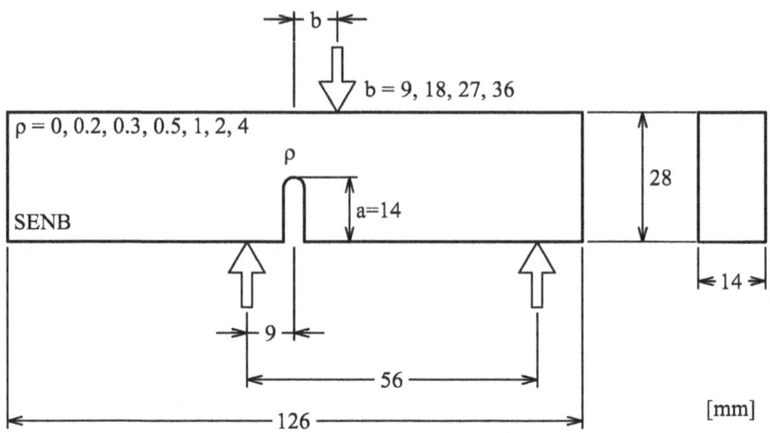

Fig. 3.36 U-notched SENB specimens (PMMA, $T = -60°$) with eccentric loading and support arrangement producing mixed mode loading conditions at the U-notch root; (Gómez et al. 2007)

conditions based on $\overline{W}_c$ and R_0, or vice versa for determining the critical SED $\overline{W}_c$ from the measured failure load F_c. The measured values of F_c deviate between −7 and +13 % (with some freak values up to 25 %) from the analytical or numerical results. A similar scatter range applies to the measured compared with the calculated initial crack angle. The values of $\overline{W}_c$ found analytically based on F_c and σ_{max} from the FE model and those values determined from averaging the SED in the crescent-shaped control volume from the FE model directly are well in agreement. The scatter range of the $\overline{W}_c$ values is substantially smaller than that in the comparable Fig. 3.8.

The conclusion from the above is that the brittle fracture strength of blunt V-notches is well described by the SED concept with crescent-shaped control volumes. Evaluating data from static fracture testing of specimens with pointed and rounded notches made of PMMA, ceramics, duraluminium and other brittle materials, a uniform horizontal scattes band of $(\overline{W}/W_0)^{1/2}$over ρ/R_0 is derived, where W_0 is the critical SED in the unnotched specimen (Berto and Lazzarin 2009).

3.5 Relation of the Local SED Concept to Comparable Concepts

3.5.1 Survey of Section Contents

The relation of the local SED concept to comparable other concepts is investigated. First the relation to the Kitagawa and Atzori diagrams is demonstrated inclusive of Taylor's comparison with notch-mechanical concepts. Then, the relation to the Neuber concept of fictitious notch rounding applied to welded joints is presented. Finally, the relation to the J-integral is derived with focus on the ratio of elastic strain energy within the control volume to the J-integral dependent on the control radius.

The expositions are primarily based on the analytical developments carried out by Atzori, Lazzarin and Berto. A list of symbols and a list of references are available at the end of Chap. 3.

3.5.2 Relation to the Kitagawa and Atzori Diagrams

Kitagawa's Original Diagram

The threshold value of cyclic stress (stress range $\Delta\sigma$ or stress amplitude σ_a), at which growth of short cracks is initiated, may be plotted over the crack length a (Kitagawa and Takahashi 1976). The diagram presents a curve, which runs horizontally first and then decreases steadily, Fig. 3.37. Below this limit curve, non-propagating cracks are observed, if there are any, above the curve fracture occurs after a finite number of cycles.

The horizontal curve section characterises the endurance limit of the material with accordingly negligible short cracks, corresponding to the microstructurally described short cracks in the diagram. The endurance limit $\Delta\sigma_E = 2\sigma_A$ can be approximately equated with two times the value of the cyclic yield limit, $2\sigma'_Y$. The following increasing slope of the curve comprises the range of short cracks described according to continuum-mechanics. Finally, the constant further slope characterising the threshold stress intensity factor ΔK_0 of long cracks is reached, which is valid below $2\sigma'_Y/3$ in linear-elastic fracture mechanics according to the limit condition on the extension of the plastic zone at the crack tip. Above the considered limit curve, crack propagation occurs, which can be described differentiating between the microstructural, elastic-plastic and linear-elastic ranges.

The limit curve of crack propagation in the Kitagawa diagram is asymptotically defined by the endurance limit $\Delta\sigma_E$ of the material and the threshold SIF ΔK_0 of long cracks within this material. The crack closure effect observed with long cracks (in contrast to the observation with short cracks) enlarges the threshold SIF ΔK_0 relative to its effective value, thus shifting the inclined part of the limit curve to higher values.

A compilation of test results from the literature (up to 1980), related to the crack length dependency of the threshold cyclic stress, is set into comparison with theoretical predictions (Smith 1977; Tanaka et al. 1981) in Fig. 3.38. Actual surface crack sizes are converted to an equivalent through thickness crack length a_{eq} (with identical SIF) and related to the 'intrinsic crack length' defined by El Haddad et al., Eq. (3.69). The threshold cyclic stresses $\Delta\sigma_0$ are normalised by the endurance limit $\Delta\sigma_E$.

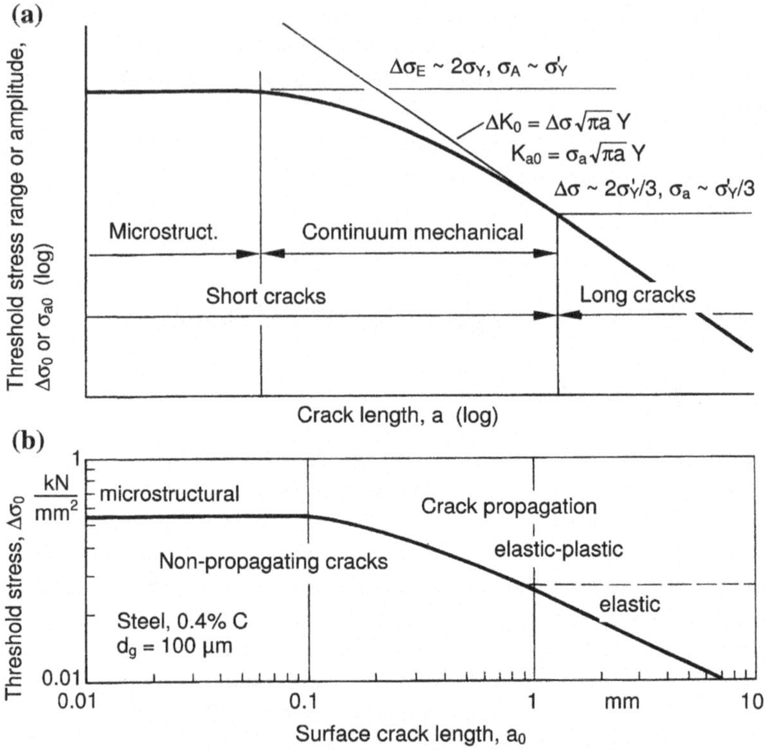

Fig. 3.37 Threshold cyclic stress range or amplitude dependent on crack length (Kitagawa diagram); definition of 'short' and 'long' cracks (**a**) and kinds of crack propagation (**b**); grain size d_g; (Radaj and Vormwald 2007 following Brown 1986)

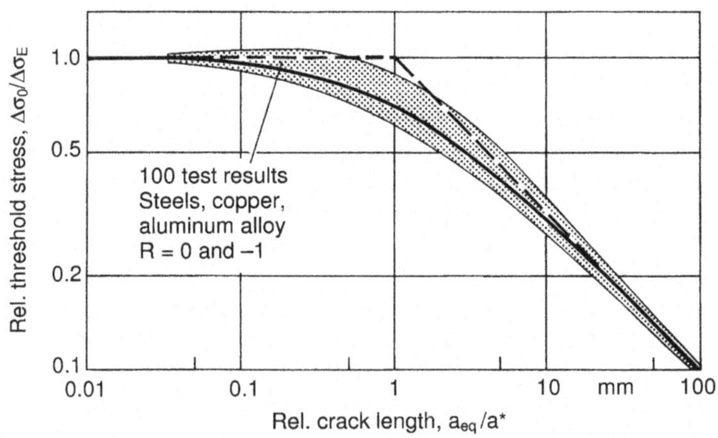

Fig. 3.38 Threshold cyclic stress range dependent on relative crack length; slip band blocking model according to Tanaka (full line curve), approach of Smith and El Haddad (broken line curve) and scatter band of test results (half tone band); equivalent through-thickness crack length a_{eq} and material-dependent length parameter a^* according to Eq. (3.69); (Radaj and Vormwald 2007, following Tanaka et al. 1981)

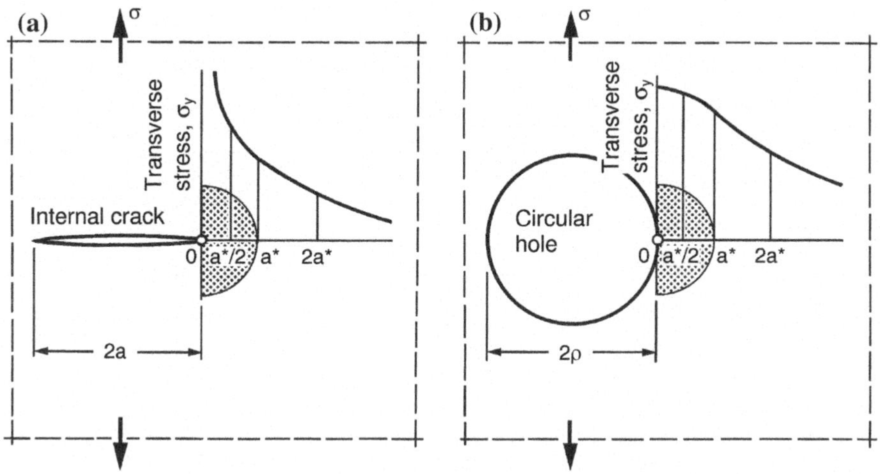

Fig. 3.39 Internal crack (**a**) and circular hole (**b**) subjected to transverse tension with plotted ligament transverse stress; evaluation of this stress at $a^*/2$, averaged over the length $2a^*$ and averaged over the semicircle area with radius a^*; microstructural material parameter a^* for high-cycle fatigue; (Taylor 1999)

Taylor's Comparison with Notch-Mechanical Concepts

The dependency of the nominal endurance limit on short and longer crack length (limit curve of crack propagation in the Kitagawa diagram) or on the comparable diameter of a circular hole (slightly below the above limit curve) has been uniformly presented according to different notch-mechanical support concepts and compared with test results from the literature (Taylor 1999; Taylor and Wang 2000). Using the example of a thin plate with a short crack or a small hole, subjected to tensile loading, the following concepts are contrasted, Fig. 3.39:

- Concept of critical stress at the short distance $a^*/2$ from the crack tip or notch root.
- Concept of critical average stress over the short distance $2a^*$ ahead of the crack tip or notch root.
- Concept of critical average stress over the small semicircular area with radius a^* ahead of the crack tip or notch root.

Additionally, the link to the concept of El Haddad, Smith and Topper has been established by the material-characteristic length parameter a^* ('intrinsic crack length') defined on the basis of the endurance limit $\Delta\sigma_E$ of the material and the threshold SIF ΔK_0 of long cracks in this material (El Haddad et al. 1979):

$$a^* = \frac{1}{\pi}\left(\frac{\Delta K_0}{\Delta \sigma_E}\right)^2 \qquad (3.69)$$

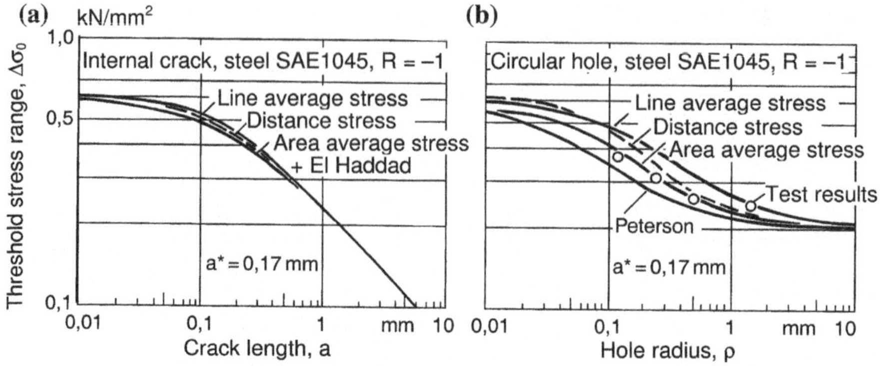

Fig. 3.40 Calculated high-cycle fatigue strength (threshold cyclic stress range $\Delta\sigma_0$) as a function of the crack semilength a (**a**) or the hole radius ρ (**b**) for carbon steel SAE 1045; different theoretical concepts and comparison with some test results for the hole; (Taylor 1999)

The crack length a^* denotes the intersection point of the two asymptotes in the Kitagawa diagram, compare Fig. 3.38. It can be interpreted as a material parameter characterising the defect sensitivity.

Also, the short distance stress concept of Peterson is included. It has to be noted, that Neuber's microstructural support length ρ^* corresponds to $2a^*$ in Taylor's investigation, and Peterson's microstructural distance a^* to $a^*/2$ in Taylor's evaluation. An inconsistent element is the fact, that the values of ρ^* given by Neuber and the values of a^* given by Peterson, do not reproduce the necessary condition $\rho^* \approx 4a^*$; instead, $\rho^* < a^*$ is mostly found (Radaj and Vormwald 2007).

The result of a mainly theoretical comparative investigation for a tensile loaded plate with internal crack or hole, made of carbon steel SAE 1045 ($\sigma_Y = 470$ N/mm^2, $\sigma_U = 745$ N/mm^2) is shown in Fig. 3.40. For the internal crack, the results according to the different concepts are largely identical, but for the circular hole, certain deviations occur. The solutions of El Haddad et al. and of Peterson can be considered as experimentally verified. The test result for the circular hole are taken from the literature (DuQuesnay et al. 1986). Obviously, the calculated endurance limit with circular hole may be smaller than the endurance limit with an internal crack of comparable size. This is an indication, that cracks initiated at the hole edge may remain arrested.

The limit curves for the internal hole converge asymptotically to a bottom value for large hole radii (hole in the infinite plate), which corresponds to the theoretical stress concentration factor $K_t = 3.0$. The microstructural support is negligible in this case.

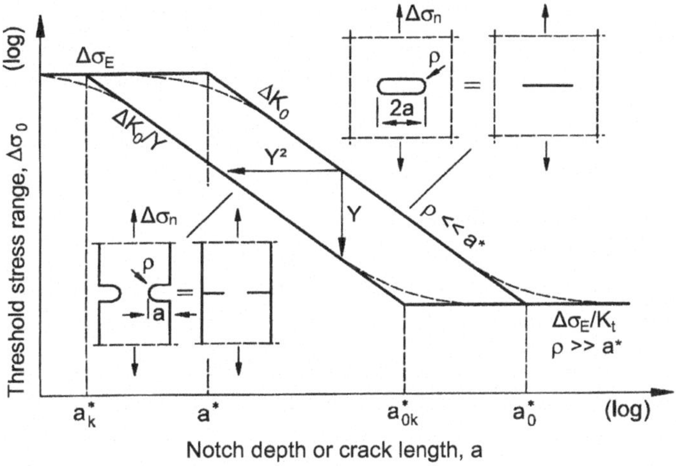

Fig. 3.41 Schematic Kitagawa diagram extended in respect of notch mechanics; upper curve corresponding to reference model, lower curve corresponding to structural member; (following Atzori et al. 2003)

Atzori's Notch-Mechanical Extension of the Kitagawa Diagram

Taylor's conceptual link to notch mechanics has been further elaborated by Atzori and his team (Atzori and Lazzarin 2001; Atzori et al. 2001, 2003). The question is answered, under what conditions a notch (or notch-like defect) can be analysed as a crack, which is a precondition for achieving identical endurance limit values. The notch can be treated as a crack (or as a re-entrant corner in cases of non-zero notch opening angles), provided the length parameter a^* according to Eq. (3.69) is substantially larger than the notch radius ρ. The influence of the dimensional parameters of the structural member (with exclusion of the notch radius ρ) enters into the relevant SIF values. On the other hand, the theoretical SCF K_t is fully effective, provided the length parameter a^* is small compared with the notch radius ρ. The influence of the dimensional parameters (with inclusion of the notch radius ρ) is expressed by the theoretical SCF K_t. The equivalent crack model is not anymore valid in this range.

The authors above have quantified and illustrated the effects just mentioned in a Kitagawa diagram with notch-mechanical extensions, Fig. 3.41. The extended diagram has been substantiated with numerous test results from the literature for structural notches and notch-like defects. The curve sequence marked $\Delta\sigma_E$, ΔK_0, $\Delta\sigma_E/K_t$ for the tensile loaded infinite plate with internal notch or crack (reference model) is contrasted with the curve sequence marked $\Delta\sigma_E$, $\Delta K_0/Y$, $\Delta\sigma_E/K_t$ for the tensile loaded plate specimen with external notches (representing the structural member). The distance between the lines ΔK_0 and $\Delta K_0/Y$ corresponds to the geometry factor Y (in vertical direction) and Y^2 (in horizontal direction. The geometry factor Y is derived from $K_I = \sigma_n(\pi a)^{1/2} Y$ with the nominal stress σ_n in the gross cross-section. In the considered case, $Y > 1.0$, but for semielliptical surface cracks $Y < 1.0$. The derivations have been carried out for two-dimensional models only. Rounded curve transitions (broken lines) correspond to the conditions in reality.

The corner points of the curves are characterised by the following special crack lengths or notch depths (compare Fig. 3.41):

- Intrinsic crack length a^* of the material, Eq. (3.69).
- Intrinsic crack length $a_k^* = a^*/Y^2$ of the specimen.
- Crack length $a_0^* = K_t^2 a^*$ at the intersection of lines ΔK_0 and $\Delta\sigma_E/K_t$.
- Crack length $a_{0k}^* = a_0^*/Y^2$ at the intersection of lines $\Delta K_0/Y$ and $\Delta\sigma_E/K_t$.

The crack lengths a^* and a_k^* are correlated with the defect sensitivity, the crack lengths a_0^* and a_{0k}^*, on the other hand, with the notch sensitivity. One may term this: 'Two sides of the same medal'. Ample experimental data related to different materials, geometries and loading conditions confirm the validity of Atzori's diagram (Atzori et al. 2003).

Based on the diagram above, the endurance limit of the specimen or structural member can be predicted with reference to three ranges of notch depth or crack length:

- Range $a \le a_k^*$: The endurance limit $\Delta\sigma_E$ is relevant, the notch is negligible.
- Range $a_k^* \le a \le a_{0k}^*$: The threshold value $\Delta K_0/Y$ is relevant, the notch depth has to be taken into account.
- Range $a \ge a_{0k}^*$: The endurance limit $\Delta\sigma_E/K_t$ is relevant, the notch severity a/ρ being decisive.

Lazzarin's SED-Based Version of the Kitagawa Diagram

A SED-based version of the original Kitagawa diagram has been presented. Once more, the tensile loaded infinite plate with an internal crack is analysed. Two different control radii for determining the average SED are considered, $R_0 = 0.02$ and 0.2 mm, corresponding to two different materials. Linear-elastic conditions are assumed.

For generating a Kitagawa-like diagram, the inverse average SED $1/\overline{W}_1$ in the control volume is introduced. Its value is decreasing with increasing crack length under the condition of a constant control volume radius R_0 and a constant tensile load σ_0. Therefore, it characterises the decrease of strength relative to the specimen without a crack, provided the strength is described by a constant critical SED value, which is the basis of the local SED approach.

The result of a numerical investigation based on the FE method, inverse average SED plotted over crack length, is shown in Fig. 3.42. When the crack length is very small, the inverse SED of the specimen without a crack is relevant. It is expressed by $1/\overline{W}_1 = 2E/(1-\nu^2)\sigma_0^2$ under plane strain conditions independent

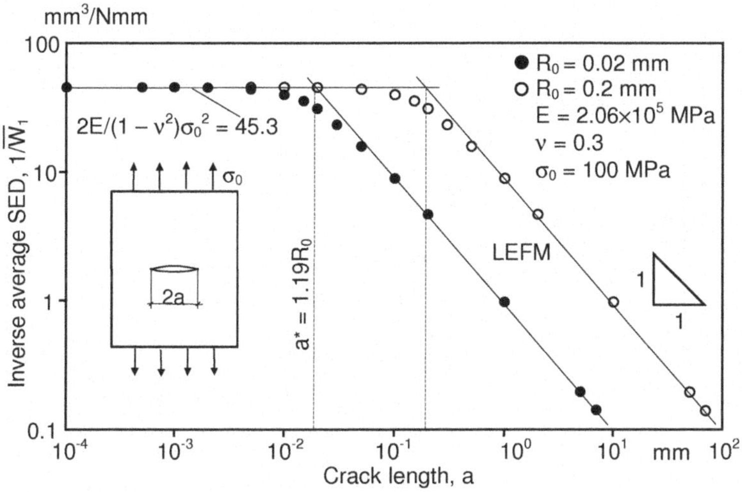

Fig. 3.42 Kitagawa-analogue crack-mechanical diagram of inverse average local SED; internal crack in tensile-loaded infinite plate under plane strain conditions; smaller and larger control volume of radius R_0; point symbols for FE analysis results; LEFM linear-elastic fracture mechanics; (Lazzarin and Berto 2005 [(2)]

of the control volume radius. When the crack gets larger while the loading σ_0 remains constant, then the inverse SED curve will linearly drop with the slope exponent $\kappa = -1$. The drop starts theoretically at $a^* = 1.19R_0$, compare Eq. (3.69) with Eq. (3.43) for $2\alpha = 0, f_1 = 1.936$ and $\lambda_1 = 0.5$. The two straight lines are in agreement with linear-elastic fracture mechanics (LEFM) considering the average SED $\overline{W}_1 = f(K_I/R_0)$.

The above numerical investigation has been extended in respect of the bottom value $1/\overline{W}_{10}$ given by notch mechanics in analogy to the Taylor and Atzori diagrams. The geometry of the analysed two-sided U-notch is characterised by $a/\rho = 5$ with the crack semilength a being varied. The reference curves gained for two values of R_0 (i.e. two materials) are plotted in Fig. 3.43. The curve for a specimen or structural member may be derived therefrom using the geometry factor Y, as shown in Fig. 3.41.

3.5.3 Relation to the Neuber Concept of Fictitious Notch Rounding

The fictitious notch rounding (FNR) concept and its derivative, the reference notch concept, are widely used for assessing the fatigue strength of welded joints (Radaj et al. 2006). Endurable values of the notch stress range are given in the IIW design

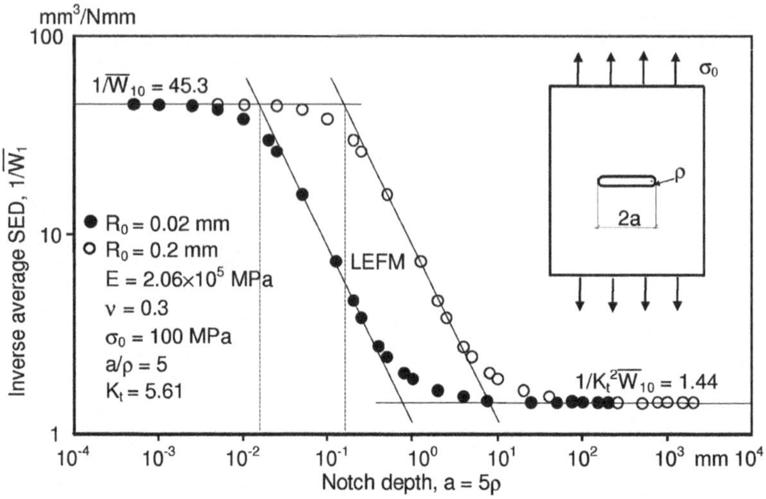

Fig. 3.43 Kitagawa-analogue notch-mechanical diagram of inverse local average SED; internal U-notch in tensile-loaded infinite plate under plane strain conditions; smaller and larger control volume of radius R_0; point symbols for FE analysis results; LEFM: linear-elastic fracture mechanics

recommendations (Hobbacher 2009). The original FNR concept proposed by Neuber refers to non-welded sharply notched specimens (Neuber 1968).

According to the reference notch concept for welded joints, the weld toe and weld root notches are rounded by the reference radius $\rho_r = 1$ mm, which is derived from the fictitious notch radius ρ_f according to Neuber's expression:

$$\rho_f = \rho + s\rho^* \tag{3.70}$$

with the real radius ρ, the microstructural support length ρ^* and the support factor s representing the multiaxiality conditions. The reference radius $\rho_r = 1.0$ results from Eq. (3.70) with $\rho = 0$ (worst case), $\rho^* = 0.4$ mm (cast iron) and $s = 2.5$ (representative conventional value). The parameter ρ^* in the original Neuber concept is given for different material groups (e.g. steels or aluminium alloys) dependent on the yield limit σ_Y (ultimate strength σ_U would have been a better founded choice). This expresses the dependency of ρ^* on the fatigue strength $\Delta\sigma_E$ of the material, as far as σ_Y or σ_U and $\Delta\sigma_E$ are correlated.

A theoretical fatigue notch factor $K_{f,N}$ (subscript N for Neuber) may be derived considering a cross-sectional model of the welded joint with the notches really rounded by the fictitious radius:

$$K_{f,N} = \frac{\Delta\bar{\sigma}_{k\,max}}{\Delta\sigma_n} \tag{3.71}$$

where $\Delta\overline{\sigma}_{k\,max}$ is the maximum notch stress range after notch rounding and $\Delta\sigma_n$ the nominal stress range. The fatigue notch factor allows to assess the fatigue strengths of basic welded joint types inclusive the size effect in relation to one another. For an assessment in absolute terms, the fatigue strength $\Delta\sigma_E$ of the as-welded material in unnotched specimens is equated with $\Delta\overline{\sigma}_{k\,max}$, so that the endurable nominal stress range is $\Delta\sigma_{nE} = \Delta\sigma_E/K_{f,N}$. The problem is the value $\Delta\sigma_E$, which cannot be generally stated.

The local SED concept can be set into comparison to the FNR or reference notch concept based on the following relationships. The control volume radius according to the Eq. (3.43) is $R_0 = 0.28$ mm, determined with $\Delta K_{1E} = 211$ MPa mm$^{0.326}$ and $\Delta\sigma_E = 155$ MPa ($P_f = 50$ %, $N_E = 5 \times 10^6$ cycles). In order to make a direct comparison of the ΔK_f values gained with the two concepts easier, the control volume radius $R_0 = 0.15$ mm is introduced which corresponds to $\Delta\sigma_E = 190$ MPa. The same fatigue strength is predicted based on $R_0 = 0.28$ mm combined with $\Delta\sigma_E = 155$ MPa and based on $R_0 = 0.15$ mm combined with $\Delta\sigma_E = 190$ MPa. A factor of 1.226 is acting both on the $K_{f,N}$ and $\Delta\sigma_E$ values.

A theoretical fatigue notch factor $K_{f,W}$ may be derived based on the local average SED $\overline{W}$ over the control volume in the following form:

$$K_{f,W}(R_0) = \sqrt{\frac{\Delta\overline{W}(R_0)}{\Delta W_n}} \tag{3.72}$$

where ΔW_n is the nominal SED in the remote cross-section under plane strain conditions. The fatigue notch factor $K_{f,W}$ depends via R_0 on the fatigue strength $\Delta\sigma_E$ relative to ΔK_{1E}. In contrast, $K_{f,N}$ is independent of $\Delta\sigma_E$.

When assuming a control volume radius $R_0^* > R_0$, e.g. $R_0^* = 1$ mm, with the aim to reduce the number of degrees of freedom in the FE model, the local average SED obtained directly from the FE model can be reconverted to the correct control volume by using the following expression:

$$\overline{W}(R_0 = 0.15\,\text{mm}) = \overline{W}(R_0^* = 1\,\text{mm}) \times \left(\frac{R_0}{R_0^*}\right)^{2(\lambda_1 - 1)} \tag{3.73}$$

A precondition for acceptable accuracy of such a procedure is that the radius R_0^* remains small in relation to other geometrical parameters such as plate or throat thickness. The influence of the *T*-stress and of the higher order stress terms must also remain negligible.

A comparison of the theoretical fatigue notch factors has been performed for the cross-sectional models of welded joints shown in Fig. 3.44 for the FNR or reference notch approach (toe and root notches rounded by $\rho_r = 1$ mm) in contrast to the SED approach, where the notches remain pointed. The models comprise tensile and bending loaded cruciform joints with the slits transverse and parallel to the external loading direction (load-carrying and non-load-carrying fillet welds) as well as one bending loaded T-joint. The tensile force results from prescribed boundary displacements. The corresponding support is achieved by suppressed

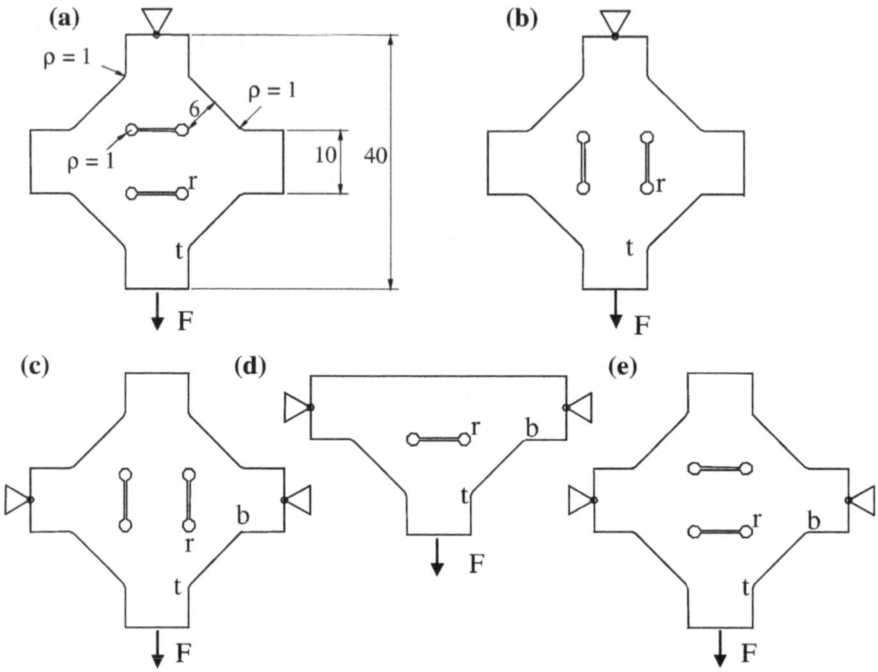

Fig. 3.44 Geometries of the welded joint cross-sectional models analysed in respect of their theoretical fatigue notch factors, first according to the FNR concept with (real) reference notches of $\rho = 1$ mm (boundary element method) and then according to the local SED concept with pointed V-notches and slits (FE method); tensile toe notch at t, bending toe notch at b and root notch at r; (Lazzarin et al. 2007)

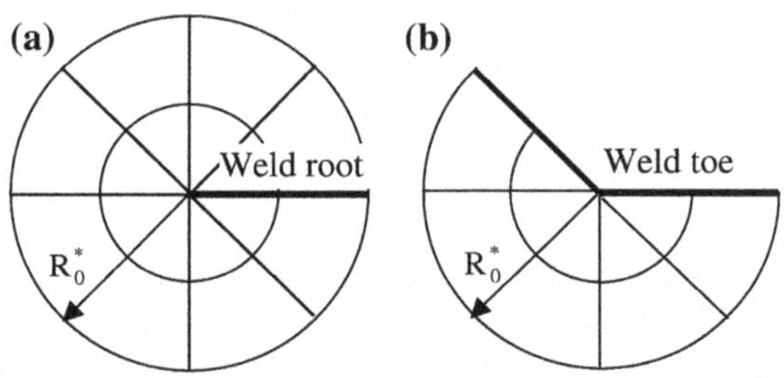

Fig. 3.45 Coarse FE mesh in the control volume at the weld root (**a**) and the weld toe (**b**) used in one variant of the FE evaluations; radius of outer circle, $R_0^* = 1$ mm; (Lazzarin et al. 2007)

boundary displacements. The side supports act without creating a bending moment in the support (therefore the hinge symbol). The hinge support is chosen in order to produce identical bending moments in the toe point cross-section.

The fatigue notch factors $K_{\text{f,N}}$ according to the FNR or reference notch concept are found by a boundary element (BE) analysis with high accuracy. The factors $K_{\text{f},W}^{(\text{f})}$ are determined by FE-analysis with very fine meshes. On the other hand, the factors $K_{\text{f},W}^{(\text{c})}$ result from very coarse meshes, Fig. 3.45 ($R_0^* = 1$ mm), after reconversion to

Table 3.3 Comparison of SED-based fatigue notch factors $K_{\mathrm{f},W}^{(\mathrm{f})}$(fine mesh) and $K_{\mathrm{f},W}^{(\mathrm{c})}$ (coarse mesh) with FNR-based fatigue notch factors $K_{\mathrm{f,N}}$ (BE analysis Values from coarse mesh ($R_0^* =$ 1 mm) reconverted to $R_0 = 0.15$ mm; compare Fig. 3.44; (Lazzarin et al. 2007)

Model	Weld toe, point t ('tensile')			Weld toe, point b ('bending')			Weld root, point r	
	$K_{\mathrm{f},W}^{(\mathrm{f})}$	$K_{\mathrm{f},W}^{(\mathrm{c})}$	$K_{\mathrm{f,N}}$	$K_{\mathrm{f},W}^{(\mathrm{f})}$	$K_{\mathrm{f},W}^{(\mathrm{c})}$	$K_{\mathrm{f,N}}$	$K_{\mathrm{f},W}^{(\mathrm{c})}$	$K_{\mathrm{f,N}}$
a	3.17	3.13	3.30	–	–	–	3.95	4.03
b	2.41	2.40	2.45	–	–	–	1.65	1.77
c	2.66	2.67	2.68	4.42	4.31	4.51	1.66	1.56
d	3.12	3.15	3.14	5.61	5.60	5.79	2.16	2.33
e	3.08	3.09	3.13	4.40	4.34	4.50	2.50	2.45

$R_0 = 0.15$ mm according to Eq. (3.73). The fatigue notch factors above for the five welded joint models at the toe and root notch points are listed in Table 3.3.

The results for the weld toe are first considered. Fine and coarse meshes give fatigue notch factors corresponding to one another within 2.5 %. Therefore, rather coarse meshes are suitable for local average SED evaluations. These meshes are not suitable for NSIF evaluations. Very satisfactory is also the correspondence of the $K_{\mathrm{f},W}$ and $K_{\mathrm{f,N}}$ values, the maximum difference being 5 %. Then, the data related to the weld root are considered. The differences between $K_{\mathrm{f},W}$ and $K_{\mathrm{f,N}}$ are only slightly larger than 5 %. Therefore, the two methods, FNR or reference notch and SED, give closely related predictions, provided $\Delta\sigma_{\mathrm{E}} = 190$ MPa ($P_{\mathrm{f}} = 50$ %, $N_{\mathrm{E}} = 5 \times 10^6$ cycles) is used for determining R_0.

In conclusion, the derived theoretical fatigue notch factors $K_{\mathrm{f,N}}$ and $K_{\mathrm{f},W}$ are well suited for the assessment of the fatigue strength of welded joints in relative terms, but they cannot establish an absolute assessment. The following uncertainties or inconsistencies occur in the latter case:

- The reference material and its fatigue strength are uncertain: parent or filler material, non-welded or as-welded, dependency on specimen size, contradictory strength data in codes and literature.
- The high-cycle fatigue strength at $N_{\mathrm{E}} = 2 \times 10^6$ compared with 5×10^6 cycles is uncertain: identical values for non-welded steels, decreasing values for non-welded aluminium alloys and for as-welded materials in general (Haibach 2003).
- The scatter range of the high-cycle fatigue data is uncertain: non-welded materials in unnotched specimens may have a substantially larger scatter range than sharply notched specimens in cases where the material and surface quality is not well defined.

In order to illustrate part of the uncertainties just mentioned, the reference fatigue strength $\Delta\sigma_{\mathrm{E}}$ for the common structural steels is presented from different sources, all data relating to $R = 0$, $P_{\mathrm{f}} = 50$ % and $N = 5 \times 10^6$ cycles:

- $\Delta\sigma_{\mathrm{E}} = 155$ MPa for flush-ground butt welds (Atzori and Dattoma 1983).

- $\Delta\sigma_E = 160$ MPa for non-welded structural members of steel (Eurocode 3 2005).
- $\Delta\sigma_E = 182$ MPa for non-welded (FAT160) structural steels (IIW, Hobbacher 2009).
- $\Delta\sigma_E = 190$ MPa for flush-ground butt welds (Lazzarin et al. 2007).
- $\Delta\sigma_E = 227$ MPa for weld notches (FAT225) of structural steels (IIW, Hobbacher 2009).
- $\Delta\sigma_E = 300$ MPa for mill-finished mild steel (Radaj 1990, 1998).
- $\Delta\sigma_E = 312$ MPa for weld notches of high-strength steel (Olivier et al. 1989, 1994).

The following conversion factors on the endurable stresses were applied where appropriate. They result from the scatter range index for welded joints, $T_\sigma = 1 : 1.5$ referring to the failure probability $P_f = 10\ \%$ compared with 90 %.

- Factor 1.25 on stress for $P_f = 50\ \%$ compared with $P_f = 10\ \%$.
- Factor 1.35 on stress for $P_f = 50\ \%$ compared with $P_f = 2.3\ \%$.

Considering the large differences in the $\Delta\sigma_E$ values, it is evident that these values are suitable reference parameters only in connection with the special approach, for which they are defined. The problem comes up with comparisons of different approaches on the basis of fatigue notch factors. The reference parameter in the fatigue notch factor should have the same value in the compared approaches.

3.5.4 Relation to the J-Integral

The J-integral according to Rice is a convenient parameter for characterising the local elastic-plastic fields at crack tips and sharp notches in respect of crack initiation and propagation (Rice 1968).

The linear version of the J-integral for crack tips is presented in Sect. 2.2.6, comprising the pure and mixed mode loading cases, just as the contribution of the T-stress. The J-integral referring to K_I, K_{II} and K_{III} does not depend on the selected integration path, but the component referring to T does.

The J-integral concept has been extended to pointed V-notches, at first under linear-elastic and then under elastic-plastic conditions (Lazzarin et al. 2002), Sect. 2.3.8. The J_V-integral of V-notches in linear-elastic materials subjected to mode 1 and 2 loading depends on the integration path and the notch opening angle, but can be made path-independent in the pure loading modes 1 and 2 by $J_L = J_V/R^{2\lambda_1-1}$ or $J_L = J_V/R^{2\lambda_2-1}$ where R is the radius of a circular integration path around the V-notch tip. The integrals J_V and J_L have also been defined for pure mode 3 loading with inclusion of elastic-plastic material behaviour (Lazzarin and Zappalorto 2008). A different line integral applicable to V-notches subjected to mode 3 loading has been proven to be path-independent, provided the notch flanks are free of loads (Quian and Hasebe 1997).

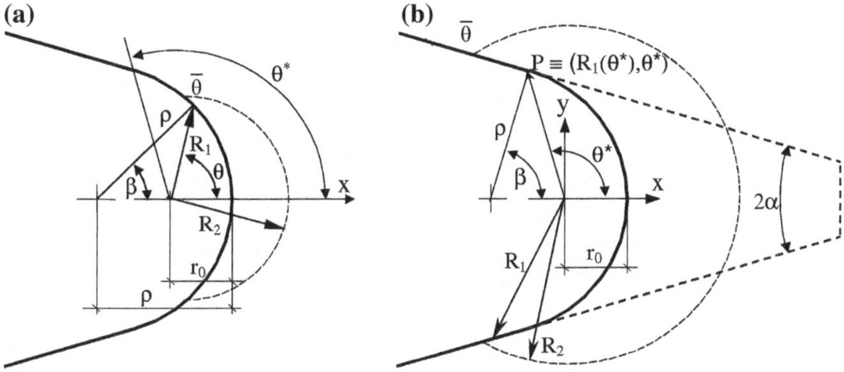

Fig. 3.46 Fundamental and analysis-related geometrical parameters of blunt V-notch consisting of a circular arc and rectilinear extensions (transition point P); crescent-shaped control areas with radius $R_0 = R_2 - r_0$ ending at the circular arc (**a**) or at the rectilinear extension (**b**); (Berto and Lazzarin 2007)

Expressions for the local strain energy at crack tips and pointed notches have been derived in Sect. 3.2.1, followed by the corresponding analysis for blunt notches in Sect. 3.4.1. For establishing a general link between the local strain energy and the *J*-integral, the following extension to U-notches and rounded V-notches under mode 1 loading has been carried out (Livieri 2003; Berto and Lazzarin 2007). Reference is made to the geometrical and stress field parameters for these notches shown in Fig. 3.25. Some additional parameters are introduced, which describe the geometrical situation at the rounded V-notch in combination with the crescent-shaped control area in more detail, Fig. 3.46. The notch contour consisting of a circular arc and rectilinear extensions is plotted in Fig. 3.46a for the crescent-shaped control area meeting the circular arc, and plotted in Fig. 3.46b for the control area ending on the rectilinear extensions. The control radius $R_0 = R_2 - r_0$ is chosen differently in the two plots. The radial distance R_1 in the polar coordinate system is a function of θ and ρ, both in the circular and rectilinear sections of the notch contour. The transition point between the two sections is described by the angle θ^*.

The *J*-integral is written in the following form:

$$J = \int_{\Gamma} \left(W\mathrm{d}y - \boldsymbol{T}\frac{\partial \boldsymbol{u}}{\partial s}\mathrm{d}s \right) \tag{3.74}$$

where Γ is the integration path, *W* the strain energy density, ***T*** the traction vector perpendicular to Γ and ***u*** the displacement vector.

Following Livieri, the path Γ is chosen along the load-free edge of the notch where the vector ***T*** is zero and the stresses are restricted to the σ_v component in the curvilinear *u–v* coordinate system. The expression for *J* can be written in the following form:

$$J = 2\int_0^{\overline{\theta}} W\,\mathrm{d}(R_1(\theta)\sin\theta) = 2\int_0^{\overline{\theta}} \frac{\sigma_v^2}{2E'}\left(\frac{\mathrm{d}R_1(\theta)}{\mathrm{d}\theta}\sin\theta + R_1(\theta)\cos\theta\right)\mathrm{d}\theta \quad (3.75)$$

where the end points of the crescent-shaped control area are characterised by the angle $\overline{\theta}$ and $E' = E$ for plane stress or $E' = E/(1 - \nu^2)$ for plane strain conditions. Linear-elastic material behaviour is presumed.

For tensile loaded hyperbolic notches, the stress σ_v at the notch edge is given by the polar stress components σ_r, σ_θ, $\tau_{r\theta}$ (Neuber 1958):

$$\sigma_v = \frac{1}{2}(\sigma_r + \sigma_\theta) - \frac{1}{2}(\sigma_r - \sigma_\theta)\cos\frac{2\theta}{q} + \tau_{r\theta}\sin\frac{2\theta}{q} \quad (3.76)$$

where $q = (2\pi - 2\alpha)/\pi$ characterises the notch opening angle 2α.

For hyperbolic notches, $R_1(\theta\ )$ has the following form (Livieri 2003):

$$R_1(\theta) = \rho\frac{q-1}{q}\frac{1}{\cos^q(\rho/q)} \quad (3.77)$$

For the parabolic notch, $2\alpha = 0$ and $q = 2$ have to be introduced, resulting in :

$$R_1(\theta) = \frac{\rho}{2}\frac{1}{\cos^2(\rho/2)} \quad (3.78)$$

In the analytical investigation referenced below (Berto and Lazzarin 2007), the elastic strain energy in the crescent-shaped control area is compared with the appertaining J-integral on an appropriate path. Since the J-integral is equal to zero when evaluated on a closed path within the elastic field, the contribution to J evaluated on a path described by $R_1(\theta)$ (the notch edge) is equal to the contribution evaluated on a path described by R_2 (the contour of the control volume). The analytical calculation is easier to perform along the notch edge, where only the stress component σ_v parallel to the edge is different from zero, whereas the traction vector $\boldsymbol{T}$ is equal to zero.

As mentioned by the authors above, J is dependent only on the two points of the notch edge selected to define the integration path, whereas the local strain energy E_1 within the control volume area A is dependent on the geometry of this area:

$$E_1 = \int_A W_1 \mathrm{d}A \quad (3.79)$$

The results for J and E_1/J presented below refer to the rounded V-notch inclusive of the U-notch (circular arc with rectilinear extensions) combined in Eq. (3.75) with the notch stress σ_v of the corresponding hyperbolic notch for identical values of θ. The stresses on the hyperbolic notch edge are assumed to act on the edge of the actual rounded V-notch for the purpose of solving the J-integral.

The level of the stresses σ_v in an actual V-notched specimen used for the validation of the theory is quantified by the NSIF $K_{1\rho}$ or the corresponding maximum notch stress σ_{max}. The parameter $K_{1\rho}^2$ is occurring in the expressions for the *J*-integral. These parameters must be determined by FE analysis. Any analytical curve shown in the diagrams below has this information as its basis.

Typical results of the quoted analytical investigation (Berto and Lazzarin 2007) referring to pointed V-notches, to U-notches and to rounded V-notches subjected to tensile loading are presented in the following.

For pointed V-notches, a linear relationship is found for the ratio E_1/J as a function of the control volume radius R_0, which depends on the notch opening angle 2α and Poisson's ratio:

$$\frac{E_1}{J} = k(2\alpha, v)R_0 \tag{3.80}$$

For the crack ($2\alpha = 0$) under plane strain conditions, the relationship reads:

$$\frac{E_1}{J} = \frac{5 - 8v}{8(1 - v)} R_0 \tag{3.81}$$

Under plane stress conditions, the corresponding relationship is:

$$\frac{E_1}{J} = \frac{5 - 3v}{8} R_0 \tag{3.82}$$

One result of the derivation for pointed V-notches, the ratio E_1/J as a function of the control volume radius R_0 for $v = 0.4$ (PMMA) for different notch opening angles 2α, analytical expression compared with FE analysis results, is plotted in Fig. 3.47.

For U-notches, the expression for the *J*-integral is based on the generalised NSIF $K_{1\rho}$ derived from the Creager–Paris stress field solution for the parabolic V-notch ($2\alpha = 0$). The NSIF $K_{1\rho}$ is correlated with the maximum notch stress σ_{max} (Irwin 1957, 1958; Glinka 1985):

$$K_{1\rho} = \frac{\sigma_{max}}{2}\sqrt{\pi\rho} \tag{3.83}$$

The maximum value of *J* is obtained for the path between the transition points between the semicircular arc and the rectilinear extensions of the notch edge:

$$J_{max} = 0.815\frac{(K_{1\rho})^2}{E'} \tag{3.84}$$

The factor 0.815 is close to the FE analysis result 0.785 (Livieri 2003), whereas a path along the parabolic notch edge according to Eq. (3.78) results in the factor 1.009.

The *J*-integral of a tensile loaded PMMA specimen with U-notches is plotted in Fig. 3.48 dependent on the control volume radius R_0. The analytical expression is

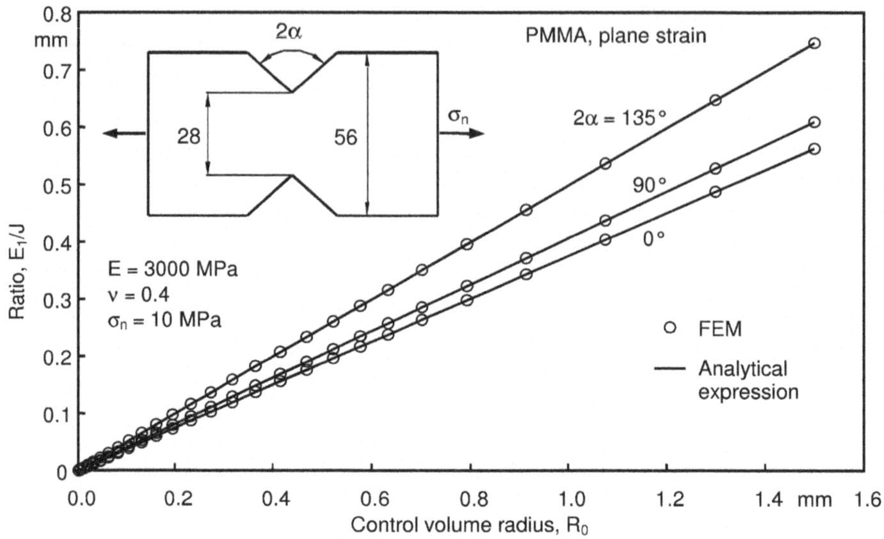

Fig. 3.47 Ratio of elastic strain energy E_1 in control volume to the *J*-integral (determined along the notch edge) dependent on control volume radius R_0 for different notch opening angles; tensile loaded pointed V-notch; K_1-based analytical compared with numerical solution; (Berto and Lazzarin 2007)

compared with FE analysis results. It is well visible from the plot that *J* increases to a maximum value related to the transition points and then remains constant.

For tensile loaded rounded V-notches, the expression for the *J*-integral is based on the generalised NSIF $K_{1\rho}$ derived for these notches (Lazzarin and Filippi 2006):

$$K_{1\rho} = \frac{\sigma_{\max}\sqrt{2\pi}r_0^{1-\lambda_1}}{1+\tilde{\omega}_1} \tag{3.85}$$

The *J*-integral including the arc and rectilinear extension contributions (if present) is given in the following form:

$$J = C\frac{K_{1\rho}^2}{E'}\rho^{2\lambda_1-1} \tag{3.86}$$

where the coefficient *C* depends on the notch opening angle 2α and on the path end angle $\overline{\theta}$, but not on Poisson's ratio ν.

The *J*-integral of a tensile loaded PMMA specimen with rounded V-notches ($2\alpha = 135°$) is plotted in Fig. 3.49 dependent on the control volume radius R_0. The analytical expression is compared with FE analysis results. The curve of the corresponding hyperbolic notch, path of *J* according to Eq. (3.77), is contrasted. Its values are substantially larger. A similar plot with the notch opening angles 2α being varied is shown in Fig. 3.50. It is well visible that only the curve for the U-notch ($2\alpha = 0$) exhibits a constant maximum value of *J* where the path follows the rectilinear notch edges.

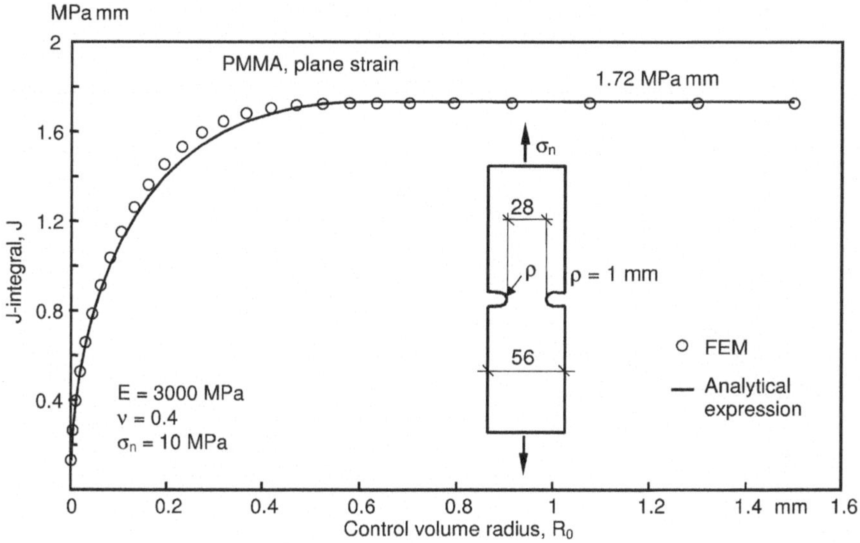

Fig. 3.48 *J*-integral (determined along the notch edge) of tensile loaded U-notch dependent on control volume radius R_0; $K_{1\rho}$-based analytical compared with numerical solution; (Berto and Lazzarin 2007)

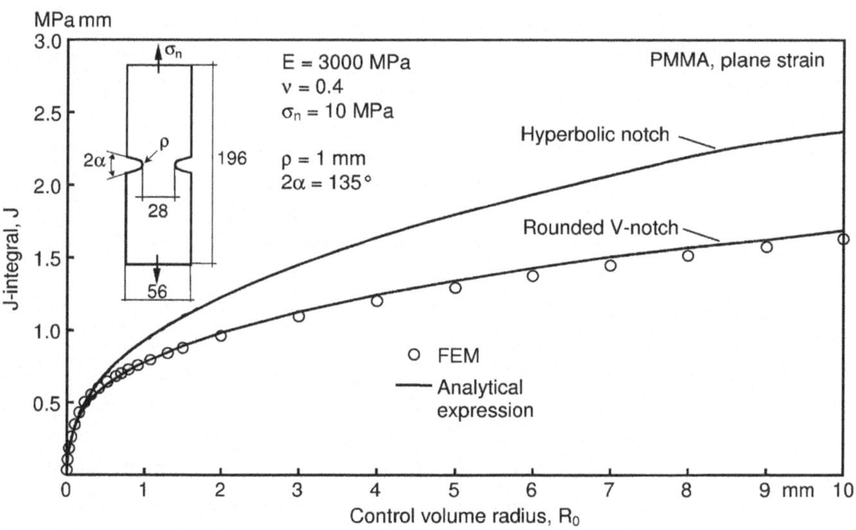

Fig. 3.49 *J*-integral (determined along the notch edge) of tensile loaded rounded V-notch ($\rho = 1$ mm, $2\alpha = 135°$) and corresponding hyperbolic notch dependent on control volume radius R_0; $K_{1\rho}$-based analytical compared with numerical solution; *J* given per unit of plate thickness; (Berto and Lazzarin 2007)

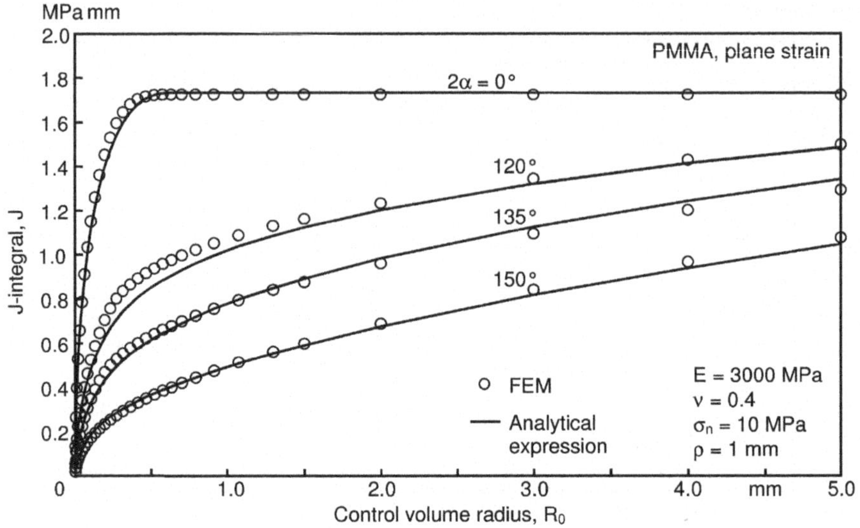

Fig. 3.50 *J*-integral (determined along the notch edge) of tensile loaded blunt V-notch ($\rho = 1$ mm, different values of 2α) dependent on control volume radius R_0; $K_{1\rho}$-based analytical compared with numerical solution; *J* given per unit of plate thickness; (Berto and Lazzarin 2007)

J-integral and local strain energy E_1 are now considered in comparison, at first separately for the U-notch PMMA specimen, Fig. 3.51. Obviously, the ratio E_1/J cannot be expected to be linearly dependent on the control volume radius R_0, as it is the case for pointed V-notches. The plot of E_1/J over R_0 for different notch radii, Fig. 3.52, shows that only slight oscillations occur in comparison to straight line curves.

Finally, the ratio E_1/J is plotted over the control volume radius R_0 for different notch opening angles 2α with the notch radius $\rho = 1$ mm remaining constant, Fig. 3.53, and for different notch radii ρ with the notch opening angle 2α remaining constant, Fig. 3.54. Once more, the oscillating deviations from a strictly linear behaviour are visible.

In conclusion, approximately linear relationships between the ratio E_1/J and the control volume radius R_0 are valid for tensile loaded rounded V-notches. These depend on the notch opening angle 2α, the notch radius ρ and the multiaxiality condition with an influence of Poisson's ratio ν. From the local strain energy E_1 in the control volume area A, the strength-relevant average SED $\overline{W}_1 = E_1/A$ can easily be determined, which should be independent of the geometrical parameters 2α and ρ. The influence of the material expresses itself in the critical value of $\overline{W}_1$ together with the control volume radius R_0. Critical values of J_V are applicable independently of 2α when $\rho = 0$.

Successful application of the J_V integral has been demonstrated using mainly PMMA specimens (besides duraluminium) with pointed V-notches under static loading (Livieri 2008). The application of the J_V-integral to welded joints under fatigue loading is reviewed in Sect. 2.3.8.

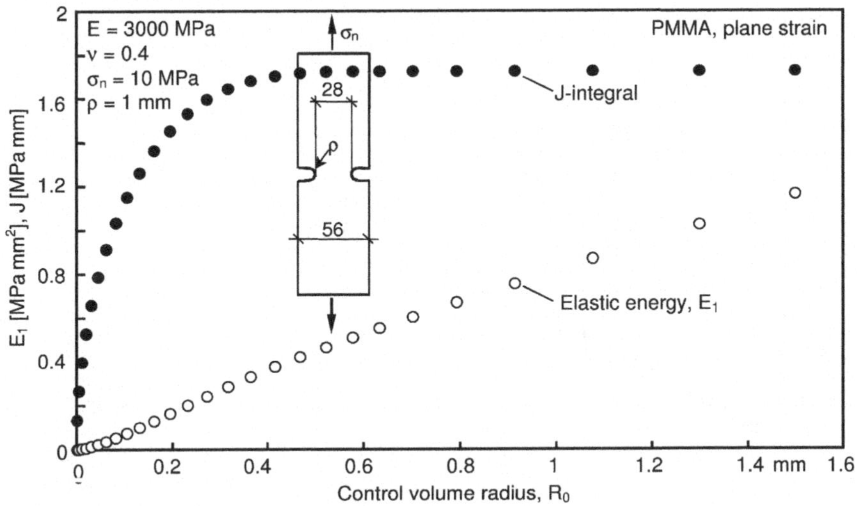

Fig. 3.51 J-integral (determined along the notch edge) and elastic strain energy E_1 in control volume for tensile loaded U-notch ($\rho = 1$ mm) dependent on control volume radius R_0; $K_{1\rho}$-based analytical solution; J and E_1 given per unit of plate thickness; (Berto and Lazzarin 2007)

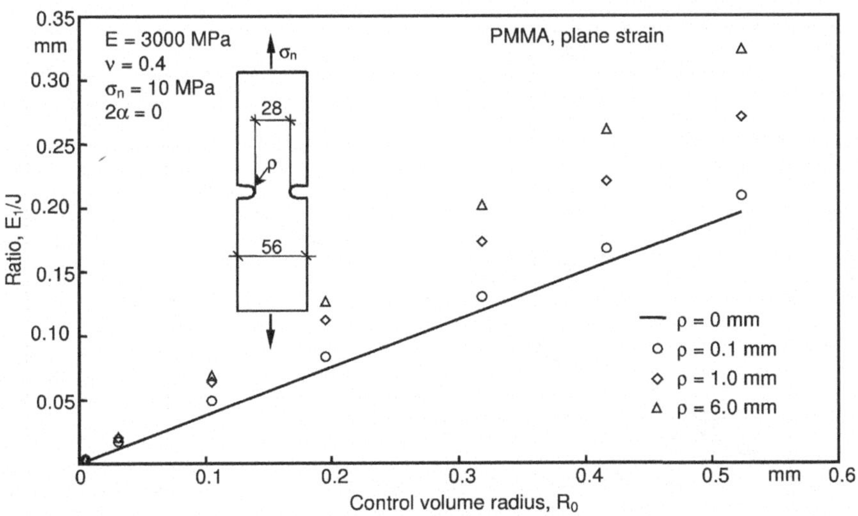

Fig. 3.52 Ratio of elastic strain energy E_1 in control volume to the J-integral (determined along the notch edge) dependent on control volume radius R_0; tensile loaded U-notch with different notch radii ρ; $K_{1\rho}$-based analytical solution; (Berto and Lazzarin 2007)

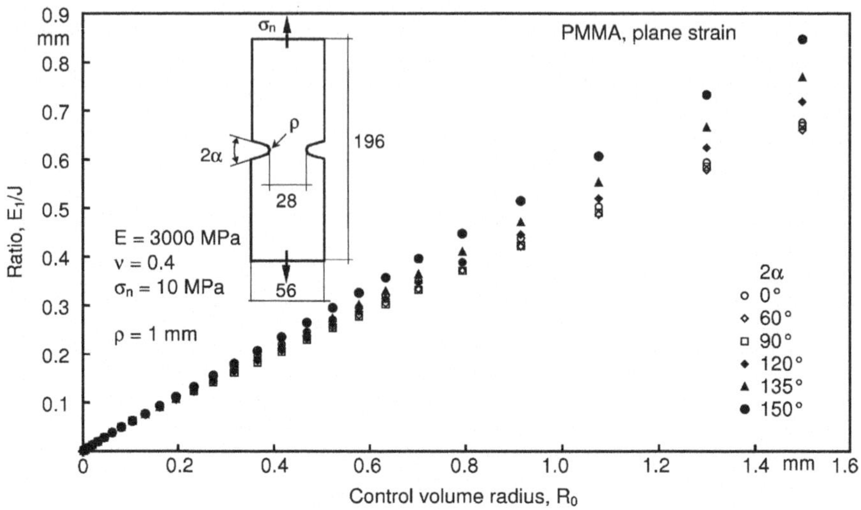

Fig. 3.53 Ratio of elastic strain energy E_1 in control volume to the J-integral (determined along the notch edge) dependent on control volume radius R_0; tensile loaded blunt V-notch ($\rho = 1$ mm) with different notch opening angles 2α; $K_{1\rho}$-based analytical solution; (Berto and Lazzarin 2007)

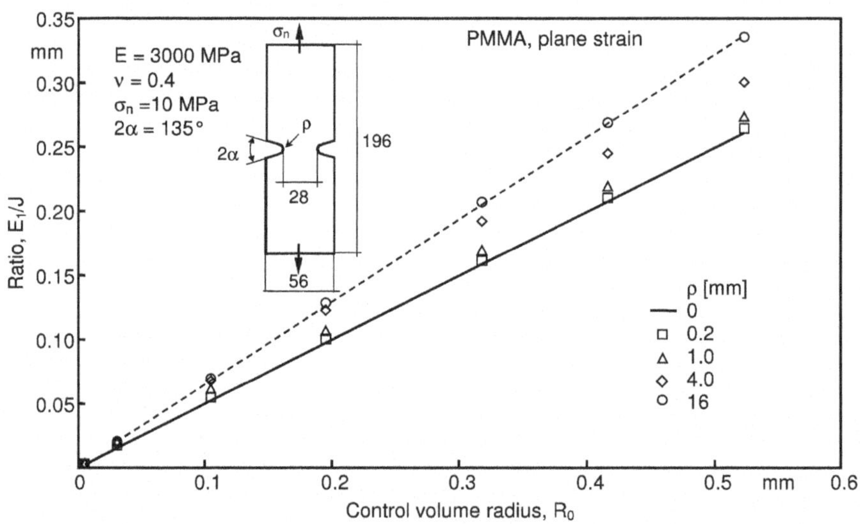

Fig. 3.54 Ratio of elastic strain energy E_1 in control volume to the J-integral (determined along the notch edge) dependent on control volume radius R_0; tensile loaded blunt V-notch ($2\alpha = 135°$) with different notch radii ρ; $K_{1\rho}$-based analytical solution; (Berto and Lazzarin 2007)

3.6 Special Issues of the SED Concept

3.6.1 Survey of Section Contents

This section addresses three special issues which are of relevance for the application of the local SED concept to the assessment of brittle fracture and fatigue failure.

The first issue (Sect. 3.6.2) is related to the question whether other shapes of the control volume, deviating from the full sector or full circle, might be appropriate under special circumstances. The idea behind the question is the observation that crack propagation is mainly controlled by processes ahead of the crack tip. The conditions in the wake of the propagating crack seem to be less important.

The second issue of investigation (Sect. 3.6.3) are microrounded reference notches which have found some application for assessing the fatigue strength of thin sheet welded joints and of very small laser beam welded components.

The third considered issue (Sect. 3.6.4) is slit-parallel loading of a keyhole notch in comparison to a pointed slit. A notch-effect occurs in the former case which is absent in the latter case. The widely used reference notch concept does not work well with slit-parallel loading.

The issues mentioned above are further clarified by numerical evaluation procedures (Sect. 3.6.5) directed to models of practical relevance: a slit or V-notch plate model under tensile and in-plane shear loading (Sect. 3.6.6), cruciform welded joint models under tensile loading (Sect. 3.6.7) and finally non-load-carrying welds (Sect. 3.6.8).

3.6.2 Semicircle Control Volume SED

Within the local SED concept, as developed by Lazzarin and his colleagues, a uniform critical value of the average SED, $\overline{W}_{cr}$, in the control volume of radius R_0 is assumed. The control volume has the shape of a full sector ($2\alpha > 0°$) or of a full circle ($2\alpha = 0°$), Fig. 3.55a. All the derivations and applications presented above in this chapter use this convention. The concept is well proven for brittle fractures of PMMA specimens with pointed or rounded notches as well as for the high-cycle fatigue behaviour of fillet-welded joints, primarily in respect of fractures.

The following arguments can be stated in favour of the full sector or full circle approach. From a numerical point of view it is convenient to have the starting points of the contour line defining the control volume on the load-free edges. Also the link to the *J*-integral is thus established. Further, the full circular arcs can easily be adopted to the local SED analysis of blunt notches. Another advantage for applications is that the direction of crack propagation must not be known in advance.

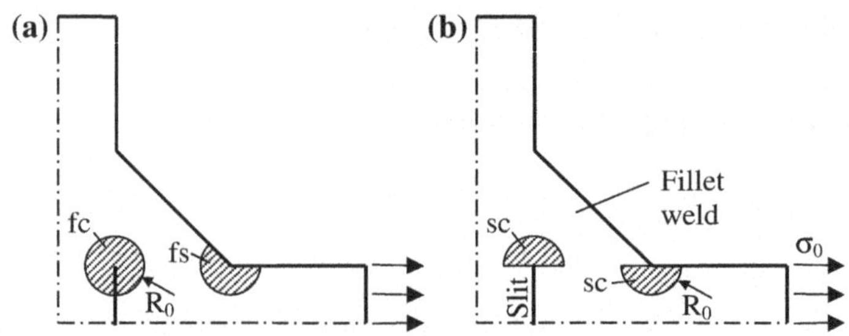

Fig. 3.55 Local SED evaluation in control volume (radius R_0) at weld toe and weld root; full circle and full sector SED (**a**) compared with semicircle SED (**b**); symmetry quarter of fillet-welded cruciform joint model; (Radaj et al. 2009[(1)])

The procedure is theoretically well founded by its link to the *J*-integral whose path independence was originally proven for slit or crack tips under non-reversed loading where the slit or crack flanks remain load-free. The path independence in cyclic loading can be established for materials with a cyclic Masing behaviour (hysteresis loops generated by doubling the σ–ε curve of the cusp points) excluding crack closure effects. The missing path independence in the case of V-notches even under non-reversed elastic conditions can be overcome by using a defined circular path shape.

Using only circular arcs (of radius R_0) in the procedure is another means to facilitate the numerical evaluations. A theoretically better choice would be to limit the averaging volume by SED isolines (Lazzarin and Berto 2005[(1)]; Zappalorto and Lazzarin 2011), but this necessitates more numerical effort. Another argument against an excessive differentiation of shape for the control volume is the influence of the mostly unknown residual stresses on fatigue failure.

The question after the appropriate control radius R_0 has been answered for brittle fracture and fatigue failure based on statistical evaluations. Corresponding data are given in the preceding sections of this chapter. It is important for application that any dependency of R_0 on geometrical parameters such as the notch opening angle should be suppressed as far as possible. For example, weld root fatigue fractures are evaluated with $R_0 \approx 0.3$ mm. (the correct value for toe fractures) instead of the correct root value $R_0 = 0.36$ mm.

The alternative use of a semicircle instead of a full circle (at slit tips) or a full sector (at V-notches) for averaging the local SED, Fig. 3.55b, is based on the hypothesis that the process zone of crack formation will always be positioned ahead of the propagating crack with its centre line in the direction of the expected crack path, whereas the conditions in the wake have no effect. The direction of the crack path can roughly be estimated or more accurately be determined according to the maximum tangential stress criterion (Erdogan and Sih 1963). There is no major difference between the average SED values of the semicircle and full sector volume shape in the case of weld toe notches ($2\alpha = 135°$), but for weld root notches ($2\alpha = 0°$), the difference is larger, at least under mode 1 loading conditions. The difference is neglected in the investigations reviewed below. Identical values of R_0 are applied to full circles, full sectors and semicircles. Stress averaging (not SED averaging) in a semicircle area has already been used for the evaluation of fatigue data (Sheppard 1991; Taylor 1999).

Initially, it seemed possible that by investigating different directional arrangements of the semicircle, the most probable direction of crack propagation inclusive of the critical condition could be determined. The idea turned out to be unrealistic. Also, the SED averaged over a narrow sector along the centre line of the semicircle could not be confirmed as a local fatigue strength parameter.

Basic general conclusions with regard to the influence of the shape of the control volume (semicircle compared with full sector or full circle) can be deduced from the angular distribution of dimensionless SED parameters at slit tips (polar SED plots) derived on the basis of stress intensity factors (Radaj et al. 2009[(1)]). Dimensionless equivalent stress parameters at slit tips and corresponding polar plots are known from the literature (Radaj and Zhang 1993, 1994). Equivalent stresses σ_{eq} are considered there based on the total SED (Beltrami criterion), the distortional SED (von Mises criterion) and the dilatational SED. The dimensionless parameters have the form $\sigma_{\mathrm{eq}}(2\pi r)^{1/2}/K_{\mathrm{res}}$ with $K_{\mathrm{res}} = (K_{\mathrm{I}}^2 + K_{\mathrm{II}}^2)^{1/2}$. Similar dimensionless SED parameters are defined and used further below.

In the case of slit notches ($2\alpha = 0°$), these parameters are:

$$\omega = \frac{1}{2}\frac{2\pi r}{K_{\mathrm{i}}^2}\sigma_{\mathrm{B}}^2 = \frac{2\pi r}{K_{\mathrm{i}}^2} WE \quad (\mathrm{i} = \mathrm{I,\ II,\ III}) \tag{3.87}$$

$$\omega_{\mathrm{d}} = \frac{1+\nu}{3}\frac{2\pi r}{K_{\mathrm{i}}^2}\sigma_{\mathrm{vM}}^2 = \frac{2\pi r}{K_{\mathrm{i}}^2} W_{\mathrm{d}}E \quad (\mathrm{i} = \mathrm{I,\ II,\ III}) \tag{3.88}$$

With ω referring to the total SED W represented by the Beltrami equivalent stress σ_{B}, with W_{d} referring to the distortional SED W_{d} represented by the von Mises equivalent stress σ_{vM} and with the modulus of elasticity E.

In the case of V-notches ($2\alpha \neq 0$), these parameters are:

$$\omega = \frac{1}{2}\frac{2\pi r^{2(1-\lambda_{\mathrm{i}})}}{K_{\mathrm{i}}^2}\sigma_{\mathrm{B}}^2 = \frac{2\pi r^{2(1-\lambda_{\mathrm{i}})}}{K_{\mathrm{i}}^2} WE \quad (\mathrm{i} = 1, 2, 3) \tag{3.89}$$

$$\omega_{\mathrm{d}} = \frac{1+\nu}{3}\frac{2\pi r^{2(1-\lambda_{\mathrm{i}})}}{K_{\mathrm{i}}^2}\sigma_{\mathrm{vM}}^2 = \frac{2\pi r^{2(1-\lambda_{\mathrm{i}})}}{K_{\mathrm{i}}^2} W_{\mathrm{d}}E \quad (\mathrm{i} = 1, 2, 3) \tag{3.90}$$

with λ_{i} being the eigenvalues of the one-term asymptotic solutions of the elastic V-notch problem. The eigenvalues for $2\alpha = 0°$ (weld root) are $\lambda_1 = \lambda_2 = \lambda_3 = 0.5$, those for $2\alpha = 135°$ (weld toe) are $\lambda_1 = 0.674$, $\lambda_2 = 1.302$ and $\lambda_3 = 0.800$. The stresses at the pointed notch tips are singular in all cases with the exception of $\lambda_2 = 1.302$ for mode 2 loading and $2\alpha = 135°$. Non-singular states are treated analytically in the same manner as singular states.

The angular plots of ω and ω_{d} for the loading modes 1, 2, 3 and the notch opening angles $2\alpha = 0$ and $135°$, are shown in Fig. 3.56. Reference circles are added for indicating a definite numerical value (0.5 or 1.0). The distortional SED is always smaller or equal to the total SED, $\omega_d \leq \omega$ because distortional and dilatational SED add up to the total SED. This condition presumes that the dilatational SED is introduced with its absolute value, not differentiating between tensile dilatation (positive) and compressive dilatation (negative). The curve shapes are similar for $2\alpha = 0$ and 135°. Comparing the ω_d curves with the ω curves, the following general statements are possible. In mode I with $2\alpha = 0°$ a pronounced difference in curve shape and curve size exists between ω and ω_d which decreases for $2\alpha = 135°$. In mode II with $2\alpha = 0°$, both curves extend substantially more in the horizontal direction with less difference between the two curves. These features are enhanced for mode 2 with $2\alpha = 135°$. In mode III with $2\alpha = 0°$ and $2\alpha = 135°$, one single circular curve occurs, indicating $\omega_3 = \omega_{d3} = 1.3$.

The SED values $\overline{W}$ or $\overline{\omega}$ averaged over the control volume (full circle, full sector and semicircle variants) are obtained by angular integration of $\omega(\theta)$ with θ being the polar angle. This is shown by the following derivation, using the example of the total SED W for $2\alpha \neq 0$ and a full sector up to the notch edges:

$$W = \frac{K_\mathrm{i}^2}{2\pi r^{2(1-\lambda_\mathrm{i})}} \frac{1}{E} \omega \quad (\mathrm{i} = 1, 2, 3) \tag{3.91}$$

$$\begin{aligned} \overline{W} &= \frac{1}{(\pi-\alpha)R_0^2} \int_0^{R_0} \int_{-\pi+\alpha}^{\pi-\alpha} W r \,\mathrm{d}\theta \,\mathrm{d}r \\ &= \frac{1}{(\pi-\alpha)R_0^2} \frac{K_\mathrm{i}^2}{2} \int_0^{R_0} r^{2\lambda_\mathrm{i}-1} \,\mathrm{d}r \int_{-\pi+\alpha}^{\pi-\alpha} \omega \,\mathrm{d}\theta \\ &= \frac{K_\mathrm{i}^2}{2\pi\lambda_\mathrm{i} R_0^{2(1-\lambda_\mathrm{i})}} \frac{1}{E} \overline{\omega} \end{aligned} \tag{3.92}$$

The conclusion is that $\overline{W} \propto \overline{\omega}$, so that angular integration of ω can substitute the conventional area integration of W. It is also obvious that direct comparisons between the $\overline{\omega}$ values of different loading modes are reasonable only with identical λ_i values, i.e. only for slit notches ($2\alpha = 0°$, $\lambda_\mathrm{i} = 0.5$):

$$\frac{\overline{W}_1}{\overline{W}_\mathrm{k}} = \frac{\overline{\omega}_1}{\overline{\omega}_\mathrm{k}} \frac{\lambda_\mathrm{k}}{\lambda_1} \left(\frac{K_1}{K_\mathrm{k}}\right)^2 R_0^{2(\lambda_1-\lambda_\mathrm{k})} \quad (1 = 1, 2, 3;\ \mathrm{k} = \mathrm{I}, \mathrm{II}, \mathrm{III}) \tag{3.93}$$

In the case of V-notches ($2\alpha \neq 0°$), the corresponding relationship reads:

$$\frac{\overline{W}_1}{\overline{W}_\mathrm{k}} = \frac{\overline{\omega}_1}{\overline{\omega}_\mathrm{k}} \frac{\lambda_\mathrm{k}}{\lambda_1} \left(\frac{K_1}{K_\mathrm{k}}\right)^2 R_0^{2(\lambda_1-\lambda_\mathrm{k})} \quad (1 = 1,\ 2,\ 3;\ \mathrm{k} = \mathrm{I},\ \mathrm{II},\ \mathrm{III}) \tag{3.94}$$

It is now possible to obtain and compare the $\overline{\omega}$ values plotted in Fig. 3.56, considering the loading modes 1, 2 and 3, the notch opening angles $2\alpha = 0$ and 135°, full circle or full sector (fc), semicircle (sc) and narrow sector (ns) SED, total

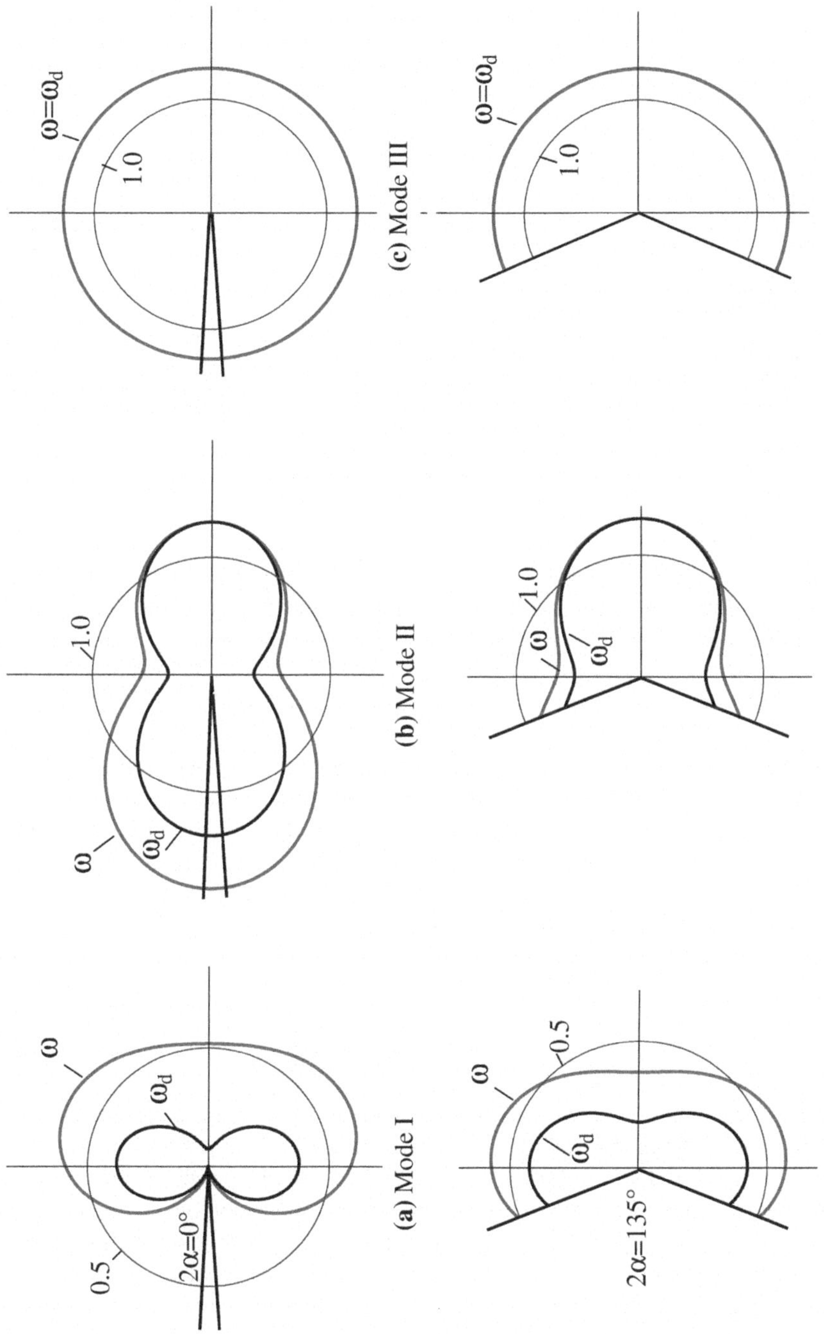

Fig. 3.56 Polar plots of the dimensionless total and distortional SED values, $\overline{\omega}$ and $\overline{\omega}_d$, at slit-tip ($2\alpha = 0$) and at V-notch ($2\alpha = 135°$) under tensile loading (mode I or 1), in-plane shear loading (mode II or 2) and antiplane shear loading (mode III or 3); plane strain condition in mode I or 1 and mode II or 2 loading; analytical solution; (Radaj et al. 2009[1])

Table 3.4 Dimensionless average SED parameters, $\overline{\omega}$ and $\overline{\omega}_d$ (total and distortional), for slit notch ($2\alpha = 0°$) and V-notch ($2\alpha = 135°$) under mode I or 1 and mode II or 2 loading; full circle SED (fc), full sector SED (fs), semicircle SED (sc) and narrow-section SED (ns) with centreing angle θ_0; analytical solution for infinite plate under plane strain conditions; (Radaj et al. 2009[(1)])

Slit notch, $2\alpha = 0°$										
	Mode I, $\theta_0 = 0°$				Mode II, $\theta_0 = 70.6°$			Ratios $\overline{\omega}_2/\overline{\omega}_1$		
	fc	sc	ns		fc	sc	ns	fc2/fc1	sc2/sc1	ns2/ns1
$\overline{\omega}_1$	0.42	0.59	0.52	$\overline{\omega}_2$	1.07	1.02	0.61	2.54	1.72	1.17
$\overline{\omega}_{d1}$	0.20	0.22	0.07	$\overline{\omega}_{d2}$	0.85	0.84	0.46	4.29	3.83	6.58
V-notch, $2\alpha = 135°$										
	Mode I, $\theta_0 = 0°$				Mode II, $\theta_0 = 44.7°$			Ratios $\overline{\omega}_2/\overline{\omega}_1$		
	fc	sc	ns		fc	sc	ns	fc2/fc1	sc2/sc1	ns2/ns1
$\overline{\omega}_1$	0.50	0.48	0.38	$\overline{\omega}_2$	0.92	1.00	0.95	1.85	2.06	2.53
$\overline{\omega}_{d1}$	0.32	0.30	0.18	$\overline{\omega}_{d2}$	0.85	0.92	0.91	2.61	3.03	5.07

and distortional SED, Table 3.4. The semicircle is centred by the expected crack propagation direction. This direction is found by application of the Erdogan–Sih maximum tangential stress criterion, resulting in $\theta_0 = 0°$ for mode 1 loading or in $\theta_0 = 70.6°$ ($2\alpha = 0°$) and $\theta_0 = 44.7°$ ($2\alpha = 135°$) for mode 2 loading. The narrow sector has an inscribed angle of four degrees. One finds $\overline{\omega}_3 = \overline{\omega}_{d3} = 1.3$ in mode 3 loading.

The data in Table 3.4 are first evaluated with regard to the question whether the total SED or the distortional SED is better suited for the fatigue assessment of welded joints. The question is answered based on theoretically founded and/or empirically confirmed equivalent SIF formulae ($2\alpha = 0°$). Equivalent NSIF formulae and data for the weld toe notch ($2\alpha = 135°$) are also available (Lazzarin et al. 2004, 2008[(2)]). The following formula is in general use for kinking crack initiation and propagation in general or in fatigue assessments for spot welded joints:

$$K_{eq} = \sqrt{K_I^2 + \beta K_{II}^2 + \gamma K_{III}^2} \tag{3.95}$$

The coefficients β and γ can be approximated based on theoretical and/or experimental findings documented in the literature and reviewed in Sect. 2.2.9, Table 3.5. These β and γ values refer partly to the crack initiation threshold, SIF K_0, partly to crack propagation (Paris equation) and only occasionally to the fatigue strength directly. When comparing these values with the relevant ratios of $\overline{\omega}$ ($2\alpha = 0°$) in Table 3.4, it is obvious (despite a rather large variation in the β and γ values) that $\overline{\omega}_d$ cannot be a realistic choice whereas $\overline{\omega}$ seems to be acceptable. Therefore, the distortional SED is further recorded only for comparison, not for application. The preference of the total SED to the distortional SED expressed by the above consideration stands against the engineering convention to relate total

Table 3.5 Coefficients β of K_{II}^2 and γ of K_{III}^2 in expression for K_{eq}, Eq. (3.95), extracted from various references; (Radaj et al. 2009[(1)])

Reference	β	γ	Reference	β	γ
Irwin (1957)	1.0	1.39	Tanaka (1974)	1.45	1.74
J-integral	1.0	1.43	Radaj et al. (2001)	1.82	1.16 or 1.34[a]
Sih (1975)	1.03	2.27	Kurath (1992)	2.0 or 3.0[b]	–
Erdogan–Sih (1963)	1.15	–			

[a] Beltrami or von Mises criterion applied to keyhole
[b] Derived with questionable K_{I} and K_{II} values for spot-welded joints, low-carbon or high-strength steels

SED to brittle behaviour and distortional SED to ductile behaviour. Actually, high-cycle fatigue fractures show a rather brittle behaviour.

The question after the most appropriate shape of the control volume (fc, sc, ns) can also be answered based on Table 3.4. The most convincing results are derived with the semicircle approach, i.e. $\beta = 1.02/0.59 = 1.72$ and $\gamma = 1.3/0.59 = 2.20$. With the full circle approach, the corresponding values $\beta = 1.07/0.42 = 2.55$ and $\gamma = 1.3/0.42 = 3.10$, are too large. But the reason for this is solely the too low $\overline{\omega}$ value for the slit notch ($2\alpha = 0°$) in mode I loading. Therefore, taking into account that the full circle approach is much easier to use than the semicircle approach (no information on the probable crack path is needed) and has the merit to have the starting points of the contour line defining the control area on the load-free slit edges, away from the point of singularity, the recommendation to use semicircles should be restricted to slit notches under more or less pure mode I loading and other exceptional cases with a markedly uneven angular distribution of the SED, like, for example, tensile shear loaded lap joints. The narrow sector approach is deleted because $\overline{\omega} = 0.61$ in mode II loading is unrealistic. Speaking more generally, the narrow-sector SED values depend too much on the assumed crack path direction. Therefore, the narrow sector approach is not further investigated. The consequence is that both the distortional SED variant and the narrow sector variant are only occasionally treated in the investigation below. Only the total SED combined with full circles or full sectors and semicircles, respectively, are further considered.

Finally, the average SED values of mode I and II loading at the slit notch and of mode 1 and 2 loading at the V-notch may be compared. For this purpose, the $\overline{\omega}$ values alone are not sufficient, see Eq. (3.94). The reason is that the degree of singularity expressed by the eigenvalue λ_{i} is different in these cases. For the comparison, it is necessary to introduce certain assumptions relating to the notch stress intensity factors K_1 and K_2. One may consider these factors in a form related to the notch depth a and the gross cross-sectional nominal stresses σ_{n} or τ_{n}:

$$K_1 = Y_1 \sigma_{\mathrm{n}} a^{1-\lambda_1} \tag{3.96}$$

$$K_2 = Y_2 \tau_{\mathrm{n}} a^{1-\lambda_2} \tag{3.97}$$

Table 3.6 Ratios of average total SED values in control volume ($R_0 = 0.28$ mm), comparing mode II or 2 with mode I or 1 ($\overline{W}_{II}/\overline{W}_I, \overline{W}_2/\overline{W}_1$) and V-notch ($2\alpha = 135°$) with slit notch ($2\alpha = 0°$) ($\overline{W}_1/\overline{W}_I, \overline{W}_2/\overline{W}_{II}$); semicircle SED with centreing angle θ_0; dependency on R_0/a for $a = 1, 10, 100$ mm; (Radaj et al. 2009[(1)])

Notch depth a [mm]	Ratio of the average total SED values ($R_0 = 0.28$ mm)			
	$\overline{W}_{II}/\overline{W}_I$	$\overline{W}_1/\overline{W}_I$	$\overline{W}_2/\overline{W}_1$	$\overline{W}_2/\overline{W}_{II}$
1	1.72	0.38753	0.21799	0.04887
10	1.72	0.17390	0.01209	0.00122
100	1.72	0.07804	0.00067	0.00003

with the geometry factors Y_1 and Y_2 taking load-free plate edge conditions into account. Assuming $Y_1 = Y_2$ and $\sigma_n = \tau_n$, Eq. (3.94) adopts the following form:

$$\frac{\overline{W}_1}{\overline{W}_k} = \frac{\overline{\omega}_1}{\overline{\omega}_k} \frac{\lambda_k}{\lambda_1} \left(\frac{R_0}{a} \right)^{2(\lambda_1 - \lambda_k)} \quad (k = I, II, III;\ 1 = 1, 2, 3) \tag{3.98}$$

The SED ratios $\overline{W}_1/\overline{W}_k$ for three ratios R_0/a, indicating the field of values relevant for engineers, are presented in Table 3.6. In contrast to the $\overline{W}_{II}/\overline{W}_I$ relationship for slit tips, the other three relationships are highly dependent on R_0/a. It has to be noted that $\overline{W}_2$ belongs to a non-singular state.

The following conclusions can be drawn from the above comparison with regard to the procedures of fatigue assessment of welded joints. The local SED approach should be based on the full circle or full sector evaluation of the total SED combined with the control radius $R_0 = 0.28$ mm (in case of steels). The distortional SED is not appropriate as is not the narrow sector of evaluation. In exceptional cases where the angular distribution of the total SED is markedly unilateral, the SED evaluation over a semicircle centred by the provisional crack path is more appropriate than the full circle evaluation.

3.6.3 Microrounded Notch SED

In recent years, the microrounded notch approach has found application to thin-sheet lap joints ($t = 0.7$–3.0 mm), either resistance spot-welded or laser beam seam-welded (Zhang and Richter 2000; Eibl et al. 2003). A microkeyhole or a correspondingly slim U-shaped notch with a reference radius ρ_r in the grain size range, usually $\rho_r = 0.05$ mm, is introduced when modelling the slit tip at the searm weld root or spot weld edge. The maximum elastic notch stress is evaluated and used to characterise endurable stress ranges dependent on number of load cycles in the high-cycle to medium-cycle fatigue range.

Applying circular or semicircular control volumes at the microrounded notch in the reviewed investigation (Radaj et al. 2009[(1)]), the centre of the hole is chosen to be identical with the former notch tip, Fig. 3.57. Also the circular or semicircular

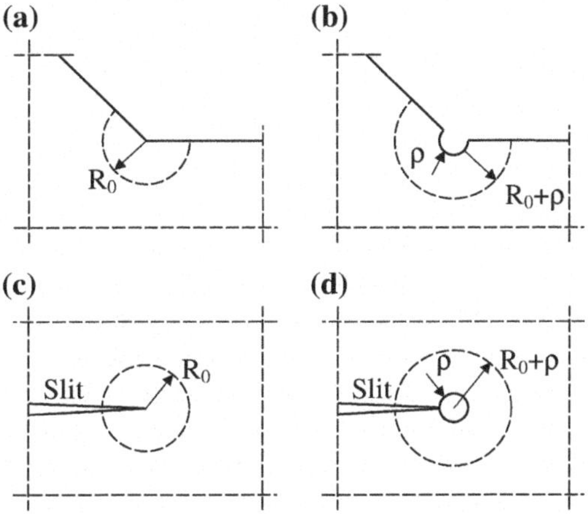

Fig. 3.57 Full sector control volumes at the weld toe without and with microrounded notch (**a**, **b**) as well as full circle control areas at the weld root with the same conditions (**c**, **d**); details from transverse attachment joint; control volume radii R_0 and $R_0 + \rho$; (Radaj et al. 2009[(1)])

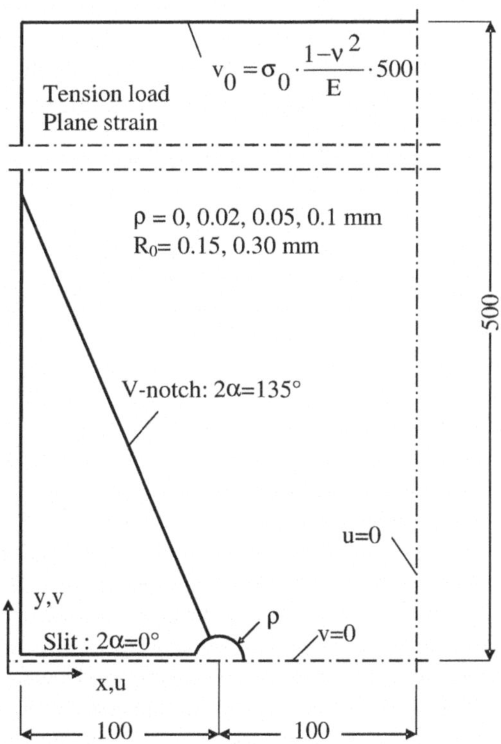

Fig. 3.58 V-notch plate model ($2\alpha = 0$ and 135°) under remote tensile loading by constant boundary displacements v_0; symmetry quarter with symmetry conditions; notch radius ρ and control volume radius R_0; (Radaj et al. 2009[(1)])

control volume with radius R_0 has the same centre. In order to readjust the control volume for the purpose of SED comparisons, $(R_0 + \rho)$ is used as the external radius of the control volume. This choice deviates in a simplifying manner from the better substantiated procedure for rounded notches (Lazzarin and Berto 2005[(1)]) reviewed in Sect. 3.4.2.

The question to be answered is, to what extent the average local SED is changed by microrounding. The answer is given based on the FE analysis results for a tensile-loaded slit or V-notch plate model, Fig. 3.58. Remote tensile loading is generated by prescribed constant boundary displacements v_0. Plane strain conditions are assumed for the analysis. The remote reference tensile stresses σ_0 correspond to the homogeneous plate without the notch. As a consequence, the actual remote tensile stresses are not constant, i.e. they are reduced to some extent above and below the slit or V-notch.

The SED ratios $\overline{W}_{1,\rho}/\overline{W}_{1,0}$ and $\overline{W}_{1d,\rho}/\overline{W}_{1d,0}$ respectively, are plotted in Fig. 3.59 dependent on the notch radius ρ for $2\alpha = 0$ and 135° combined with $R_0 = 0.15$ and 0.30 mm. Decreasing SED values occur in all cases. The decrease is stronger for $2\alpha = 0°$ than for $2\alpha = 135°$, stronger for $R_0 = 0.15$ mm than for $R_0 = 0.30$ mm and stronger for the total SED than for the distortional SED in the case of $2\alpha = 0°$.

Because it is not self-evident that the same SED value is gained for the pointed notch with $\rho = 0$ mm and the rounded notch in the limit $\rho \to 0$, an analytical comparison has been performed between the two total SED values using $\rho = 0.01$ mm to approximate $\rho \to 0$. Without specifying the formula and its parameters here, one can write according to Eqs. (3.10) and (3.65):

$$\frac{\overline{W}_{1,\rho}}{\overline{W}_{1,0}} = \frac{(H/E)K_{1,\rho}^2 \rho^{2(\lambda_1-1)}}{(e_1/E)K_1^2 R_0^{2(\lambda_1-1)}} = \frac{H}{e_1}\left(\frac{K_{1,\rho}}{K_1}\right)^2 \left(\frac{\rho}{R_0}\right)^{2(\lambda_1-1)} \tag{3.99}$$

Inserting $2\alpha = 0°$, $R_0 = 0.3$ mm and $\rho = 0.01$ mm and setting $K_{1,\rho} \approx K_1$, the result is $\overline{W}_{1,\rho}/\overline{W}_{1,0} = 1.007$. Other choices of R_0 down to 0.02 mm result also in values close to 1.0. This means that identical SED values result for $\rho = 0$ mm and $\rho \to 0$.

The SED ratios for $\rho = 0.05$ mm combined with $R_0 = 0.30$ mm (total SED) or $R_0 = 0.15$ mm (distortional SED), where R_0 is confirmed by fatigue-testing of welded joints considering $\rho = 0$ mm, are listed in Table 3.7. Supplemented in the table are the semicircle SED ratios. They are larger than the full circle or full sector SED values. They may be even larger than 1.0, i.e. not a decrease but an increase is observed for increasing ρ values. The semicircle approach provides an advantage over the full circle and full sector approach insofar as the SED ratios remain closer to 1.0. The decrease or increase in the SED value may be relevant for engineering applications. A reference notch radius $\rho_r = 0.01$ mm would be more appropriate if the endurable SED values for $\rho = 0$ mm are used with the microrounded notch approach because the SED ratios are once more closer to 1.0. In any case, the assumption $\rho = 0$ mm of the NSIF approach, as formalised and applied to welded joints by Lazzarin and Tovo, is a better choice for engineering applications than the microrounded notch concept.

The different gradients of decrease in SED for different parameters and conditions shown in Fig. 3.59 can principally be explained by the stress relief caused

Table 3.7 Ratios of average total and distortional SED values for microrounded V-notches ($2\alpha = 135°$) and slit notches ($2\alpha = 0°$) with $\rho = 0.05$ mm; evaluation of FE analysis results for tensile loaded plate model; two different control volume shapes; (Radaj et al. 2009(1))

Control volume shape	Ratio $\overline{W}_{1,\rho}/\overline{W}_{1,0}$ ($R_0 = 0.30$ mm)		Ratio $\overline{W}_{1d,\rho}/\overline{W}_{1d,0}$ ($R_0 = 0.15$ mm)	
	$2\alpha = 0°$	$2\alpha = 135°$	$2\alpha = 0°$	$2\alpha = 135°$
Full circle, full sector	0.813	0.932	0.758	0.859
Semicircle	0.972	1.071	1.114	1.066

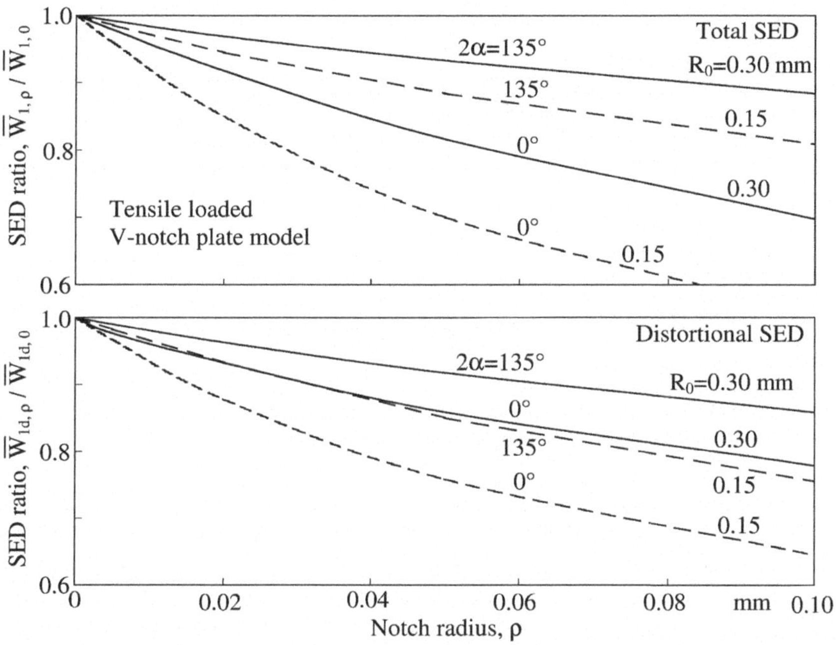

Fig. 3.59 Tensile loaded V-notch plate model: total and distortional SED ratios, $\overline{W}_{1,\rho}/\overline{W}_{1,0}$ and $\overline{W}_{1d,\rho}/\overline{W}_{1d,0}$ dependent on notch radius ρ for notch opening angles $2\alpha = 0$ and 135° (full circle and full sector evaluation, respectively) with control volume radii $R_0 = 0.15$ and 0.30 mm; (Radaj et al. 2009(1))

by the rounded in comparison to the pointed notch. This effect is greater for a higher original stress intensity per unit load and a smaller control volume. One obvious influence is the change in the multiaxiality condition between pointed and rounded notches. The latter influence is considered in more detail. For characterising the degree of multiaxiality, the ratio of hydrostatic stress σ_h to von Mises stress σ_{vM} is defined as the multiaxiality index:

$$\chi_h = \frac{\sigma_h}{\sigma_{vM}} \qquad (3.100)$$

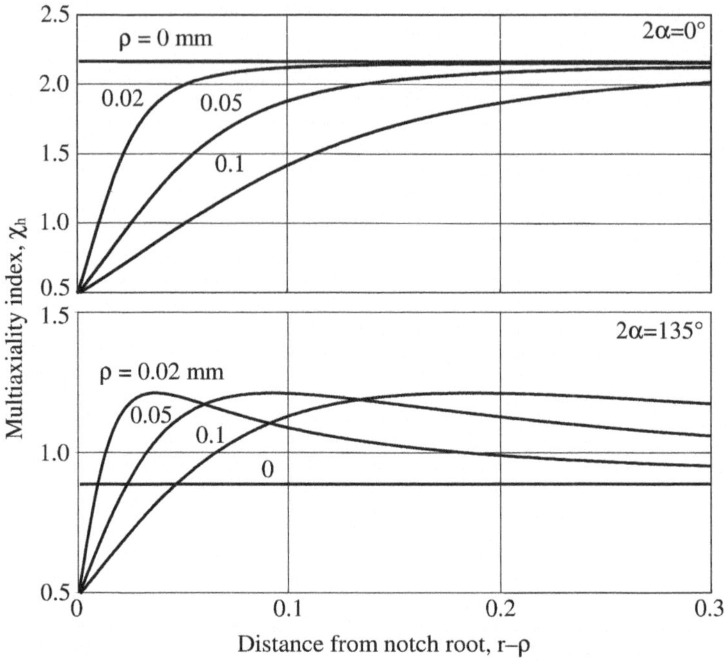

Fig. 3.60 Multiaxiality index $\chi_h = \sigma_h/\sigma_{vM}$ (hydrostatic related to von Mises stress) plotted over ligament line of V-notch plate model under tensile loading; slit notch ($2\alpha = 0°$) (**a**) and V-notch ($2\alpha = 135°$) (**b**), different notch radii ρ; FE analysis results; (Radaj et al. 2009[(1)])

The multiaxiality index in the ligament or bisector plane for the tensile loaded slit or V-notch plate model is plotted for rounded notches ($\rho = 0.02$, 0.05, 0.10 mm, $2\alpha = 0$ and 135°) in comparison to the constant value of the pointed notch ($\rho = 0$ mm), Fig. 3.60. For the slit notch ($2\alpha = 0°$), the multiaxiality index is substantially reduced by notch rounding for $(r - \rho) \leq 2.5\rho$. For the V-notch ($2\alpha = 135°$), it is reduced for $(r - \rho) \leq 0.5\rho$ and enlarged thereafter. The position and shape of the curves will change to some extent for polar angles other than $\theta = 0$ (ligament or bisector), but the principal behaviour will probably persist. The multiaxiality effect is less pronounced in mode 2 loading and vanishes in mode 3 loading.

3.6.4 Slit-Parallel Loading SED

Microrounded as well as macrorounded slit tips in comparison to pointed slit tips pose a problem in cases of substantial slit-parallel loading, expressed by the *T*-stress in the limit case $\rho \to 0$. The problem and its solution is treated in Sect. 2.4.8 referring to the generalised NSIF. It is a disturbing element also within

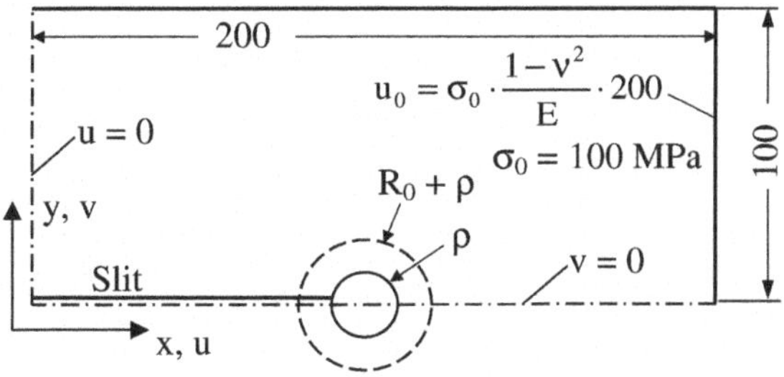

Fig. 3.61 Slit-plate model under slit-parallel tensile loading; quarter plate with prescribed or suppressed boundary displacements, u and v; plane strain conditions; with notch radius ρ and control volume radius R_0; (Radaj et al. 2009[(1)])

the fictitious notch rounding concept, as shown in Sect. 1.5.4. The problem, expressed in terms of the SED concept, is that slit-parallel loading increases the average local SED at the rounded slit tip above the value of the homogeneous tensile field applicable to the pointed slit tip.

Slit-parallel tensile or compressive loading causes no stress concentration at the pointed slit tip, but generates appreciable notch stresses as soon as the notch tip gets rounded. This effect is described by a theoretical SCF $K_t = 3.0$ for the circular hole in the infinite plate. Actually, $K_t = 3.02$ is determined by FE analysis for the considered rectangular plate model with the reference notch radius $\rho_r = 1$ mm (corresponding to the IIW design recommendations), the slight increase above 3.0 is caused by cross-sectional weakening. The SCF is much lower, if the fatigue-effective stresses are considered instead of the theoretical stresses, resulting in $\overline{K}_t \leq 1.1$ for hole diameters $2\rho \leq 0.1$ mm, if the usual microstructural support conditions are applied. The effect is now quantified based on the SED approach.

The SED-based SCFs, $K_{t,W}$ and $K_{t,Wd}$, are defined in Sect. 3.6.5. They are compared with the effective SCFs, $\overline{K}_{t,B}$ and $\overline{K}_{t,vM}$, gained according to Neuber by averaging the equivalent notch stress over the microstructural support length ρ^* along the provisional crack path and relating it to the equivalent nominal stress under plane strain conditions:

$$\overline{K}_{t,B} = \frac{\overline{\sigma}_{max,B}}{\sigma_{n,B}}, \quad \sigma_{n,B} = \sigma_0\sqrt{1-\nu^2} \tag{3.101}$$

$$\overline{K}_{t,vM} = \frac{\overline{\sigma}_{max,vM}}{\sigma_{n,vM}}, \quad \sigma_{n,vM} = \sigma_0\sqrt{1-\nu+\nu^2} \tag{3.102}$$

with the maximum average notch stresses $\overline{\sigma}_{max,B}$ and $\overline{\sigma}_{max,vM}$, the slit-parallel remote tensile stress σ_0 and the subscripts B and vM for the Beltrami and von Mises equivalent stresses.

The FE analysis is performed for the slit plate model shown in Fig. 3.61 together with the relevant boundary conditions. The remote tensile stress σ_0 is generated by a constant displacement u_0 of the remote boundary. The expected crack path in the upper and lower apex points of the keyhole is directed perpendicular to σ_0, that is in the vertical direction ($\theta = 90°$). Full circle and semicircle SED values are identical, provided the semicircle is centred by the vertical crack

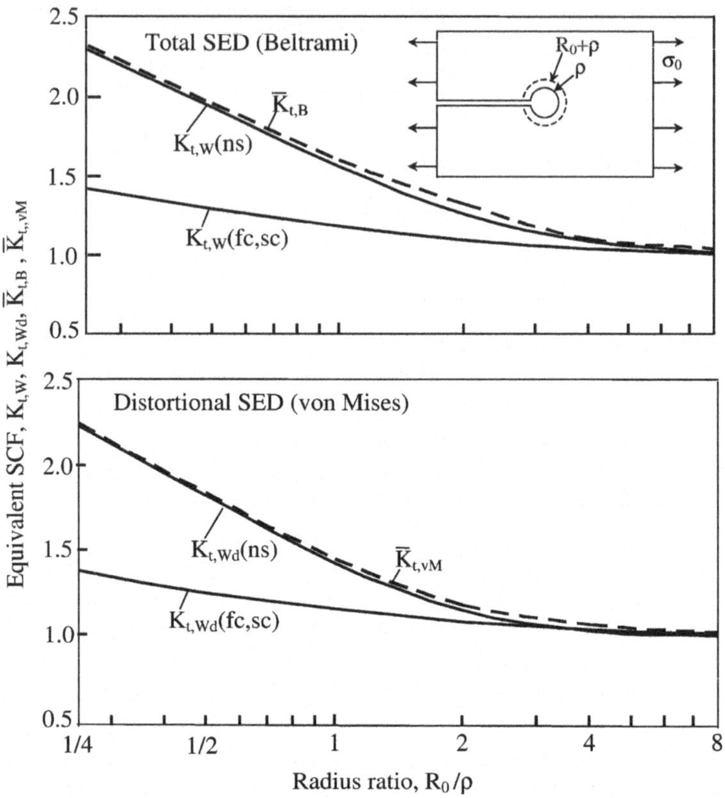

Fig. 3.62 Equivalent SCF values, $K_{t,W}$ (SED) and $\overline{K}_t$ (Neuber), for microrounded notch under slit-parallel tensile loading, total and distortional SED, dependent on radius ratio R_0/ρ; full circle (fc) or semicircle (sc) compared with narrow sector (ns) evaluation; results from FE analysis of slit-plate model; (Radaj et al. 2009[(1)])

path. The narrow sector SED is gained for the sector $88° \leq \theta \leq 92°$ and compared with Neuber's line-averaged stress at $\theta = 90°$ with the microstructural support length $\rho^* = R_0$.

The SED-based SCF values, $K_{t,W}$ and $K_{t,Wd}$ and the effective SCF values, $\overline{K}_{t,B}$ and $\overline{K}_{t,vM}$, are plotted in Fig. 3.62 dependent on the ratio R_0/ρ with the control radius R_0 and the notch radius ρ. All SCF values decrease with increasing R_0/ρ and converge to $K_t = 1.0$ for very large values of R_0/ρ (or $\rho \rightarrow 0$). The starting values at $R_0/\rho = 0$ (or $R_0 \rightarrow 0$) are approximately:

- $K_{t,W}$(fc, sc) $\approx$ $K_{t,Wd}$(fc, sc) $\approx$ 1.7
- $K_{t,W}$(ns) $\approx$ $K_{t,Wd}$(ns) $\approx \overline{K}_{t,B} \approx \overline{K}_{t,vM} \approx 3.0$

The narrow sector SED values (ns) are markedly higher than the full circle or semicircle values (fc, sc). The $\overline{K}_t$ values are slightly higher than the $K_{t,W}$(ns) or $K_{t,Wd}$(ns) values. These trends are naturally explained by the stress field characteristics of a tensile-loaded plate with a circular hole. A more detailed discussion has to take the following secondary effects in account:

- The considered keyhole model corresponds only approximately to a circular hole model. The introduction of the slit removes the transverse compressive notch stresses in that region. The keyhole SCF is 3.06 (Radaj et al. 2001).
- The SCF values referring to σ_0 from prescribed displacements at the remote boundary may be slightly lower than from prescribed stresses there because of the hole-induced unloading effect (more severe with rising ρ values). This effect may be counterbalanced or even reversed by the multiaxiality of the stress field near the hole.
- The SCF limit values for $R_0/\rho \rightarrow 0$ are actually determined for $R_0/\rho = 0.02$.

For the SED-based fatigue assessment methods under discussion for welded joints, the following SCF values are appropriate:

- $K_{t,W}(\mathrm{fc}) \approx 1.04$ (Beltrami, $\rho = 0.05$ mm, $R_0 = 0.28$ mm)
- $K_{t,Wd}(\mathrm{fc}) \approx 1.08$ (von Mises, $\rho = 0.05$ mm, $R_0 = 0.15$ mm)

The relative low $K_{t,W}$ and $K_{t,Wd}$ values indicate that the effect of slit-parallel tensile loading may be negligible in many cases of microrounded reference notches, but it is an important effect with larger reference notches.

3.6.5 Numerical Evaluation Procedures

The numerical evaluations in the following Sects. 3.6.6–3.6.8 have been performed using the FE method (FE programme Ansys, version 11.0). Extremly fine meshes are applied at the notch tip where the SED is evaluated (12 angular sections per quadrant, radial element length about 1/3 of circumferential element length, resulting in about 10000 elements per quarter circle sector of 1 mm radius). Isoparametric quadrilateral elements (PLANE 82) are used with a linear distribution of the stresses or strains along the edges of the element.

The total strain energy of the element (following Beltrami's definition) is found from the nodal point displacements and the stiffness matrix of the element using an easy-to-call function provided by the Ansys code. By adding up the strain energies of all elements within the control volume characterised by the control radius R_0, the local strain energy is obtained. Finally, the SED value is computed by dividing the local strain energy by the control volume. A different procedure has to be used to determine the distortional SED, because no such function is directly provided by Ansys. The von Mises equivalent stress is evaluated for all elements within the control volume as mean value of the nodal point stresses and the corresponding SED values are obtained from these stresses according to Eq. (3.106). The SED value for each element is multiplied by the relevant element volume to obtain the distortional strain energy of the element. Summing up the energies of all elements and dividing by the control volume, the average distortional SED is obtained. The distortional SED values determined based on element stresses are less accurate than the total SED values calculated based solely on nodal point displacements (see Sect. 3.7.2).

For modelling slit closure in relevant cases, the ANSYS contact element CONTA172 (2D 3-node) combined with target surface TARGE169 is used, thus allowing for contact at deformable surfaces. The friction coefficient of the contact element is set to zero.

The (elastic) total strain energy density W (in-plane loading) and its distortional component W_{d} are expressed by the stress components in a Cartesian frame in the following form:

$$W = \frac{1}{2E}\left[\sigma_{\mathrm{xx}}^2 + \sigma_{\mathrm{yy}}^2 + \sigma_{\mathrm{zz}}^2 - 2\nu\left(\sigma_{\mathrm{xx}}\sigma_{\mathrm{yy}} + \sigma_{\mathrm{xx}}\sigma_{\mathrm{zz}} + \sigma_{\mathrm{yy}}\sigma_{\mathrm{zz}}\right) + 2(1+\nu)\tau_{\mathrm{xy}}^2\right] \tag{3.103}$$

$$W_{\mathrm{d}} = \frac{1+\nu}{3E}\left[\sigma_{\mathrm{xx}}^2 + \sigma_{\mathrm{yy}}^2 + \sigma_{\mathrm{zz}}^2 - \left(\sigma_{\mathrm{xx}}\sigma_{\mathrm{yy}} + \sigma_{\mathrm{xx}}\sigma_{\mathrm{zz}} + \sigma_{\mathrm{yy}}\sigma_{\mathrm{zz}}\right) + 3\tau_{\mathrm{xy}}^2\right] \tag{3.104}$$

The total and distortional SED values are expressed in terms of the Beltrami equivalent stress σ_{B} and the von Mises equivalent stress σ_{vM} as follows:

$$W = \frac{1}{2E}\sigma_{\mathrm{B}}^2 \tag{3.105}$$

$$W_{\mathrm{d}} = \frac{1+\nu}{3E}\sigma_{\mathrm{vM}}^2 \tag{3.106}$$

For the purpose of comparison with the Neuber–IIW reference notch approach, the SED-based stress concentration factors, $K_{\mathrm{t},W}(R_0)$ and $K_{\mathrm{t},W\mathrm{d}}(R_{0\mathrm{d}})$, are introduced in the following form:

$$K_{\mathrm{t},W}(R_0) = \sqrt{\frac{\overline{W}(R_0)}{W_{\mathrm{n}}}} \tag{3.107}$$

$$K_{\mathrm{t},W\mathrm{d}}(R_{0\mathrm{d}}) = \sqrt{\frac{\overline{W}_{\mathrm{d}}(R_{0\mathrm{d}})}{W_{\mathrm{d,n}}}} \tag{3.108}$$

where $\overline{W}(R_0)$ or $\overline{W}_{\mathrm{d}}(R_{0\mathrm{d}})$ is the total or distortional SED averaged over the control volume at the notch tip (control volume radius R_0 or $R_{0\mathrm{d}}$) and W_{n} or $W_{\mathrm{d,n}}$ is the nominal total or distortional SED remote from the notch, all parameters for plane strain conditions.

The nominal total and distortional SED values under plane strain conditions for tensile loading (remote normal stress σ_0) and pure shear loading (remote shear stress τ_0), respectively, are:

$$W_{\mathrm{n}} = \frac{(1-\nu^2)\sigma_0^2}{2E} \tag{3.109}$$

$$W_{\mathrm{d,n}} = \frac{(1+\nu)(1+\nu^2-\nu)\sigma_0^2}{3E} \tag{3.110}$$

$$W_{\mathrm{n}} = W_{\mathrm{d,n}} = \frac{\tau_0^2}{2G} = \frac{(1+\nu)\tau_0^2}{E} \tag{3.111}$$

The control volume radius R_0 is introduced as follows:

- $R_0 = 0.28$ mm for total SED evaluations
- $R_0 = 0.15$ mm for distortional SED evaluations

This corresponds to the value of R_0 derived in Sect. 3.2.4 for welded joints of steel, with the extension to $R_{0\mathrm{d}} = 0.54R_0$ stated in Sect. 3.2.3.

The comparisons of the SED approach with the Neuber–IIW reference notch approach in Sects. 3.6.6–3.6.8 below refer to the slit plate model and to welded joint models, both with a reference notch radius $\rho_{\mathrm{r}} = 1$ mm. The fatigue-effective SCF of these models is defined as follows:

$$K_{\mathrm{t}}(\rho_{\mathrm{r}}) = \frac{\sigma_{\max}(\rho_{\mathrm{r}})}{\sigma_{\mathrm{n}}} \quad (\rho_{\mathrm{r}} = 1\,\mathrm{mm},\ \sigma_{\mathrm{n}} = \sigma_0) \tag{3.112}$$

where $\sigma_{\max}$ is the maximum notch stress and σ_{n} the nominal stress, both stresses without multiaxiality consideration. With the endurance limit stress $\sigma_{\mathrm{n}} = \sigma_{\mathrm{E}}$ of the unnotched specimen, the fatigue notch factor results $K_{\mathrm{f}} = K_{\mathrm{t}}(\rho_{\mathrm{r}}, \sigma_{\mathrm{E}})$.

For a more convincing comparison the SCF values, $\sigma_{\max}$ and σ_{n} are defined based on the Beltrami or von Mises equivalent stresses, respectively, under plane strain conditions:

$$K_{\mathrm{t,B}}(\rho_{\mathrm{r}}) = \frac{\sigma_{\max,\mathrm{B}}(\rho_{\mathrm{r}})}{\sigma_{\mathrm{n,B}}}, \qquad \sigma_{\mathrm{n,B}} = \sigma_0\sqrt{1-\nu^2} \tag{3.113}$$

$$K_{\mathrm{t,vM}}(\rho_{\mathrm{r}}) = \frac{\sigma_{\max,\mathrm{vM}}(\rho_{\mathrm{r}})}{\sigma_{\mathrm{n,vM}}}, \qquad \sigma_{\mathrm{n,vM}} = \sigma_0\sqrt{1-\nu+\nu^2} \tag{3.114}$$

Once more, $\rho_{\mathrm{r}} = 1$ mm has been introduced uniformly. The reference notch approach is well suited for the pointed and sharply rounded notches under tensile and in-plane shear loading considered in Sects. 3.6.6 and 3.6.7, but it is questionable in the case of slit-parallel loading of such notches further considered in Sect. 3.6.8.

Closing this subsection on numerical evaluation procedures, the reader's attention is drawn once more to the unconventional definition of the SCF values under multiaxial stress conditions: both the maximum and nominal stresses or SED values are considered under plane strain conditions in combination with the appropriate strength criterion (Beltrami or von Mises). Another peculiarity of the models, to be kept in mind, are the remote boundary conditions in terms of prescribed displacements. Thus, the remote boundary stresses (and SED values) of notched components are changed to some extent and mostly not constant along the boundary. Thereby, the SCF values are modified to some extent.

The stress intensity factors K_{I} and K_{II} as well as the T-stress are evaluated in the investigation on slit-parallel loading. They are determined in the close neighbourhood of the stress singularity. The SIF values are evaluated from the expressions $\sigma_y(2\pi\, r)^{1/2}$ and $\tau_{xy}(2\pi r)^{1/2}$ with r being the distance from the slit tip on the ligament ($\theta = 0°$), whereas the T-stress is equal to the difference ($\sigma_x - \sigma_y$) detected on the ligament.

The equivalent SIF K_{eq} is determined according to Irwin without taking the T-stress into account:

$$K_{\mathrm{eq}} = \sqrt{K_{\mathrm{I}}^2 + K_{\mathrm{II}}^2} \tag{3.115}$$

The corresponding SIF K_{eq}^* incorporating the T-stress is determined according to Eq. (2.21b) with $J_{\mathrm{tot}} = K_{\mathrm{eq}}^{*2}/E'$ (Gdoutos 1990):

$$K_{\mathrm{eq}}^* = \left[K_{\mathrm{I}}^4 + 2K_{\mathrm{I}}^2K_{\mathrm{II}}^2 + K_{\mathrm{II}}^4 + 4K_{\mathrm{II}}^2\left(K_{\mathrm{I}} + T\sqrt{\frac{8}{\pi}R}\right)^2\right]^{1/4} \tag{3.116}$$

where R is the radius of the selected circular integration path.

3.6.6 Evaluation for Slit Plate Model

The slit plate model under plane strain conditions already described in Sect. 3.6.3 is used here to investigate the following influencing conditions on the SED values expressed by the corresponding SCF values:

- Uniaxial tensile loading (mode 1) compared with pure shear loading (mode 2).
- Total SED compared with distortional SED evaluation.
- Pointed slit tip compared with micro keyhole notch.
- Full circle compared with semicircle SED evaluation.

The slit plate model, Fig. 3.63, has the overall dimension 400 × 200 mm^2. Only one quarter of the plate needs actually to be modelled when using appropriate symmetry or antimetry conditions. The slit semi length is 100 mm. The slit tip radius is $\rho = 0$ mm and $\rho_{\mathrm{r}} = 0.05$ mm, respectively, in combination with the control volume radius $R_0 = 0.28$ mm for the total SED and $R_{0\mathrm{d}} = 0.15$ mm for the distortional SED. In comparison, the reference notch radius according to Neuber–IIW ($\rho_{\mathrm{r}} = 1$ mm) is introduced for fatigue-effective notch stress evaluations. Mode 1 tensile loading and mode 2 shear loading are considered. Full circle and semicircle SED values are evaluated under plane strain conditions. The semicircle is arranged with its base transverse to the slit tip in the case of tensile loading. It is oriented transverse to the crack propagation direction (angle θ_0

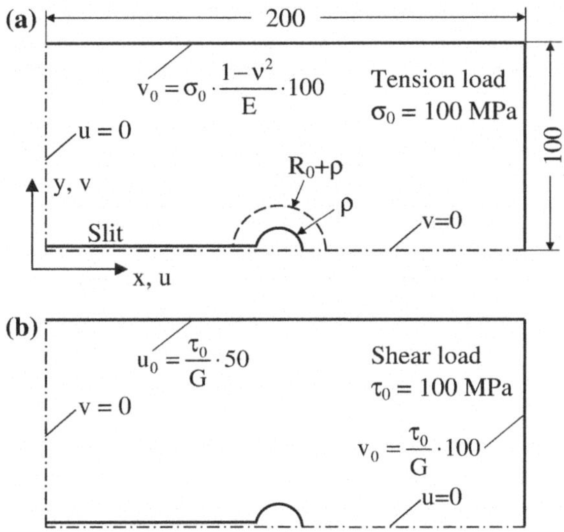

Fig. 3.63 Slit plate model under mode 1 tensile loading (**a**) and mode 2 shear loading (**b**); quarter plate with prescribed or suppressed boundary displacements, u, v; plane strain conditions; with notch radius ρ and control volume radius R_0; (Radaj et al. 2009[(1)])

according to the Erdogan–Sih criterion) in the case of shear loading. The SED-based SCF values, Eqs. (3.107) and (3.108), are used for the purpose of comparisons.

Prescribed constant boundary displacements are applied to generate the remote loading of the model. The reference remote stresses σ_0 or τ_0, respectively, result for the homogeneous plate without a slit or notch. As a consequence, the actual remote stresses are not constant, i.e. they are reduced to some extent above and below the slit or notch. Prescribed boundary displacements are easier to implement in FE programmes (no equilibrium or constraint problems). They approximate identical net cross-section stresses independent of the slit length.

The results of the FE analysis of the slit plate model under tensile and shear loading, respectively, are summarised in Table 3.8. The SED-based SCF values, $K_{t,W}(R_0 = 0.28$ mm) of the total energy and $K_{t,Wd}(R_{0d} = 0.15$ mm) of the distortional energy, are shown for the pointed and keyhole-rounded slit tips. Full circle (fc) control volumes can be compared with semicircle (sc) control volumes. The semicircle is arranged with its basis perpendicular to the slit line in tensile loading or transverse to the out-of-bisector crack propagation direction in shear loading. Semicircles with their bases in the direction of the ligament give results which are identical with those for full circles because of the symmetry conditions of the SED field. The SED-based SCF values are set into comparison with the Neuber–IIW SCF values, $K_{t,B}$ (Beltrami) and $K_{t,vM}$ (von Mises), using the uniform reference notch radius $\rho_r = 1$ mm. The parameters R_0, R_{0d} and ρ_r correspond to data recommended for the fatigue assessment of welded joints of steel.

The following SCF values are predicted based on the analytical solution derived for the keyhole in the infinite plate (Radaj et al. 2001), here with the notch

Table 3.8 SED-based and Neuber–IIW-based SCF values for slit plate under tensile (mode 1) and shear (mode 2) loading; total SED (W) and distortional SED (W_d), pointed compared with microrounded (keyhole) notch, full circle (fc) compared with semicircle (sc) evaluation; sc centred by $\theta_0 = 0°$ in case of σ_0 and $\theta_0 = 71°$ or $67°$ in case of τ_0 ($\rho = 0$ or 0.05 mm); results from FE analysis under plane strain conditions; with control volume radius R_0 or R_{0d}, actual notch radius ρ and reference notch radius ρ_r; (Radaj et al. 2009[(1)])

	SCF $K_{t,W}$ ($R_0 = 0.28$ mm)				SCF $K_{t,Wd}$ ($R_{0d} = 0.15$ mm)				Neuber–IIW SCF	
	$\rho = 0$ mm		$\rho = 0.05$ mm		$\rho = 0$ mm		$\rho = 0.05$ mm		$\rho_r = 1$ mm	
	fc	sc	fc	sc	fc	sc	fc	sc	$K_{t,B}$	$K_{t,vM}$
σ_0	10.35	12.27	9.27	12.07	11.35	12.12	9.86	12.79	12.73	12.73
τ_0	15.89	15.48	14.84	16.25	19.29	19.19	16.23	18.71	15.65	13.57

depth $t = 100$ mm and the notch radius $\rho = 1$ mm. In mode 1 tensile loading, $K_t = 2.12(t/\rho)^{1/2} = 21.2$ is nearly identical with Inglis' or Neuber's elliptical hole solution, $K_t = 1 + 2(t/\rho)^{1/2} = 21.0$. It can be concluded from Fig. 62 in the SCF compendium (Radaj and Schilberth 1977) that under remote displacement loading with the actual finite plate width, $K_t = \sigma_{max}/\sigma_0 = 0.6 \times 21.2 = 12.7$ (about twice this value occurs under remote stress loading). In mode 2 shear loading, $K_t = \sigma_{max}/\tau_0 = 2.86(t/\rho)^{1/2} = 28.6$ (at $\theta_0 = 67.5°$), with an unknown reduction factor.

Considering tensile and shear loading of the slit plate, the SCF results presented in Table 3.8 can be discussed as follows:

- Total SED ($R_0 = 0.28$ mm) compared with distortional SED ($R_{0d} = 0.15$ mm) evaluation shows that the SCF values are nearly identical in mode 1 loading. But this is not the case with mode 2 loading where the distortional SED is substantially larger. The reason is that the control volume radius $R_{0d} = 0.15$ mm is adapted only to the mode 1 conditions. In mode 2 loading, the difference between the local total and distortional SED values is small, so that an identical control volume radius would be more appropriate.
- Pointed slit tip compared with microrounded keyhole notch SCF evaluation shows that the micronotch values deviate up to about ±10 % from the pointed notch values.
- Full circle compared with semicircle evaluations for $\rho = 0$ mm show that the behaviour documented by Table 3.4 is reproduced in tendency, the deviations being larger for mode 1 than for mode 2 loading.
- The Neuber–IIW SCF values for $\rho_r = 1$ mm show a high degree of correspondence to the SED-based SCF values for $\rho = 0$ mm with the expected larger deviation of the distortional SCF values in mode 2 loading. The $K_{t,B}$ value in mode 1 is identical with that derived from the above simple SCF formula. In mode 2, the unknown reduction factor results as 15.65/28.6 = 0.55 which is close to the mode 1 value of 0.6.

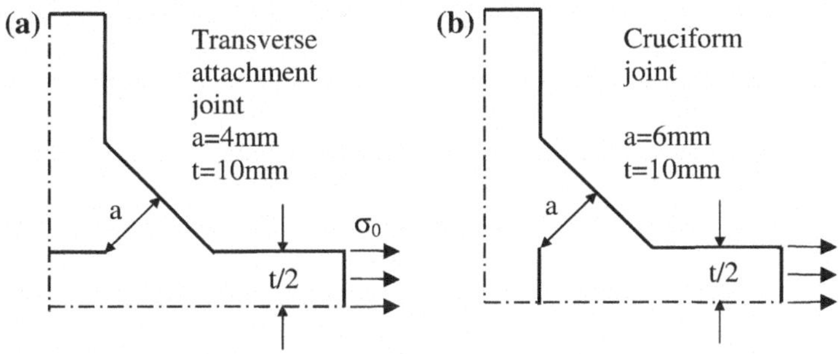

Fig. 3.64 Shape and dimensions of FE models (symmetry quarters) of transverse attachment fillet-welded joint (**a**) and of cruciform fillet-welded joint (**b**); with throat thickness a, plate thickness $t = 10$ mm and remote stress $\sigma_0 = 100$ MPa; plate ends are longer than depicted; (Radaj et al. 2009[(1)])

The general conclusions from the discussion above of the SCF results in Table 3.8 are the following:

- The inconsistencies of the distortional SED concept are once more confirmed.
- The micronotch concept has no advantage over the NSIF concept, as has already been stated.
- The findings on the relationships between full circle and semicircle values are confirmed.
- The Neuber–IIW SCF values for $\rho_r = 1.0$ mm are well in correspondence with the SED-based SCF values, provided the total SED and the Beltrami criterion are considered.

3.6.7 Evaluation for Two Cruciform Welded Joint Models

The application-related investigation comprises two fillet-welded joint models with the usual simplified geometry which are both of cruciform shape, a transverse attachment joint (non-load-carrying fillet welds) and an original cruciform joint (load-carrying fillet welds), both models under tensile loading, Fig. 3.64. Any deviation from the simplified geometry originating from the welding process is not taken into account. The throat thickness in the attachment joint, $a = 4$ mm, is smaller than in the cruciform joint, $a = 6$ mm, because of its lower load transfer. Only one quarter of the joint is modelled with introduction of the corresponding symmetry conditions. The plate ends are much longer than shown in the figure, in order to eliminate any influence from these loaded or non-loaded ends. A constant nominal stress, $\sigma_n = \sigma_0$, is applied to the remote plate ends. The notch tip or slit tip radius is $\rho = 0$ mm and $\rho_r = 0.05$ mm, respectively, in combination with the control volume radius $R_0 = 0.28$ mm for the total SED and $R_{0d} = 0.15$ mm for the distortional SED. Slit closure at the weld root which may occur in the transverse attachment joint is not simulated by the FE model. Semicircle and full circle

Table 3.9 SED-based and Neuber–IIW-based SCF values for transverse attachment joint; total SED evaluation with pointed compared with rounded notches, full circle (fc) or full sector (fs) compared with semicircle (sc) evaluation (see Fig. 3.65 for sc1 and sc2); results from FE analysis under plane strain conditions; ρ, ρ_r, R_0 given in mm; fine mesh (fm) and coarse mesh (cm); variants V1 and V2; (Radaj et al. 2009[(1)])

Stress concentration factors	Transverse attachment joint					Cruciform joint			
	Toe		Root			Toe		Root	
	fs	sc	fc	sc1	sc2	fs	sc	fc	sc
$K_{t,W}$ ($\rho = 0$, $R_0 = 0.28$), fm	1.90	1.90	0.81	0.72	0.89	2.55	2.50	2.74	3.25
$K_{t,W}$ ($\rho = 0$, $R_0^* = R_0$), V1 cm	1.92	1.91	0.82	0.73	0.90	2.58	2.58	2.78	3.28
$K_{t,W}$ ($\rho = 0$, $R_0^* = 1 \rightarrow R_0$), V2 cm	1.94	1.97	1.27	1.25	1.28	2.57	2.50	2.91	3.39
$K_{t,W}$ ($\rho = 0.05$, $R_0 = 0.28$)	1.84	1.93	0.79	0.70	0.87	2.47	2.56	2.50	3.27
$K_{t,B}$ ($\rho_r = 1$), FE analysis	2.39		2.06			3.33		4.07	
$K_{t,B}$ ($\rho_r = 1$), BE analysis	2.39		2.04			3.34		4.03	

or full sector evaluations are performed in comparison. Reference notch tip rounding is applied in a further step.

The results of the FE analysis of the two fillet-welded joint specimens, transverse attachment joint and cruciform joint, under tensile loading are summarised in Table 3.9. The SCF values based on total SED, $K_{t,W}(R_0 = 0.28$ mm), are shown for the weld toe and the weld root, full sector or full circle compared with semicircle SED and pointed notch compared with microrounded notch. The semicircle at the weld toe is positioned with its baseline in the plate edge. The semicircle at the slit tip of the transverse attachment joint is positioned on the plate side (sc1) or on the attachment side (sc2) with its baseline in the slit line, Fig. 3.65. The semicircle at the slit tip of the cruciform joint (sc) is positioned with its base line perpendicular to the slit line, Fig. 3.66. The SED-based SCF values are set into comparison with the Neuber–IIW-based SCF values $K_{t,B}$ (Beltrami), applying the reference notch radius $\rho_r = 1.0$ mm to the weld toe and weld root. The parameters R_0 and ρ_r correspond to data recommended for the fatigue assessment of welded joints of steel. Reference notch rounding at the weld toe is performed with the circular arc being tangent to the plate and weld edges, i.e. without producing an undercut. The solution referenced in the first line uses the FE model. The solution referenced in the second line corresponds to a boundary element (BE) model (Radaj and Helmers 1997; Radaj et al. 2006, ibid. pp. 169–171).

It has been shown that accurate SED values can be derived based on extremely coarse FE meshes, see Sect. 3.6.8 and Fig. 3.45. The coarse mesh option is applied below to the welded joint models in comparison with the fine mesh option. The number of elements within a circular area of 1 mm radius is reduced by a factor of 2500. Actually four angular sections are applied per quadrant instead of two, but with only minor influence on the results.

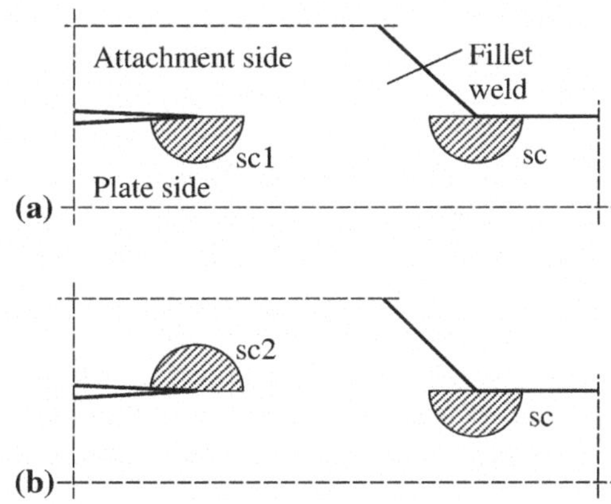

Fig. 3.65 Semicircle control volume positions at weld root (sc1, sc2) and weld toe (sc) of transverse attachment fillet-welded joint model; (Radaj et al. 2009[(1)])

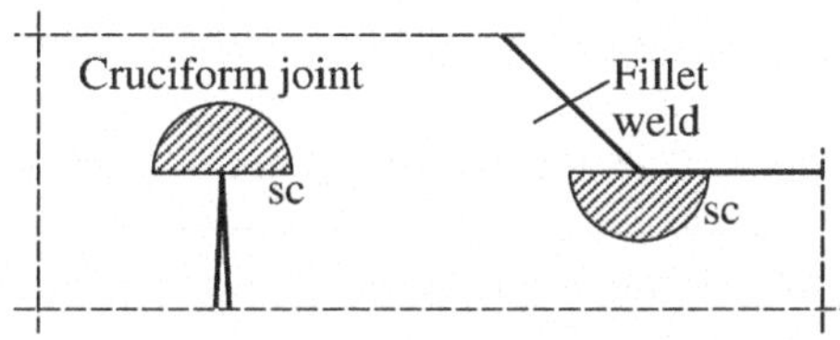

Fig. 3.66 Semicircle (sc) control volume positions at weld root and weld toe of cruciform fillet-welded joint model; (Radaj et al. 2009[(1)])

There are two variants of using the rough mesh of Fig. 3.45: variant V1 with $R_0^* = R_0$ and variant V2 with $R_0^* = 1$ mm. The transformation of the average total SED value for $R_0^* = 1$ mm to that for $R_0 = 0.28$ mm, and therefrom the transformation of the corresponding SCF values, is expressed by the following relationships:

$$\overline{W}(R_0^* \to R_0) = \overline{W}(R_0^*)\ \left[R_0/R_0^*\right]^{2(\lambda_1-1)} \tag{3.117}$$

$$K_{t,W}(R_0^* \to R_0) = K_{t,W}(R_0^*)\ \left[R_0/R_0^*\right]^{(\lambda_1-1)} \tag{3.118}$$

with $\lambda_1 = 0.674$ at the weld toe ($2\alpha = 135°$) and $\lambda_1 = 0.5$ at the weld root ($2\alpha = 0°$). The term $(R_0/R_0^*)^{2(\lambda_1-1)}$ is 1.226 in the former case and 1.367 in the latter case. The relationships above are valid for singular mode 1 slit or notch tip loading. They can also be used with pure mode 2 loading (with λ_1 substituted by λ_2) as far as singular stresses occur ($2\alpha \leq 102.6°$). In the presence of major nonsingular stresses, these relations become invalid. The R_0^* value should remain within the range where the stress field is described solely by the NSIF.

The stress concentration factor values presented in Table 3.9 can be compared and discussed as follows:

- The transverse attachment joint compared with the cruciform joint shows substantially lower SCF values at the weld toe and weld root. This is explained from the fact that different loads are transferred by the fillet welds in the two cases, or alternatively, from slit-parallel loading compared with slit-transverse loading.

- The full sector or full circle approach gives almost the same $K_{t,W}$ values as the semicircle approach with the exception of the slit tip in the cruciform joint which is subjected to nearly pure mode 1 loading. In the latter case, the semicircle value is more representative as shown in Sect. 2.6.3.
- The ratio of $K_{t,W}$ values gained from full circle (fc) compared with semicircle (sc) evaluations, $K_{t,W}(\text{fc})/K_{t,W}(\text{sc}) = 2.74/3.25 = 0.843$ is identical with the corresponding SED ratio $(0.42/0.59)^{1/2}$ resulting from Table 3.4.
- The pointed notch tip compared with the microrounded notch shows that the deviation from one another is within the 10 % range for $\rho_r = 0.05$ mm.
- The $K_{t,B}$ values according to the Neuber–IIW reference notch approach ($\rho_r = 1$ mm) are substantially larger than the $K_{t,W}$ values according to the SED approach (factor 1.25–1.33 with the exception of the root sc1 case where the factor 2.85 occurs because the effect of slit-parallel loading is overrated by the reference notch). It has been shown (Lazzarin et al. 2006 and Sect. 3.5.3) that most of the discrepancy can be removed by using $R_0 = 0.15$ mm instead of $R_0 = 0.28$ mm, resulting in a factor 1.226 on $K_{t,W}$ at the weld toe ($2\alpha = 135°$) and of 1.367 at the weld root ($2\alpha = 0°$). This corresponds to introducing a higher endurable stress range, $\Delta\sigma_E = 190$ MPa instead of $\Delta\sigma_E = 155$ MPa at $N_E = 5 \times 10^6$ cycles.
- The coarse mesh variants V1 and V2 give excellent results compared with the fine mesh, but the variant V2 fails in the case of the attachment joint weld root ($K_{t,W} = 1.25-1.28$) because of predominant non-singular effects which are not covered by Eqs. (3.117) and (3.118).

The application-relevant result of the investigation above is now validated by assessing the fatigue strength of the two welded joints, the transverse attachment joint (TAJ) and the cruciform joint (CJ), based on the SED approach, on the Neuber–IIW reference notch approach on the IIW nominal stress approach and on *S–N* data from the literature. The endurable basic stress $\Delta\sigma_E$ at $N_E = 2 \times 10^6$ cycles for $P_s = 97.7$ % (two standard deviations from the mean) and $R = 0$ (zero-to-tension loading) is considered, together with the predicted toe fractures in the TAJ and root fractures in the CJ. The details of the selection of $\Delta\sigma_E$ in the two approaches are presented in Sect. 3.6.8.

The derived $\Delta\sigma_{nE}$ values for the TAJ and the CJ are listed in Table 3.10. It cannot be expected that the different approaches show more or less full correspondence. The experimental basis and its evaluation are different. Therefore, the results are not identical. Only the two versions of the SED approach ($K_{t,W}$) and ($\Delta\overline{W}$), should converge. It is an noteworthy fact that for the CJ slit tip, the SED (fc) results seem to be more 'on the line' of the IIW and *S–N* results than the SED (sc) results.

Table 3.10 Endurable nominal stress ranges $\Delta\sigma_{nE}$ ($P_s = 97.7$ %, $N = 2 \times 10^6$ cycles) for transverse attachment joint (TAJ, toe fractures) and cruciform joint (CJ, root fractures) resulting from different approaches; (Radaj et al. 2009[(1)])

Fatigue strength	Fatigue assessment procedure				
	SED($K_{t,W}$,$\Delta\sigma_E$)[a]	SED $\Delta\overline{W}_E$[b]	IIW-FAT225(K_t)[c]	IIW-FAT80/54[c]	S-N[d]
TAJ: $\Delta\sigma_{nE}$ [MPa]	84.2	85.2	94.1	80.0	95.7
CJ: $\Delta\sigma_{nE}$ [MPa]	49.2, 58.4[e]	49.9, 59.1[e]	55.6, 55.3[f]	63.0	58.9

[a] $\Delta\sigma_E = 160$ MPa
[b] $\Delta\overline{W}_E = 0.058$ Nmm/mm^3
[c] Hobbacher (2009)
[d] Olivier and Ritter (1979–1985)
[e] sc, fc
[f] Keyhole, U-notch

3.6.8 Evaluation for Butt Weld Joint with Backing Plate Model

Survey on the Investigation

The investigation refers to slit-parallel loading of welded joints. Typical examples are fillet-welded transverse attachment and cover plate joints, fillet-welded, laser-beam-welded or resistance-spot-welded lap joints and butt weld joints with backing plate. Slit-parallel loading is mostly connected in these joints with a markedly one-sided distribution of the stresses or strain energy. Also the *T*-stress at pointed slit tips or its counterpart at rounded slit tips may be substantial.

The conventional reference notch approach shows severe deficiencies when applied to models with slit-parallel loading. The positioning of the reference notch is subjected to ambiguity. Centreing the circular hole in the slit tip causes cross-sectional weakening and produces the notch stresses of slit-parallel loading which do not occur at the pointed slits. The slit length is simultaneously enlarged by the hole radius. Engineering stress analysts use an eccentric circular hole with a smooth transition from the circular arc into the rectilinear slit line, the hole being positioned on the load-free side of the slit. The strength assessment then tends to be non-conservative and the ambiguity with the slit length is not removed.

The more recently developed SED approach has not the drawbacks of the reference notch approach just mentioned. There is no overrating of the notch effect, no ambiguity of circular hole positioning and no cross-sectional weakening. And it is applicable based on extremely coarse FE meshes without a major loss of accuracy. An open question was, to what extent the semicircle variant of the SED approach is more appropriate than the full circle variant for assessing the fatigue strength of the special joints mentioned above. The question has been answered by a comprehensive investigation with comparisons of the SED and reference notch approaches (Radaj et al. 2009[(2)]).

The investigation refers to three types of welded joints: the butt weld joint with backing plate, the single side fillet-welded attachment joint and the fillet-welded lap joint. Finally, a welded sandwich panel joint is considered, a novel option in ship structural design which is of the first mentioned joint type. In the following,

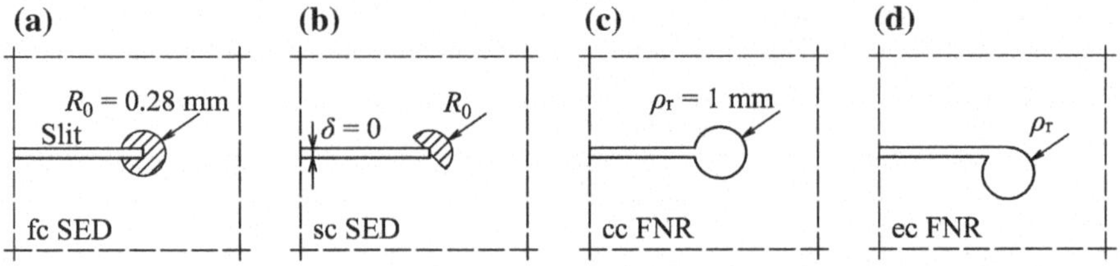

Fig. 3.67 Full circle (fc) and semicircle (sc) control volumes for SED evaluation at the pointed slit tip (**a**, **b**) as well as concentric (cc) and eccentric (ec) circular reference notch at the slit tip (**c**, **d**); (Radaj et al. 2009[(2)])

only the numerical investigation referring to the butt weld joint with backing plate is reviewed. All essential points can be demonstrated by this one example.

Local SED Compared with Reference Notch Approach

The circular and semicircular control volumes at the pointed slit tip for the SED evaluation are shown in Fig. 3.67a, b. In all models analysed, the gap width δ is set equal to zero. The semicircle is centred by the expected crack path which is determined by the position of the maximum tangential stress $\sigma_{\theta\ \mathrm{max}}$ on the circle with control volume radius R_0. This $\sigma_{\theta\ \mathrm{max}}$ criterion resembles to the Erdogan–Sih maximum tangential stress criterion with which it coincides for $R_0 \rightarrow 0$. The advantage of evaluating $\sigma_{\theta\ \mathrm{max}}$ on the circular contour of the control volume with $R_0 = 0.28$ mm (for steels) is that the influence of the T-stress on early crack propagation (a major part of the fatigue life) is thus taken into account whereas the Erdogan-Sih criterion refers to early crack initiation solely under the singular stress components.

The semicircle variant is expected to give better-grounded results than the full circle variant in cases of markedly one-sided angular SED distributions. This should also be true in all cases of predominant mode 2 loading where the zones with negative principal stresses are excluded from the outset by the semicircle SED approach. The control radius R_0 of the semicircle is assumed here to be equal to the value provided from the full sector or full circle evaluations under predominant mode 1 loading conditions. To find a new, more accurate R_0 value for the semicircle SED approach would require not only a re-evaluation of the bulk of fatigue test data with the semicircle model, but also a number of not yet available fatigue test data from welded joints with markedly one-sided angular SED distributions. The above engineering assumption of an unchanged R_0 value has the consequence that the semicircle results will be more conservative than the full circle results.

The concentric and eccentric circle notches ($\rho_r = 1$ mm) for the reference notch evaluation are shown in Fig. 3.67c, d. In these models, the gap width δ is always set to zero. The eccentric circle notch is tangentially arranged in the more

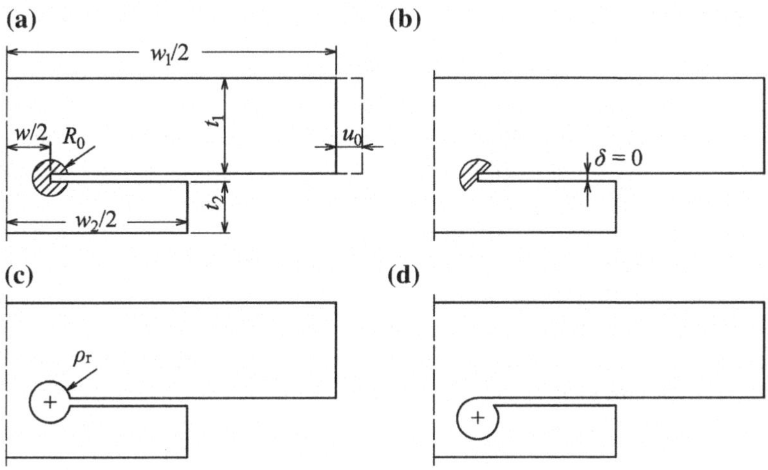

Fig. 3.68 Butt weld joint with backing plate: cross-sectional plane strain model (symmetry half), dimensional parameters and prescribed displacement u_0 of the remote plate end; full circle (fc) and semicircle (sc) control volumes for the SED approach (**a**, **b**); concentric circle (cc) and eccentric circle (ec) hole for the reference notch approach (**c**, **d**); (Radaj et al. 2009[(2)])

or less unloaded lower plate where it does not disturb the force flow in the upper base plate. Also cross-sectional weakening of the base plate is avoided. Therefore only the eccentric circle notch in the unloaded plate has a chance to describe the fatigue behaviour of the butt weld joint with backing plate sufficiently well.

Modelling Features

The geometry and its characteristic parameters of the plane strain cross-sectional model of the considered butt weld joint with backing plate is shown in Fig. 3.68. The geometry is simplified in the usual manner, idealising the slits and weld profiles. Any deviation from the simplified geometry originating from the welding process is not taken into account. A uniform base plate thickness of $t_1 = 10$ mm is introduced. The dimensional variants refer to this thickness. Main influencing parameters are the ratio of plate thicknesses t_2/t_1 and the ratio of joint face width w to base plate thickness t_1. Plane strain modelling is performed for steel with the elastic modulus $E = 2.06 \times 10^5$ MPa and Poisson's ratio $\nu = 0.3$. The influence of variations of the base plate thickness is treated separately at the end of this subsection.

The remote loading of the models is produced by a rotation-free prescribed constant displacement u_0 of the end section in the horizontal x direction:

$$u_0 = \sigma_0 \frac{1-\nu^2}{E} \frac{w_1}{2} \tag{3.119}$$

with the remote nominal stress $\sigma_0 = 100$ MPa and the tensile-loaded base plate length w_1. The relationship above presumes an unnotched plate strip under plane strain conditions subjected to the tensile load σ_0. As the prescribed displacement u_0 acts on the notched model, the actual remote tensile stresses in the FE model will deviate from σ_0. For plate ends sufficiently remote from the geometrical inhomogeneities, the remote stresses will be linearly distributed, described by the membrane and bending components $\overline{\sigma}_0$ and $\overline{\sigma}_{0b}$. In general, $\overline{\sigma}_0 \geq \sigma_0$, because the notched model reacts stiffer than the corresponding unnotched plate strip. In the considered case, $\overline{\sigma}_0 \approx \sigma_0$ is found with a few exceptions where $\overline{\sigma}_0 \geq 1.05\sigma_0$. The bending stress component $\overline{\sigma}_{0b} \ll \overline{\sigma}_0$ depends on the extent of one-sided stiffening by the backing plate. It enlarges the membrane stress on the slit side of the base plate.

The remote end of the model can freely move up or down without rotation. There is no vertical support of the end in order to avoid a transverse shear force which would complicate the loading system. The prescribed constant displacements at the model ends are in better agreement with the conditions in welded structures than prescribed constant stresses.

Only one symmetry half of the butt-weld joint has to be analysed. The model is supported in such a way that it can deform without any constraint from the supports.

The slit flanks are free to move with the consequence that the ideal slit flanks will penetrate each other in cases of local mode 1 compressive load components. This condition may be realistic insofar as real world slit flanks of welded joints have often a small (or even larger) gap between them, resulting from the production conditions, so that they can move against one another without penetration. Also the endurable structural or nominal stress ranges $\Delta\sigma$ in the IIW design recommendations (Hobbacher 2009) are given for residual tensile prestresses (corresponding to the load ratios $R > 0.5$) which may keep the slit permanently open. On the other hand, such movements of the slit flanks are limited by the gap width so that it is of interest what changes in the fatigue-relevant parameters occur when the slit flanks contact each other. This slit closure is modelled by inserting contact elements into the gap-free slit.

Selection of Endurable Basic Stresses

The unexpressed assumption behind the SCF definitions according to Eqs. (3.107–3.114) is that the endurable reference stresses $\Delta\sigma_E$, are identical in the SED and reference notch approach. Only then, it is possible to compare the fatigue-effective SCF values directly. The problem with comparisons between the SED and reference notch approach is, that the two approaches use rather different $\Delta\sigma_E$ values, $\Delta\sigma_E = 162$ MPa in the SED approach, derived from notched specimens, in contrast to $\Delta\sigma_E = 225$ MPa in the reference notch approach applicable to notched specimens (both endurable stress values for $N = 2 \times 10^6$ cycles and $P_s = 97.7$ %).

In the SED approach, the uniform *W–N* curve (Lazzarin et al. 2003; Livieri and Lazzarin 2005) is relevant which is statistically well founded based on a large bulk of published fatigue test results relating to welded joints under predominant mode 1 loading. This curve provides $\Delta\overline{W}_E = 0.058$ Nmm/mm^3 for $N = 2 \times 10^6$ cycles and $P_s = 97.7$ % based on evaluations using sectorial or full circles with radius $R_0 = 0.28$ mm marking the boundary of the control volume. The corresponding endurable stress is gained from the following relationship:

$$\Delta\sigma_E = \sqrt{\frac{2E}{1-\nu^2}\Delta\overline{W}_E} \tag{3.120}$$

with $E = 2.06 \times 10^5$ MPa and $\nu = 0.3$ resulting in $\Delta\sigma_E = 162$ MPa. To what extent the reference fatigue strength $\Delta\overline{W}_E$ may actually be found in an unnotched specimen, is questionable, see Sect. 3.5.3. It has to be noted that the *W–N* scatter band does not include data from unnotched specimens. Therefore, $K_{t,W}$ is only an expedient which indicates the fatigue-effective stress concentration at the weld notches in analogy to the reference notch approach. Unnotched welded specimens, however, are of major importance only for finding the R_0 value. Actually, the butt weld ground flush to plate was used for that purpose ($\Delta\sigma_E = 155$ MPa for $N = 5 \times 10^6$ cycles and $P_s = 50$ %).

In the reference notch approach (Neuber–IIW version, $\rho_r = 1$ mm), the reference fatigue strength at $N = 2 \times 10^6$ cycles and $P_s = 97.7$ % is $\Delta\sigma_E = 225$ MPa, defined by the IIW code developers for notched welded joints with $K_t(\rho_r = 1\text{ mm}) \geq 1.6$ whereas $\Delta\sigma_E = 160$ MPa is assigned to the unnotched base material close to the weld notches. This means that the fatigue-effective SCF values in the reference notch approach cannot be directly compared with those in the SED approach. Only the endurable stresses $\Delta\overline{\sigma}_{0E} = \Delta\sigma_E/K_t$, inserting reference-notch-relevant or SED-relevant values of $\Delta\sigma_E$ and K_t, can directly be compared.

Numerical Analysis Results

The analysed model of the butt weld joint with permanent backing plate has already been shown in Fig. 3.68. Such joints are used where plates are butt-welded without accessibility of the rear side. The backing plate supports the weld pool during welding in a controllable manner. It may also result from a supporting web, e.g. in the welded sandwich panel joint mentioned in the subsection survey. The thickness t_2 of this backing member may then be rather large. The width w of the joint face may vary between small values, e.g. in laser beam welding, and larger values, e.g. in conventional multipass welding.

Table 3.11 Butt weld joint with backing plate subjected to prescribed displacement u_0 of the remote base plate end ($\sigma_0 = 100$ MPa); SIFs K_{I}, K_{II}, K_{eq} and K^*_{eq} at weld root dependent on w/t_1 and t_2/t_1 for $t_1 = 10$ mm; K_{eq} without and K^*_{eq} with T-stress effect; (Radaj et al. 2009[(2)])

SIF values K_{I}, K_{II}, K_{eq} [MPa mm], t_1=10 mm

	$t_2/t_1 = 0.2$				$t_2/t_1 = 1.0$				$t_2/t_1 = 5.0$			
w/t_1	K_{I}	K_{II}	K_{eq}	K^*_{eq}	K_{I}	K_{II}	K_{eq}	K^*_{eq}	K_{I}	K_{II}	K_{eq}	K^*_{eq}
0.2	2.8	37.2	37.3	63.0	0.0	42.0	42.0	62.9	0.0	38.5	38.5	57.4
									1.5	43.7[a]	43.8[a]	66.2[a]
1.0	38.3	58.2	69.7	108.6	1.5	79.8	79.8	98.5	−6.0	90.1	90.3	101.0
									2.0[a]	95.1[a]	95.1[a]	111.1[a]
5.0	49.9	71.0	86.8	133.7	22.2	103.7	106.1	139.3	−78.2	161.9	179.8	184.7
									−1.7[a]	178.0[a]	178.0[a]	189.8[a]

[a] With contact elements

Nine dimensional variants of the joint geometry are considered: $t_1 = 10$ mm, $t_2/t_1 = 0.2$, 1.0, 5.0, $w/t_1 = 0.2$, 1.0, 5.0, $(w_1 - w)/2 = 5t_1$, $(w_2 - w)/2 = 5t_2$, combined with $R_0 = 0.28$ mm and $\rho_{\mathrm{r}} = 1$ mm.

The SIF values K_{I}, K_{II}, K_{eq} without T-stress, Eq. (3.115), and K^*_{eq} with T-stress, Eq. (3.116), for $\sigma_0 = 100$ MPa are recorded in Table 3.11 dependent on w/t_1 and t_2/t_1 for $t_1 = 10$ mm. The actual membrane stress $\overline{\sigma}_0$ ranges from 91.0 to 116.0 MPa, with the latter value being valid for ($w/t_1 = t_2/t_1 = 5.0$). Suppressing the penetration of the slit flanks (occurring in cases of a negative K_{I} value) by contact elements, thus simulating slit closure, rises both K_{II} and the T-stress to some extent, so that K_{eq} and K^*_{eq} increase slightly, provided the suppressed negative K_{I} value was small enough. Otherwise a decrease occurs, as expected.

The markedly one-sided angular distributions of the SED ratio $W/W_{\max}$ on the outer circle ($R_0 = 0.28$ mm) of the control volume at the pointed slit tip and the notch stress ratio $\sigma_\theta/\sigma_{\theta\ \max}$ at the rounded ($\rho_{\mathrm{r}} = 1$ mm) slit tip with concentric or eccentric hole position are plotted in Fig. 3.69 for the considered welded joint ($w/t_1 = 1.0$, $t_2/t_1 = 1.0$, $t_1 = 10$ mm).

The average local SED values $\overline{W}$ and the fatigue-effective SCF values $K_{\mathrm{t},W}$ are recorded in Table 3.12 dependent on w/t_1 and t_2/t_1 for $t_1 = 10$ mm. They refer to $\overline{\sigma}_0$ which is slightly larger than σ_0 (see above). The semicircles are centred by the $\sigma_{\theta\ \max}$ direction, $\theta = 78$–$90°$. Semicircles being aligned with the slit direction, $\theta = 90°$, result in $K_{\mathrm{t},W,\mathrm{sc}}$ values which are slightly larger (up to 3 %). The values $K_{\mathrm{t},W,\mathrm{fc}} \leq 1.0$ from full circle SED indicate that $\overline{W}$ at the slit tip is smaller than $\overline{W}$ in the tensile-loaded base plate remote from the slit tip region. This means that the fatigue strength of the welded member will be equal to the plain material strength, as it is the case in the Kitagawa diagram and in its extension by Atzori and Lazzarin (Sect. 3.5.2) for all notches located on the left-hand side of the intersection between the horizontal curve of the plain specimen fatigue strength and the inclined curve of the threshold SIF K_0.

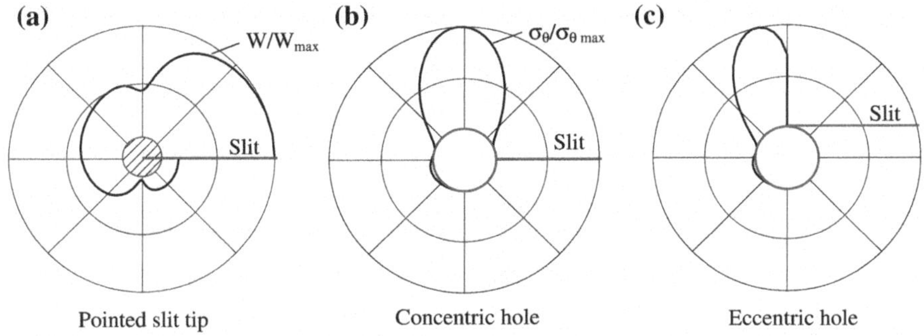

Fig. 3.69 Butt weld joint with backing plate ($w/t_1 = 1.0$, $t_2/t_1 = 1.0$, $t_1 = 10$ mm): angular distribution of the normalised SED W/W_{max} on the outer circle ($R_0 = 0.28$ mm) of the control volume at the pointed slit tip (**a**) and of the normalised notch stress $\sigma_\theta/\sigma_{\theta\ max}$ of the rounded ($\rho_r = 1$ mm) slit tip with concentric or eccentric hole position (**b**, **c**); (Radaj et al. 2009[(2)])

Table 3.12 Butt weld joint with backing plate subjected to prescribed displacement u_0 of the remote base plate end ($\sigma_0 = 100$ MPa): average SED values $\overline{W}$, $\overline{W}_1$, $\overline{W}_2$ ($R_0 = 0.28$ mm) at weld root and corresponding fatigue-effective SCF values $K_{t,W}$ ($\Delta\sigma_E = 162$ MPa) dependent on w/t_1 and t_2/t_1 for $t_1 = 10$ mm; full circle (fc) and semicircle (sc) SED evaluation (sc centred by $\sigma_{\theta\ max}$); (Radaj et al. 2009[(2)])

SED $\overline{W}(R_0 = 0.28\text{mm})[10^{-3}\text{Nmm/mm}^3]$, SCF $K_{t,W}$, $t_1 = 10$ mm												
	$t_2/t_1 = 0.2$				$t_2/t_1 = 1.0$				$t_2/t_1 = 5.0$			
w/t_1	$\overline{W}_{fc}$	$\overline{W}_{sc}$	$K_{t,W,fc}$	$K_{t,W,sc}$	$\overline{W}_{fc}$	$\overline{W}_{sc}$	$K_{t,W,fc}$	$K_{t,W,sc}$	$\overline{W}_{fc}$	$\overline{W}_{sc}$	$K_{t,W,fc}$	$K_{t,W,sc}$
0.2	17.3	29.1	0.89	1.15	17.8	30.4	0.91	1.19	17.6	30.0	0.91	1.18
									19.2[a]	32.9[a]	0.94[a]	1.22[a]
1.0	33.8	53.0	1.23	1.53	44.1	62.3	1.40	1.67	53.2	70.4	1.53	1.77
									59.4[a]	78.4[a]	1.63[a]	1.88[a]
5.0	50.2	78.3	1.44	1.80	74.8	106.6	1.73	2.07	170.0	187.8	2.42	2.54
									193.9[a]	230.6[a]	2.56[a]	2.80[a]

[a] With contact elements

The fatigue-effective SCF values K_t according to Eq. (3.112) with $\sigma_n = \overline{\sigma}_0$ from the reference notch approach ($\rho_r = 1$ mm) are recorded in Table 3.13 dependent on w/t_1 and t_2/t_1 for $t_1 = 10$ mm. One $K_{t,cc}$ value in the table is confirmed by data from the literature.

Endurable membrane stress ranges $\Delta\overline{\sigma}_{0E}$ at $N = 2 \times 10^6$ cycles and $P_s = 97.7$ % for $t_1 = 10$ mm have been determined according to the SED and reference notch approaches based on the fatigue-effective SCFs and the appertaining endurable reference stresses $\Delta\sigma_E$, Table 3.14. The $\Delta\overline{\sigma}_{0E}$ values determined with semicircles being aligned with the slit direction are slightly smaller (up to 3 %) than those from semicircles centred by $\sigma_{\theta\ max}$ which are recorded in the table. There are three cases in the table where the IIW condition $K_t > 1.6$ is not fulfilled ($w/t_1 = 0.2$), but the resulting $\Delta\overline{\sigma}_{0E}$= 163 MPa is more or less identical with the base metal value

Table 3.13 Butt weld joint with backing plate subjected to prescribed displacement u_0 of the remote base plate end ($\sigma_0 = 100$ MPa): fatigue-effective SCF values $K_t(\rho_r = 1$ mm, $\Delta\sigma_E = 225$ MPa) at the weld root dependent on w/t_1 and t_2/t_1 for $t_1 = 10$ mm; concentric circle (cc) and eccentric circle (ec) variants of the reference notch approach; (Radaj et al. 2009[(2)])

	SCF K_t ($\rho_r = 1$ mm) for reference notch, $t_1 = 10$ mm					
	$t_2/t_1 = 0.2$		$t_2/t_1 = 1.0$		$t_2/t_1 = 5.0$	
w/t_1	$K_{t,cc}$	$K_{t,ec}$	$K_{t,cc}$	$K_{t,ec}$	$K_{t,cc}$	$K_{t,ec}$
0.2	2.86	1.38	2.86	1.38	2.86	1.38
1.0	3.28	1.95	3.30	1.97	3.31	1.98
		1.99[b]		2.02[a]		2.12[a]
5.0	3.77	2.29	3.93	2.49	3.90	2.47
				2.62[a]	4.21[a]	3.16[a]

[a] With contact elements

[b] Radaj et al. 2006, ibid. Fig. 4.48c

Table 3.14 Butt weld joint with backing plate subjected to prescribed displacement u_0 of the remote base plate end ($\sigma_0 = 100$ MPa): endurable values of the remote structural stress range $\Delta\bar{\sigma}_{0E}$ (membrane portion, root fractures) dependent on w/t_1 and t_2/t_1 for $t_1 = 10$ mm; determined based on the SED approach (variants fc and sc, the latter centred by $\sigma_{\theta\ max}$) and the reference notch approach ($\rho_r = 1$ mm, variants cc and ec); (Radaj et al. 2009[(2)])

	$\Delta\bar{\sigma}_{0E}(N = 2 \times 10^6, P_s = 97.7\ \%)$ [MPa], SED (fc, sc), reference notch (cc, ec)											
	$t_2/t_1 = 0.2$				$t_2/t_1 = 1.0$				$t_2/t_1 = 5.0$			
w/t_1	fc	sc	cc	ec	fc	sc	cc	ec	fc	sc	cc	ec
0.2	183	141	79	163	179	137	79	163	180	137	79	163
									174[a]	132[a]		
1.0	132	105	69	115	115	98	68	114	105	92	68	114
									101[a]	88[a]		106[a]
5.0	113	91	60	98	95	79	57	90	67	64	58	91
								86[a]	64[a]	58[a]	53[a]	71[a]

[a] With contact elements

$\Delta\sigma_E = 160$ MPa in the IIW design recommendations and can therefore be accepted. Fatigue test data are not available for the considered butt weld joint with backing plate, but there are endurable values given in relevant codes: $\Delta\bar{\sigma}_{nE}$= 71 MPa in the IIW design recommendations (Hobbacher 2009) and $\Delta\bar{\sigma}_{nE}$= 68 MPa in the British Standard (BS 7608 1993), identical value in BS 5400, *S–N* curve F (Maddox 1991, ibid. p. 101).

The following conclusions are drawn from the endurable membrane (or nominal) stresses $\Delta\bar{\sigma}_{0E}$ listed in Table 3.14:

- $\Delta\bar{\sigma}_{0E}$ depends strongly on w/t_1 and t_2/t_1, no uniform value can be assigned corresponding to the IIW or BS classifications, which specify a substantially lower endurable stress.

- $\Delta\overline{\sigma}_{0E}$ based on the full circle SED needs a comparison with the base material strength in cases of extremely one-sided angular SED distribution and might then be substituted by the semicircle SED variant which gives always conservative results.
- $\Delta\overline{\sigma}_{0E}$ based on the concentric circle reference notch is too low because of the overrated notch effect under slit-parallel loading and because of the effect of cross-sectional weakening; $\Delta\overline{\sigma}_{0E}$ based on the eccentric circle reference notch is more realistic, but probably on the non-conservative side.
- $\Delta\overline{\sigma}_{0E}$= 140 MPa may be used in cases of a small joint face width, e.g. $w/t_1 = 0.2$, occurring with laser beam welded joints.
- $\Delta\overline{\sigma}_{0E}$ is substantially lowered by increasing w/t_1 and t_2/t_1; the value in the IIW design recommendations, $\Delta\overline{\sigma}_{0E} \approx 70$ MPa, occurs with large values of w/t_1 and t_2/t_1.
- $\Delta\overline{\sigma}_{0E}$ is reduced by less than 10 % by slit closure under the considered remote plate end condition.
- $\Delta\overline{\sigma}_{0E}$ is reduced by less than 3 % when applying the semicircle SED approach with the semicircle being aligned with the slit direction instead of being centred by the $\sigma_{\theta\ \max}$ direction.

Influence of Plate Thickness

The base plate thickness is uniformly chosen in the investigation above, reference thickness $t_0 = 10$ mm. So the question arises, how the results can be applied to welded joints with another base plate thickness.

According to the NSIF concept supporting the SED approach, the fatigue strength ratio σ_n/σ_{n0} for V-notched members subjected to mode 1 loading is correlated with the plate thickness ratio t/t_0 (Lazzarin and Tovo 1998):

$$\frac{\sigma_{nE}}{\sigma_{nE0}} = \frac{\kappa_{1,0}}{\kappa_1}\left(\frac{t}{t_0}\right)^{-n_1}, \quad n_1 = 1 - \lambda_1 \tag{3.121}$$

where σ_{nE} and σ_{nE0} are endurable nominal stresses, κ_1 and $\kappa_{1,0}$ are nondimensional shape factors depending on the dimensional ratios of the joint geometry and n_1 is an exponent depending on the eigenvalue λ_1 of the elastic singularity problem and on the notch opening angle 2α. The index 0 refers to the reference joint, the index 1 to mode 1 loading. For $2\alpha = 135°$, representative for weld toes, $n_1 = 0.326$; for $2\alpha = 0°$, representative for weld roots, $n_1 = 0.5$. At the weld toe, Eq. (3.121) is an approximation neglecting the (non-singular) mode 2 loading contribution in the bisector plane. At the weld root, it is an approximation because of the neglected T-stresses. Both approximations are sufficiently accurate, as far as the singular terms in the solution are dominating. The exponent $n_1 = 0.5$ at the weld root has also been determined based on a fracture mechanics approach related to fillet-welded cruciform joints (Frank and Fisher 1979).

Under mode 2 loading conditions, as prevailing in the considered welded joints, the relationship is expected to be given in terms of endurable nominal shear stresses τ_{nE}/τ_{nE0}, with $\kappa_{2,0}/\kappa_2$ and $n_2 = 0.5$ (at the weld root). Nominal shear stresses cannot be defined for the considered joint type, but τ_{nE}/τ_{nE0} can certainly be expressed by σ_{nE}/σ_{nE0}. Therefore, Eq. (3.121) remains valid with $n_1 = n_2 = 0.5$ without the shape factors in cases of self-similar size changes and with modified shape factors ($\kappa_2 \neq \kappa_{2,0}$) in cases where the other dimensional ratios (other than t/t_0) are kept constant in comparison to the reference joint and its endurable stress. These shape factors can easily be determined from the mode 2 shear stress distribution (Lazzarin and Tovo 1998).

According to the reference notch approach based on fatigue-effective SCFs, similar considerations apply, but $n_1 = n_2 = 0.5$ at the weld root is not appropriate. For demonstration, the result of a SCF analysis for the tensile loaded cruciform joint with load-carrying fillet welds (Radaj et al. 2006, ibid. pp. 115–116, 187–189) is evaluated with respect to the thickness effect. After reducing the four parameter problem there to a simpler two parameter problem here (by introducing $\rho_1 = \rho_2 = \rho$, $t_2/t_1 = 1$ and $g/t_1 = 1$) the SCFs at the weld toe and weld root, K_{t1} and K_{t2}, are:

$$K_{t1} = 1.192 \left(\frac{a}{t_1}\right)^{-0.311} \left(\frac{\rho}{t_1}\right)^{-0.392} \tag{3.122}$$

$$K_{t2} = 1.155 \left(\frac{a}{t_1}\right)^{-0.720} \left(\frac{\rho}{t_1}\right)^{-0.371} \tag{3.123}$$

where t_1 is the base plate thickness, a the throat thickness of the fillet weld and ρ the notch radius at the weld toe and weld root.

In the reference notch approach, the thickness effect expresses itself by the ρ/t_1 term. If we consider $\rho = \rho_r = 1$ mm as a fraction of the reference thickness $t_{1,0}$, ρ/t_1 can be substituted by $t_{1,0}/t_1$ (and a factor). Thus we get the form of Eq. (3.121). We then find by comparison $n_1 = 0.392$ (weld toe) and $n_1 = 0.371$ (weld root), but only for a constant a/t_1. The influence of a/t_1 on the SCF expressed by the exponents is equally strong (weld toe) or even stronger (weld root) than that of ρ/t_1 substituted by $t_{1,0}/t_1$. Based on the above consideration and in view of lacking investigations under mode 2 loading conditions, $n_1 = n_2 = 0.4$ is recommended for weld toe and weld root fractures in cases of self-similar size change or of thickness change with constant other dimensional ratios.

According to the IIW design recommendations, a strength reduction by $(t_0/t)^n$ with $n = 0.2$ or 0.3 is prescribed for $t > t_0$, $t_0 = 25$ mm, referring to butt welds and fillet welds, respectively, in the as-welded condition. In both cases root fractures are excluded. The exponents above are derived from fatigue test data, where other fatigue-relevant dimensional ratios have been varied together with t/t_0, so that these other parameter changes are also included in the exponent. The IIW formula is not applicable to the welded joints considered in the investigation above where root fractures are prevailing.

3.7 Coarse Mesh Options

3.7.1 *Survey of Section Contents*

The coarse mesh option for the FE analysis of sharp notches comprises two different procedures, the SED-based coarse mesh evaluation (Sect. 3.7.2) and the peak stress method (Sect. 3.7.3).

The SED-based coarse mesh evaluation benefits from the fact that the requirements on mesh refinement for an accurate determination of the local strain energy are much lower than the corresponding requirements in the case of direct stress field evaluations. The procedure is first applied to pointed V-notches subjected to mode 1, mode 2 and mixed mode loading, respectively, inclusive of weld toe notches, resulting in accurate values of K_1 and K_2. The procedure is then applied to rounded notches with the aim to get an accurate value of the maximum notch stress.

The peak stress method proves that the peak stresses determined in a plane FE mesh consisting of linear quadrilateral elements of globally even element size d (e.g. $d = 1$ mm) are a measure for the NSIFs K_1 and K_2 of pointed V-notches. The link to the SED-based procedure is established with definition of 'weighted peak stresses'. The method is first applied to cruciform joints, rescaling the statistically founded design scatter band. The method is then used to show that recently established fatigue data for welded lap joints and cover plates are well in correspondence with this scatter band.

3.7.2 *Coarse Mesh SED Evaluation*

Theoretical Basis of the Coarse Mesh Procedure

The average SED at pointed or rounded notches can be evaluated with high accuracy based on coarse FE meshes. The NSIF or maximum notch stress values, respectively, derived therefrom are sufficiently accurate for technical applications. The extremely fine meshes required for determining the NSIF values based on the usual limit value $r \to 0$ procedure are avoided, as well as the correspondingly fine meshes for an accurate maximum stress analysis. This is shown by the following expositions which review work done mainly by Lazzarin and Berto, referring to linear-elastic material behaviour with an extension into the elastic-plastic range (Lazzarin et al. 2007, 2010; Lazzarin and Berto 2008).

The reason for the excellent performance of SED evaluations within the widely used displacement (or stiffness) method of finite element analysis is the fact that the nodal point displacements of the FE structural system are the primary unknown parameters which are determined based on a variational formulation of the potential Π leading to the principle of virtual work. Within this approach, the element strains are fully compatible whereas the stresses are not equilibrated at the element boundaries. Therefore, the evaluation of the strain energy from the nodal point displacements without using stresses is superior to a stress evaluation. The procedural steps are shortly summarised in the following (Lazzarin et al. 2010; Cook 1995; Bathe 1996).

The displacements within a finite element, column vector $\{u\}$, are expressed by the nodal point displacements $\{d\}$ by means of the displacement interpolation matrix $[N]$ (representing the 'shape functions') which depends on the element type:

$$\{u\} = [N]\{d\} \tag{3.124}$$

The strains $\{\varepsilon\}$ within the element follow according to well-known relations:

$$\{\varepsilon\} = \frac{\partial\{u\}}{\partial x_i} = [B]\,\{d\} \tag{3.125}$$

with the strain–displacement matrix $[B]$ whose elements are obtained by appropriately differentiating the elements of the matrix $[N]$.

Under linear-elastic conditions, the stresses are linked to the strains by means of the elasticity matrix $[E]$:

$$\{\sigma\} = [E]\{\varepsilon\} = [E][B]\{d\} \tag{3.126}$$

The strain energy density W in the points of the finite element can be written in the following form:

$$W = \frac{1}{2}\{\varepsilon\}^{t}\{\sigma\} = \frac{1}{2}\{d\}^{t}[B]^{t}[E]\,[B]\,\{d\} \tag{3.127}$$

The total strain energy E_{fe} in the finite element of volume V is:

$$E_{fe} = \int_V W\mathrm{d}V = \frac{1}{2}\{\varepsilon\}^{t}\left(\int_V [B]^{t}[E]\,[B]\,\mathrm{d}V\right)\{d\} = \frac{1}{2}\{d\}^{t}[K]\,\{d\} \tag{3.128}$$

with the stiffness matrix $[K]$ given by:

$$[K] = \int_V [B]^{t}[E]\,[B]\,\mathrm{d}V \tag{3.129}$$

It is evident from Eq. (3.128) that the elastic strain energy E_{fe} is directly determined from the nodal point displacements, without determining any stresses. The same holds true for the average strain energy density, $\overline{W} = E_{fe}/V$.

'For monotonic convergence (to the exact solution), the elements must be complete and the elements and mesh must be compatible. The requirement of completeness of an element means that the displacement functions of the element must be able to represent the rigid body displacements and the constant strain states. The requirement of compatibility means that the displacements within the

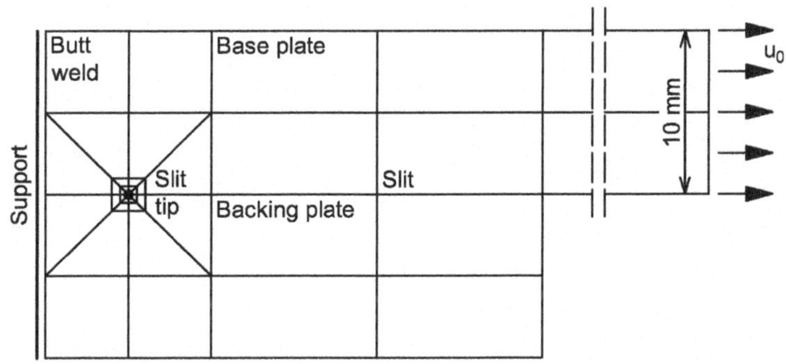

Fig. 3.70 Coarse FE mesh of tensile loaded (by remote displacement u_0) butt weld joint with backing plate (symmetry half); model for comparison with a fine mesh SED evaluation; parabolic isoparametric finite elements; (Lazzarin et al. 2010)

elements and across the element boundaries must be continuous.' (Bathe 1996, ibid. pp. 229 and 231).

The continuity of displacements across the element boundaries does not imply that the stresses are also continuous. Actually, the stresses along the element boundaries or at the element nodal points meeting in one mesh nodal point may be rather different, especially in the case of coarse meshes. Therefore, the requirements on mesh refinement for an accurate determination of the strain energy are much lower than the corresponding requirements in the case of stress field evaluations.

Coarse Mesh SED Evaluation for Welded Joint

Fine and coarse mesh evaluations in comparison have been performed for the tensile loaded butt weld joint with backing plate, already analysed in Sect. 3.6.8. The coarse finite element mesh surrounding the slit tip area is shown in Fig. 3.70. The fine and coarse meshes in the slit tip area are plotted in Fig. 3.71 with registration of the number of elements within the circle of control volume radius R_0 and in the total model (symmetry quarter). The appertaining local average SED values are recorded, first from a full circle and then from a semicircle evaluation (the semicircle extending in the base plate). Considering the fine mesh results as 'exact', the deviations of the coarse mesh results are rather small (0.2–0.7 %, further reduced for the corresponding NSIF values which depend on the square root of $\overline{W}$). The application of the SED-based coarse mesh option to welded joints in general is shown in a special paper (Lazzarin et al. 2008 [(1)]).

Slit tip R_0=0.28 mm	Slit tip 2×2 mm^2	Number of elements in circle	in model	$\overline{W}_{fc}$ [Nmm/mm^3]	$\overline{W}_{sc}$	Δ_{fc} [%]	Δ_{sc}
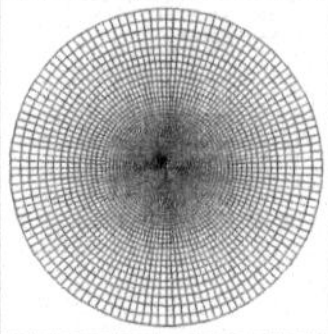	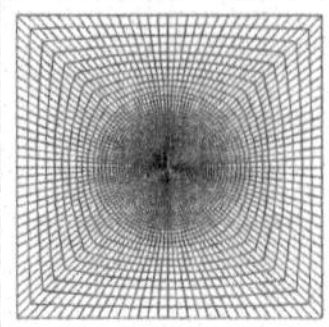	17546	23751	0.0437	0.0649		
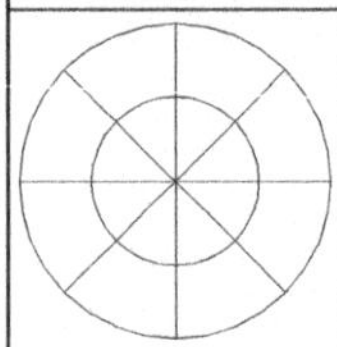	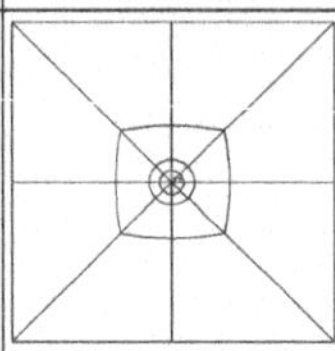	16	54	0.0450	0.0657	+2.7	+1.3
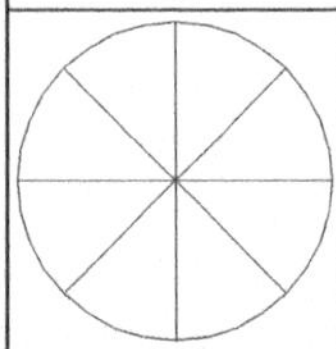	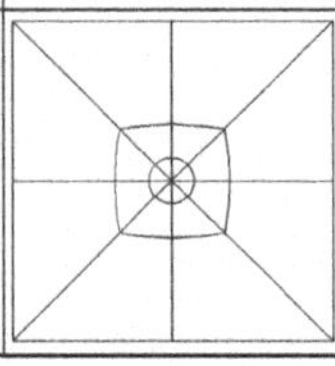	8	46	0.0447	0.0650	+2.1	+0.2

Fig. 3.71 Average local SED values (fc full circle, sc semicircle) at slit tip of tensile loaded butt weld joint with backing plate; coarse FE meshes in comparison to fine FE mesh; parabolic isoparametric finite elements; (Lazzarin et al. 2010)

Coarse Mesh SED Evaluation for Pointed V-Notches

First, the applied formulae from preceding sections are recapitulated. The NSIFs K_1 and K_2 characterising the stress singularity at the V-notch tip under in-plane tensile and shear (mode 1 and 2) loading conditions have the following form (Williams 1952; Gross and Mendelson 1972), Eqs. (2.77) and (2.78):

$$K_1 = \lim_{r \to 0} \sqrt{2\pi} r^{1-\lambda_1} \sigma_\theta(r, 0), \qquad K_2 = \lim_{r \to 0} \sqrt{2\pi} r^{1-\lambda_2} \tau_{r\theta}(r, 0) \tag{3.130}$$

where the exponents $(1 - \lambda_1)$ and $(1 - \lambda_2)$ depend on the notch opening angle 2α.

Considering the first order terms of Williams' solution, the total elastic strain energy density averaged over a small sector area of radius R, symmetric with respect to the V-notch bisector line, can be written as follows (Lazzarin and Zambardi 2001), Eq. (3.10):

$$\overline{W} = \frac{e_1}{E}\left(\frac{K_1}{R^{1-\lambda_1}}\right)^2 + \frac{e_2}{E}\left(\frac{K_2}{R^{1-\lambda_2}}\right)^2 \tag{3.131}$$

The parameters e_1 and e_2 depend on the multiaxiality condition (plane stress or plane strain), on the notch opening angle 2α and on Poisson's ratio v. For plane strain conditions, $v = 0.3$ and 2α given in degrees, the following expressions can be used, Eqs. (3.56) and (3.57):

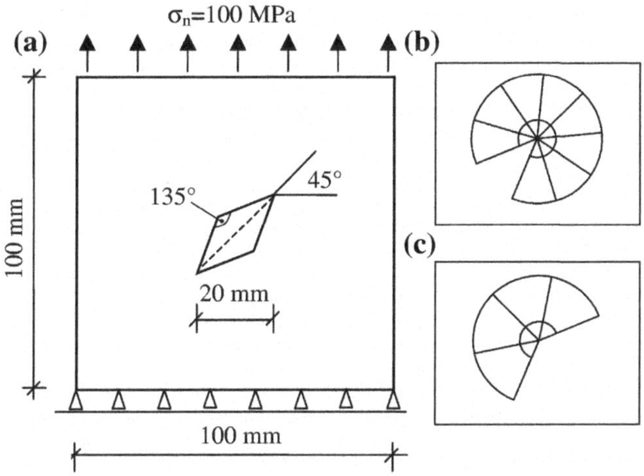

Fig. 3.72 Diamond-shaped inclined hole with pointed corner notches in tensile loaded quadratic plate (**a**) with globally even coarse mesh details at the corner notches (**b**, **c**); (Lazzarin et al. 2010)

$$e_1 = 0.133 + 6.151 \times 10^{-4}(2\alpha) - 5.373 \times 10^{-6}(2\alpha)^2 \tag{3.132}$$

$$e_2 = 0.340 - 2.346 \times 10^{-3}(2\alpha) + 4.809 \times 10^{-6}(2\alpha)^2 \tag{3.133}$$

Under pure mode 1 or, alternatively, pure mode 2 loading conditions, the appertaining NSIFs K_1 and K_2 can be determined from Eq. (3.131) in the following form:

$$K_1 = R^{1-\lambda_1}\sqrt{\frac{E\overline{W}}{e_1}}, \quad K_2 = R^{1-\lambda_2}\sqrt{\frac{E\overline{W}}{e_2}} \tag{3.134}$$

Under mixed mode loading conditions, K_1 and K_2 can be determined from Eq. (3.131) by evaluation of $\overline{W}$ for two different radii R resulting in two equations with the two unknown parameters K_1^2 and K_2^2. Solving these equations for K_1 and K_2 is straightforward.

In the case of slit tips ($2\alpha = 0$), appreciable T-stresses may be acting. The average SED $\overline{W}$ is then enlarged or diminished by two terms which are related to T^2 and $K_{\mathrm{I}}T$, Eq. (3.26). The $K_{\mathrm{I}}T$-related term is often negligibly small. When using Eq. (3.131) for the determination of K_1 and K_2, the T-stress related terms must first be subtracted from or added to the K_1 and K_2 related average SED $\overline{W}$. After establishing the K_1 and K_2 stress fields, the T-stress has to be reinserted.

As an example of application (without T-stress), a diamond-shaped inclined hole with pointed corner notches in a tensile loaded quadratic plate is analysed using fine and coarse meshes in comparison, Fig. 3.72. The corner notches have different opening angles, $2\alpha = 45$ and 135°. Both notches are subjected to mixed mode loading conditions. At the notch with $2\alpha = 45°$, the stress distributions from

Table 3.15 NSIF values K_1 and K_2 at corner notches of diamond-shaped inclined hole in tensile loaded quadratic plate; SED-based evaluation using fine and coarse meshes of parabolic isoparametric finite elements; number of elements n_1 and n_2 in smaller and larger control volume; (Lazzarin et al. 2010)

Corner notch			Fine mesh		Coarse FE mesh, $\sigma_n = 100$ MPa					
2α	λ_1	λ_2	K_1	K_2	n_1	n_2	K_1	K_2	Δ_1	Δ_2
			[MPa mm$^{1-\lambda}$]		Elements		[MPa mm$^{1-\lambda}$]		[%]	
45°	0.505	0.624	355	325	8	16	366	315	3.09	−3.1
135°	0.674	1.302	127	−97	4	8	128	−101	0.78	−4.1

mode 1 and mode 2 loading are both singular. At the notch with $2\alpha = 135°$, only the mode 1 stresses are singular (proportional to $1/r^{1-\lambda_1} = 1/r^{0.326}$), whereas the mode 2 stresses are non-singular (proportional to $1/r^{1-\lambda_2} = r^{0.302}$). The result of the comparative investigation with fine and coarse meshes using the radii $R = 0.3$ mm and $R = 1$ mm for differentiating between K_1 and K_2 is listed in Table 3.15. The deviations Δ from the 'exact' values when using coarse meshes are sufficiently small for technical applications.

Coarse Mesh SED Evaluation for Rounded Notches

The average SED evaluation in the crescent-shaped control volume of rounded notches has been performed with the aim to determine the maximum notch stresses using fine and coarse meshes in comparison (Lazzarin et al. 2010). The basic relationships are given in Sect. 3.4.2. They are applied here to the tensile loaded circular hole and to in-plane tensile and shear loaded U-notches (with a possible extension to rounded V-notches). The centre of the outer arc of the control volume is chosen at one half of the hole or U-notch radius.

The relationship between average local SED and maximum notch stress for the circular hole in a tensile loaded quadratic plate is derived based on the stress field solution for the hole in the infinite plate (Kirsch 1898), Fig. 3.73. The remote tensile stress is denoted by σ_0, the notch radius by a. The polar coordinate system is positioned with $\theta = 0$ in the loading direction. The stress components are:

$$\begin{aligned}
\sigma_r &= \frac{\sigma_0}{2}\left[1 + 3\left(\frac{a}{r}\right)^4 - 4\left(\frac{a}{r}\right)^2\right]\cos 2\theta + \frac{\sigma_0}{2}\left[1 - \left(\frac{a}{r}\right)^2\right] \\
\sigma_\theta &= -\frac{\sigma_0}{2}\left[1 + 3\left(\frac{a}{r}\right)^4\right]\cos 2\theta + \frac{\sigma_0}{2}\left[1 + \left(\frac{a}{r}\right)^2\right] \\
\tau_{r\theta} &= -\frac{\sigma_0}{2}\left[1 - 3\left(\frac{a}{r}\right)^4 + 2\left(\frac{a}{r}\right)^2\right]\sin 2\theta
\end{aligned} \tag{3.135}$$

The maximum notch stress occurs for $r = a$ and $\theta = 90°$, $\sigma_{\theta\,\max} = 3.0\sigma_0$.

The crescent-shaped control volume has the hole edge as boundary at the one side (distance $R_1 = a$ from the hole centre) and a circle arc around point P (at $a/2$)

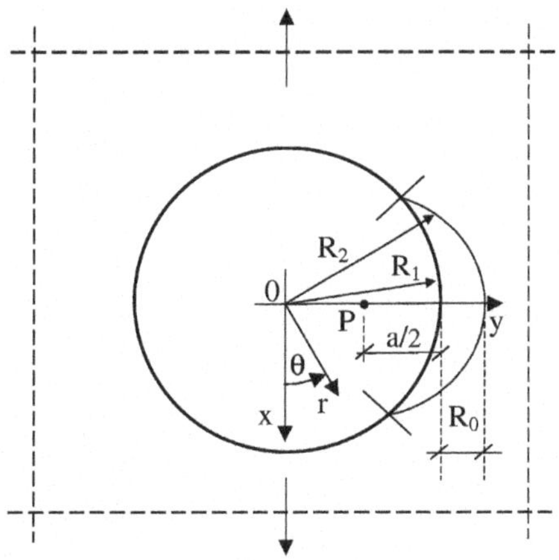

Fig. 3.73 Circular hole of radius *a* in tensile loaded plate; coordinate systems and crescent-shaped control volume; (Lazzarin et al. 2010)

on the other side (distance R_2 from the hole centre). The distance R_2 varies according to the following expression (Lazzarin et al. 2010):

$$R_2(\theta) = \frac{1}{2}\left(a \sin\theta + \sqrt{4R_0^2 + 4aR_0 + a^2 \sin^2\theta}\right) \tag{3.136}$$

The apex of the crescent-shaped control volume corresponds to the angle $\overline{\theta}$ according to:

$$\overline{\theta} = \arccos\left(1 - \frac{R_0}{a} - \frac{R_0^2}{a^2}\right) + \frac{\pi}{2} \tag{3.137}$$

The strain energy density under the condition of linear-elastic material behaviour and in-plane loading is given by:

$$W(r,\theta) = \frac{1}{2E}\left[\sigma_r^2 + \sigma_\theta^2 + \sigma_z^2 - 2\nu(\sigma_r\sigma_\theta + \sigma_\theta\sigma_z + \sigma_z\sigma_r) + 2(1+\nu)\tau_{r\theta}^2\right] \tag{3.138}$$

with $\sigma_z = 0$ for plane stress conditions and $\sigma_z = \nu(\sigma_r + \sigma_\theta)$ for plane strain conditions.

The average SED in the control volume (area A) is determined by:

$$\overline{W} = \frac{2}{A}\left[\int_{+\pi/2}^{+\overline{\theta}} \mathrm{d}\theta \int_{R_1}^{R_2} W(r,\theta)\, r\, \mathrm{d}r\right] \tag{3.139}$$

The solution of this integral can be written in the following form:

$$\overline{W} = \frac{(3\sigma_0)^2}{2E} H(R_0/a) \tag{3.140}$$

Table 3.16 Parameter H dependent on R_0/a for tensile loaded plate with circular hole; plane strain conditions, $\nu = 0.3$; (Lazzarin et al. 2010)

R_0/a	0.01	0.03	0.05	0.1	0.2
H	0.881	0.826	0.775	0.663	0.496

Table 3.17 Maximum notch stress σ_{max} at circular hole of radius a in tensile loaded quadratic plate, 80 × 80 mm^2; SED-based evaluation using fine and coarse meshes of parabolic isoparametric finite elements; number of elements in symmetry quarter, in total and in control volume; (Lazzarin et al. 2010)

Circular hole		Fine mesh	Coarse FE mesh, $\sigma_n = 100$ MPa				
a [mm]	R_0 [mm]	σ_{max} [MPa]	Elements in total	Elements in control volume	$\overline{W}$ [Nmm/mm^3]	σ_{max} [MPa]	Δ [%]
10	0.1	330.4	29	1	0.237	333.3	0.9
10	0.3	330.4	35	3	0.218	330.2	−0.1
10	0.5	330.4	32	3	0.203	328.2	−0.7
10	1	330.4	32	6	0.173	327.9	−0.7
10	2	330.4	31	6	0.129	326.9	−1.0

The equation is exact for the infinite plate, but can be used with a small error, provided that σ_{max} substitutes the term $3\sigma_n$:

$$\overline{W} = \frac{\sigma_{max}^2}{2E} H(R_0/a) \tag{3.141}$$

The parameter H is listed in Table 3.16 dependent on R_0/a for plane strain conditions and $\nu = 0.3$.

As an example of application, the maximum notch stress σ_{max} is determined for a circular hole ($a = 10$ mm) in a tensile loaded quadratic plate (80 × 80 mm^2). The SED-based evaluation using fine and coarse meshes is set into comparison, Table 3.17. The deviations of σ_{max} evaluated from the coarse mesh is not larger than 1 %.

Analytically derived values from the literature are $\sigma_{max} = 325$ MPa for the hole in a plate of finite width but infinite length (Howland 1929/30) and $\sigma_{max} \approx 338$ MPa (rough interpolation) for the hole in a quadratic plate (Savin 1970), see (Radaj and Schilberth 1977, ibid. p. 65).

The SED-based procedure for determining the maximum notch stress can also be applied to U-notches and V-notches with a circular root shape based on Eq. (3.64) for blunt V-notches:

$$\overline{W} = F(2\alpha)H(2\alpha, R_0/\rho)\frac{\sigma_{max}^2}{E} \tag{3.142}$$

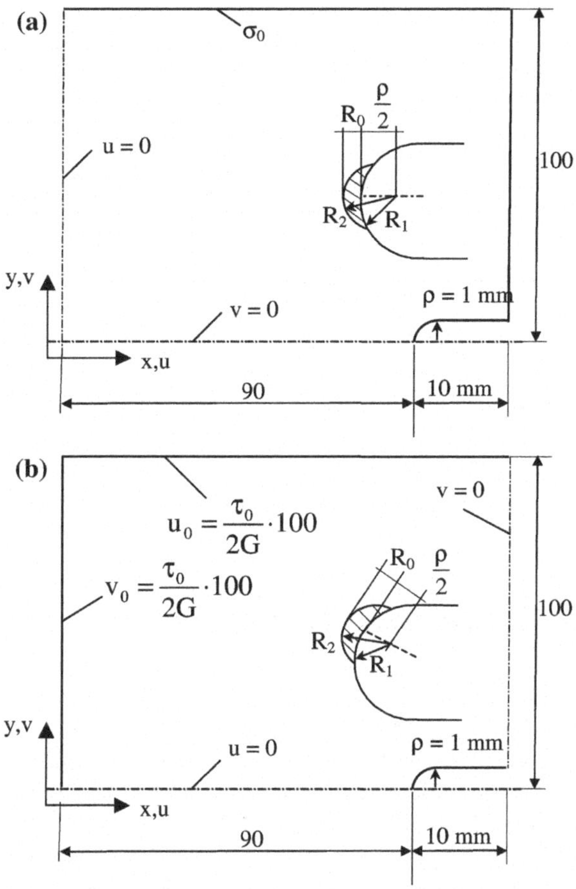

Fig. 3.74 Two-sided external U-notch in quadratic plate (symmetry quarter) subjected to tensile loading σ_0 (**a**) and two-sided internal U-notch in quadratic plate subjected to in-plane shear loading τ_0 (**b**), the latter produced by prescribed displacements v_0 and u_0 on the plate edges; geometrical parameters of the control volume; (Lazzarin et al. 2010)

For the U-notch, the maximum notch stress derived from this equation inserting $2\alpha = 0°$ reads:

$$\sigma_{\max} = \sqrt{\frac{4E}{\pi}\frac{\overline{W}}{H}} \qquad (3.143)$$

with the parameter H listed in Table 3.2.

The Eq. (3.143) is applied to the U-notch with semicircular root shape with the centre point of the crescent-shaped control volume at $\rho/2$ in the direction of $\sigma_{\max}$ ($\theta = 0°$ in mode 1 and $\theta \approx 70°$ in mode 2 loading).

A quadratic plate with two-sided external U-notch subjected to tensile loading (σ_0) and a corresponding plate with two-sided internal U-notch subjected to shear loading (τ_0) are analysed, Fig. 3.74. Fine and coarse meshes are applied in comparison. The results are listed in Table 3.18. The comparison values from an analytical solution for the U-shaped internal notch in the infinite plate (Radaj and

Table 3.18 Maximum notch stress σ_{max} at U-notch of radius ρ in rectangular plate subjected to tensile (σ_0) and shear (τ_0) loading; SED-based evaluation using fine and coarse meshes of parabolic isoparametric finite elements; number of elements in total and in control volume; (Lazzarin et al. 2010)

	U-notch		Fine mesh	Coarse FE mesh, $\sigma_n = 100$ MPa				
	ρ [mm]	R_0 [mm]	σ_{max} [MPa]	Elements in total	Elements in control volume	$\overline{W}$ [Nmm/mm^3]	σ_{max} [MPa]	Δ [%]
σ_0	1	0.3	816	33	3	0.740	796	−2.5
τ_0	1	0.3	880	34	4	1.186	830	−5.6

Schilberth 1977, ibid. pp. 22 and 26) with the remote reference stresses reduced by a factor of 1.1 on σ_0 or 1.05 on τ_0 (stiffness reduction by the hole) are $\sigma_{max} \approx 808$ MPa for σ_0 and $\sigma_{max} \approx 882$ MPa for τ_0. These values are well in correspondence with the fine mesh results.

3.7.3 Peak Stress Method

It has been shown, at first for crack tips (Nisitani and Teranishi 2004) and then for V-notch tips (Meneghetti and Lazzarin 2007), that the SIF K_I or NSIF K_1, respectively, can be approximated based on the elastic maximum or peak stress, $\sigma_{max} = \sigma_p$, calculated at the crack or notch tip by means of the FE method using a mesh pattern with a defined element type and a globally even element size. The global evenness of the element size is the feature of meshes generated by a special algorithm in the FE programme ANSYS. The peak stress under mode 1 loading is defined as the maximum elastic principal stress in the considered slit tip or notch tip nodal point of the mesh. This nodal point stress results as the arithmetic mean of the stress components in the two ($2\alpha > 90°$) or four elements ($2\alpha < 90°$) meeting in the nodal point. It is proven that the ratio K_I/σ_p or K_1/σ_p depends mainly on the element size and type, modified solely by the notch opening angle 2α. In the following, linear quadrilateral elements representing constant stresses within the element are referred to.

The ratio K_1/σ_p as a function of the global element size d and the notch opening angle 2α (represented by λ_1) is derived based on the elastic maximum notch stress for the blunt V-notch (Lazzarin and Filippi 2006), Eq. (2.135):

$$\sigma_p = \sigma_{max} = (1 + \tilde{\omega}_1)\frac{K_{1\rho}}{\sqrt{2\pi} r_0^{1-\lambda_1}} \tag{3.144}$$

where r_0 is the distance of the notch tip from the origin of the curvilinear coordinates and $\tilde{\omega}_1$ is an auxiliary parameter which depends on 2α. The distance r_0 varies as a function of 2α and ρ (notch radius), Eq. (2.115):

$$r_0 = \frac{q-1}{q}\rho = \frac{\pi - 2\alpha}{2\pi - 2\alpha}\rho \quad (\rho = \rho^*) \tag{3.145}$$

where ρ^* is an auxiliary notch tip radius which is considered as independent on notch depth a and notch angle 2α.

In order to include the element size d, the dimensionless ratio κ_{fe} is introduced. The generally weak influence of ρ/a is taken into account by $K_{1\rho} \approx \varphi K_1$ (Sect. 2.4.5). Based on the above equations, the following expression is gained:

$$\kappa_{\mathrm{fe}} = \frac{K_1}{\sigma_{\mathrm{p}} d^{1-\lambda_1}} = \frac{1}{\varphi(1+\tilde{\omega}_1)}\sqrt{2\pi}\left(\frac{q-1}{q}\frac{\rho^*}{d}\right)^{1-\lambda_1} \tag{3.146}$$

Numerical evaluations for various element sizes d, notch depths a and notch angles 2α gave the following relationship for the considered linear quadrilateral elements (Meneghetti and Lazzarin 2007):

$$\kappa_{\mathrm{fe}} = \frac{K_1}{\sigma_{\mathrm{p}} d^{1-\lambda_1}}, \quad \kappa_{\mathrm{fe}} = 1.38 \tag{3.147}$$

The relationship expresses the fact that for a defined global element size, e.g. $d \approx 1$ mm, the ratio K_1/σ_{p} is also defined. The influence of ρ/a is neglected in the following ($\varphi \approx 1.0$). Any other choice of the element type, e.g. parabolic quadrilateral elements, necessitates a new calibration resulting in another value of κ_{fe}.

Application to Cruciform Joints

An element mesh with the mentioned properties (linear quadrilateral elements, globally even element size, element size $d \approx 1$ mm) is applied to a cruciform joint model with toe and root notches, Fig. 3.75. The mesh is generated using the 'free mesh algorithm' of the FE programme ANSYS, providing a globally even element size ($d = 1$ mm). It is essential that two elements share the node located at the weld toe, while four elements must meet at the weld root. The scatter band of test results for fillet-welded attachment joints (weld toe angle $\phi = 45°$) failing from the weld toe, presented in previous publications in terms of the NSIF ΔK_1 (Lazzarin and Tovo 1998; Lazzarin and Livieri 2001), could thus be recalibrated in terms of the peak stress $\Delta\sigma_{\mathrm{p}}$ (Meneghetti and Lazzarin 2007; Meneghetti 2008).

Relationship Between Peak Stress and Local SED

The relationship between the average local SED and the NSIFs K_1 and K_2 is given by Eq. (3.131). The relationship to the peak stress can be derived therefrom. First, the case of pure K_1 loading is dealt with. The mode 2 stress contribution is considered to be non-singular or simply negligible in comparison to the singular

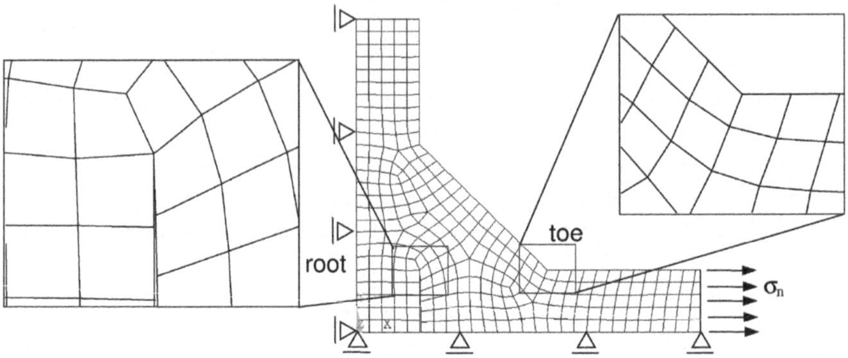

Fig. 3.75 Globally even FE mesh for application of the peak stress method to a cruciform welded joint model with toe and root notches (symmetry quarter); quadrilateral elements of even size $d \approx 1$ mm; (Meneghetti and Lazzarin 2010)

mode 1 contribution. Then, the relationship restricted to mode 1 loading has the following form:

$$\Delta\overline{W} = \frac{e_1}{E}\left(\frac{\Delta K_1}{R_0^{1-\lambda_1}}\right)^2 \tag{3.148}$$

A comparison between Eq. (3.148) and the K_1 term in Eq. (3.146) leads to the following equation referring to plane strain conditions (Meneghetti and Lazzarin 2010):

$$\Delta\overline{W} = \frac{e_1}{E}\left[\kappa_{\text{fe}}\Delta\sigma_{\text{p}}\left(\frac{d}{R_0}\right)^{1-\lambda_1}\right]^2 = \frac{1-\nu^2}{2E}\Delta\sigma_{\text{eq,p}}^2 \tag{3.149}$$

The SED-related equivalent peak stress $\sigma_{\text{eq,p}}$ can therefore be expressed in terms of the original peak stress σ_{p} (introducing $\kappa_{\text{fe}} = 1.38$):

$$\Delta\sigma_{\text{eq,p}} = 1.38\Delta\sigma_{\text{p}}\sqrt{\frac{2e_1}{1-\nu^2}}\left(\frac{d}{R_0}\right)^{1-\lambda_1} = f_{\text{w}}\Delta\sigma_{\text{p}} \tag{3.150}$$

The SED-based weight factor f_{w} modifies the peak stress to some extent, which is now termed 'weighted peak stress'. Some values of λ_1, e_1 and f_{w} which are useful for the evaluation of Eq. (3.150) are listed in Table 3.19.

Using the weighted peak stress $f_{\text{w}}\Delta\sigma_{\text{p}}$, the scatter band of fatigue test data referring to fillet-welded attachment joints mentioned above can be converted to a modified scatter band, Fig. 3.76, which is now valid for any weld toe angle ϕ (substituting $\phi \approx 45°$) and also for welded joints with root fractures. About 200 fatigue test data referring to weld root fractures have been plotted into the scatter band of the toe fractures, Fig. 3.77. The different colours of the triangular symbols cannot be distinguished in the black and white plot. The welded joints are made of

Table 3.19 Values of parameters in Eq. (3.150) dependent on notch angle 2α; (Meneghetti and Lazzarin 2010)

2α	λ_1^a	e_1^a	f_w^b	2α	λ_1^a	e_1^a	f_w^b
0°	0.500	0.133	1.410	125°	0.633	0.126	1.159
90°	0.544	0.145	1.392	135°	0.674	0.118	1.064
110°	0.586	0.136	1.278	145°	0.723	0.109	0.961
120°	0.616	0.129	1.198	150°	0.752	0.104	0.905

[a] Values according to (Lazzarin and Zambardi 2001)
[b] Values calculated with $R_0 = 0.28$ mm, $d = 1$ mm, $\nu = 0.3$

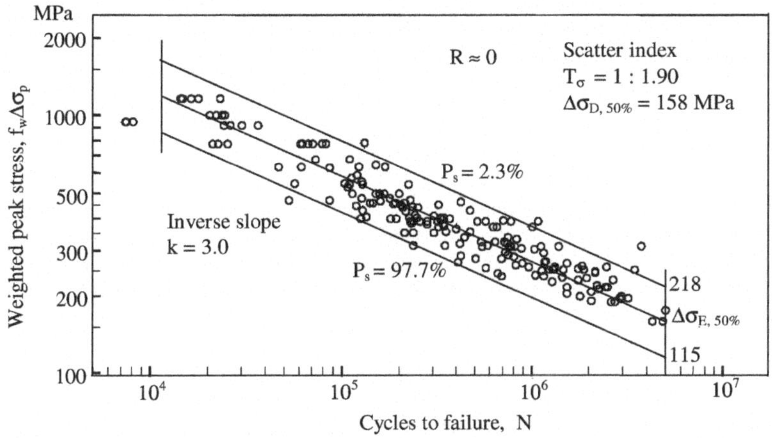

Fig. 3.76 Fatigue test results in terms of the weighted peak stress; cruciform attachment joints of steel with non-load-carrying fillet welds, weld toe angle $\phi = 45°$, element size $d \approx 1$ mm; design curve inverse slope exponent $k = 3.0$; (Meneghetti and Lazzarin 2010, Livieri and Lazzarin 2005)

steels with a yield limit ranging from 252 to 690 MPa. They were fatigue-tested in the as-welded condition applying the load ratio $R \approx 0$. The same bulk of data has already been evaluated in terms of the local SED (Livieri and Lazzarin 2005), Figs. 3.11 and 3.12.

Extension to Mode 2 and Mixed Mode 1 and 2 Loading

The peak stress method reviewed above for mode 1 loading relates the NSIF K_1 to the maximum principal stress $\sigma_p = \sigma_{11,p}$ (see Fig. 3.78 for $\sigma_{11,p}$):

$$\kappa_{1,fe} = \frac{K_1}{\sigma_{11,p} d^{1-\lambda_1}}, \quad \kappa_{1,fe} \approx 1.38 \tag{3.151}$$

A corresponding relationship is established for mode 2 loading:

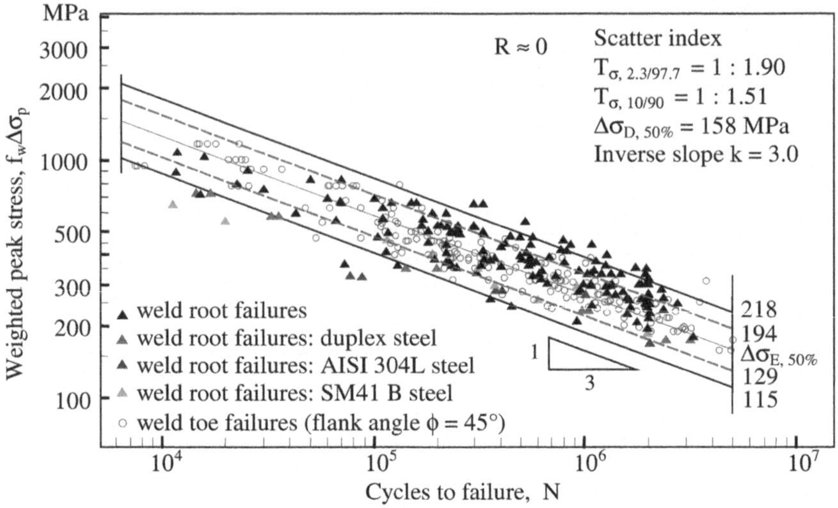

Fig. 3.77 Fatigue test results in terms of weighted peak stress; cruciform welded joints of steel with inclusion of the root failures of load-carrying welds; circle symbol for toe fractures and limit curves identical with Fig. 3.76, triangular symbols for root fractures; solid line curves for $P_s = 2.3$ and 97.7 %, broken line curves for $P_s = 10$ and 90 %; (Meneghetti and Lazzarin 2010)

$$\kappa_{2,\text{fe}} = \frac{K_2}{\tau_{xy,\text{p}} d^{1-\lambda_2}}, \quad \kappa_{2,\text{fe}} \approx 3.38 \tag{3.152}$$

The value $\kappa_{2,\text{fe}} \approx 3.38$ is found by numerical evaluations for various globally even element sizes d applied to the slit notch ($2\alpha = 0$) with different notch depths a. The value above is valid for $a/d > 14$. It decreases for $a/d < 14$. Linear quadrilateral elements are considered. Any other choice of the element type, e.g. parabolic quadrilateral elements, necessitates a new calibration resulting in other values of $\kappa_{1,\text{fe}}$ and $\kappa_{2,\text{fe}}$.

The mixed mode relationship in analogy to Eq. (3.149) referring only to mode 1 has the following form (see Fig. 3.78 for $\sigma_{yy,\text{p}}$ and $\tau_{xy,\text{p}}$):

$$\begin{aligned} \Delta\overline{W} &= \frac{e_1}{E}\left[\kappa_{1,\text{fe}}\Delta\sigma_{yy,\text{p}}\left(\frac{d}{R_0}\right)^{1-\lambda_1}\right]^2 + \frac{e_2}{E}\left[\kappa_{2,\text{fe}}\Delta\tau_{xy,\text{p}}\left(\frac{d}{R_0}\right)^{1-\lambda_2}\right]^2 \\ &= \frac{1-\nu^2}{E}\Delta\sigma_{\text{eq,p}}^2 \end{aligned} \tag{3.153}$$

Solving Eq. (3.153) for $\sigma_{\text{eq,p}}$ has the following result:

$$\Delta\sigma_{\text{eq,p}} = \sqrt{f_{\text{w1}}^2\Delta\sigma_{yy,\text{p}}^2 + f_{\text{w2}}^2\Delta\tau_{xy,\text{p}}^2} \tag{3.154}$$

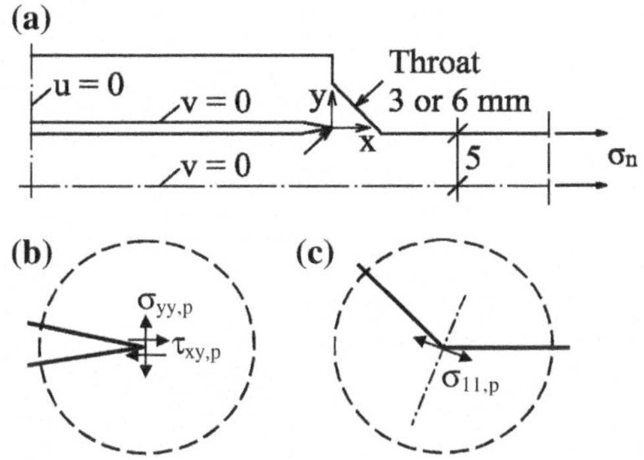

Fig. 3.78 Model of a fillet-welded cover plate joint subjected to tensile loading (**a**), symmetry quarter; lap joint model generated by cutting the base plate in the left hand middle section; peak stresses at weld root (**b**) and weld toe (**c**); (Meneghetti and Lazzarin 2011)

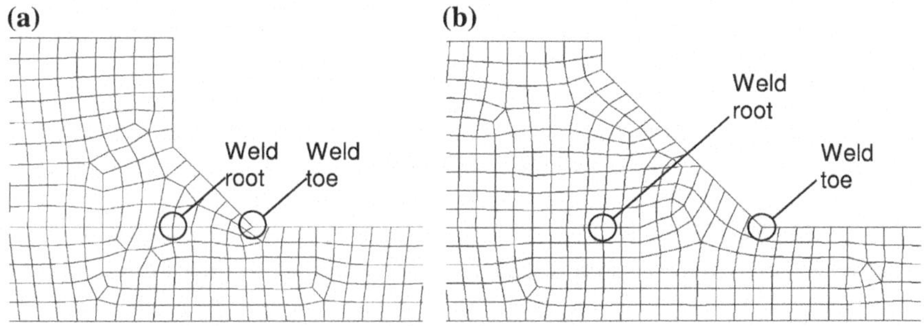

Fig. 3.79 Globally even FE mesh for application of the peak stress method to the model of the welded cover plate and lap joint; smaller (**a**) and larger (**b**) throat thickness; quadrilateral elements of even size $d \approx 1$ mm; (Meneghetti and Lazzarin 2011)

$$f_{w1} = \kappa_{1,fe}\sqrt{\frac{2e_1}{1-\nu^2}}\left(\frac{d}{R_0}\right)^{1-\lambda_1}, \quad f_{w2} = \kappa_{2,fe}\sqrt{\frac{2e_2}{1-\nu^2}}\left(\frac{d}{R_0}\right)^{1-\lambda_2} \tag{3.155}$$

Introducing $\kappa_{2,fe} = 3.38$ for the weld root ($2\alpha = 0$) together with $\lambda_2 = 0.5$, $d = 1$ mm and $R_0 = 0.28$ mm, one obtains $f_{w2} = 5.52$ (compared with $f_{w1} = 1.410$, see Table 3.19).

Application to Lap Joints and Cover Plates

Recent fatigue test results for fillet-welded lap joints and cover plates subjected to tensile loading (Meneghetti and Lazzarin 2011) are compared below with the statistically well founded scatter band for cruciform joints (as-welded condition, $R \approx 0$). The model of the lap joint and cover plate (symmetry quarter) is shown in

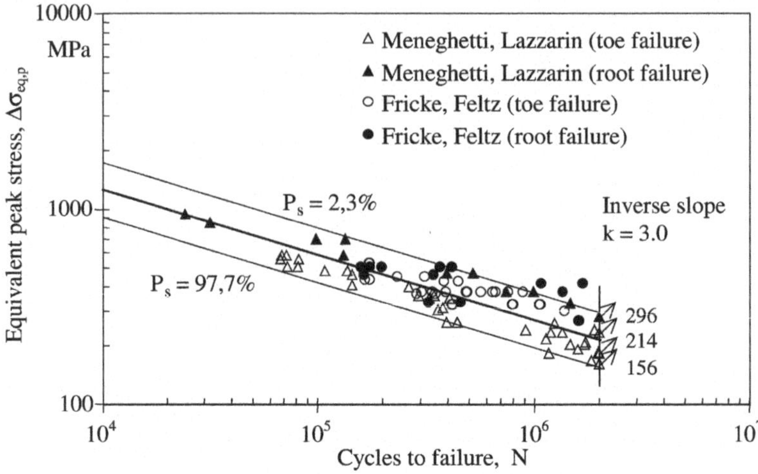

Fig. 3.80 Fatigue test results in terms of the equivalent peak stress; welded cover plate and lap joints made of steel; globally even element size $d \approx 1$ mm; scatter band transferred from Fig. 3.76; (Meneghetti and Lazzarin 2011)

Fig. 3.78 together with the peak stress components at the weld root and weld toe. Two throat thicknesses are considered. Toe and root fractures are observed in the lap joint tests, but only toe fractures in the cover plate tests.

The applied FE mesh with the globally even element size $d \approx 1$ mm is plotted in Fig. 3.79. The two meshes represent the smaller and larger throat thickness. The test results are assigned to the design scatter band of cruciform joints considered above resulting in excellent correspondence, Fig. 3.80. Relevant new data from the literature (Fricke and Feltz 2010) don't have this high degree of correspondence. Unexplained is the too large inverse slope expressed by these data. The deviations occur mainly in the high-cycle fatigue range where the influence of the residual stresses is strong. Unknown compressive stresses may have caused the deviations.

3.8 Conclusions from Chapter Contents

The local strain energy density approach provides powerful tools to solve the mechanical field problem at sharp notches, thus facilitating the assessment of brittle fracture and fatigue failure of structural members. The size effect is naturally included. Finite element procedures based on coarse meshes are possible without a major loss in accuracy.

The main part of the local SED concept refers to linear-elastic conditions. Such conditions are directly applicable to brittle fractures and high-cycle fatigue failures. The concept has also been extended to elastic-plastic local or global conditions, thus establishing a link between elastic and plastic field properties.

Multiaxial local loading conditions are easily taken into account (presuming linear-elastic material behaviour). The concept is applicable to pointed notches as well as to rounded notches. In the former case, the link to the notch stress intensity factors is established, in the latter case the link to the maximum notch stresses.

The local SED concept is well in correspondence with related notch-mechanical approaches such as the Kitagawa and Atzori diagrams, the Neuber concept of fictitious notch rounding and the J-integral.

Special issues are easily analysed and discussed on the basis of the SED concept. Semicircular control volumes, microrounded reference notches and slit-parallel loading are among them.

Most important for technical application is the fact that accurate results can be gained with coarse FE meshes based on the SED concept. A statistically well-founded uniform design W–N curve for welded joints of steel or aluminium alloy provides the basis for a reliable fatigue assessment of welded joints.

List of Symbols

A	Control volume area
a	Notch depth, hole radius, crack length
a^*, a_0^*, a_{0k}^*	Crack length parameters
a_i	Initiated crack length
B_W	Parameter, dependent on 2α and n
$[B]$	Strain–displacement matrix
C	Coefficient in J-integral
c_W	Prestress coefficient of SED
d	Globally even FE size
d_g	Grain size
$\{d\}$	Vector of nodal point displacements
$[E]$	Elasticity matrix
E_{fe}, E_l	Strain energy in finite element, in local volume
E, E'	Modulus of elasticity
e_1, e_2, e_3	Total SED coefficient, mode 1, 2, 3
e_{d1}, e_{d2}, e_{d3}	Distortional SED coefficient, mode 1, 2, 3
F	Parameter, dependent on 2α
F, F_c	Force, critical force
f_1, f_2	Angle-dependent function related to W_1, W_2
f_{d1}, f_{d2}, f_{d3}	Angle-dependent function related to W_{d1}, W_{d2}, W_{d3}
f_w, f_{w1}, f_{w2}	Peak stress coefficient, mode 1, 2
G	Shear modulus
H	Parameter, dependent on 2α and R_0/ρ
H, H_τ	Hardening coefficient, tensile and shear loading
I_1, I_2	Integral over f_1, f_2
I_{d1}, I_{d2}, I_{d3}	Integral over f_{d1}, f_{d2}, f_{d3}
I_e, I_p	Integral over elastic and plastic function f_{1e}, f_{1p}

J	J-integral
J_V, J_L	J-integral for V-notches
K_1, K_2, K_3	NSIF, mode 1, 2, 3
K_{1e}, K_{3e}	Elastic NSIF, mode 1, 3
K_{1p}, K_{3p}	Plastic NSIF, mode 1, 3
K_{1E}, K_{3E}	Endurable NSIF, mode 1, 3
$K_{1\rho}$, $K_{2\rho}$	Generalised NSIF, mode 1, 2
K_{1c}	Critical NSIF, mode 1
K_I, K_{II}, K_{III}	SIF, mode I, II, III
K_{eq}, K_{eq}^*	Equivalent SIF, without and with T-stress
K_{Ic}, K_c	Fracture toughness
K_0	Threshold SIF
$K_t, \overline{K}_t$	Theoretical SCF, fatigue-effective SCF
K_{t1}, K_{t2}	SCF at weld toe, weld root
$K_{t,B}$, $K_{t,vM}$	SCF of equivalent stress, Beltrami, von Mises
$\overline{K}_{t,B}$, $\overline{K}_{t,vM}$	Fatigue-effective value of $K_{t,B}$, $K_{t,vM}$
K_W, $K_{t,W}$, $K_{t,Wd}$	SED-based SCF, total, distortional
$K_{f,N}$, $K_{f,W}$	Fatigue notch factor, based on Neuber, on SED
k	Inverse slope exponent, S–N or W–N curve
m	Elastic-plastic antiplane shear exponent
N, N_i, N_E	Number of cycles, crack initiation, endurable value
$[N]$	Interpolation matrix
n	Hardening exponent
n_1, n_2	Exponent of plate thickness effect
P_f, P_s	Probability of failure, of survival
q	Factor on $\pi/2$ for notch internal angle
R	Load ratio, nominal stress ratio
R	Radius
R_0	Radius of integration path
R_0, R_{0d}	Radius of control volume for W, W_d
R_{01}, R_{03}	Radius of control volume, mode 1, 3
R_0, R_0^*	Radius of control volume, original, enlarged
R_1, R_2	Radii of crescent-shaped control volume
r	Polar coordinate, radial distance
r_0	Notch root distance
r_p	Radius of plastic zone
s	Elastic-plastic eigenvalue dependent on λ_1 and n
s	Microstructural support factor
T	Temperature or T-stress
T_σ, T_W	Scatter range index, related to σ, to W
$\boldsymbol{T}$	Traction vector in J-integral
t, t_0, t_1, t_2	Plate thickness
t	Notch depth
u_0	Remote boundary displacement, x direction
$\{u\}$, $\boldsymbol{u}$	Displacement vector
V	Volume

v_0	Remote boundary displacement, *y* direction
W, W_c	Total SED, critical value
W_1, W_2, W_3	Total SED, mode 1, 2, 3
W_d, W_{d1}, W_{d2}	Distortional SED, mode 1, 2
W_n, $W_{n,n}$, $W_{n,g}$	Nominal SED, net and gross cross-section
$W_{1\ max}$	Maximum total SED, mode 1
$W_{1,\rho}$, $W_{1,0}$	Total SED, with and without microrounding
$W_{1d,\rho}$, $W_{1d,0}$	Distortional SED, with and without microrounding
$\overline{W}$, $\overline{W}_1$, $\overline{W}_2$	Average total SED, mode 1, 2
$\overline{W}_d$	Average distortional SED
$\overline{W}_I$, $\overline{W}_{II}$	Average total SED, mode I, II
$\overline{W}_{1e}$, $\overline{W}_{1p}$	Average total SED, elastic, plastic, mode 1
$\overline{W}_{3e}$, $\overline{W}_{3p}$	Average total SED, elastic, plastic, mode 3
$\overline{W}_c$, $\overline{W}_{dc}$	Critical average SED, total, distortional
W_0	Critical SED in unnotched specimen
$\overline{W}_E$	Endurable average total SED
w	Width of cross-section
w_1, w_2	Geometric parameters
x, y, z	Cartesian coordinates
Y, Y_1, Y_2	Geometry factor, mode 1, 2
α	Notch opening semi-angle
β	Notch arc related angle, coefficient in σ_{eq} formula
γ	V-notch internal semi-angle, coefficient in σ_{eq} formula
Δ, Δ_{fc}, Δ_{sc}	Relative deviation, full circle, semicircle
δ	Gap width
δ_{ij}	Kronecker delta
ε, ε_{ij} $\{\varepsilon\}$	Strain, strain tensor, strain vector
$\varepsilon_{\theta\theta}$, ε_{rr}, $\varepsilon_{r\theta}$	In-plane strain components in polar coordinates
ε_{zz}	Out-of-plane strain component
θ, $\overline{\theta}$, θ^*	Polar angle, crescent shape angle, transition angle
θ_0	Crack propagation angle
κ	Slope exponent
κ_1, $\kappa_{1,0}$	Dimensionless shape factor
κ_{fe}	Dimensionless ratio, K_1 or K_2 related to σ_p
λ_1, λ_2, λ_3	Eigenvalue at V-notch, mode 1, 2, 3
λ	Stress amplitude ratio τ_a/σ_a
ρ, ρ_c	Notch radius, notch curvature radius
ρ_f, ρ_r	Radius of fictitious notch, of reference notch
ρ^*	Microstructural support length
ρ^*	Auxiliary notch tip radius
σ, σ_{ij}, $\{\sigma\}$	Stress, stress tensor, stress vector
$\sigma_{\theta\theta}$, σ_{rr}, $\sigma_{r\theta}$	In-plane stress components in polar coordinates
σ_θ, σ_r, $\tau_{r\theta}$	In-plane stress components in polar coordinates
σ_{zz}, σ_z	Out-of-plane stress component
$\tilde{\sigma}_{\theta\theta}$, $\tilde{\sigma}_{rr}$, $\tilde{\sigma}_{r\theta}$	Angular functions of in-plane stresses
$\tilde{\sigma}_{zz}$	Angular functions of out-of-plane stress

σ_{kk}	Sum of principal stresses
σ_n, $\sigma_{n,g}$, $\sigma_{n,n}$	Nominal stress, in net and gross cross-section
σ_l, σ_u	Lower and upper nominal stress of load cycle
$\sigma_{k\ max}$	Maximum notch stress
$\sigma_{\theta\ max}$, $\sigma_{t\ max}$	Maximum tangential stress
σ_a, σ_A	Stress amplitude, endurable value
σ_0	Substitute yield limit, remote nominal stress
$\overline{\sigma}_0$, $\overline{\sigma}_{0b}$	Remote stress, membrane, bending
σ_Y, $\sigma_Y{}'$	Yield limit, monotonic, cyclic
σ_U	Ultimate tensile strength
σ_E, σ_{nE}	Endurance limit stress, endurable nominal stress
σ_{eq}, σ_B, σ_{vM}	Equivalent stress, Beltrami, von Mises
σ_h	Hydrostatic stress
σ_p, $\sigma_{eq,p}$	Peak stress in FE mesh, equivalent peak stress
τ	Shear stress
τ_a	Shear stress amplitude
$\tau_{xy\infty}$	Remote shear stress
τ_0	Substitute shear yield limit, remote nominal shear stress
τ_{max}	Maximum notch shear stress
τ_E	Endurance limit shear stress
ϕ	Phase shift angle, weld toe angle
φ	Coefficient of K_1
χ_h	Multiaxiality index
ψ	Inclination angle
ω, ω_d	Dimensionless SED, total, distortional
$\overline{\omega}$, $\overline{\omega}_d$	Dimensionless average SED, total, distortional
$\tilde{\omega}_1$	Auxiliary parameter, dependent on 2α
BE	Boundary element
BS	British Standard
CJ	Cruciform joint
FAT	Fatigue strength class
FE	Finite element
FEM	Finite element method
FNR	Fictitious notch rounding
IIW	International Institute of Welding
LEFM	Linear-elastic fracture mechanics
NSIF	Notch stress intensity factor
SCF	Stress concentration factor
SED	Strain energy density
SIF	Stress intensity factor
TAJ	Transverse attachment joint
cc, ec	Concentric, eccentric (reference notch)
fc, sc	Full circle, semicircle (average SED)
fm, cm	Fine mesh, coarse mesh
ns	Narrow section (average SED)

References

Atzori B, Dattoma V (1983) A comparison of the fatigue behaviour of welded joints in steels and aluminium alloys. IIW Doc XIII-1089–1083

Atzori B, Lazzarin P (2001) Notch sensitivity and defect sensitivity: two sides of the same medal. Int J Fract 107:L3–L8

Atzori B, Lazzarin P, Tovo R (1999) From a local stress approach to fracture mechanics: a comprehensive evaluation of the fatigue strength of welded joints. Fatigue Fract Engng Mater Struct 22:369–381

Atzori B, Lazzarin P, Filippi S (2001) Cracks and notches – analogies and differences of the relevant stress distributions and practical consequences in fatigue limit predictions. Int J Fatigue 23:355–362

Atzori B, Lazzarin P, Meneghetti G (2003) Fracture mechanics and notch sensitivity. Fatigue Fract Engng Mater Struct 26:257–267

Atzori B, Berto F, Lazzarin P, Quaresimin M (2006) Multiaxial fatigue behaviour of a severely notched carbon steel. Int J Fatigue 28:485–493

Bathe KJ (1996) Finite element procedures. Prentice Hall, Englewood Cliffs

Beltrami E (1885) Sulle condizioni di resistenza dei corpi elastici. Rendic R Istituto Lombardo di Scienze, Lettere e Arti 18, No. 2

Berto F, Lazzarin P (2007) Relationships between *J*-integral and the strain energy evaluated in a finite volume surrounding the tip of sharp and blunt V-notches. Int J Solids Struct 44:4621–4645

Berto F, Lazzarin P (2009) A review of the volume-based strain energy density approach applied to V-notches and welded structures. Theor Appl Fract Mech 52:183–194

Berto F, Lazzarin P (2011) Fatigue strength of structural components under multiaxial loading in terms of local energy density averaged on a control volume. Int J Fatigue 33:1055–1065

Berto F, Lazzarin P, Radaj D (2008) Fictitious notch rounding concept applied to sharp V-notches: Evaluation of the microstructural support factor. Part I: Basic stress equations. Engng Fract Mech 75:3060–3072

Berto F, Lazzarin P, Radaj D (2009) Fictitious notch rounding concept applied to sharp V-notches: Evaluation of the microstructural support factor. Part II: Microstructural support analysis. Engng Fract Mech 76:1151–1175

Berto F, Lazzarin P, Yates JR (2011) Multiaxial fatigue of V-notched steel specimens: a non-conventional application of the local energy method. Fatigue Fract Engng Mater Struct 33:921–943

Boukharouba T, Tamine T, Nui L, Chehimi C, Pluvinage G (1995) The use of notch intensity factor as a fatigue crack initiation parameter. Engng Fract Mech 52:503–512

Brown M W (1986) Interfaces between short, long, and non-propagation cracks. In: The behaviour of short fatigue cracks (eds Miller KJ, de los Rios ER), pp. 423–439. Mech Engng Publ, London

BS 7608 (1993) Fatigue design and assessment of steel structures – Code of practice. BSI, London

Carpinteri A (1987) Stress singularity and generalised fracture toughness at the vertex of re-entrant corners. Engng Fract Mech 26:143–155

Cook RD (1995) Finite element modeling for stress analysis. J Wiley, New York

Costa JD, Abreu LM, Pinho AC, Ferreira JA (2005) Fatigue behaviour of tubular AlMgSi welded specimens subjected to bending-torsion loading. Fatigue Fract Engng Mater Struct 28: 399–407

Dunn ML, Suwito W, Cunningham SJ, May CW (1997) Fracture initiation at sharp notches under mode I, mode II, and mild mixed mode loading. Int J Fract 84:367–381

DuQuesnay DL, Topper TH, Yu MT (1986) The effect of notch radius on the fatigue notch factor and the propagation of short cracks. In: Behaviour of short fatigue cracks (eds Miller KJ, de los Rios ER), pp. 323–335. Mech Engng Publ, London

Eibl M, Sonsino CM, Kaufmann H, Zhang G (2003) Fatigue assessment of laser-welded thin sheet aluminium. Int J Fatigue 25:719–731

El Haddad MH, Smith KN, Topper TH (1979) Fatigue crack propagation of short cracks. J Engng Mater Technol 101:42–46

Erdogan F, Sih GC (1963) On the crack extension in plates under plane loading and transverse shear. J Basic Engng 85:519–527

Eurocode 3 (2005) Design of steel structures, Part 1-9, Fatigue. EN 1993-1-9, Brussels, CEN

Frank KH, Fisher JW (1979) Fatigue strength of fillet-welded cruciform joints. J Struct Div ASCE 105(ST9):1727–1740

Fricke W, Feltz O (2010) Fatigue tests and numerical analyses of partial load and full load carrying fillet welds at cover plates and lap joints. Welding in the World 54:R225–R233

Gdoutos EE (1990) Fracture mechanics criteria and applications. Kluwer Acad Publ, Dordrecht

Gimperlein D (1998) *S-N* curve slope for welded structures. Welding in the World 41:399–406

Glinka G (1985) Energy density approach to calculation of inelastic strain-stress near notches and cracks. Engng Fract Mech 22:485–508

Gómez FJ, Elices M (2003$^{(1)}$) Fracture of components with V-shaped notches. Engng Fract Mech 70:1913–1927

Gómez FJ, Elices M (2003$^{(2)}$) A fracture criterion for sharp V-notched samples. Int J Fract 123:163–175

Gómez FJ, Elices M (2004) A fracture criterion for blunted V-notched samples. Int J Fract 127:239–264

Gómez FJ, Elices M, Valiente A (2000) Cracking in PMMA containing U-shaped notches. Fatigue Fract Engng Mater Struct 23:795–803

Gómez FJ, Elices M, Berto F, Lazzarin P (2007) Local strain energy density to assess the static failure of U-notches in plates under mixed mode loading. Int J Fract 145:29–45

Gross R, Mendelson A (1972) Plane elastostatic analysis of V-notched plates. Int J Fract Mech 8:267–272

Gurney TR (1991) The fatigue strength of transverse fillet-welded joints. Abington Publ, Cambridge

Haibach E (ed) (^{5}2003) FKM Guideline – Analytical strength assessment of components in mechanical engineering. FKM, Frankfurt/M

Hobbacher A (ed) (2009) Fatigue design of welded joints and components. IIW-Doc XIII-2151-07/XV-1254-07. Welding Research Council, Bulletin 520, New York

Howland RC (1929/30) On the stresses in the neighbourhood of a circular hole in a strip under tension. Phil Trans Roy Soc (London) A229:49–86

Irwin GR (1957) Analysis of stresses and strains near the end of a crack traversing a plate. J Appl Mech 24:361–364

Irwin GR (1958) Fracture. In: Handbuch der Physik, Vol 4, pp 551–590. Springer, Berlin

Kihara S, Yoshii A (1991) A strength evaluation method of a sharply notched structure by a new parameter, the equivalent stress intensity factor. JSME Int J 34:70–75

Kihl D, Sarkani S (1997) Thickness effects on the fatigue strength of welded steel cruciforms. Int J Fatigue 19:S311–S316

Kirsch G (1898) Die Theorie der Elasitzität und die Bedürfnisse der Festigkeitslehre. Z VDI 42:797–807

Kitagawa H, Takahashi S (1976) Applicability of fracture mechanics to very small cracks. In: Proc 2nd Int Conf Mech Behaviour of Materials, pp. 627–631. Boston Mass

Küppers M, Sonsino CM (2003) Critical plane for the assessment of the fatigue behaviour of welded aluminium under multiaxial loading. Fatigue Fract Engng Mater Struct 26:507–513

Kurath P (1992) Multiaxiality fatigue criteria for spot welds. SAE Techn Paper 920668. Warrendale Pa

Lazzarin P, Berto F (2005$^{(1)}$) Some expressions for the strain energy in a finite volume surrounding the root of blunt V-notches. Int J Fract 135:161–185

Lazzarin P, Berto F (2005$^{(2)}$) From Neuber's elementary volume to Kitagawa and Atzori's diagrams: an interpretation based on local energy. Int J Fract 135:L33–L38

Lazzarin P, Berto F (2008) Control volumes and strain energy density under small and large scale yielding due to tension and torsion loading. Fatigue Fract Engng Mater Struct 31:95–107

Lazzarin P, Filippi S (2006) A generalized stress intensity factor to be applied to rounded V-shaped notches. Int J Solids Struct 43:2461–2478

Lazzarin P, Livieri P (2001) Notch stress intensity factors and fatigue strength of aluminium and steel welded joints. Int J Fatigue 23:225–232

Lazzarin P, Tovo R (1998) A notch stress intensity factor approach to the stress analysis of welds. Fatigue Fract Engng Mater Struct 21:1089–1103

Lazzarin P, Zambardi R (2001) A finite volume energy based approach to predict the static and fatigue behaviour of components with sharp V-shaped notches. Int J Fract 112:275–298

Lazzarin P, Zambardi R (2002) The equivalent strain energy density approach reformulated and applied to sharp V-shaped notches under localized and generalized plasticity. Fatigue Fract Engng Mater Struct 25:917–928

Lazzarin P, Zappalorto M (2008) Plastic notch stress intensity factors for pointed V-notches under antiplane shear loading. Int J Fract 152:1–25

Lazzarin P, Livieri P, Zambardi R (2002) A *J*-integral based approach to predict the fatigue strength of components weakened by sharp V-shaped notches. Int J Comp Appl Techn 15:202–210

Lazzarin P, Lassen T, Livieri P (2003) A notch stress intensity approach applied to fatigue life predictions of welded joints with different local toe geometry. Fatigue Fract Engng Mater Struct 26:49–58

Lazzarin P, Sonsino CM, Zambardi R (2004) A notch stress intensity approach to assess the multiaxial fatigue strength of welded tube-to-flange joints subjected to combined loadings. Fatigue Fract Engng Mater Struct 27:127–140

Lazzarin P, Berto F, Radaj D (2006) Uniform fatigue strength of butt and fillet welded joints in terms of local strain energy density. In: Proc 9th Int Fatigue Congress, ICF9. Elsevier, Oxford, CD Rom

Lazzarin P, Berto F, Radaj D (2007) Fatigue design of welded joints by local approaches: comparison between fictitious notch rounding and strain energy averaging. In: Advances in fracture and damage mechanics VI, Key Engng Mater 248/249:449–452

Lazzarin P, Berto F, Gómez FJ, Zappalorto M ($2008^{(1)}$) Some advantages derived from the use of the strain energy density over a control volume in fatigue strength assessments of welded joints. Int J Fatigue 30:1345–1357

Lazzarin P, Livieri P, Berto F, Zappalorto M ($2008^{(2)}$) Local strain energy density and fatigue strength of welded joints under uniaxial and multiaxial loading. Engng Fract Mech 75: 1875–1889

Lazzarin P, Berto F, Radaj D (2009) Fatigue-relevant stress field parameters of welded lap joints: pointed slit tip compared with keyhole notch. Fatigue Fract Engng Mater Struct 32:713–735

Lazzarin P, Berto F, Zappalorto M (2010) Rapid calculations of notch stress intensity factors based on averaged strain energy density from coarse meshes: Theoretical bases and applications. Int J Fatigue 32:1559–1567

Livieri P (2003) A new path-independent integral applied to notched components under mode I loadings. Int J Fract 123:107–125

Livieri P (2008) Use of *J*-integral to predict static failures in sharp V-notches and rounded U-notches. Engng Fract Mech 75:1779–1793

Livieri P, Lazzarin P (2005) Fatigue strength of steel and aluminium welded joints based on generalised stress intensity factors and local strain energy values. Int J Fract 133:247–276

Maddox SJ (1987) The effect of plate thickness on the fatigue strength of fillet-welded joints. Abington Publ, Cambridge

Maddox SJ (^{2}1991) Fatigue strength of welded structures. Abington Publ, Cambridge

Meneghetti G (2008) The peak stress method applied to fatigue assessments of steel and aluminium fillet-welded joints subjected to mode 1 loading. Fatigue Fract Engng Mater Struct 31:346–369

Meneghetti G, Lazzarin P (2007) Significance of the elastic peak stress evaluated by FE analyses at the point of singularity of sharp V-notched components. Fatigue Fract Engng Mater Struct 30:95–106

Meneghetti G, Lazzarin P (2010) The peak stress method for fatigue strength assessment of welded joints with weld toe or weld root failures. IIW Doc XIII-2322-10; Welding in the World (2011), 55:22–29

Meneghetti G, Lazzarin P (2011) The use of peak stress for fatigue strength assessments of welded lap joints and cover plates with toe and root failures. IIW Doc XIII-2375–2311

Moftakhar A, Buczynski A, Glinka G (1995) Calculation of elastoplastic strains and stresses in notches under multiaxial loading. Int J Fract 70:357–373

Neuber H ([2]1958, [3]1985) Kerbspannungslehre. Springer Verlag, Berlin

Neuber H (1968) Über die Berücksichtigung der Spannungskonzentration bei Festigkeitsberechnungen. Konstruktion 20:245–251

Nisitani H, Teranishi T (2004) The K_{I} value of a circumferential crack emanating from an ellipsoidal cavity obtained by the crack tip stress method in FEM. Eng Fract Mech 71: 579–585

Olivier R, Ritter W (1979–1985) Wöhlerlinienkatalog für Schweißverbindungen aus Baustählen. DVS-Report 56, DVS-Verlag, Düsseldorf

Olivier R, Köttgen VB, Seeger T (1989) Schweißverbindung I – Schwingfestigkeitsnachweise für Schweißverbindungen auf der Grundlage örtlicher Beanspruchungen. FKM-Forschungsheft 143, FKM, Frankfurt/M

Olivier R, Köttgen VB, Seeger T (1994) Schweißverbindung II – Untersuchung zur Einbindung eines neuartigen Zeit- und Dauerfestigkeitsnachweises von Schweißverbindungen aus Stahl in Regelwerke. FKM-Forschungsheft 180, FKM, Frankfurt/M

Park J, Nelson D (2000) Evaluation of an energy-based approach and a critical plane approach for predicting constant amplitude multiaxial fatigue life. Int J Fatigue 22:23–39

Person NL (1971) Fatigue of aluminium alloy welded joints. Welding Res Suppl 50:77s–87s

Pluvinage G (1997) Rupture and fatigue initiated from notches: application of the notch intensity factor (in French). Revue Francaise de Mécanique 1997-1:53–61

Quian J, Hasebe N (1997) Property of eigenvalues and eigenfunctions for an interface V-notch in antiplane elasticity. Engng Fract Mech 56:729–734

Radaj D (1990) Design and analysis of fatigue resistant welded structures. Abington Publ, Cambridge

Radaj D (1998) Abminderungsfaktor der Schwingfestigkeit bei Schweißverbindungen. Konstruktion 50:55–62

Radaj D, Helmers K (1997) Bewertung von Schweißverbindungen hinsichtlich Schwingfestigkeit nach dem Kerbspannungskonzept. Konstruktion 49:21–27

Radaj D, Schilberth G (1977) Kerbspannungen an Ausschnitten und Einschlüssen. DVS-Verlag, Düsseldorf

Radaj D, Vormwald M ([3]2007) Ermüdungsfestigkeit – Grundlagen für Ingenieure. Springer Verlag, Berlin

Radaj D, Zhang S (1993) Analogies between crack tip and rigid line tip stresses and displacements. Engng Fract Mech 44:913–919

Radaj D, Zhang S (1994). Stress distribution at bimaterial crack tips. Engng Fract Mech 47: 613–617

Radaj D, Lehrke HP, Greuling S (2001) Theoretical fatigue-effective notch stresses at spot welds. Fatigue Fract Engng Mater Struct 24:293–308

Radaj D, Sonsino CM, Fricke W ([2]2006) Fatigue assessment of welded joints by local approaches. Woodhead Publ, Cambridge and CRC Press, Boca Raton Fla

Radaj D, Berto F, Lazzarin P (2009[(1)]) Local fatigue strength parameters for welded joints based on strain energy density with inclusion of small-size notches. Engng Fract Mech 76: 1109–1130

Radaj D, Lazzarin P, Berto F (2009[(2)]) Fatigue assessment of welded joints under slit-parallel loading based on strain energy density or notch rounding. Int J Fatigue 31:1490–1504

Razmjoo GR (1996) Fatigue of load-carrying fillet-welded joints under multiaxial loadings. TWI Ref 7309.02/96/909, Cambridge

Rice JR (1968) A path-independent integral and the approximate analysis of strain concentration by notches and cracks. J Appl Mech 35:379–386

Savin GN (1970) Stress distribution around holes. NASA Techn Translation F-607

Seweryn A (1994) Brittle fracture criterion for structures with sharp notches. Engng Fract Mech 47:673–681

Seweryn A, Poskrobko S, Mróz Z (1997) Brittle fracture in plane elements with sharp notches under mixed mode loading. J Engng Mech 123:535–543

Sheppard SD (1991) Field effects in fatigue crack initiation: long life fatigue strength. J Mech Design (ASME) 113:118–194

Sih GC (1975) A three-dimensional strain energy density factor theory of crack propagation. In: Three-dimensional crack problems. Noordhoff, Leyden, pp 15–80

Siljander A, Kurath P, Lawrence FV (1992) Non-proportional fatigue of welded structures. In: ASTM STP 1122, pp 319–338, Philadelphia Pa

Smith RA (1977) On the short crack limitations of fracture mechanics. Int J Fract 13:717–720

Sonsino CM (1995) Multiaxial fatigue of welded joints under in-phase and out-of-phase local strains and stresses. Int J Fatigue 17:55–70

Tanaka K (1974) Fatigue propagation from a crack inclined to the cyclic tensile axis. Engng Fract Mech 6:493–507

Tanaka K, Nakai Y, Yamashita M (1981) Fatigue growth threshold of small cracks. Int J Fract 17:519–533

Taylor D (1999) Geometrical effects in fatigue – A unifying theoretical model. Int J Fatigue 21:413–420

Taylor D, Wang G (2000) The validation of some methods of notch fatigue analysis. Fatigue Fract Engng Mater Struct 23:387–395

Verreman Y, Nie B (1996) Early development of fatigue cracking at manual fillet welds. Fatigue Fract Engng Mater Struct 19:669–681

Williams ML (1952) Stress singularities resulting from various boundary conditions in angular corners of plates in tension. J Appl Mech 19:526–528

Yosibash Z, Bussiba AR, Gilad I (2004) Failure criteria for brittle elastic materials. Int J Fract 125:307–333

Yung JY, Lawrence FV (1986) Predicting the fatigue life of welds under combined bending and torsion. Rep no 125, UIL-ENG86-3602. Univ of Illinois, Urbana Ill

Zappalorto M, Lazzarin P (2011) Strain energy based evaluations of plastic notch stress intensity factors at pointed V-notches under tension. Engng Fract Mech 78:2691–2706

Zhang G, Richter B (2000) A new approach to the numerical fatigue life prediction of spot-welded structures. Fatigue Fract Engng Mater Struct 23:499–508

Chapter 4
Elastic-Plastic Fatigue Crack Growth

M. Vormwald

Abstract Conventional fatigue crack propagation approaches rely on similitude arguments and relationships between the stress intensity factor range and crack growth rate. The application limit of this approach is specified by small-scale yielding conditions. Still within these limits, an explanation and the straightforward modelling of the mean stress influence and the influence of variable amplitudes requires consideration of cyclic plasticity. Plasticity-induced crack closure greatly influences the crack growth rate. Modelling tools and algorithms are presented. Outside the small-scale yielding limits, the stress intensity factor range must be substituted by a crack driving force parameter of elastic-plastic fracture mechanics. Various proposals are presented and discussed with a focus on the ΔJ-integral. Together with an adequate consideration of crack closure, advances in simulating fatigue crack growth in this regime more realistically are presented. Multiaxial and mixed mode loading are a continuing challenge for actual research. These topics are discussed against the background of current expertise and available computational resources.

Contents

D. Radaj and M. Vormwald, *Advanced Methods of Fatigue Assessment*,
DOI: 10.1007/978-3-642-30740-9_4,

4.1 Survey of Chapter Contents

For many applications, fatigue crack propagation can be modelled by describing the growth rate by an empirical equation and integrating this equation to arrive at crack lengths as functions of the number of applied cycles. Such a detailed view at the fatigue process provides more information and insight than conventional fatigue approaches using correlations of stress or strain ranges with the number of cycles to failure.

The crack growth rate is determined by stresses and deformations at the crack tip or crack front. According to the linear theory of elasticity, a special and unique type of stress singularity appears at pointed crack tips. Its intensity is closely related to the near-tip stress field. It is self-evident that—in a first-order approximation—crack growth rates are usually linked to the range of the corresponding elastic stress intensity factor. In detail, however, irreversible plastic and disruptive processes at the crack tip are responsible for crack propagation. The limits of the applicability of a parameter derived from a linear theory to describe highly nonlinear phenomena are specified.

Remaining within these limits of applicability, certain observable phenomena of fatigue crack growth cannot be explained without considering cyclic plasticity. The influence of mean stress on the fatigue crack growth rate is discussed against the background of plasticity-induced crack closure. The crack closure argument also provides an insight into the various phenomena of load history effects, which may accelerate or decelerate the growth rate depending on a variety of conditions.

The strip-yield model and a finite element based nodal release technique are the two main approaches for modelling plastic deformation effects in fatigue crack growth. Actual feasibilities and limitations are shown.

Beyond the application limit of the stress intensity factor based life calculation, a substitute procedure has to be defined. The pros and cons of various proposals for a crack driving force parameter in elastic-plastic fracture mechanics are discussed. A preference in favour of the cyclic ΔJ-integral is elaborated, keeping its theoretical limitations in mind. Crack closure is also an important issue under large-scale yielding conditions. Experience in this subject cannot be transferred from the small-scale to the large-scale cyclic yielding regime or vice versa.

The chapter ends with a discussion on multiaxial and mixed mode aspects in the large-scale cyclic yielding regime. It is concluded with a list of symbols followed by a list of references.

4.2 Crack Growth under Small-Scale Yielding Conditions

4.2.1 Basic Modelling Features

The conditions for a crack or a crack-like defect to grow in a stable way under cyclic loading are of major engineering and scientific interest. Engineering fracture

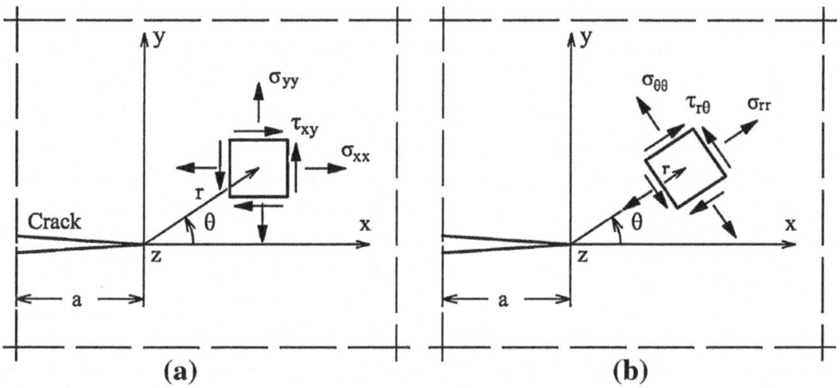

Fig. 4.1 Cartesian (**a**) and polar (**b**) coordinate systems with correspondingly defined stresses at the crack tip; (Lazzarin and Tovo 1998)

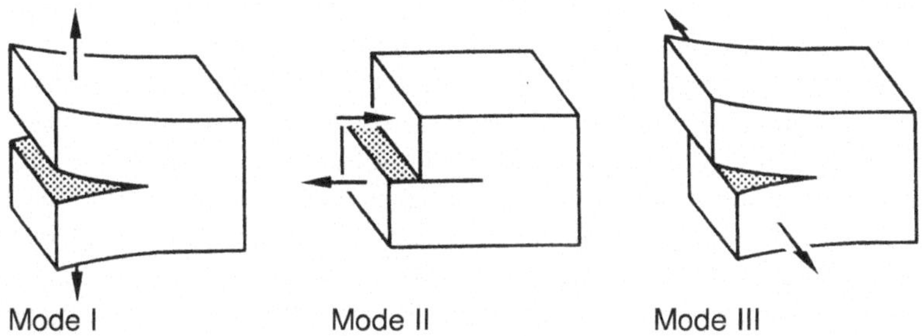

Fig. 4.2 Basic loading modes with stress singularity at crack front: mode I (tension load normal to crack plane), mode II (shear load normal to crack front), mode III (shear load parallel to crack front); (Radaj and Vormwald 2007)

mechanics describes the behaviour of macroscopically long cracks based on the principles of continuum mechanics. The approach is therefore limited to a macroscopic scale where the real process—irreversible slip and dislocation movement in a highly inhomogeneous and anisotropic microstructure in the vicinity of a crack front—is not considered explicitly. In general, a simple isotropic, linear-elastic material behaviour is assumed, leading to the theory of linear-elastic fracture mechanics. Within this theory, the stresses in the close vicinity of the crack tip in the Cartesian and polar coordinate systems depicted in Fig. 4.1 can be determined according to the following equation (Irwin 1957):

$$\sigma_{\mathrm{ij}} = \frac{1}{\sqrt{2\pi r}}\left[K_{\mathrm{I}} f_{\mathrm{I,ij}}(\theta) + K_{\mathrm{II}} f_{\mathrm{II,ij}}(\theta) + K_{\mathrm{III}} f_{\mathrm{III,kz}}(\theta)\right] + \mathrm{O}(r^0)$$
$$(\mathrm{i,j} = x, y \text{ and } \mathrm{k} = x, y \text{ or } \mathrm{i,j} = r, \theta \text{ and } \mathrm{k} = r, \theta) \qquad (4.1)$$

The stress singularity at crack fronts or tips is of type $r^{-1/2}$. The mode-related stress intensity factors K_{I}, K_{II} and K_{III}, Fig. 4.2, depend on the magnitude of the load, the crack length and further geometrical parameters of the considered configuration. The mode-related angle-dependent functions $f_{I,ij}, f_{II,ij}, f_{III,kz}$ describe the angular distribution of the stresses at the crack tip. The relationship above is strictly valid for $r \to 0$ and approximately valid for values of r which are small in relation to the crack length and other geometrical parameters of the configuration.

The higher order terms, $O(r^0)$, gain only importance in cases where the leading terms on the right hand side of Eq. (4.1) do not explain the phenomena of the fracture process under consideration. For furher details, see the standard text books on fracture mechanics (e.g. Anderson 1995; Broek 1988; Sanford 2003; Gross and Seelig 2011).

In fatigue failure, mode I loading with tensile stress transverse to the crack plane dominates. The experience from growing fatigue cracks shows that the cracks seem to search for an orientation which provides pure mode I loading along its front.

The occurence of theoretically infinite stresses clearly violates physical evidence. In real materials and structures, the stresses at crack tips remain finite, mainly for the following reasons. Even the theory of linear elasticity indicates that the geometry of the crack tip changes from a mathematically sharp crack to a rounded notch. However, the equilibrium condition is formulated in the undeformed reference configuration. Formulating the equilibrium condition in the deformed configuration—as it would be the correct way—would on the one hand cause the singularity to disappear and on the other hand increase the numerical expense to an extent which is unacceptable in engineering applications today. Additionally, in the theory of linear elasticity, the strains are considered to be small. According to the assumption of linearised kinematics of deformation the strains at crack tips become infinitely high. This result is obviously incompatible with the assumption of small strains. Only by improving the theory allowing for large deformations and large strains, will keep the crack tip stresses finite.

To this point, only arguments based on equilibrium and deformation kinematics have been brought forward to explain why infinite stresses do not exist in contrast to the theoretical results. The most important explanation is provided by the material behaviour. All materials show nonlinearities in their stress-strain relationship which, at least for metals, are due to plastic deformation. It is necessary to consider the conditions under which fatigue crack growth can be uniquely assigned to the stress intensity factor. In fact, this is only guaranteed if the criteria of similitude and embedding are not violated. This means that a limited amount of plastic deformation can be allowed at the tip of real cracks. However, this plastic zone must remain embedded in an elastic stress field which is governed by the stress intensity factor. The stress intensity factor dominated fields at crack tips are similar for different structures. Therefore, failure-relevant information can be transferred between structures of the same material.

The fatigue crack growth is characterised by its rate, da/dN, where a is the crack length and N is the number of load cycles. With the assumption of linear elasticity, the crack growth rate is controlled by the cyclic stress intensity factor ΔK which is defined as the stress intensity factor range, $\Delta K = K_{max} - K_{min}$, of the load cycle. The crack growth rate is correlated to the cyclic stress intensity factor.

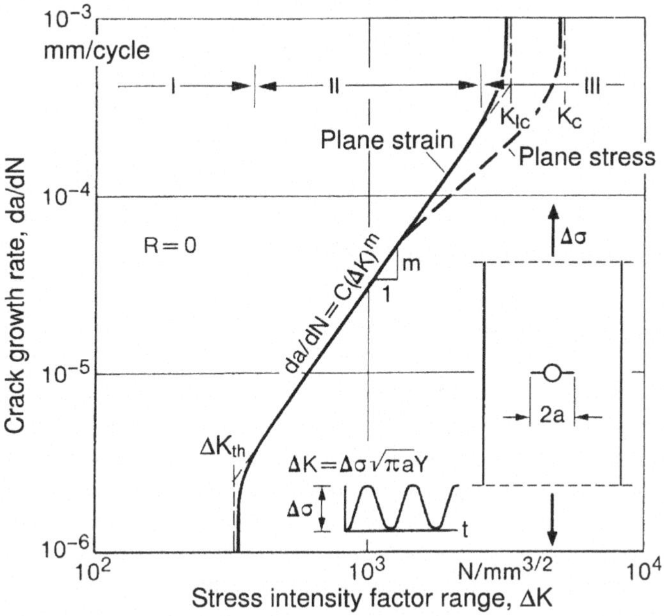

Fig. 4.3 Crack growth rate of long cracks dependent on cyclic stress intensity factor range; threshold region (I), Paris law region (II), stable tearing region (III); schematic representation; (Radaj and Vormwald 2007)

A typical relationship between da/dN and ΔK is schematically illustrated in Fig. 4.3. Three regions of crack growth can be identified. Region I is associated with small growth rates and finally crack arrest, when the stress intensity factor ranges are lower than a threshold value ΔK_{th}. Region III is characterised by very high growth rates with $K_{\max}$ close to the static fracture toughness indicating stable tearing (K_{c}) or brittle fracture (K_{Ic}). In region II, the relationship between crack growth rate and stress intensity factor range is linear in logarithmic scales. The simplest, for many applications sufficient relationship is given by the Paris power law (Paris 1960; Paris and Erdogan 1963):

$$\frac{\mathrm{d}a}{\mathrm{d}N} = C\Delta K^{m} \tag{4.2}$$

where the material constants C and m are dependent on the stress intensity ratio, $R = K_{\min}/K_{\max}$. Often used values for metallic materials are $m = 2.25$–4.0.

Many empirical equations have been proposed for correlating the fatigue crack growth rate and the stress intensity factor range with the aim to include the behaviour in regions I and III. The equation published by Forman and Mettu is widely accepted for its ability to describe fatigue crack growth from the threshold to the monotonic failure region (Forman and Mettu 1992):

$$\frac{da}{dN} = C\left[\frac{1-f}{1-R}\Delta K\right]^m \frac{(1-\Delta K_{th}/\Delta K)^p}{(1-K_{max}/K_c)^q} \quad (4.3)$$

where the material parameters C, m, p and q are empirically derived constants. An expression for the crack opening function f is given by Eq. (4.15).

4.2.2 Load Ratio Effects on Crack Growth

Basic Phenomena and Effective Stress Intensity

Dealing with small-scale yielding at the crack tip, Irwin assumed elastic perfectly-plastic material behaviour with the yield stress σ_Y and a small plastic zone size compared to the other dimensions of the structure and to the crack length a. For this small-scale yielding zone, he presumes the stresses to decrease outside the plastic zone in the same way as under elastic conditions. Taking the equilibrium condition in the ligament additionally into account, he arrives at a first estimate for the plastic zone size ω under plane stress conditions (Irwin 1956). Under the (positive) maximum stress intensity K_{max} of the cycle, this parameter reads:

$$\omega_{max} = \frac{1}{\pi}\left(\frac{K_{max}}{\sigma_Y}\right)^2 \quad (4.4)$$

The following unloading process from K_{max} to K_{min}, expressed by the range ΔK, is treated as a loading in the negative direction. It is assumed that all material elements of the structure exhibit simultaneously a reversal of the load. During the unloading process they show a material behaviour approximately described by the hysteresis loop branch of the elastic perfectly-plastic material. This means that yielding under compressive load starts at a stress range equal to twice the monotonic yield stress (ranging from plus σ_Y to minus σ_Y). The cyclic plastic zone size can therefore be estimated according to:

$$\omega_{cyc} = \frac{1}{\pi}\left(\frac{\Delta K}{2\sigma_Y}\right)^2 \quad (4.5)$$

The above equation holds only theoretically with the assumption that for $R \geq -1$ the crack faces do not come into contact. The special case of $R = 0$, for example, results in the cyclic plastic zone size ratio, $\omega_{cyc}/\omega_{max} = 1/4$. For $R \geq -1$ the size ratio becomes $\omega_{cyc}/\omega_{max} = (1 - R)^2/4$. The Eqs. (4.4) and (4.5) give the plastic zone size under plane stress conditions at the crack tip. The corresponding plane strain plastic zone size may be gained by replacing the yield stress σ_Y by $\sigma_Y/(1 - 2\nu)$.

Application of linear elastic fracture mechanics for assessing the fracture limit under monotonic loading according to ASTM E1820 (2007) or BS 7910 (2005) is based on predominantly plane strain conditions. These are guaranteed by geometrical parameters such as specimen thickness B, crack length a and ligament width $(W - a)$ according to:

$$[B,\ a,\ (W - a)] \geq 2.5 \left(\frac{K_{max}}{\sigma_Y}\right)^2 \tag{4.6}$$

This equation may be interpreted as the requirement that the geometrical dimensions exceed the plastic zone size under plane strain conditions by a factor of 40–50. In cyclic loading, a first estimate of the limit of crack growth description based on linear-elastic fracture mechanics, e.g. by Eq. (4.3), may be defined in analogy with the monotonic case. The geometrical dimensions should exceed the cyclic plastic zone size under plane strain conditions by the same factor expressed by the following condition:

$$[B,\ a,\ (W - a)] \geq 2.5 \left(\frac{\Delta K}{2\sigma_Y}\right)^2 \tag{4.7}$$

During cyclic loading, the stresses at the crack tip are limited by $+\sigma_Y$ at the tensile reversal point. During unloading, these crack tip stresses turn to $-\sigma_Y$. At a pointed crack tip and under the above assumptions, the material undergoes cyclic stress loading with $R = -1$ between the positive and negative yield stress, independently from the applied load ratio R. This simplified view on the mechanics of a stationary (not growing) crack cannot explain the influence of the load ratio R on the crack growth rate. An example of an experimentally found load ratio influence is shown in Fig. 4.4.

The mechanisms of fatigue crack growth have been summarised (Wang 1993). The widely accepted blunting and resharpening model was proposed earlier (Forsyth and Ryder 1960). Variants of the model have also been presented (McMillan and Pelloux 1967; Laird 1967; Schijve 2009). Experimental observations by Pippan et al. (2011) will be shown further below. During an ascending-load half cycle the crack tip material yields. In the descending-load half cycle, the direction of yielding is reversed. The crack surface created under tensile loading cannot be completely rejoined by cold welding. The new surfaces come into contact, thus resharpening the crack tip for the next ascending half cycle which will soon act on a crack which has grown by an amount Δa. It is important to emphasise that the plastic elongations of the material in the forward plastic zone are not completely removed during the descending-load half cycle. Plastically elongated material remains at the crack faces. It gives rise to the plasticity-induced crack closure phenomenon. Naturally growing fatigue cracks will remain closed during part of the applied load cycle. While the crack surfaces are in contact during a descending half cycle, the compressive plastic deformation at the crack tip still continues, however to a drastically reduced extent.

The crack closure phenomenon was observed and published for the first time by Elber who attributed the crack growth rates to the effective stress range defined as the difference between the maximum stress and the stress at crack closure σ_{cl} or at crack

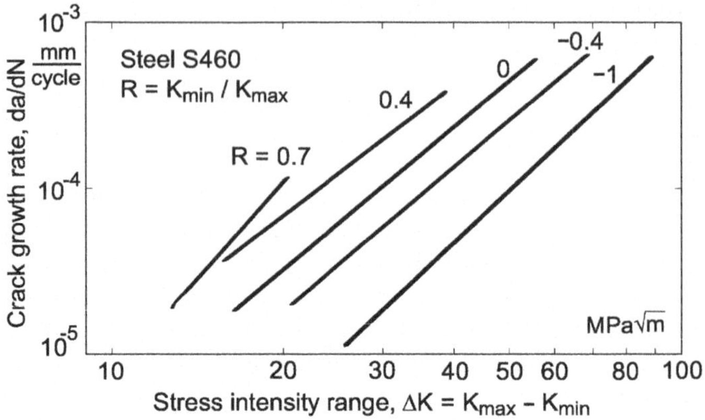

Fig. 4.4 Crack growth rate of long cracks dependent on cyclic stress intensity factor range for various load ratios *R*; (Vormwald 2011)

opening σ_{op} (Elber 1970, 1971). Crack tip plastic deformation during crack contact is thus neglected. Crack opening stresses may be measured during fatigue tests. Figure 4.5 shows the data of Fig. 4.4 where the crack growth rate is replotted in terms of the effective stress intensity range, $\Delta K_{eff} = K_{max} - K_{op}$. It should become clear that K_{op} and also K_{cl} are fictitious load-proportional quantities of linear-elastic fracture mechanics which have no simple relation to the actual crack tip field of stresses and strains at the moment of opening and closing of a fatigue crack.

Cyclic plastic deformation at the crack tip drives the fatigue crack. The effective range of the stress intensity factor, ΔK_{eff}, is only a measure of cyclic crack tip plastic deformation which is applicable within the range of linear-elastic fracture mechanics. Again, it is proposed that the main geometric dimensions of the structures are by a factor of 50 larger than the cyclic plastic zone under plane strain conditions:

$$[B,\ a,\ (W-a)] \geq 2.5\left(\frac{\Delta K_{eff}}{2\sigma_Y}\right)^2 \tag{4.8}$$

As the relation $\Delta K_{eff} < \Delta K$ is usually valid, Eq. (4.8) reduces the requirements with regard to the geometrical dimensions in comparison to Eq. (4.7). The crack growth analysis may then be performed based on effective stress intensity factor ranges, replacing Eq. (4.2) by the following expression:

$$\frac{da}{dN} = C_{eff}\Delta K_{eff}^m \tag{4.9}$$

where the constant C_{eff} is only material-dependant and not any more dependent on the mean stress.

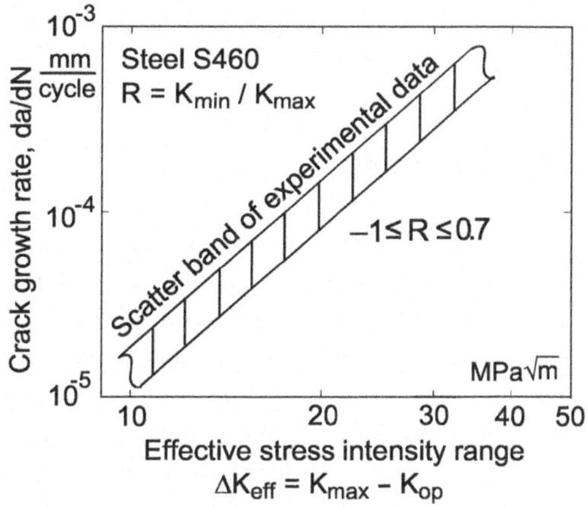

Fig. 4.5 Crack growth rate of long cracks dependent on effective stress intensity factor range for various load ratios R; (Vormwald 2011)

The crucial point of the approach using effective stress intensity factor ranges is the determination of the opening and closure point loads. There is no easy way to get this information.

Guided by experimental findings, several authors proposed approximation formulae for the crack opening ratio U, which allows to determine ΔK_{eff} from ΔK:

$$U = \frac{\Delta\sigma_{\text{eff}}}{\Delta\sigma} = \frac{\Delta K_{\text{eff}}}{\Delta K} \tag{4.10}$$

The first of these formulas was given for the aluminium alloy 2024-T3 (Elber 1970, 1971):

$$U = 0.5 + 0.4R \qquad (-0.1 \leq R \leq 0.7) \tag{4.11}$$

This formula was modified for the aluminium alloy AlZnMgCu0.5 (Schwalbe 1980):

$$U = 0.6 + 0.5R \qquad (0 \leq R \leq 0.8) \tag{4.12}$$

The approximation was later on extendend (Schijve 1981, 2009):

$$U = 0.55 + 0.33R + 0.12R^2 \qquad (-1 \leq R \leq 1) \tag{4.13}$$

The following alternative expression is also attributed to Schijve (Heitmann 1983):

$$U = \frac{3.72}{(3 - R)^{1.74}} \tag{4.14}$$

The Eqs. (4.11) to (4.14) have in common that the crack opening ratio U is independent of the geometry, especially the crack length, the loading, $\sigma_{\max}$ and $\sigma_{\min}$, and the material. Overviews still valid have been given by Schijve and McClung, the latter with the conclusion that a unique approximation has not been established (Schijve 1988; McClung 1991[2]). Some further approximations have been discussed (Kumar and Singh 1995; Thumser 2009).

Strip-Yield Model

A deeper insight into the mechanics of crack tip plastic deformation is achieved by applying the Dugdale–Barenblatt model (Dugdale 1960; Barenblatt 1962). The model confines the plastic deformation to a narrow strip embedding the crack and ligament line. Therefore, the name strip-yield model is used. The simplification above allows for solving the nonlinear mechanical problem by superposition of solutions gained by applying the theory of linear elasticity. This is a tremendous advantage, keeping in mind that thousands or even millions of cycles have to be simulated—depending on the actual engineering problem. Up to now, only two-dimensional models have become known.

The extension for treating cyclic loading cases started in the seventies (Seeger 1973; Dill and Saff 1976; Dill et al. 1980; Führing 1977; Führing and Seeger 1979). Further contributions enhanced the theoretical basis and the applicability of the model (Budiansky and Hutchinson 1978; Newman 1981; Ibrahim et al. 1986; Ibrahim 1989; Wang and Blom 1991).

The cyclic strip-yield model, according to Newman, is available as software for general use, (NASGRO 2002) and ESACRACK (http://esacrack.dorea.eu). The model is illustrated in Fig. 4.6. Three regions can be distinguished: a region behaving linear-elastically containing a crack with length $(a + \omega)$, a plastic zone of size ω and a zone of irreversible plastic deformation on the crack flanks. The plastic zone size in compression ω_c under a constant amplitude load ratio $R \geq -1$ is identical with the cyclic plastic zone size ω_{cyc}. It includes material elements of both plastic regions, i.e. in the wake and ahead of the crack tip.

The stress singularity in the ligament line is removed by plastic deformation. This is the main condition for constructing a solution of the mechanical problem. For the elastic region, the stress intensities and crack flank displacements due to the external load have to be provided by solving the linear-elastic problem of the cracked structure (crack length $a + \omega$). Also the stress intensities and crack flank displacements under the crack flank loads, i.e. the stresses $\sigma_i(x_i)$ in a small bar of width w_i at position x_i, must be known. In the ligament plastic zone, these bar element stresses are limited by the tensile yield stress $+\alpha\sigma_Y$ and compressive yield stress $-\beta\sigma_Y$. Adjusting the constraint factor α allows for interpolating between plane stress ($\alpha = 1$) and plane strain ($\alpha = 3$) conditions. At the crack flanks, the bar element stresses are limited by zero and $-\beta\sigma_Y$. Penetrating crack flanks are not allowed. The length l_i of the bars has to be equal or smaller (if broken) to the flank displacement v_i.

Plastic zone sizes, when calculated by the strip-yield model, are no longer subject to the small-scale yielding condition. Provided the small-scale yielding condition is met, the model provides plastic zone sizes similar to the Irwin

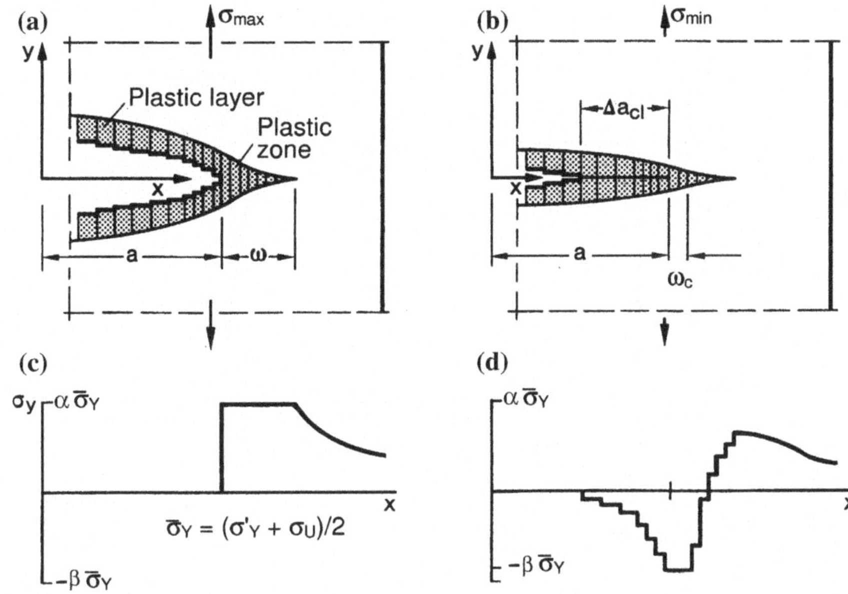

Fig. 4.6 Strip-yield model with plastic layers on crack flanks, produced by cyclic plastic deformation during crack growth: loading with crack opening (**a**), load relief with crack closure (**b**), corresponding stresses at the crack tip (**c, d**); with closed crack length Δa_{cl}, extension ω and ω_{c} of tensile and compressive plastic zones at crack tip, substitute yield limit $\bar{\sigma}_{\mathrm{Y}}$ and constraint factors α and β; schematic presentation; (Newman 1981, 1982)

approximation, however the factor $1/\pi \approx 0.32$ in Eqs. (4.4) or (4.5) has to be replaced by $\pi/8 \approx 0.39$.

The strip-yield model is usually applied to calculate the crack opening stresses. A crack length increment is produced by changing the boundary condition of the crack tip bar from intact to broken. Then, the next load cycle is applied and the crack opening stresses are identified. Based on the effective stress range, the effective stress intensity factor range ΔK_{eff} is calculated. Inserting ΔK_{eff} into Eq. (4.9) together with the crack tip bar width equal to the crack length increment, the required number of cycles to break the bar is calculated. Driving a crack in this way step by step through the cross-section, gives the crack opening stresses, Fig. 4.7. Typically, a rapidly increased (from $\sigma_{\min}$) crack opening stress is observed, followed by stabilisation (crack length independence) at a level characteristic for the stress ratio R, the applied maximum stress $\sigma_{\max}$, and the constraint factor α. An approximation for calculating the stabilised level of the crack opening stress reads as follows (Newman 1984):

$$f = \frac{\sigma_{\mathrm{op}}}{\sigma_{\max}} = \begin{cases} A_0 + A_1 R + A_2 R^2 + A_3 R^3 & R \geq 0 \\ A_0 + A_1 R & R < 0 \end{cases} \tag{4.15}$$

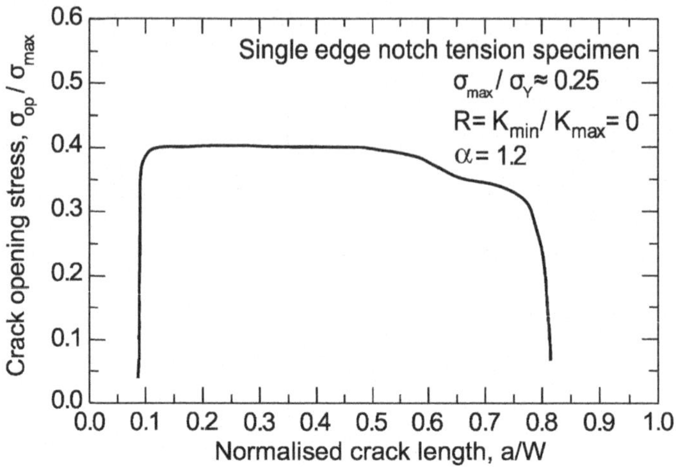

Fig. 4.7 Crack opening stress dependent on crack length, calculation results according to the strip-yield model; (Buschmöhle et al. 1996)

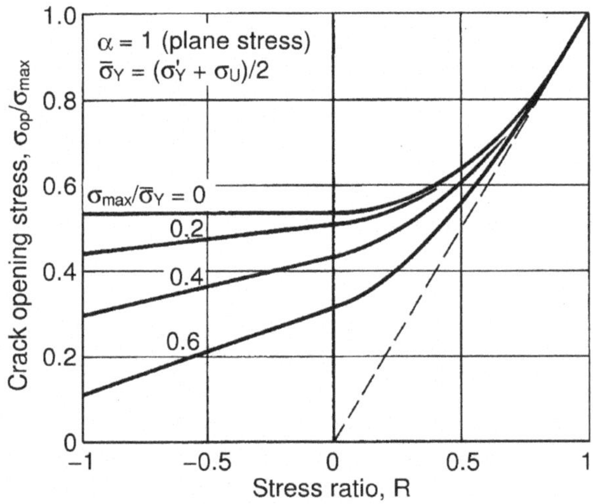

Fig. 4.8 Crack opening stress dependent on load ratio R for various maximum stresses $\sigma_{\max}$; plane stress condition; calculation results based on the strip-yield model; (Newman 1981, 1982)

where the constants A_0 to A_3 are defined as follows:

$$\begin{aligned} A_0 &= (0.825 - 0.24\alpha + 0.05\alpha^2)(\cos(\pi\sigma_{\max})/(2\sigma_Y))^{1/\alpha} \\ A_1 &= (0.415 - 0.071\alpha)\sigma_{\max}/\sigma_Y \\ A_2 &= 1 - A_0 - A_1 - A_3, \quad A_3 = 2A_0 + A_1 - 1 \end{aligned} \tag{4.16}$$

The Fig. 4.8 shows a graphical representation of the influence of the stress ratio R on the crack opening stress under plane stress conditions ($\alpha = 1$), as described by Eqs. (4.15) and (4.16). The two additional parameters σ_Y and α allow consideration of the load level and material strength as well as of geometrical (plate

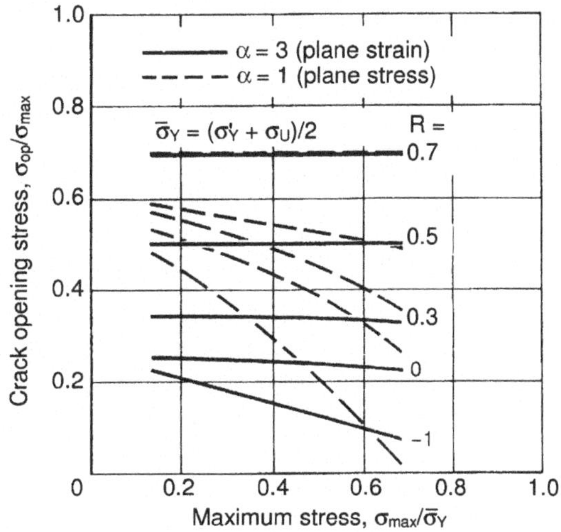

Fig. 4.9 Crack opening stress dependent on maximum stress for various load ratios R; plane strain and plane stress conditions; calculation results based on the strip-yield model; (Newman 1981)

thickness) influences on the effective ranges and crack growth rates. The magnitude of these influences can be anticipated from Fig. 4.9.

Cyclic plastic deformation at the crack tip and in consequence irreversible plastic deformation at the crack flanks producing plasticity-induced crack closure may explain the stress ratio influence on fatigue crack growth to a large extent. This view of the mechanism may be traced in the literature over the past few decades until today (Ziegler et al. 2008; Thumser et al. 2008). Others (e.g. McEvily 1988; Sadananda et al. 2009) emphasise that the role of plasticity-induced crack closure is much less important than considered by Elber and his successors, or that the 'extrinsic closure arguments' are even irrelevant for modelling the fatigue process compared to the intrinsic material resistance against static and cyclic failure mechanisms. The latter arguments led to the two-parameter criterion (Sadananda et al. 2009). The stress ratio influence is explained in this philosophy by superimposed monotonic modes of damage on cyclic damage.

Other crack closure causes besides crack tip plastic deformation have to be taken into consideration, debris and corrosion products among them. Debris phenomena are especially important for fatigue fractures under torsional loads. They may produce beneficial crack closure effects.

It is clear that not all the observable fatigue related phenomena can be explained by closure arguments alone. On the other hand, a two parameter criterion based on the maximum value of the cycle and on the range of this cycle can—from a mathematical point of view—always be transformed into a criterion based on that maximum value and a second load parameter which defines the crack opening level. The two latter quantities lead to effective ranges. Therefore, this philosophy has the potential to produce life calculations in good agreement with experimental evidence—the major concern of engineers responsible for the reliability of fatigue-loaded structures.

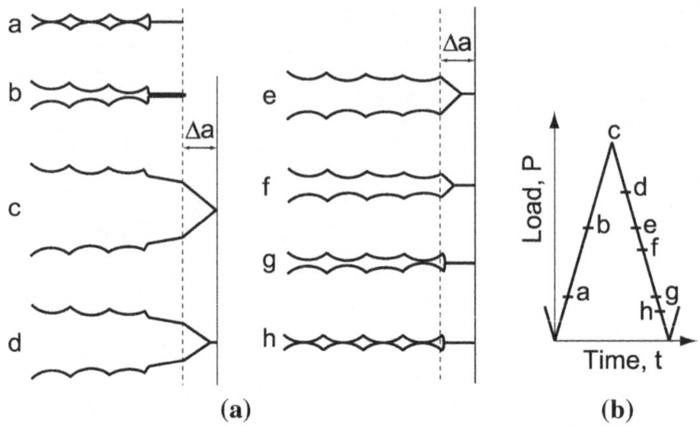

Fig. 4.10 Micromechanism of fatigue crack growth with pointed crack tip and re-established crack closure; crack flank shaping (**a**) and reference points on load cycle (**b**); (Pippan et al. 2011)

Plasticity-induced crack closure and cyclic crack tip plastic deformation are measurable quantities. However, high-precision techniques are required and not all of the published data satisfy the demand for producing crack tip relevant results with sufficiently low error margins. The direct observation methods have been summarised (Schijve 1988). Crack closure is also the result of mechanical modelling of the fatigue crack growth process. The dispute of crack closure phenomena will continue as long as only a few rigorous comparisons of measured and modelled crack closure are available—both tasks performed by applying highly sophisticated methods. Regarding experimental evidence, major progress has been achieved by applying a special fractographic technique in combination with a stereophotogrammetric reconstruction of the fracture surface (Pippan et al. 2011). These authors have been able to reconstruct the fatigue fracture surface at different load levels during one cycle. They were able to summarise their findings in a simplified model of fatigue crack growth, Fig. 4.10, and to describe the crack formation process in the following form (Pippan et al. 2011):

> At the minimum of the load in a constant load amplitude experiment, the crack is sharpened and closed. At about 25 % of K_{max} at $R = 0.05$ the crack tip opens. Additional loading causes a blunting and forms a V-shaped micro-notch. The plastic flow is mainly concentrated at the tip of this V-shaped micro-notch. The blunting process continues till K_{max} is reached. The new crack surface was produced by the formation of this V-shaped micro-notch. Load reversal causes resharpening, i.e. shearing of the activated slip bands in opposite direction occurs. This deformation is again concentrated at the tip of the V–notch. This continues until the crack is fully closed. The fissure is formed between the last two striations. During loading, a corner is formed between the last striation and the new created fracture surface in the 'blunting' region. During unloading the shearing is not strictly reversible, the material on the corner between the previous formed striation and the partly closed micro V-notch generates a fold, which results in a fissure.

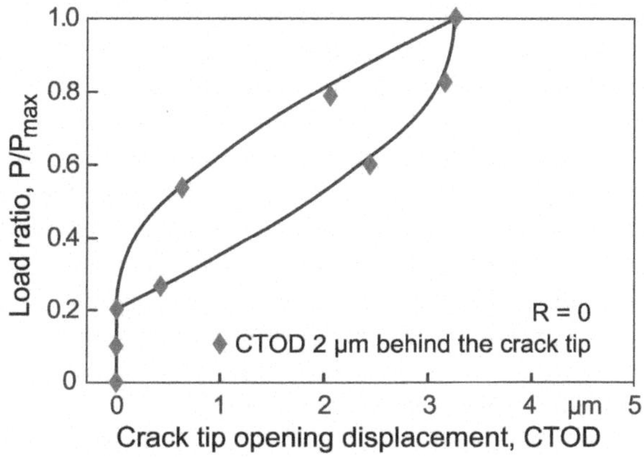

Fig. 4.11 Hysteresis loop of load cycle over crack opening displacement (CTOD) with crack closure; measurement results; (Pippan et al. 2011)

The basic micromechanism of fatigue crack growth with a pointed crack tip and re-established crack closure was already published much earlier (Forsyth and Ryder 1960; McMillan and Pelloux 1967).

The findings above strongly support the crack closure argument. The closure level of 20 % of the maximum load for $R = 0$, measured by observing crack tip opening displacements (CTOD) in close vicinity of the crack tip, Fig. 4.11, can be well explained by modelling the deformation mechanics as described above. As cyclic plastic deformation is the main determinant of the ongoing physical process, it is slightly surprising that such realistic descriptions are gained, although the cyclic plastic deformation model used in Newman's version of the strip-yield model is extremely simple, especially too with regard to the perfectly plastic material behaviour in the plastic zone. Introducing additional hardening capabilities in the strip-yield model (e.g. Wang 1993) may improve the assessment accuracy to some extent.

The strip-yield model is rather restricted with respect to the geometry and the material behaviour. Especially during cyclic loading, the stress–strain relationship of the material is far from being perfectly-plastic. First, it might be argued that the originally recommended use of the static yield stress σ_Y should be replaced by the cyclic yield stress σ'_Y. Keeping in mind that the permissible maximum load is equal to the chosen yield stress and that locally higher stresses, e.g. at notches, are found, a substitute yield stress is often recommended according to the formula $\overline{\sigma}_Y = (\sigma'_Y + \sigma_U)/2$ where σ_U is the ultimate tensile strength.

Finite Element Modelling of Crack Growth

The restriction to a two-dimensional structure and a one-dimensional crack in a slim strip of perfectly-plastic material can be overcome by modelling the crack growth with finite elements. Finite element based calculations of crack opening levels have early been presented (Newman 1974, 1976; Ohji et al. 1974). This method was applied to investigate the crack closure phenomenon under plane strain conditions (Fleck and Newman 1988). The calculus was accelerated considerably by other authors (McClung and Sehitoglu 1996). As a consequence, these authors were able to perform a rigorous study on the effects of the modelling techniques on the accuracy of the calculated results. They were able to conclude that several of the previous researchers may have fallen victim to model deficiencies. The progress that had been achieved until the mid-nineties was summarised (Sehitoglu and Gall 1996; McClung 1999). The influence of the constitutive material model was emphasised as well as the roughness-induced crack closure and the influence of biaxial loading.

Being aware of the vital importance of the material model, Jiang and Feng applied the plasticity model according to Jiang in their crack propagation simulations (Jiang and Feng 2004; Jiang 1995). Andersson et al. used the viscoplastic material model according to Bodner and Partom in the version of Moreno and Jordan (Andersson et al. 2004; Moreno and Jordan 1986). They confirmed the significant influence of the material model by comparing the results to those obtained by applying simple models with linear kinematic hardening. The hardening model proposed by Chaboche was also used (Zhao et al. 2004; Chaboche 1986, 1989).

Both Jiang's plasticity model and Chaboche's model apply kinematic hardening rules of the Armstrong–Frederick type. However, the evolution of ratchetting, as predicted by all these models, is different. Therefore, different crack opening levels will arise. Thanks to increasing computer resources, investigations on the three-dimensional problem of semicircular and semi-elliptical surface cracks have been started (Chermahini et al. 1993). Investigations on this problem are continuing (Zhang and Bowen 1998; Skinner and Daniewicz 2002; Hou 2004). Recent publications provide literature overviews (Lei 2008; Antunes and Rodrigues 2008).

Two main items have to be properly addressed in the finite element modelling of cyclic crack tip plastic deformation and fatigue crack closure: the mesh geometry, especially the element size near the crack tip and the plasticity-model, especially the kinematic hardening rule. A very fine mesh is needed near the crack tip in order to enable a realistic representation of the strongly inhomogeneous strain field. The size of the crack tip element also determines the crack length increment in the generally applied nodal release technique.

The nodal release technique received its name because, in the finite element analysis, the crack growth is performed by releasing the boundary condition of the crack tip node from a fixed support to a free support (in case of only one symmetry half of the structure being modelled). In case of both crack flanks being modelled, two nodes at the crack tip are provided. Initially their boundary condition is specified by enforcing identical displacements of both crack tip nodes by constraint equations. The crack growth is again simulated by releasing this constraint condition between the crack tip nodes. The history of plastic deformation is preserved in the finite element model. The procedure is repeated according to a numerical scheme which has to be properly defined in order to achieve converging results.

The application of the procedure is limited to cases for which the geometry of the growing crack is known a priori. As part of this scheme, it is recommended that at least three elements should be positioned in the cyclic plastic zone (McClung 1991[(1)]). As the cyclic plastic zone size is unknown a priori, an iterative procedure is usually required to properly arrive at converging results. The convergence with respect to the crack tip element size depends on the plasticity model selected.

A load cycle with plasticity-induced crack closure gives rise to non-proportional multiaxial loading at the crack tip. As this loading rarely leads to locally mean stress free conditions, it is of great importance how multiaxial ratchetting is modelled. An early attempt to apply the Mróz plasticity model (Mróz 1967, 1969) was disappointing (Anthes et al. 1996, Anthes 1997). The closure level is strongly dependent on the number of cycles applied without node release, i.e. without numerical crack growth. The severe non-proportional ratchetting leads to vanishing crack closure. Similarly strong non-proportional ratchetting rates are modelled by the kinematic hardening rules proposed by Ziegler and Chaboche (Ziegler 1959; Chaboche 1989). These older plasticity models cannot be recommended for application in finite element based calculations of fatigue crack growth. Stable, converging and non-zero crack closure levels are calculated when applying plasticity models which are adjusted to describe the ratchetting behaviour of the material in a realistic way (Ohno and Wang 1993; Jiang 1993; Jiang and Sehitoglu 1996[(1)], 1996[(2)]; Döring et al. 2003).

The numerical crack length increment—equal to the crack tip element size—is generally much larger than the physical crack length increment per real load cycle. Ratchetting, however, is associated with the real cycle number applied to the material. Therefore, a sufficient number of cycles should be provided for the cyclic plastic zone to develop its ratchetting deformation before the next node at the crack tip is released. No final recommendation can be given. Again, an iterative study of the development of the crack closure level is necessary. On the one hand, its value is believable only if it does not change any more with increasing number of cycles without a node release. On the other hand, if this stabilised value tends to vanish as it is the case for strong-ratchetting plasticity models, again these results are not trustworthy. Against the background of information which can be obtained with finite element based fatigue crack modelling, considering all interacting influence parameters in respect of element size, plastic deformation model and nodal release scheme, it is once again surprising that the simple strip-yield model gives realistic crack closure estimates.

Cyclic plastic deformation, even under small-scale yielding conditions, is the origin of plasticity induced crack closure. The closure level depends on the load ratio. This effect on the fatigue crack growth rate can be explained by plasticity-induced crack closure.

4.2.3 Load Sequence Effect on Crack Growth

Sequence-Independent Fatigue Life

The Eqs. (4.2) or (4.9), can be the basis of a fatigue crack growth calculation without taking sequence effects into account. A block of n_i cycles with stress range $\Delta\sigma_i$ acting on a structure with the initial crack length a_i and the stress intensity factor range $\Delta K_i = \Delta\sigma_i(\pi a)^{1/2}Y(a)$ will produce fatigue crack growth according to the following integral:

$$\int_{a_i}^{a_{i+1}} \frac{da}{a^{m/2}(Y(a))^m} = C_i \Delta\sigma_i^m \pi^{m/2} n_i \tag{4.17}$$

A further loading block of n_{i+1} cycles with stress range $\Delta\sigma_{i+1}$ results in:

$$\int_{a_{i+1}}^{a_{i+2}} \frac{da}{a^{m/2}(Y(a))^m} = C_{i+1} \Delta\sigma_{i+1}^m \pi^{m/2} n_{i+1} \tag{4.18}$$

Performing this calculation for all blocks of a given load sequence gives the crack length as function of the applied number of cycles. If the calculation starts at an initial crack length a_0 and ends at a final crack length a_f the following expression results:

$$\int_{a_0}^{a_1} \frac{da}{a^{m/2}(Y(a))^m} + \int_{a_1}^{a_2} \frac{da}{a^{m/2}(Y(a))^m} + \ldots + \int_{a_i}^{a_{i+1}} \frac{da}{a^{m/2}(Y(a))^m} + \ldots \int_{a_k}^{a_f} \frac{da}{a^{m/2}(Y(a))^m} =$$
$$= C_0 \Delta\sigma_0^m \pi^{m/2} n_0 + C_1 \Delta\sigma_1^m \pi^{m/2} n_1 + \ldots + C_i \Delta\sigma_i^m \pi^{m/2} n_i + \ldots C_k \Delta\sigma_k^m \pi^{m/2} n_k \tag{4.19}$$

where k is the counter for the last block. Eq. (4.19) can be written as follows:

$$\int_{a_0}^{a_f} \frac{da}{a^{m/2}(Y(a))^m} = \pi^{m/2} \sum_{i=1}^{k} C_i \Delta\sigma_i^m n_i \tag{4.20}$$

Constant amplitude loading with crack growth from a_0 to a_f gives

$$\int_{a_0}^{a_f} \frac{da}{a^{m/2}(Y(a))^m} = \pi^{m/2} C \Delta\sigma^m N \tag{4.21}$$

where N is the number of cycles between initial and final crack size.

The Eq. (4.21) can be written with insertion of $\Delta\sigma = \Delta\sigma_i$ and $N = N_i$ where the subscript i refers to a constant amplitude block as part of a variable amplitude load sequence as in Eq. (4.17). Dividing Eq. (4.17) by this modified Eq. (4.21) leads to

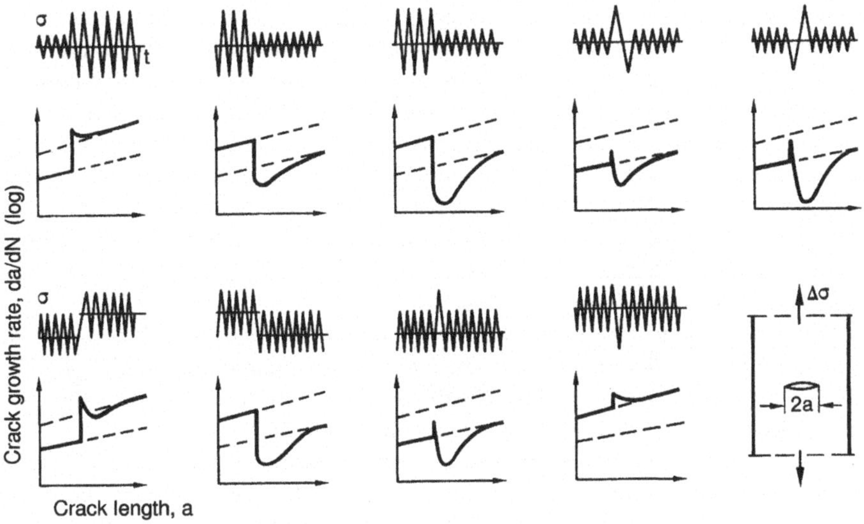

Fig. 4.12 Fatigue crack growth rate increasing with crack length; retardation and acceleration effects at changes of load amplitude or mean stress; (Führing 1982)

a hypothetical total damage contribution n_i/N_i. Summing up these damage contributions for all blocks of a variable amplitude sequence leads to

$$\sum_{i=1}^{k} \frac{n_i}{N_i} = 1 \tag{4.22}$$

This result—the linear damage accumulation rule—is obtained independently of whether the calculation starts with Eq. (4.2) or with Eq. (4.9). The sum in Eq. (4.22) shows that the above simplified fatigue crack growth calculation results in sequence-independent fatigue lives. The underlying assumption is that the growth rates, Eq. (4.2) or Eq. (4.9), besides being described by a power law, are a function of the actually applied stress range only. For crack growth described by Eq. (4.2), this assumption is often strongly violated.

Crack Growth Retardation and Acceleration

Various crack growth acceleration and retardation phenomena have been reported in the literature. An overview is provided in Fig. 4.12 (Führing 1982). For simplicity, only two different amplitudes or mean stresses are considered. Under constant amplitude (and mean stress) the crack growth rates would be described by the dashed lines. Switching loads from one block to the next without any interaction between the blocks would simply produce a switching between the two dashed lines. Such a situation and its consequences are mathematically modelled by Eqs. (4.17) to (4.22). In Fig. 4.12 the various deviations are shown as solid lines. A single tensile peak load causes gradually developing retardation with a minimum growth rate after some crack length increments (delayed retardation), followed by approaching the constant amplitude growth rate. The amount of retardation expressed in terms of additionally applicable cycles strongly depends on the overload ratio K_p/K_{max} where K_p is the stress intensity factor at the single tensile peak load and K_{max} is the stress intensity factor at the maximum load of the constant amplitude block load. The fatigue life increases with the increase of this ratio. In rare beneficial cases, crack arrest and unlimited life may be achieved (Alzos et al. 1976; Bernard et al. 1976; Jacoby et al. 1976; Lankford and Davidson 1976; Matsuoka et al. 1976).

The crack length range influenced by the retardation effect depends on the size of the plastic zone (Petrak 1974). Low yield stresses enlarge the retardation effect. Also, a decreasing sheet thickness will increase retardation because the plane stress portion at the crack front increases (Mills and Herzberg 1975).

Single compressive overloads cause an acceleration of fatigue crack growth which acts without delay. However, the effect is much weaker than the retardation effect. In thick plates, the acceleration effect nearly vanishes (Buschermöhle et al. 1996). Repeated tensile peak loads may produce ample retardation, especially when the peak is applied at the crack length with minimum growth rate caused by the previous peak. An amplification of acceleration with multiple compressive overload peaks does not occur (Trebules et al. 1973).

The effects are similar when switching from one block to the next. The difference is that after a switch the constant amplitude growth rate of the next block is approached—be it after retardation (high–low sequence) or acceleration (low–high sequence).

For more complex sequences, it is difficult to make a judgement on interaction effects without any modelling (Schijve 1973, 1976$^{(1)}$, 1976$^{(2)}$, 1992, 2009; Schijve et al. 1985). For identical load histograms, crack growth life under random loading is shorter than under blocked loading. However, the shorter the blocks are, the more are the random sequence lives approached. In blocked loading, a descending sequence produces longer lives than an ascending sequence, Fig. 4.13 (Schijve 1973). A variety of data is gathered in another overview (Singh et al. 2011).

Plastic Zone Size Related Retardation Models

Load sequence effects as described above, cannot be explained by only considering crack tip parameters of linear elasticity. Obviously, information on previous loading conditions remains stored in the structure. They exert an influence on crack growth in subsequent loading. The most important means of storing such information in a mechanically loaded structure are plastic deformations. Load sequence effects are mainly a consequence of cyclic plastic deformation at the crack tip. Describing load sequence effects based on the ongoing retardation and acceleration mechanisms requires considering cyclic plastic deformation in the structure, even if the constant amplitude crack growth behaviour is advantageously characterised by the stress intensity factor range as in Eqs. (4.2) or (4.3).

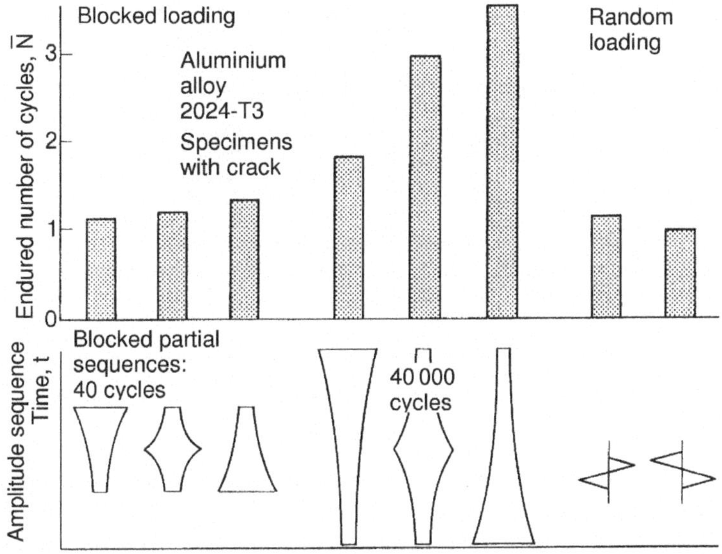

Fig. 4.13 Crack growth lives of aluminium alloy specimens ($a_0 = 12$ mm, $a_f = 50$ mm) under blocked sequence loading and random loading in comparison (gust spectra); blocked sequence with small or large partial sequence subdivision; random loading with tension–compression cycles compared with compression-tension cycles; (Schijve 1973)

Under small-scale yielding conditions the main portion of cyclic plastic deformation occurs in the forward plastic zone ahead of the crack tip. Therefore, the first approaches considering sequence effects in fatigue crack growth were focused on the size of the plastic zone. In the model proposed by Wheeler, a retardation factor C_p is introduced into the crack growth rate equation (Wheeler 1972):

$$\frac{da}{dN} = C_p f(\Delta K, R) \tag{4.23}$$

The factor C_p is smaller than one, as long as the tensile plastic zone size ω_i of cycle i is smaller than the plastic zone size ω_0 left from a previous peak load cycle, Fig. 4.14:

$$C_p = \begin{cases} [\omega_i/(a_0 + \omega_0 - a_i)]^p & \omega_i < (a_0 + \omega_0 - a_i) \\ 1 & \omega_i \geq (a_0 + \omega_0 - a_i) \end{cases} \tag{4.24}$$

The exponent p has to be chosen to fit the experimental results. The plastic zone sizes are estimated based on the small-scale yielding assumption, i.e. they are proportional to $(K/\sigma_Y)^2$. In particular, for plane strain conditions, an equation proposed by Williams may be used (Williams 1957):

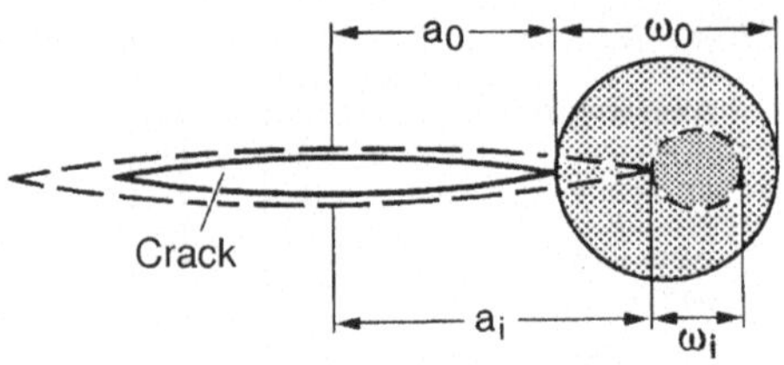

Fig. 4.14 Crack lengths a_0 and a_i as well as plastic zone sizes (schematised as circles) at peak load (subscript 0) and at constant amplitude load (subscript i), for use in crack growth retardation models; (Willenborg et al. 1971; Wheeler 1972)

$$\omega = \frac{1}{4\sqrt{2\pi}}\left(\frac{K}{\sigma_Y}\right)^2 \tag{4.25}$$

Under plane stress conditions, Eq. (4.4) can be applied. The tensile plastic zone ω_i of cycle i is calculated without considering the stress and strain of the prior peak cycle. The model does not consider crack arrest, delayed retardation, amplified retardation due to multiple overloads, and crack acceleration. Also, interaction of multiple overloads in complex load sequences cannot be taken into account.

Gray and Gallagher propose that crack growth should be retarded as long as the secondary plastic zone is enclosed in the primary plastic zone produced by a peak load (Gray and Gallagher 1976). The retardation factor definition, Eq. (4.24), is then replaced by the following equation:

$$C_p = \left(\frac{K_{i\,max}}{K^*_{i\,max}}\right)^p \tag{4.26}$$

where $K^*_{i\,max}$ is a fictitious stress intensity factor necessary to create the following plastic zone size:

$$\omega = a_0 + \omega_0 - a_i \tag{4.27}$$

The fictitious stress intensity factor $K^*_{i\,max}$ is determined from K_p according to the following expression:

$$K^*_{i\,max} = K_p\sqrt{1 - \frac{a_i - a_0}{\omega_0}} \tag{4.28}$$

Again, the interaction of multiple overloads in complex load sequences cannot be simulated.

Willenborg interpreted the fictitious stress intensity factor, Eq. (4.28), as caused by residual stresses introduced by the peak load (Willenborg et al. 1971). Responsible for retardation according to the Willenborg model is the difference between the actual and the fictitious stress intensity factor:

$$K_{resp} = K^{*}_{i\,max} - K_{i\,max} \tag{4.29}$$

A fictitious mean stress intensity shift in the negative direction (for producing retardation) by the amount K_{resp} is now introduced leading to:

$$\begin{aligned} K^{(W)}_{i\,max} &= 2K_{i\,max} - K^{*}_{i\,max} \\ K^{(W)}_{i\,min} &= K_{i\,min} + K_{i\,max} - K^{*}_{i\,max} \end{aligned} \tag{4.30}$$

The corresponding Willenborg stress ratio reads as follows:

$$R^{(W)} = \frac{K^{(W)}_{i\,min}}{K^{(W)}_{i\,max}} \tag{4.31}$$

The Willenborg stress ratio $R^{(W)}$ is smaller than the original load ratio R. The fatigue crack growth is calculated using $R^{(W)}$ instead of R in the crack growth rate equation, e.g. Eq. (4.2). This leads to a reduced growth rate as long as $(a_i + \omega_i)$ is smaller than $(a_0 + \omega_0)$. Directly after application of the peak load, the fictitious stress intensity factor $K^{*}_{i\,max}$ is equal to K_p. The fictitious Willenborg maximum stress intensity factor $K^{(W)}_{i\,max}$ will become zero as soon as K_p is twice as large as $K_{i\,max}$, see Eq. (4.30). For such a constellation, crack arrest is predicted.

Variants of plastic zone size related crack propagation models have been presented (Gallagher 1974; Eidinoff and Bell 1977). All these models are based on quantities of the forward plastic zone. Their prediction qualities have not been convincing in general.

As has been the case with explaining the load ratio effect, the plasticity-induced crack closure offers further insight into the life-relevant physical processes. Considering the crack opening stress—which may vary from cycle to cycle—the effective stress intensity factor range, ΔK_{eff}, can be calculated. As has been mentioned on the occasion of Eq. (4.22), the integration of the corresponding crack growth relationship, Eq. (4.9), leads to the linear damage accumulation rule. This means that sequence effects are captured in analysing the sequence-dependent development of crack opening and crack closure. The damage accumulation or the crack growth in a narrower sense is then sequence-independent.

Various approximative approaches have been presented, differing in the way of considering previous amplitudes (Baudin and Robert 1984; de Koning 1981; Aliaga et al. 1985). Only immediate retardation, not delayed retardation or acceleration, is modelled. The progressive transition from plane strain to plane stress conditions with increasing crack length is taken into account. An extensive comparison is available (Padmadinata 1990). A satisfying accuracy of life predictions has only been achieved with factors up to five caused by closure-related retardation.

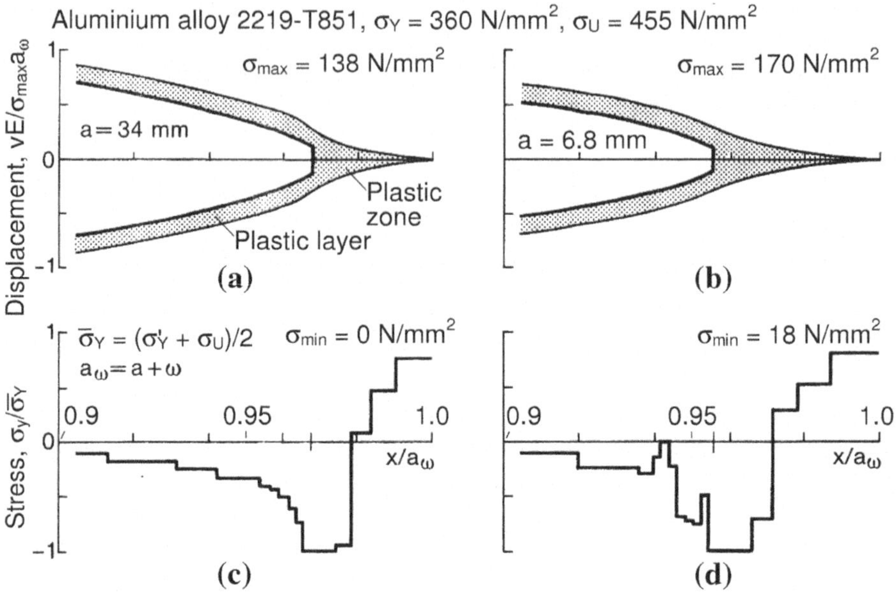

Fig. 4.15 Strip-yield model with plastic layers on crack flanks, produced by cyclic plastic deformation during crack growth: constant amplitude loading (**a**) compared with loading by a typical gust sequence with truncation at peak load (**b**); stresses at the crack tip after load relief (**c**, **d**) with plasticity-enlarged crack length, $a_\omega = a + \omega$, and with substitute yield limit $\bar{\sigma}_Y$; (Newman 1981, 1982)

Strip-Yield Based Retardation and Acceleration Model

The strip-yield model can advantageously be applied to identify both plastic zone sizes, the (residual) stress distributions connected with these zones and—more importantly—the irreversible plastic deformation in the crack wake under variable amplitude loading. Numerical solutions have been presented (Newman 1981, 1982, Newman et al. 1999[(1)], 1999[(2)]). An example of the results is shown in Fig. 4.15.

The deformation profiles in this plot are similar under constant and variable amplitude loading. Small differences in these deformation profiles, however, cause large differences in the stress profiles near the crack tip. Especially the contact stress profiles at the crack flanks are a consequence of an irregular deformation profile at the flanks.

Some simpler cases are shown in Fig. 4.16, where the sequence-dependent crack opening stresses are plotted. In Fig. 4.16a, the crack opening stress is shown as function of the crack length for two load sequences which differ only in the number of cycles with the high amplitude. With only one overload in, apart from this, constant amplitude loading, the crack opening stress initially drops sharply due to crack tip blunting. It increases with a steep slope, reaching a maximum

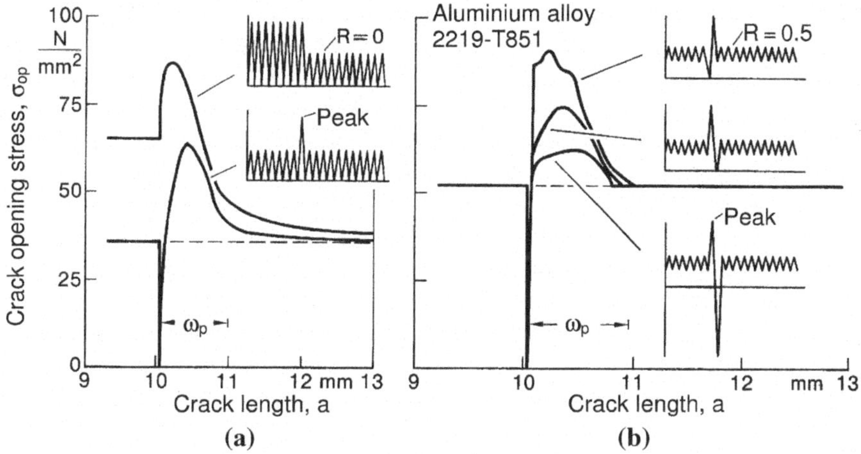

Fig. 4.16 Strip-yield model, crack opening stress dependent on crack length: two-level load sequence compared with constant amplitude loading with inserted peak load (**a**) and constant amplitude loading with inserted compression–tension compared with tension–compression peak load, zero and fully negative lower peak (**b**); extension of plastic zone, ω_p, produced by tensile peak load; $K_{OL} = 26.7$, $\Delta K_{bl} = 12.0$ MPa m$^{1/2}$, see Fig. 4.19; (Newman 1982)

approximately at half the plastic zone size caused by the peak load. The crack growth rate attains its minimum at this crack length. The crack opening stress decreases in the following and approaches its stable constant amplitude level. In the high–low load sequence, the initial decrease is less severe. The retardation effect at the lower level is therefore much more pronounced. In Fig. 4.16b, the crack opening stress is shown as a function of crack length for constant amplitude loading interrupted by a compression-tension or a tension-compression cycle. The compression load before the tension overload proves ineffective in influencing the development of the crack opening load. The compression overload after the tension overload, however, reduces the increase of crack opening stresses strongly without being able to compensate it entirely.

The fatigue crack growth calculation is usually performed on a cycle-by-cycle basis exploiting Eq. (4.9). Some numerical expense is saved when the crack opening level is first determined which is relevant for a crack increment equal to the width of the crack tip bar element. In the end, all crack length increments and associated numbers of cycles have to be added up.

Modified strip-yield models have also been developed (de Koning 1981; de Koning and Liefting 1988, 1989; Dill and Saff 1976; Dill et al. 1980; Führing and Seeger 1979, 1984; Ibrahim et al. 1986; Wang and Blom 1991).

The limitations of the strip-yield model with respect to geometry and material behaviour are substantial. Sheet thickness aspects of three-dimensional fatigue crack growth can be taken into account via the constraint factor α (Yamada et al. 2007; Ziegler et al. 2008; Beretta et al. 2009).

For complicated three-dimensional geometries, the strip-yield model can be applied by defining a two-dimensional working plane (Herz et al. 2011). For intersecting cyclically pressurised holes the bisector plane was chosen to govern the fatigue crack growth in this structure. All calculations with the strip-yield model were performed using this working plane. The constraint factor of the strip-yield model was set to $\alpha = 3$ to take into account constraint effects perpendicular to the working plane. The stress applied to the strip-yield model is adjusted in such a way that the stress intensity factor at the crack tip of the strip-yield model is equal to the stress intensity factor at the crack tip of the analysed intersecting hole specimen where the pressure is applied both to the holes and the crack faces.

In the cited application example (Herz et al. 2011) not only the geometric restrictions have been overcome. The intersecting holes are subjected to an autofrettage process during fabrication which introduces high residual stresses. Originally, the strip-yield model has been developed with the aim of taking the residual stresses and consequently the load sequence effects into account that occur under variable amplitude loading. In fact, the autofrettage process is nothing more than the first part of the load history of a pressurised component. Therefore, the influence of the autofrettage process on the propagation of fatigue cracks may be assessed by just loading the strip-yield model with a remote stress calculated based on the postulated identity of the initial stress intensity factor. However, the strip-yield model is limited to elastic perfectly-plastic material behaviour. Due to the very different deformation behaviour of the real material and the totally different geometries of the two-dimensional model and the three-dimensional pressurised holes, this approach does not lead to a residual stress distribution which satisfies the condition that the stress intensity factors from residual stresses should be equal in the real specimen and in the strip-yield model. The initial residual stress distribution therefore has to be introduced into the strip-yield model in a different way.

The main point of the extension of the strip-yield model is that the initial residual stress distribution comes from a different source and not from the strip-yield model itself. The following cyclic loading, however, occurs with load levels for which the strip-yield model should be able to provide reasonable results. It is assumed that the required initial residual stress distribution is obtained by assigning an initial length, $l_{0\,\mathrm{i}}$, to the plastically deformable bar elements of the strip-yield model, Fig. 4.17. This initial length of element i is calculated from the residual stresses $\sigma_{\mathrm{res\,j}}$ in all elements at location x_{j} and the related influence function $g(x_{\mathrm{i}}, x_{\mathrm{j}})$ (Newman 1981):

$$l_{0\,\mathrm{i}} = -\sum_{\mathrm{i}=1}^{\mathrm{k}} \sigma_{\mathrm{res\,j}}\, g(x_{\mathrm{i}}, x_{\mathrm{j}}) \tag{4.32}$$

In Fig. 4.18, an example is shown, how the original residual stresses evaluated from three-dimensional finite element calculations (or from measurements) must be modified for feeding into the two-dimensional strip-yield model in such a way that residual stress intensity factors during crack propagation are identical.

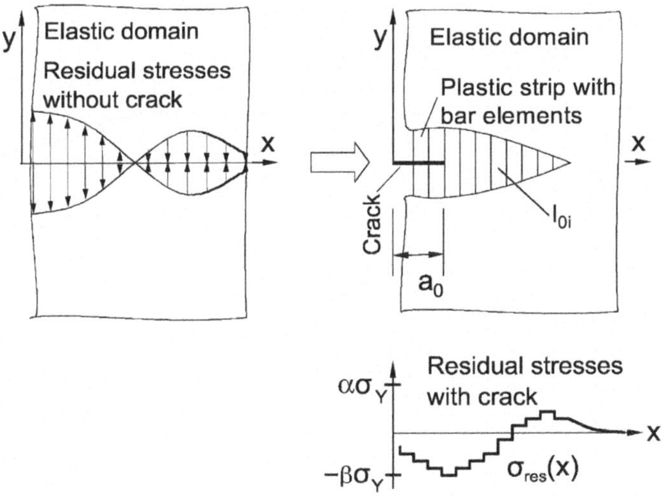

Fig. 4.17 Strip-yield model: initial residual stresses without the crack, plastic strip with bar elements and residual stresses at crack tip; schematic representation; (Herz et al. 2011)

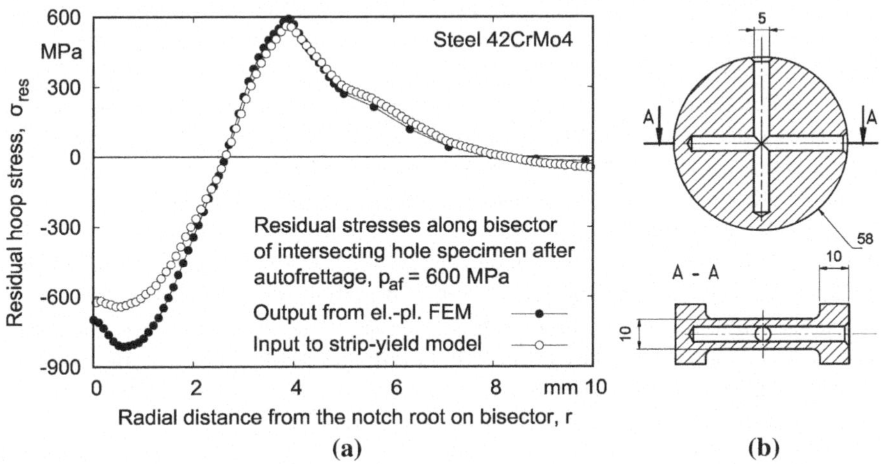

Fig. 4.18 Residual stresses along bisector (**a**) of intersecting hole specimen (**b**) after autofrettage; output from elastic-plastic finite element analysis compared with stress intensity consistent input to strip-yield model; (Herz et al. 2011)

Finite Element Modelling of Sequence Effects

For variable amplitude loading, finite element based simulations of fatigue crack growth and plasticity-induced crack closure are rare (Singh et al. 2008; Ramos et al. 2003; Park and Song 1999; Blom and Holm 1987). The numerical expense is very high.

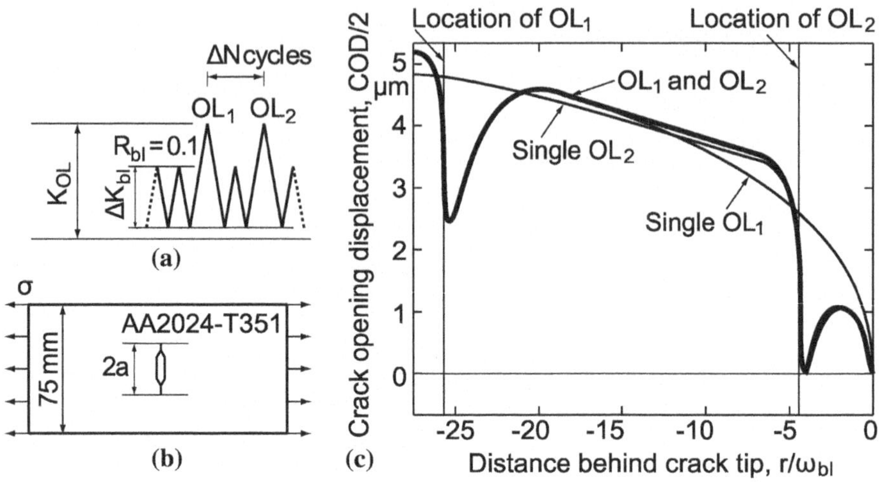

Fig. 4.19 Crack growth analysis using a finite element model of centre-cracked plate specimen; crack flank profile after single and double overload (OL); (Singh et al. 2008)

One result obtained by Singh et al. is redrawn in Fig. 4.19. The two crack flank hillocks are clearly visible. They are produced by the two overloads resulting in plastic deformation in the forward plastic zone. As the crack proceeds, some of this deformed material remains on the crack flanks, acting in a mechanically similar way to additional material injected in the crack. Consequently, the crack opening stress rises whereas the effective stress ranges and growth rates decrease until the moving crack tip leaves the crack flank hillocks far behind.

The results of the two crack growth simulation approaches—strip-yield model and finite element model—are qualitatively identical. Quantitatively, there may be differences. In the strip-yield model, the simplifications with regard to the geometry and the material behaviour are substantial. The numerical expense is acceptable even for many thousands of cycles. The application of the finite element based approach, however, still requires a confinement to rather short variable-amplitude load sequences.

Nodal Release Technique Applied to Pressurised Pipes or Intersecting Holes

The fatigue crack growth in pipes after an autofrettage process—an extremely high loading causing severe plastic deformation—was investigated based on the finite element method combined with the nodal release technique applied without remeshing and mapping (Herz et al. 2006, 2011). For pressurised Diesel injection pipes with 10 μm deep longitudinal crack-like defects originating from the deep-drawing process, the residual stresses due to the autofrettage process and the initial stress relaxation were calculated first, Fig. 4.20. A special plasticity model was applied (Döring et al. 2003). In the crack growth simulation, every node release

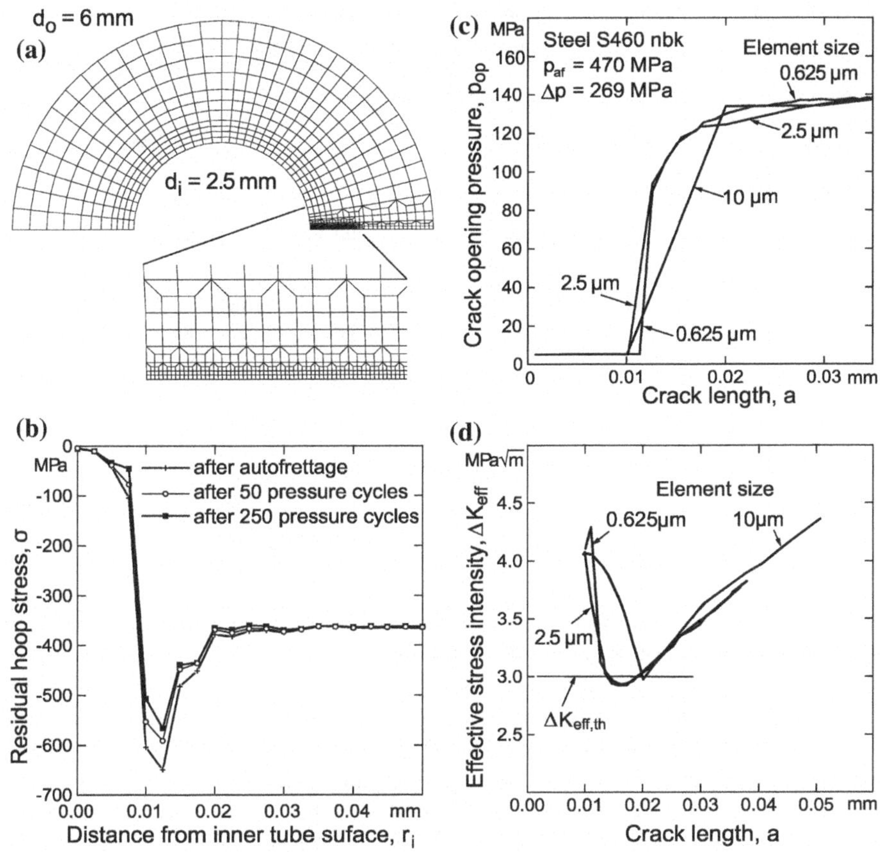

Fig. 4.20 Autofrettaged pipe specimen under cyclic internal pressure load Δp (**a**): residual hoop stresses after autofrettage pressure p_{af} and after further pressure cycles (**b**), crack opening pressure p_{op} of inner surface axial crack dependent on crack length (**c**), and effective stress intensity factor range ΔK_{eff} dependent on crack length (**d**) indicating the threshold value $\Delta K_{eff,th}$; nbk non-blank-drawn; (Herz et al. 2006)

cycle was followed by 10 relaxation cycles leading to converging results. From the crack opening stresses and available stress intensity factor solutions, the effective ranges could be obtained. Due to the steep increase in crack opening pressures, the effective stress intensity ranges initially dropped to a minimum before increasing again with rising crack length. The minimum effective stress intensity factor ranges were compared with the effective threshold range of the material. With this numerical procedure, the endurance limit values of the pipe specimens were estimated and compared to experimental results, Fig. 4.21.

The accuracy of the results is satisfactory. The finite element based simplified fatigue crack growth simulation is in the stage of conversion from a laboratory-operated descriptive tool for experimental findings to an engineering office-based

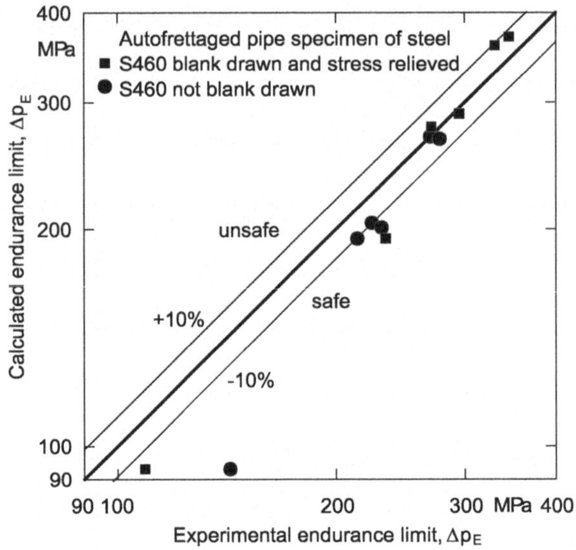

Fig. 4.21 Autofrettaged pipe specimen under cyclic internal pressure load, comparison of calculated and measured endurance limit values Δp_E associated with different crack growth fatigue lives; (Herz et al. 2006)

prediction tool for design purposes. To achieve this ambitions aim, more evidence of the prediction quality is certainly required.

For this purpose, an intersecting hole specimen, first pressurised by the autofrettage process and then cyclically loaded by the service pressure amplitudes was modelled fully three-dimensionally. The modelling is limited by the requirement that no remeshing after a crack growth increment should be necessary. Such remeshing would require that the state variables of the plastic deformation model are transferred from the old to the new mesh. Therefore, the restriction that the crack front is a straight line perpendicular to the bisector between the holes was introduced, which is not fully in agreement with the experimental findings. Both concave and convex curvatures of the crack front have been observed after sectioning unbroken specimens containing fatigue cracks (Herz et al. 2011). The edge length of the eight-node isoparametric elements with a linear shape function was set to 0.1 mm. The mesh exploiting all symmetries involved is shown in Fig. 4.22.

The simulation of constant amplitude loading with load ratio $R \approx 0$ was performed in several steps:

- Loading of the specimen by the autofrettage pressure and unloading to the minimum pressure (5 MPa) in the cyclic loading test.
- Application of 250 load cycles of pressure range Δp for stress redistribution due to combined ratchetting and stress relaxation in the whole component.
- Simulation of crack propagation along the crack front in steps of one element edge length.

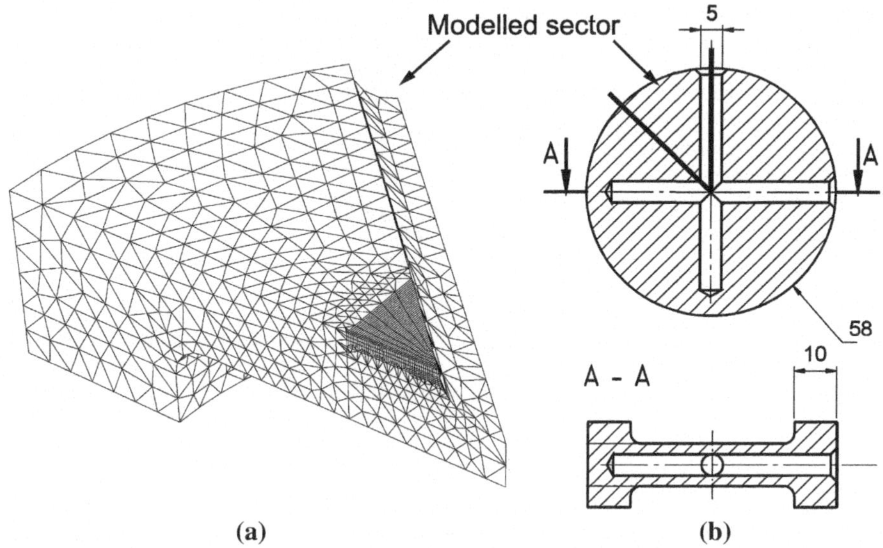

Fig. 4.22 Finite element mesh on wedge-shaped section (**a**) of intersecting-hole specimen (**b**); location where the crack initiates; minimum element size 0.1 mm; (Herz et al. 2011)

Each of the crack propagation steps is modelled as follows:

- Loading by the service pressure. The crack will get completely or partially opened.
- Unloading to 90 % of service pressure. In the same step release of the boundary condition of suppressed displacement perpendicular to the crack plane applied to the next crack front nodes on the ligament.
- Further unloading to the minimum pressure, $p_{\min} = 5$ MPa. In this step, the crack will close. This is modelled by introducing crack face contact with a rigid surface.
- Between two crack propagation cycles, ten additional load cycles are added to stabilise the mean stress in the elements ahead of the crack tip. The last of these cycles is used to obtain the displacement output for the ongoing analysis. This cycle is divided into 20 increments for loading and unloading in order to obtain enough values for accurate interpolation. The loss of contact in the last node during the loading half cycle is considered as crack opening.

In Fig. 4.23, the calculated crack opening pressures under constant amplitude loading are shown as a function of the crack length. At maximum pressure—in this example $p_{\max} = 200$ MPa—the opening pressure rises from approximately 100 MPa to a higher value of approximately 180 MPa. This is due to the residual stresses in combination with the plastic deformation due to the autofrettage process. At a crack length of 1.4 mm only the pressure range $\Delta p_{\text{eff}} = 20$ MPa is effective. The resulting effective stress intensity factor approaches a value near the effective threshold value of the material. The endurance limit is predicted.

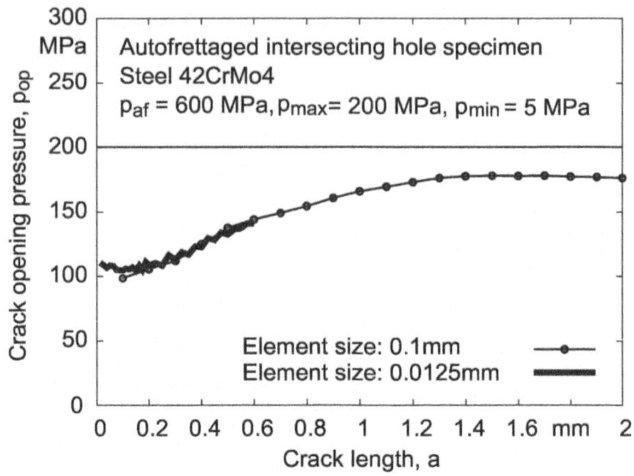

Fig. 4.23 Autofrettaged intersecting-hole specimen under constant amplitude internal pressure loading: crack opening pressure dependent on crack length for maximum pressure $p_{max} = 200$ MPa; (Herz et al. 2011)

In Fig. 4.23, also the result of a convergence study is contained. Results gained for an element edge length of 12.5 μm are included for crack lengths smaller than 0.6 mm. At that crack length the time-consuming calculations were stopped because mesh size independent results had been achieved.

The procedure has been modified and enhanced to deal with the case of variable amplitude loading. Due to the numerical limitations, it is impossible at present to subject a finely meshed finite element model of a structural component to a real-world load sequence with up to several thousands or even millions of load cycles. Nevertheless, this is necessary to correctly determine the load sequence effects. Otherwise, the high accuracy of the sophisticated crack closure model would be lost.

In a first attempt to account for load sequence effects in a finite element based simulation of fatigue crack growth, the original random load sequence was replaced by a severely shortened auxiliary load sequence. The auxiliary load sequence was defined in such a way that similar load sequence effects can be expected from both the original and auxiliary load sequences. However, there is actually no generally agreed recommendation for coming up to this requirement. In the actual application example of pressurised intersecting holes, the original variable amplitude load sequence is derived from injection pressure measurements at Diesel engines. The maximum pressure is reached frequently. The injection pressure drops at random to various minimum values depending on the operating conditions of the engine. The crack opening pressure is decreased by low pressure minima which is equivalent to large pressure ranges. The auxiliary load sequence was therefore constructed to supply information on crack opening and crack closure pressures for pressure cycles preceding and following on low pressure minima. The auxiliary load sequence is shown in Fig. 4.24b. For each level of maximum pressure of the sequence for which a life estimate has to be given, three crack growth simulations are performed according to the following procedure:

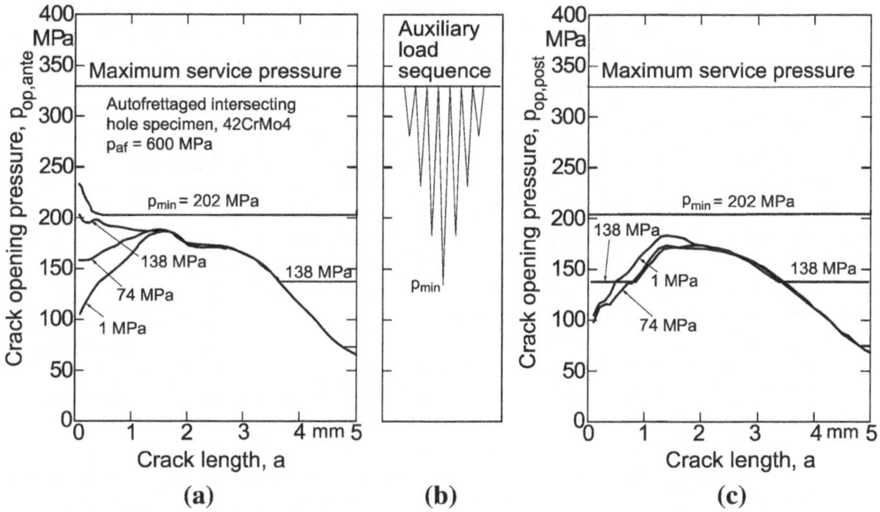

Fig. 4.24 Autofrettaged intersecting-hole specimen subjected to auxiliary load sequence (**b**): crack opening pressure p_{co} dependent on crack length for different values of minimum pressure p_{min}; $p_{op,ante}$ for the smaller cycles preceding p_{min} (**a**) and $p_{op,post}$ for the smaller cycles following p_{min} (**c**) in the auxiliary load sequence; (Herz et al. 2011)

- Simulation of constant amplitude loading.
- Simulation of 10 loading cycles to stabilise the mean stress in the elements ahead of the crack tip.
- Simulation of loading by the auxiliary load sequence.
- Repetition of the above steps until the final crack length is achieved.

It must be made sure that the largest cycle occurs at least once per crack growth increment. This is guaranteed for the slow crack growth situation in the current investigation.

Crack opening pressures are identified for each pressure cycle of the auxiliary load sequence. The analysis results are shown in Fig. 4.24. On the left hand side of the figure, the crack opening pressures $p_{op,ante}$ are shown for the smaller cycles preceding p_{min}. On the right hand side of this figure, the crack opening pressures $p_{op,past}$ for the smaller cycles following p_{min}. Using these curves, the resulting effective stress intensity factor ranges and the crack propagation rates are calculated, the latter by means of Eq. (4.9). Once again, the effective stress intensity factor range is selected to serve as the crack driving force parameter. As is the case with any strip-yield model application, the state of the art still requires simplification of the straightforward nonlinear analysis by calculating the stress intensity factors of the linear-elastic approach.

A direct observation of the cracks emanating from the hole intersection notch is impossible. As in many engineering applications, only comparisons of calculated and experimentally determined fatigue lives show whether a close approximation

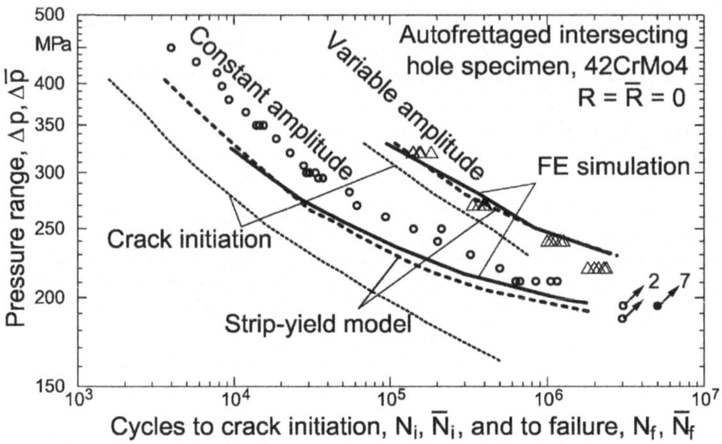

Fig. 4.25 Autofrettaged intersecting-hole specimen under constant amplitude and variable amplitude (pseudo-random) loading; fatigue life to crack initiation N_i, ($a_0 = 0.25$ mm, local strain approach) and to failure N_f ($a_f = 4.0$ mm); finite element analysis (FE) and strip-yield model (curves) compared with experimental results (point symbols); (Herz et al. 2011)

of reality is achieved. For this purpose, the crack initiation life is calculated based on the local strain approach and added to the crack growth life.

In Fig. 4.25, the lives obtained via effective stress intensity ranges are calculated based on the strip-yield model and the finite element method in comparison. In the constant-amplitude tests, $R = 0$, at pressure ranges between $\Delta p = 450$ MPa and 250 MPa, an underestimation of lives by a factor of 2 to 3 is observed. This might be due to an underestimation of the crack initiation lives. At pressure levels near the fatigue limit, the crack growth lives become more and more dominant. The fatigue limit itself is associated with crack arrest. Both numerical procedures lead to nearly identical results which correspond to the experimental findings. Although a strict proof of the quality of the crack growth models cannot be provided, it might be concluded that their accuracies in the finite life region are similar to those in the region near to the endurance limit.

The fatigue lives calculated using the crack opening pressures from the strip-yield model and from the finite element model are practically identical. Obviously, the crack opening pressures are also identical. However, there is not enough experience for generalising this result. The crack opening levels strongly depend on the cyclic deformation behaviour of the material and the multiaxial stress at the crack front. In the finite element model, both items have been considered as close to reality as currently possible. In the strip-yield model, perfect-plastic deformation without any hardening is modelled. The yield stress is simply set equal to the cyclic 0.2 % proof stress and the constraint factor to 3.

Referring to the variable amplitude results, the excellent correspondence of experimental and calculated lives is partly a matter of balancing of several inaccuracies of the models. It has been shown that variable amplitude lives are overestimated by the local strain approach (Vormwald et al. 1994). This effect annihilated the inherent life underestimation observed in simulations of constant amplitude loading.

The above arguments lead to the conclusion that the present models for crack growth simulation yield realistic results. Thus, the influence of residual stress fields on the fatigue crack growth life can be assessed by simulation of the plasticity-induced crack opening and crack closure. This is an alternative to approaches which are simply based on the stress intensity factors both from external loading and residual stresses. The simulation of plasticity-induced crack closure is nearer to the physical reality, at least for such cases where the residual stresses have been created by mechanical loading such as overloading or mechanical surface treatment. The validity for cases with residual stresses introduced by thermal treatments is excluded from the present investigation.

Despite the success of the presented simulations, the list of simplifications should always be kept in mind (Antunes and Rodriges 2008):

- The numerical crack increment is generally much larger than the crack increment per cycle in reality. One numerical increment is equivalent to many real cycles.
- The crack propagation is analytically executed by a change in the boundary conditions at a specified load. In reality, crack growth is a continuous process taking place during the loading half-cycle and a partial rejoining process by cold-welding during the unloading half-cycle, depending on the environment.
- The geometry of the crack is modelled as an ideal pointed crack tip while there is blunting and resharpening of the crack tip.
- The plastic deformation of the small material volume at the crack tip is modelled based on continuum mechanics equations derived from testing large material volumes. Their validity should be the subject of investigation when applied to describe behaviour in small plastic zones in the dimension of the microstructure.
- Crack closure phenomena other than those orginating from plastic deformation are ignored.

Mapping Technique Applied to Pressurised Pipes or Intersecting Holes

Modelling the plasticity-induced crack closure with all its implications on load ratio and load sequence effects requires that the fatigue crack path is known in advance. For the strip-yield model, this path is further simpified to a straight line, for example along the bisector for the intersecting hole specimen. For finite element modelling the generally applied nodal release algorithm requires nodes located at the expected crack tip positions. In three-dimensional modelling, the position of the subsequent crack front curves must additionally be known. In the intersecting hole example discussed above, this was the main reason for restricting the crack front geometry of a straight line. These limitations with respect to the crack geometry may not be acceptable in any case.

When applying ΔK-based fatigue crack growth calculations, which means application of linear-elastic fracture mechanics, the geometrical restriction just mentioned has been overcome. For the calculation of the stress intensity factors and crack increments including their direction, the finite element and the boundary element analysis methods are applicable. This approach is also implemented in many two-dimensional and three-dimensional, commercial and non-commercial software tools, such as FRANC2D (Swenson and Ingraffea 1988; Bittencourt et al. 1996), FRANC3D (Martha 1989; Carter et al. 2000), ZENCRACK (Zentech Int 2012; Timbrell and Cook 1997), ADAPCRACK3D (Fulland 2003; Schöllmann et al. 2003) and BEASY (Mellings et al. 2002).

For crack propagation with plastic deformation effects in a complex structure where the crack path is not known in advance, both the cited software tools and the nodal release technique are not sufficiently well suited for the analysis. The software tools perform a complete remeshing of the structure after having identified a crack increment. As the structure is supposed to behave linear-elastically, the determination of the crack driving force (i.e. the stress intensity factor range) does not require accounting for previous loadings. The situation is different if the memory of previous plastic deformation has to be transported from one mesh to the next. Such a simulation technique, an overview is shown in Fig. 4.26, has been developed (Zerres 2010; Zerres et al. 2011; Zerres and Vormwald 2012).

The main feature of this 'mapping technique' is the transportation of the plastic deformation history of the structure from a previous finite element mesh to a newly generated one. This is performed by introducing initial values for all state variables of the plasticity model at all integration points of the newly generated structure. The initial values at the integration points of the new structure must be determined from the results of the previous structure. However, as the locations of the integration points are different for the two finite element models, a mapping procedure must be implemented for the state variables. The whole crack growth simulation technique is named mapping technique after this important intermediate procedure. A prerequisite of the crack growth mapping technique is remeshing of the structure after each crack length increment.

The geometry of the structure, the parameters of the material-model applied and the load sequence must be provided as input to the procedure. The numerical simulation of the fatigue crack growth is subdivided into the following procedural steps, Fig. 4.26. First the finite element model is generated. The mesh is refined at the crack tip with regard to the high stress gradients. The structure is analysed. Within postprocessing, the crack opening and crack closure level is determined and the effective range of the crack tip parameter is calculated. With this information, the crack growth direction and the number of cycles to reach a given crack propagation (or the crack propagation for a prescribed number of cycles) can be computed by integrating an appropriate crack growth relationship. The geometry is updated by extending the crack in the calculated direction by the crack increment and the new model is generated.

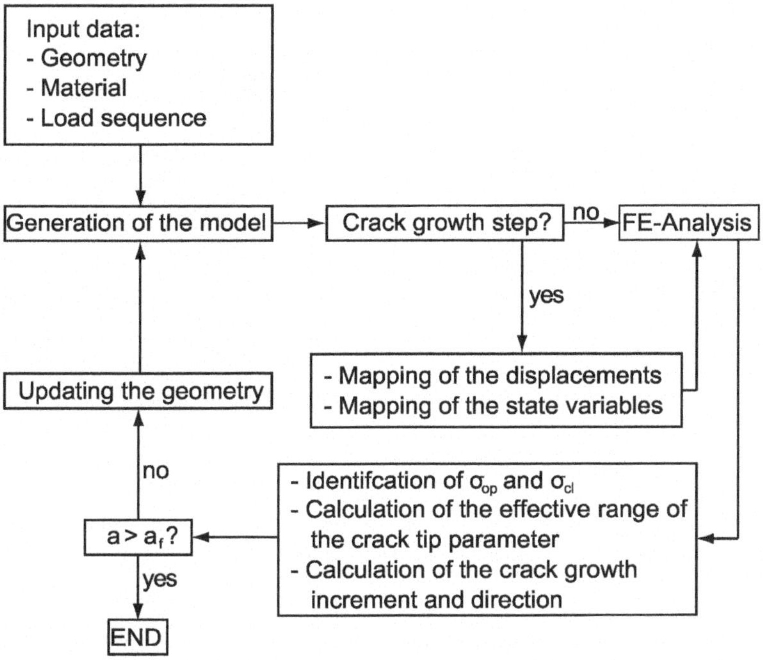

Fig. 4.26 Incremental crack growth simulation procedure based on finite element analysis comprising remeshing of the structure and mapping of the state variables of the plastic deformation model; (Zerres 2010)

In order to re-establish the stress state prior to crack propagation, the displacements and the state variables have to be mapped from the old mesh to the new one. The displacements are transferred by using the isoparametric element formulations. Afterwards all the state variables, i.e. the stresses, the strains, the components of the backstress tensors and other variables, which depend on the applied material model, are mapped from the old mesh to the new one. This is achieved by a routine, which is called up in the ABAQUS finite element software by default. Thereupon, the structure is analysed again to calculate the next increment of crack propagation. Due to the mapping of the state variables at the beginning of the analysis, illegal conditions, e.g. a stress tensor outside the yield surface, are possibly occuring. The software of the applied material model may have to be extended to take these circumstances into account, i.e. a radial return algorithm must start automatically to correct illegal conditions. The procedure is repeated until the crack length reaches a user-defined critical value.

For validation purposes, the autofrettaged pipe problem treated by Herz was reanalysed providing converged solutions achieved with the nodal release technique (Zerres 2010; Herz et al. 2006). For the nodal release analysis, half of the cross-section is modelled and symmetry boundary conditions are introduced, as shown in Fig. 4.20. To account for crack face contact, a rigid surface was introduced. In the

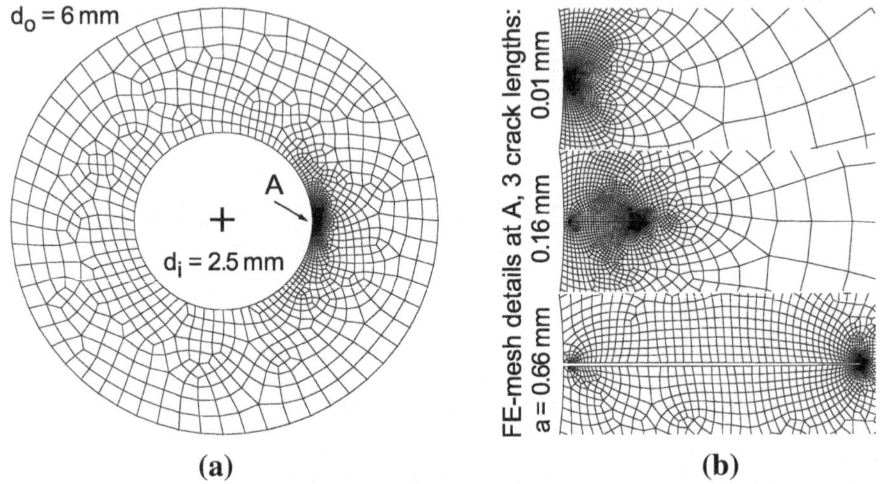

Fig. 4.27 Finite element mesh in cross-section of pipe specimen (**a**) with refined remeshing of crack tip region (**b**) for application of the mapping technique; initial and two further crack lengths; (Zerres 2010)

mapping analysis in comparison, the whole cross-section area is modelled, Fig. 4.27. Here, the crack flank contact is accomplished by introducing contact definitions between the two respective surfaces. The element size at the crack tip is set to 0.625 μm, which led to converged results. By using a propagating front algorithm and a smooth transition from the fine to the coarse mesh, a good overall mesh quality is ensured. The initial crack length is $a_0 = 10$ μm in both types of analysis. The material model was also identical in this comparison of the nodal release technique and the mapping technique.

In the nodal release analysis, the specimen is then unloaded to 90 % of the service pressure. Simultaneously the boundary condition of fixed displacements of the crack tip node is released, and a crack enlargement of one element length, $\Delta a = 2.5$ μm in this example, is thus performed. In the mapping analysis, the specimen is also unloaded to 90 % of the service pressure. At this stage, the crack is enlarged because most of the integration points are now in an elastic state, so that problems with illegal conditions are minimised. The crack growth increment which is no longer restricted to full element sizes was chosen as $\Delta a = 2.5$ μm for the comparison with the nodal release analysis and the endurance limit calculation. A larger increment $\Delta a = 0.15a$ was used for the fatigue life calculation.

The new model is generated and the mapping of the variables is performed. Then one calculation step with a constant load of 90 % of the service pressure is performed in the new model in order to establish equilibrium within the structure. After this step—for both types of analysis—the specimen is further unloaded down to $p_{\min}$. Between two crack growth steps, 10 additional load cycles are applied to stabilise the stress state at the crack tip. Within the last of these cycles the values needed for the calculation of the crack propagation, which are the displacements in

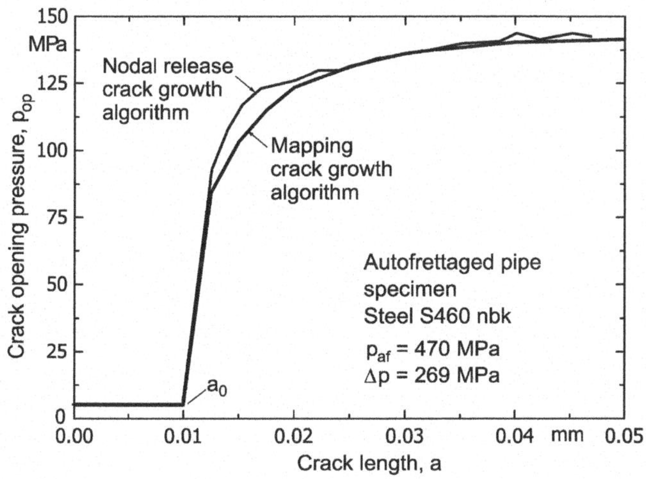

Fig. 4.28 Crack opening pressure dependent on crack length; autofrettaged pipe specimen; elastic-plastic finite element model, nodal release technique compared with remeshing and mapping technique; (Zerres 2010)

the direction perpendicular to the ligament for the nodal release and the crack opening displacement for the mapping analysis, are determined. This cycle is subdivided into at least 20 increments for both loading and unloading branch in order to obtain enough values for an accurate solution.

In this study the crack opening condition is defined as the point where the displacements or crack openings at all nodes on the crack front are greater than zero. This definition is in accordance with the previous investigation (Herz 2006). By knowing the crack opening condition, the effective pressure range Δp_{eff} can be calculated. On that basis, the effective range of the stress intensity factor ΔK_{eff} can be determined by integrating a weight function solution (Ma et al. 1994).

A comparison of the crack opening pressures is shown in Fig. 4.28. Obviously the results for both types of analysis are very similar. The conclusion is that the inaccuracies due to the mapping of the state variables are very small. The calculated endurance limits are consequently also similar to those obtained with the nodal release based effective ranges. In Fig. 4.29, the calculated and the experimental fatigue lives are compared. Two different crack growth rate equations have been applied.

The results labelled 'Paris' correspond to the following crack growth equation (Paris 1960):

$$\frac{\mathrm{d}a}{\mathrm{d}N} = 8.8 \times 10^{-6} \Delta K_{\mathrm{eff}}^{3.15} \tag{4.33}$$

with ΔK_{eff} to be inserted with the dimension (MPa $\mathrm{m}^{1/2}$) and da/dN resulting with the dimension (μm/cycle).

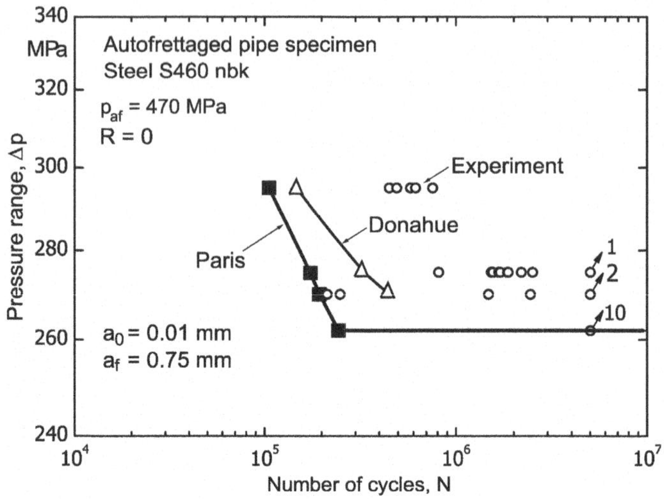

Fig. 4.29 Autofrettaged pipe specimen subjected to constant-amplitude internal pressure cycles, calculated (curves) compared with experimentally determined (circle point symbols) fatigue lives, initial crack length $a_0 = 0.01$ mm, final crack length $a_f = 0.75$ mm, crack growth rate relations according to Paris and to Donahue; (Zerres 2010)

The results labelled 'Donahue' correspond to the modified crack growth equation (Donahue et al. 1972):

$$\frac{da}{dN} = 8.8 \times 10^{-6}\left(\Delta K_{\text{eff}}^{3.15} - \Delta K_{\text{eff,th}}^{3.15}\right) \tag{4.34}$$

with ΔK_{eff} and $\Delta K_{\text{eff,th}}$ once more inserted with the dimension (MPa m$^{1/2}$).

As expected, the fatigue lives calculated by Eq. (4.34) are by a factor of about two higher than the fatigue lives calculated by Eq. (4.33). But they are by a factor of three to four smaller than the experimental results. The reasons for this underestimation of the fatigue lives—although not yet fully understood—may not only be due to the mapping of the state variables, but also to other factors. One point is the modelling as a two-dimensional structure, whereas semielliptical surface cracks grow from the inner wall in reality. Another factor may be the difference in the environmental medium (oil instead of air) and uncertainties with respect to the cyclic fluid pressure acting on the crack flanks.

The mapping technique can also be applied to three-dimensional structures. An example of freely developing crack front curves at the intersecting hole vertex point of the autofrettaged intersecting hole specimen is presented in Fig. 4.30. The simulation started with a quarter-circular starter crack with radius 0.25 mm. The crack propagation along the bisector is slower than along the bore holes due to the higher and three-axial compressive residual stresses there. Therefore, the crack front geometry changes from circular to approximately elliptical. A straight crack front as assumed in the application of the nodal release technique is not attained in

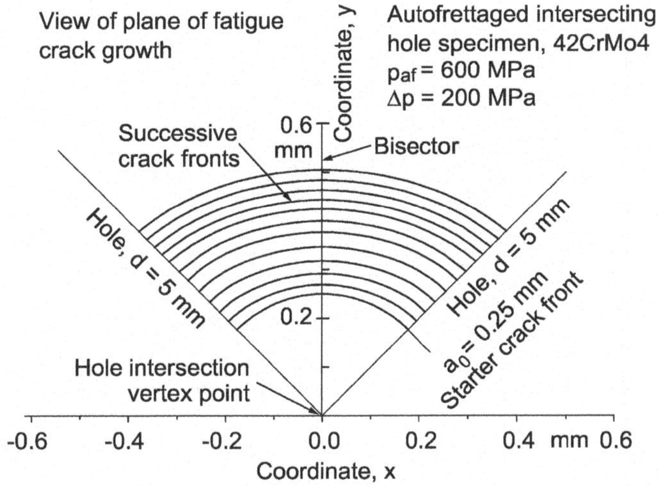

Fig. 4.30 Three-dimensional finite element simulation of initially quarter-circular crack growth; wedge-shaped section of intersecting-hole specimen; crack length $a_0 = 0.25$ mm and $a_f = 0.45$ mm; (Zerres 2010)

this example. However, the crack front geometry depends on the applied load level. Here, the assumption of a straight crack front becomes more realistic with decreasing service pressure ranges.

The fatigue crack growth rate depends on the previous load history. Neglecting this history by simply integrating rate equations based on linear-elastic fracture mechanics, relying only on the range of the stress intensity factor, may lead to large errors in life prediction. The information on the load history in respect of the actual crack growth behaviour is stored in the plastic deformations of the structure. The highest plastic deformations in cracked structures occur in the vicinity of the crack tip. Taking these cyclic plastic deformations into account—especially in terms of their influence on the crack closure—will considerably improve the accuracy of life predictions in variable amplitude loading.

4.2.4 Physically Short Cracks

Short cracks barely grow under small-scale yielding conditions. The small-scale yielding regime is defined by Eq. (4.8). A growing crack in steel, the material assumed to be homogeneous, requires an effective stress intensity factor range above $\Delta K_{\mathrm{eff,th}} \approx 3\ \mathrm{MPa(m)}^{1/2}$. Further assuming a low-strength structural steel and setting $\sigma_Y = 235$ MPa gives a minimum crack length $a_{\min} = 100\ \mu$m. However, inserting in Eq. (4.8) the relevant value for high-strength steel, $\sigma_Y = 700$ MPa, leads to a crack length of $a_{\min} \approx 11\ \mu$m. The minimum crack length increases quadratically with the applied load. This means that for high-

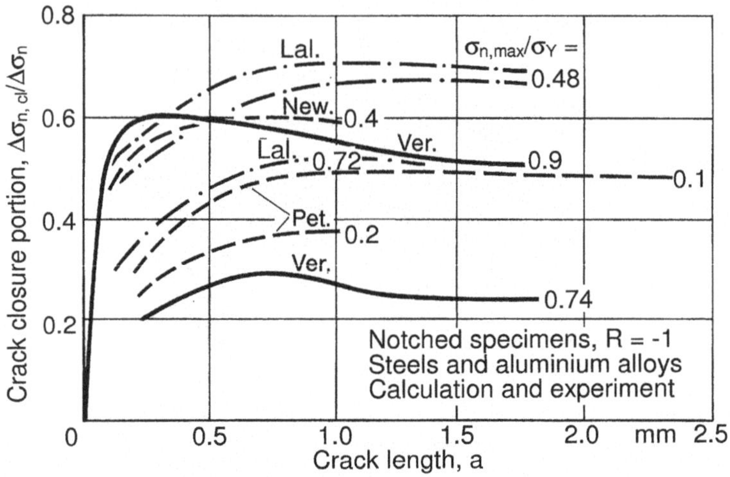

Fig. 4.31 Crack closure portion of the nominal stress range dependent on crack length; through-crack initiated at notch root; results presented by various authors: experimental results (upper curve) and calculation results (medium and lower curve) for steel AISI 1070 (Lalor et al. 1986), calculation results for steel G 50.11 (Newman 1983), experimental results for aluminium alloy AlZnMgCu (Petit and Zeghloul 1986), experimental results for steel A36 (Verreman et al. 1986); simplified diagram (Vormwald 1989)

strength steels and low applied load amplitudes, there is a small window where short fatigue crack growth might possibly be described by Eq. (4.9)—e.g. based on linear-elastic fracture mechanics. For such cases of physically short cracks, it is of vital interest to know crack opening levels and effective ranges. The effective range, however, of such short cracks cannot be described with any of the Eqs. (4.11) to (4.15).

The early closure development of short fatigue cracks has widely been investigated (Lalor et al. 1986; Petit and Zeghloul 1986; Verremann et al. 1986; Blom et al. 1986; Ritchie et al. 1987; Nicholas et al. 1988; Pang and Song 1994). The plasticity-induced crack closure phenomenon of naturally initiated surface cracks differs from the behaviour of long cracks. For these physically short cracks no plastic wake can exist at the very beginning of their growth. Crack flanks cannot get into contact; the total applied stress range is effective. Crack closure develops with increasing crack length. Under constant amplitude conditions, a steady state with crack opening loads nearly independent of the crack length develops. The steady state depends on the load ratio and on the ratio of maximum stress to yield stress. Some results are reproduced in Fig. 4.31.

In the early stage of short crack growth, the effective ranges are higher than predicted under long crack steady state conditions. Therefore, the growth rate of physically short cracks is higher than expected when estimated from the steady state and its associated relation of the crack growth rate with the stress intensity factor range ΔK. For physically short cracks, the lack of cyclic plastic deformation constitutes their special position with respect to their growth rate behaviour.

There are certainly more reasons for extraordinarily high growth rates of short cracks, which will be discussed further below.

The curve shape in the graph of developing plasticity-induced crack closure suggests the following approximation by an exponential function:

$$\Delta K - \Delta K_{\text{eff}} = (\Delta K - \Delta K_{\text{eff}})_0 (1 - e^{-\Delta a / a_{\text{cd}}}) \tag{4.35}$$

where Δa is the newly created crack length increment and the length parameter a_{cd} characterises the rate of closure development (subscript cd). The equation states that an ineffective part of a cycle, $\Delta K - \Delta K_{\text{eff}}$, is increasing from zero to a steady state value, $(\Delta K - \Delta K_{\text{eff}})_0$, the latter value being equal to the range with closed crack flanks. Such an expression has been proposed in the literature and later on modified (Gamache and McEvily 1993; Anthes 1997; McEvily et al. 2003; Zerbst et al. 2011). Knowledge on absolute values of the closure development length a_{cd} is rather limited. Not even the anticipated main dependencies on the material and the loading conditions are reported. Anthes states that usual values should amount to 10–30 μm, and more specifically for steels $a_{\text{cd}} \approx 9800$ μm divided by σ_{U} in MPa. This approximation disagrees with Gamache and McEviliy's proposal of $a_{\text{cd}} \approx 1000$ μm devided by $(0.01\sigma_{\text{U}} - 2.2)$. For several steels McEvily used $a_{\text{cd}} = 167$ μm and for an aluminium alloy Zerbst identified a value of 15 μm (McEvily et al. 2003; Zerbst et al. 2011).

4.2.5 Conclusions

In the small-scale yielding regime of fatigue crack growth, the growth rates can be conveniently related to the range of the stress intensity factor. The cyclic plastic deformation, however, must be limited and such limits are indicated.

Several experimentally observable phenomena of fatigue crack growth, even under constant amplitude loading, cannot be sufficiently understood without considering cyclic plastic deformation. Most important is that the influence of load ratio on the growth rates may be traced back to plastic deformation to a large extent. These irreversible deformations of the material at the crack tip are especially influential when the crack tip plastic zone gives rise to a crack closure behaviour which differs considerably from the behaviour of a ideal crack in purely elastic material. The plasticity-induced crack closure level depends on a variety of variables ranging from material parameters over load amplitudes, load ratios, crack lengths, to stress multiaxiality.

On the same physical basis, load sequence effects—i.e. the dependency of growth rates on the preceding load history—can be better understood. Again cyclic plastic deformation enters into the analysis mainly via crack closure. Thus, some remarkable findings of growth acceleration and retardation can be explained.

The experience with plasticity-induced crack closure is so general that its absence causes major concern as it is the case with physically short cracks. They seem to grow faster than long cracks. However, the growth, rates become more or less uniform when expressed in terms of a closure-free effective range of the stress intensity factor.

It is a challenge for modelling to identify the moments during a load cycle when the crack opens and closes. First approximations were based on experimental observations. Soon after the discovery of the phenomenon, calculations based on continuum mechanics were performed, taking into consideration plastic deformation laws and manipulations of the crack tip boundary conditions in order to simulate a propagating crack. Two modelling routes are pursued today in parallel. For a relatively fast application-relevant closure prediction, the strip-yield model is used. The restrictions and assumptions of this simple model are substantial. Plastic deformation is limited to a thin strip above and below the crack-ligament line and the material is treated as perfectly-plastic in the strip. Surprisingly enough, very reasonable results are achieved. In the scientific community, deeper insight is sought by applying the finite element method. Realistic material laws and geometries, even three-dimensional structures, may be modelled. The computational expense for a crack growth simulation can easily be raised to unacceptable levels. In a conflict between accuracy, convergence and expense, scientists nowadays try to improve their understanding of the process. A couple of applications of this ambitions modelling technique have been published. Further comparisons of simulation results and experimental results are on the agenda for the next decade.

4.3 Crack Growth under Large-Scale Yielding Conditions

4.3.1 Introductory Notes

In Sect. 4.2 on crack growth under small-scale cycling yielding conditions, Eq. (4.8) specifies a limit condition for the applicability of growth rate equations based on the stress intensity factor range. The limit is not sharp. Higher ranges may be acceptable after judgement of individual circumstances. However, as the ranges of the stress intensity factor and the extent of the plastic zone increase the stress intensity factor itself loses its physical basis. It can no longer describe the crack tip field. An alternative is required.

In fracture mechanics applications on monotonic loading, this problem was evident at a very early stage of the development of this scientific field. Therefore, most tools and algorithms, which were later on applied to fatigue loading, take advantage of what was developed previously for monotonic loading.

In essence, Eq. (4.8) limits the ratio of the cyclic plastic zone size to any other geometrical dimension of the structure to very small numbers. Therefore, a violation of the criterion means relatively large plastic zones compared with other dimensions, the crack length among them. On the one hand, this may occur shortly before final fracture of the structure. The final crack growth rates are high.

However, the number of cycles for which small-scale cyclic conditions are violated is low, when compared to the total fatigue life. On the other hand– and this is of technical relevance—the ratio of plastic zone size to crack length becomes large for short crack lengths. This is true for short cracks both in un-notched specimens and in notched specimens or structures. In both cases the local cyclic stresses and strains in the (hypothetically) uncracked material are so high that they approach and often exceed the cyclic yield limit. In recent years, a third scientific field of practical relevance has emerged: structures under cyclic deformation control instead of load control. Under deformation control, large plastic deformations are acceptable without risk of immediate collapse. Thermomechanical loading belongs to this category. As in many other fatigue problems, the thermomechanical fatigue process can mainly be treated as a matter of crack growth including the short-crack stage. Researchers are actually developing corresponding simulation procedures.

In case the cyclic plastic deformation extents more widely around the crack tip, what is then the fatigue crack driving force? Which parameter from elastic-plastic fracture mechanics is able to uniquely describe fatigue crack growth rates?

4.3.2 Elastic-Plastic Crack Driving Force Parameters

Strain Intensity Factor Approach

The historical overview on the development of elastic-plastic crack driving force parameters is well documented (McClung et al. 1999). Therefore this chapter can and does take much advantage of this fundamental work.

A first trial step from small-scale to large-scale yielding was to replace stress parameters by strain parameters (Boettner et al. 1965; McEvily 1969; Solomon 1972; El Haddad et al. 1979[(1)], 1979[(2)]; Haigh and Skelton 1978; Skelton 1982; Anthes 1997).

Originally, only the plastic strain range instead of the stress range was inserted in stress intensity factor formulae of a given geometry. Later on the total strain range was used. This makes more sense because a large-scale yielding parameter should continuously approach the small-scale yielding parameter for vanishing plastic deformations, e.g. in the following expression for the strain intensity factor range ΔK_ε:

$$\Delta K_\varepsilon = \left(\Delta\varepsilon_{\mathrm{e}} + \Delta\varepsilon_{\mathrm{p}}\right) E\sqrt{\pi\, a}\, Y \tag{4.36}$$

The above condition of the transition from the large-scale to the small-scale yielding parameter is violated by variants of definition of the strain intensity factor where the factor $\pi^{1/2}$ is omitted. The parameter ΔK_ε as defined by Eq. (4.36) does not provide a measure of the strain singularity at the crack tip. Nevertheless, it may be introduced as crack driving force which exceeds the linear-elastic value.

This points in the right direction and was shown to correlate growth rate data to a certain extent. The most important advantage of such a method is its ease of application. If applied as suggested by Eq. (4.36), not even the growth rate material constants have to be redetermined. They can be transferred directly from the small-scale yielding formulation expressed in terms of the stress intensity factor.

Crack Tip Opening Displacement Approach

The theoretical shortcomings of ΔK_ε have always been recognised. They can be overcome by using the cyclic crack tip opening displacement $\Delta\delta_t$ as crack driving force parameter which was proposed by many researchers, based on the previous application of δ_t by Wells under static loading conditions (Wells 1961; McEvily et al. 1974; Tomkins 1975, 1980; Tanaka et al. 1984; Brown et al. 1988). This measure of cyclic deformation is taken as close as possible to the location of material separation. It is generally assumed to provide a sound and unique correlation with the growth rate. However, its determination is a difficult task. It has been emphasised that knowledge of a proper driving force is of little value unless that driving force can be either exactly calculated with sufficient economy or estimated with sufficient accuracy (McClung et al. 1999). The authors are not aware of any high-resolution (with respect to element size), geometrically non-linear, elastic-plastic, finite element analysis of $\Delta\delta_t$ under cyclic loading conditions taking crack closure into account.

According to an extension of the Dugdale model of crack tip cyclic plastic deformation in thin sheets, the range of crack tip opening displacement, $\Delta\delta_t$, is given by the following expressions (Dugdale 1960):

$$\Delta\delta_t = \frac{8}{\pi} a \frac{\Delta\sigma_Y}{E} \ln\left[\sec\left(\frac{\pi}{2}\frac{\Delta\sigma}{\Delta\sigma_Y}\right)\right] \qquad (\Delta\sigma_Y = 2\sigma_Y) \qquad (4.37)$$

For semicircular surface cracks, the corresponding expression for the deepest point of the crack front under plane strain conditions, the expression reads as follows (Vormwald 1989; Vormwald and Seeger 1991):

$$\Delta\delta_t = 1.07\frac{8\,(1-\nu^2)}{\pi} a \frac{\Delta\sigma_Y}{E}\left[1-\sqrt{1-\left(\frac{\Delta\sigma}{\Delta\sigma_Y}\right)^2}\right] \qquad (\Delta\sigma_Y = 2\sigma_Y) \qquad (4.38)$$

An alternative way to determine the cyclic crack tip opening displacement is to apply the strip-yield model. Crack tip openings are naturally effective because crack face contact is considered during the simulation of the load history. In approximation formulas like Eqs. (4.37) and (4.38), the monotonic variables are simply replaced by the cyclic quantities and nothing is said about effective and ineffective ranges. As will be discussed in a later section, the crack opening and closure loads do not coincide in large-scale yielding, not even approximately.

When using $\Delta\delta_t$ together with an approximation formula and with the restriction to effective ranges, it has to be decided whether an ascending (showing opening) or a descending (showing closure) loading branch is evaluated for defining effective ranges.

A practical issue of any parameter is the availability of material data for its correlation with growth rates. Existing data are almost exclusively reported in terms of the cyclic stress intensity factor. The elastic-plastic parameter should be able to exploit this database in order to minimise the need for further testing. When going from large-scale to small-scale yielding conditions, the parameter should therefore (asymptotically) approach a functional relationship with the stress intensity factor. For the crack tip opening displacement this function is provided by the following relationship:

$$\Delta\delta_t = \frac{(\Delta K)^2}{mE\Delta\sigma_Y} \tag{4.39}$$

where $m \approx 1$ for plane stress conditions and $m \approx 2$ for plane strain conditions. If Eq. (4.39) is inserted in any crack growth rate equation the latter can be transformed from ΔK to $\Delta\delta_t$.

ΔJ-Integral Approach

The extension of the J-integral for application with cyclic loading conditions (Dowling and Begley 1976; Dowling 1977[(1)]) has created by far the greatest attention of researchers and engineers who wanted to model fatigue crack growth in the large-scale yielding region. It also gave reason to severe academic disputes. A motivation for going into some detail is seen in the immense number of successful applications. Only a minor number of early publications can be referenced here (Dowling 1977[(2)]; Mowbray 1979; Sadananda and Shahinian 1979; El Haddad et al. 1980; Huang and Pelloux 1980; Musava and Radon 1980; Reger and Remy 1982; Vardar 1982; El Haddad and Mukherjee 1983; Wang et al. 1983; Heitmann et al. 1985; Tanaka et al. 1984; Jolles 1985; Obrtlik and Polak 1985; Zheng and Liu 1986[(1)]; Grin et al. 1987; Horikawa and Cho 1987; Hoshide and Socie 1987; Rie and Schubert 1987; Hoshide et al. 1988; Bicego 1989; Hatanaka et al. 1989; Jablonski 1989; Yang and Lu 1989; Earthman 1991; Vormwald and Seeger 1991; McDowell and Berard 1992; Miura et al. 1994; Mu et al. 1996; Rie and Wittke 1996; Rahman et al. 1997; Tchankov et al. 2008).

The J-integral was originally defined for two-dimensional problems as a line integral along any contour Γ encircling the crack tip (Rice 1968):

$$J = \int_{\Gamma} \left(W\,\mathrm{d}y - T_i \frac{\partial u_i}{\partial x} \mathrm{d}s \right) \tag{4.40}$$

where W is the strain energy density, T_i are the components of the traction vector, u_i are the components of the displacement vector, y is the coordinate normal to the crack-ligament line and s is the path coordinate along the path Γ around the crack tip in a counter-clockwise direction.

The strain energy density W is given by the following integral of the stress and strain tensor components σ_{ij} and ε_{ij}:

$$W = \int_{0}^{\varepsilon_{ij}} \sigma_{ij}\, d\bar{\varepsilon}_{ij} \tag{4.41}$$

Accordingly, the cyclic J-integral ΔJ and the strain energy density range ΔW are given by the following expressions:

$$\Delta J = \int_{\Gamma} \left(\Delta W\, dy - \Delta T_i \frac{\partial(\Delta u_i)}{\partial x} ds \right) \tag{4.42}$$

$$\Delta W = \int_{0}^{\Delta\varepsilon_{ij}} \Delta\sigma_{ij}\, d(\Delta\bar{\varepsilon}_{ij}) \tag{4.43}$$

The symbol Δ preceding the stress tensor, strain tensor, traction vector and displacement vector components designates the changes of these quantities. These changes must be evaluated referring to a reference state. It serves as the new origin for defining the increments of the field variables, the latter preceded by the symbol Δ. The stress and displacement state at the moment of load reversal is a natural reference state. Then the symbol Δ designates the increments from the respective reference values. At the moment of the next reversal, these increments become the conventional cyclic ranges. However, the symbol Δ in ΔJ and ΔW does not represent changes of J and W; instead ΔJ and ΔW are functions of their arguments as defined by Eqs. (4.42) and (4.43). The symbol Δ, when used with the variables J and W, has a different meaning than usual. Therefore, the symbol Z was proposed for ΔJ (Wüthrich 1982). However, the proposal was not generally accepted (McClung et al. 1999). It was also suggested to use the name ΔJ-integral rather than range of the J-integral; the latter expression is strictly speaking wrong.

The path independence is the outstanding property of the J-integral (monotonic loading) and also of the ΔJ-integral (cyclic loading) under certain conditions which substantiates its attractiveness as a crack driving force parameter. What is path-independent must be a measure of near crack tip quantities responsible for material separation processes—on the one hand—and must be determinable by far field values, uncontaminated by numerical deficiencies—on the other hand. The most frequent objection raised against the ΔJ-integral argues that since the J-integral is based on the theory of nonlinear elasticity or (with limitation) deformation plasticity, it does not allow for unloading or non-proportional plastic deformation. However, it has been demonstrated that a properly defined ΔJ-integral maintained path independence. The path independence in applications to fatigue cracking problems could also be proven provided the ΔJ-integral is defined according to Lamba and others (Lamba 1975; Wüthrich 1982; Wüthrich and Hoffelner 1984; Tanaka 1983). More information may be found in Saxena's text book (Saxena 1998).

The ΔJ-integral has some remaining theoretical limitations. A first type of limitation has to do with the stress-strain behaviour of the material. It has been shown that the path independence is violated if the material is not completely cyclically stabilised (Yoon and Saxena 1991). Also, in the presence of temperature gradients and temperature dependent material behaviour, a strict compliance with path independence conditions cannot be achieved. Some remedies have been proposed (e.g. Blackburn 1972; Kishimoto et al. 1980; Atluri et al. 1984), which have also been reviewed (Kim and Orange 1988). These proposals for removing relevant restrictions have not yet found their way into engineering applications.

A second type of limitation has to do with crack closure. It is difficult to define a reference state as a necessary basis. There should be no stresses at the crack flanks otherwise there is no path independence. The moment of a complete loss of contact would be a natural reference state. This state has been recommended (McClung et al. 1999). However, even in this moment, the material at the various locations near the crack tip and in the far field is at different stress-strain positions on the ascending hysteresis branch. Thus, there will be no path independence during further loading. For totally open cracks the latter argument does not exist. A valid ΔJ-integral can be calculated for a reversal from minimum to maximum load. Going from maximum to minimum load the same value of the ΔJ-integral will be calculated provided that the crack length has not grown too much. Strictly speaking, for one cycle, two identical ΔJ-integrals can be calculated. The crack increment $\mathrm{d}a$, however, is usually attributed to a complete cycle, $\mathrm{d}N = 1$, and the crack driving force during this cycle is only one of two identical ΔJ-integrals. There is no reason to reject the descending ΔJ-integral in favour of the ascending ΔJ-integral. Accepting this argument so far, opens an alternative route of dealing with crack closure. Choosing the moment of the maximum load as reference state, path independence is maintained during the descending stress-strain branch until crack face contact occurs. Calculating the ΔJ-integral at the moment of first contact will provide a path independent effective ΔJ-integral, ΔJ_{eff}.

A simple functional relationship exists between the J-integral and the stress intensity factor K under linear-elastic conditions:

$$J = \frac{K^2}{E'} \tag{4.44}$$

where $E' = E$ for plane stress conditions or $E' = (1 - \nu^2)E$ for plane strain conditions. In the linear-elastic case, the corresponding relationship for cyclic loading reads:

$$\Delta J_{\mathrm{e}} = \frac{(\Delta K)^2}{E'} \tag{4.45}$$

McClung emphasises that a crack driving force parameter in fatigue should be practical, besides being theoretically valid and useful in correlating various growth rate data (McClung et al. 1999, see also McClung and Sehitoglu 1991):

> The chosen parameter should be easily and accurately calculated for a variety of materials, loads, structural geometries and crack configurations. A parameter which requires a sophisticated numerical analysis is less useful than a parameter which can be easily estimated from handbook tables or simple equations. However, the mere availability of simple estimation schemes is insufficient unless the estimation schemes used to estimate actual numerical values for the different parameters retain the theoretical distinctions between the parameters.

In the small-scale cyclic yielding regime, the governing stress intensity factor may be calculated by different methods with high accuracy. Due to Eq. (4.44), the stress intensity factor, for which many solutions are available, provides estimates of the J-integral, which are limited to the small-cale yielding regime. The J-integral increases more with the applied load than described elastically by Eq. (4.44). This increase is dependent on the hardening behaviour of the material and also on the geometry of the considered structure. In the course of a finite element based fatigue crack growth simulation, the ΔJ-integral might be determined after each increment of crack propagation. The results from several loading steps must be provided: a reference step (upper load reversal point), a step at first flank contact, an artificial newly created difference step where the field variables of the latter are subtracted from the former. An integration procedure for solving Eq. (4.42) numerically—either home-made (Kuna 2010; Brocks and Scheider 2003) or provided by a finite element postprocessor software—leads to ΔJ_{eff}, the ΔJ-value calculated with effective (closure-free) parameter ranges. However, such a so-phisticated procedure has not yet come to the authors' knowledge. In McC-lung's terminology, it would be of minor usefulness.

Cyclic *J*-Integral Approximation Formulas

For crack growth life calculation purposes relatively simple estimation formulas are required. Approximation formulas for the J-integral have been developed for monotonic failure analyses. Two of these approaches are discussed in the following. First, they are presented in the terminology for application with monotonic loading. Later, the extension to the cyclic loading case is shown.

A general engineering approach was developed at the United States Electric Power Research Institute and is therefore termed EPRI approach (Kumar et al. 1981). The J-integral is given in the following simplified form:

$$J(a,P) = J_{\text{e}}(a_{\text{eff}},P) + J_{\text{p}}(a,P) \tag{4.46}$$

where J_{e} is the linear-elastic solution according to Eq. (4.44), a_{eff} is the plastic zone corrected crack length, a is the real crack length and P is the applied load.

The following expressions are used for determining the effective crack length:

$$a_{\text{eff}} = a + \varphi r_{\text{Y}} \tag{4.47}$$

$$\varphi = \frac{1}{1 + (P/P_0(a))^2} \tag{4.48}$$

$$r_{\text{Y}} = \frac{1}{\beta\pi} \frac{n-1}{n+1} \left(\frac{K(a, P)}{\sigma_0}\right)^2 \tag{4.49}$$

where $\beta = 2$ for plane stress conditions and $\beta = 6$ for plane strain conditions. The ligament yield load P_0 is introduced as the critical reference load. The parameters n and σ_0 appear in the underlying power-law type stress-strain relationship of the material:

$$\frac{\varepsilon}{\varepsilon_0} = \frac{\sigma}{\sigma_0} + \alpha \left(\frac{\sigma}{\sigma_0}\right)^n \tag{4.50}$$

where σ_0 is a substitute yield limit parameter, $\varepsilon_0 = \sigma_0/E$, and α indicates the ratio of plastic to elastic strains for a stress value of σ_0.

The Ramberg–Osgood material law referring to (stabilised) cyclic loading is usually written as follows:

$$\varepsilon = \frac{\sigma}{E} + \left(\frac{\sigma}{K'}\right)^{1/n'} \quad (0 < n' < 1) \tag{4.51}$$

with the cyclic hardening coefficient K' and the cyclic hardening exponent n'. In applications for cyclic loading, Eq. (4.50) is used as a description of the stabilised cyclic stress-strain relationship. Therefore, a formal adjustment of the parameters is required, especially the hardening exponents n and n' are inverse values and the cyclic hardening coefficient is related to the parameters used in Eq. (4.50) by:

$$K' = (\alpha\varepsilon_0)^n \sigma_0 \tag{4.52}$$

$$n' = 1/n \tag{4.53}$$

$$E = \sigma_0/\varepsilon_0 \tag{4.54}$$

The plastic component J_{p} in Eq. (4.46) is described according to the EPRI approach as follows:

$$J_{\text{p}}(a, P) = \alpha\sigma_0\varepsilon_0 \frac{a}{W}(W - a) h_1(a/W, n) \left(\frac{P}{P_0(a)}\right)^{n+1} \tag{4.55}$$

Knowledge of the non-dimensional function h_1, a function of the ratio of crack length a to structural width W and of the hardening exponent n, is the clue to arrive at sufficiently accurate J-integral estimates. Such functions are tabulated for several structures in the EPRI handbook (Kumar et al. 1981). The handbook also provides some equations for evaluation of the ligament yield load P_0 which depends on the material parameters and on the geometry characterised by a and W among others.

An alternative J-integral estimation scheme is the reference stress method (Ainsworth 1984). This method also requires the linear-elastic solution and introduces the stress-strain relationship of the material. The J-integral estimate is written in the following form:

$$J = \frac{K^2}{E'}\left(\frac{E\varepsilon_{\mathrm{ref}}}{\sigma_{\mathrm{ref}}}\right) \tag{4.56}$$

The reference stress σ_{ref} is defined as follows:

$$\sigma_{\mathrm{ref}} = \frac{P}{P_0}\sigma_0 \tag{4.57}$$

The reference strain $\varepsilon_{\mathrm{ref}}$ is determined from the stress-strain relationship, e.g. Eq. (4.50) or any other relationship, as the strain corresponding to the uniaxially applied reference stress σ_{ref}. The Eq. (4.56) is used with the actual crack length a or with the plasticity-corrected crack length a_{eff}.

The approach is attractive because it requires only knowledge of the elastic solution, the hardening behaviour of the material and the ligament yield load P_0. Besides reference to the basic work of McClung (McClung et al. 1999), an updated collection of ligament yield load formulae is part of the FITNET compendium (Kocak et al. 2005).

In accordance with these general approaches for montonic loading, expressions for the cyclic J-integral have been derived by Dowling based on the Ramberg–Osgood material law in the form of Eq. (4.51), at first without considering crack closure (Dowling 1977(1), 1977(2), 1987):

$$\Delta J = \Delta J_{\mathrm{e}} + \Delta J_{\mathrm{p}} \tag{4.58}$$

The following general relationship has been given by Dowling:

$$\Delta J = 2\pi a\left[\left(1 - \frac{2f(n')}{n'+1}\right)\frac{(\Delta\sigma)^2}{2E} + \frac{f(n')}{n'+1}\Delta\sigma\,\Delta\varepsilon_{\mathrm{p}}\right] \tag{4.59}$$

where $f(n')$ has to be determined for the individual geometry.

Considering the application-relevant case of a semicircular surface crack inclusive of crack closure, the following two expressions are widely used. For the first one, very good agreement with finite element solutions has been shown by Dowling (Dowling 1987):

$$\Delta J = a\left[1.24\frac{(\Delta\sigma)^2}{E} + \frac{1.02}{\sqrt{n'}}\Delta\sigma\,\Delta\varepsilon_{\mathrm{p}}\right] \tag{4.60}$$

The second one is given by Heitmann (Heitmann et al. 1985):

$$\Delta J_{\mathrm{eff}} = a\left[1.45\frac{(\Delta\sigma_{\mathrm{eff}})^2}{E} + \frac{2.5n'}{n'+1}\Delta\sigma\,\Delta\varepsilon_{\mathrm{p}}\right] \tag{4.61}$$

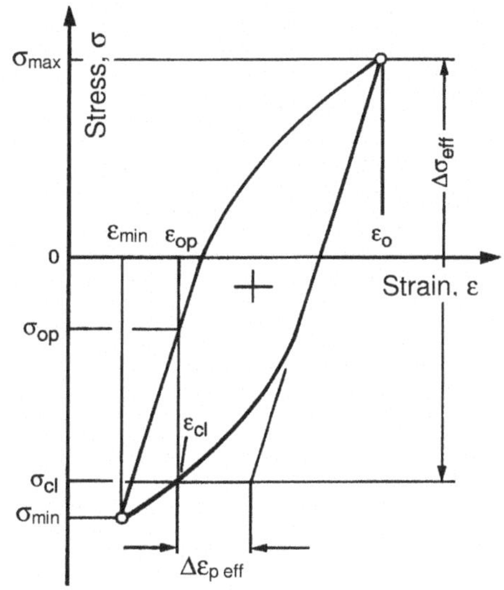

Fig. 4.32 Stress-strain hysteresis loops with crack opening (subscript op) and crack closure (subscript cl) points defining effective cyclic stress $\Delta\sigma_{eff}$ and effective plastic strain $\Delta\varepsilon_{p\ eff}$; (Savaidis and Seeger 1994)

where crack closure is taken into account, however, only in the elastic component.

Based on the descending branch of the hysteresis loop (here, ΔJ remains path-independent up to crack closure), the quantity ΔJ_{eff} can be determined based on the range between the maximum parameters σ_{max}, ε_{max} and the crack closure parameters σ_{cl}, ε_{cl} (Vormwald and Seeger 1991), Fig. 4.32:

$$\Delta J_{eff} = a\left\{1.24\frac{(\sigma_{max}-\sigma_{cl})^2}{E} + \frac{1.02}{\sqrt{n'}}(\sigma_{max}-\sigma_{cl})\left[(\varepsilon_{max}-\varepsilon_{cl}) - \frac{(\sigma_{max}-\sigma_{cl})}{E}\right]\right\} \tag{4.62}$$

In contrast to Eq. (4.61), not only the elastic component is modified by crack closure, but also the plastic component. The term in brackets is equal to the effective plastic strain range $\Delta\varepsilon_{p\ eff}$. The empirical finding that crack closure and crack opening occur at the same strain is included in Eq. (4.62) and Fig. 4.32. Further details on the treatment of crack closure are presented in Sect. 4.3.3.

4.3.3 Crack Closure under Large-Scale Yielding Conditions

Experimental Findings

The stabilised—i.e. nearly crack length independent—levels of crack opening stress decrease as the cyclic plastic deformation near the crack tip increases. This could already be observed in Fig. 4.7 (the plastic deformation increases with the crack length) and in Fig. 4.9 where the stabilised crack opening stress is plotted as

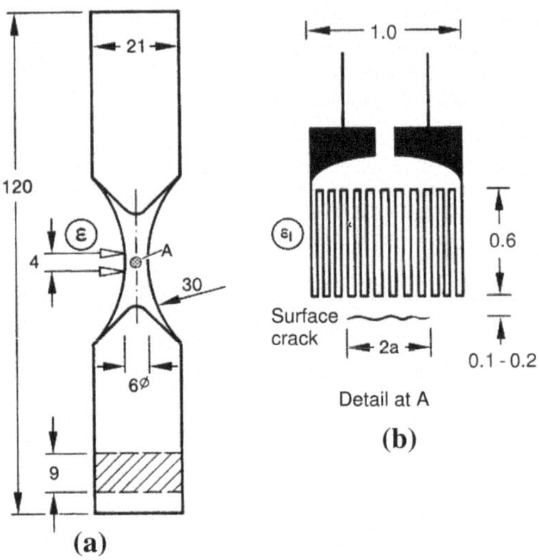

Fig. 4.33 Experimental set-up for detecting crack closure under large-scale yielding conditions: specimen (**a**) with strain gauge at a surface crack (**b**); measurement of global strain ε and local strain ε_l; (Vormwald and Seeger 1991)

a function of the applied maximum stress. There is a smooth transition from small-scale to large-scale yielding conditions. The decrease of crack opening stress means that the crack driving force parameter increases more than predicted from linear-elastic fracture mechanics. The effective ranges are also larger than would be expected from observations in the small-scale yielding region.

Measurements of the crack opening and crack closure levels under large-scale cyclic yielding conditions have been performed by many researchers (Vormwald and Seeger 1991; McClung and Sehitoglu 1988; Dowling and Iyyer 1987; Rie and Schubert 1987; DuQuesnay et al. 1992; El-Zeghayar et al. 2010, 2011). Short fatigue cracks were naturally initiated in conventional cylindrical material specimens used for low-cycle fatigue testing. The deformations in the neighbourhood of the crack were monitored—a procedure also very common with long cracks. A strain microgauge with a grid length of 0.6 mm was mounted close to the crack, Fig. 4.33. Its signal is named the local strain ε_l. While the crack is closed, its flanks can transfer stresses by contact. A nearly homogeneous uniaxial stress state is acting. The local strain ε_l equals the strain indicated by a macrogauge (global strain ε) used in the control loop of the testing machine. Upon crack opening, the tiny gauge gets into the stress and strain shadow of the crack. The local strain ε_l falls behind the global strain ε. In Fig. 4.34, the local strain ε_l and the global strain ε are plotted dependent on the global stress σ. The deviation between the global and local hysteresis loops allows to determine the crack opening and crack closure levels.

The Newman approximation formula, Eq. (4.15), for calculating crack opening stresses describes the trend correctly, even in the large-scale yielding regime (Newman 1984). A slightly simpler formula is also available (DuQuesnay et al.1992):

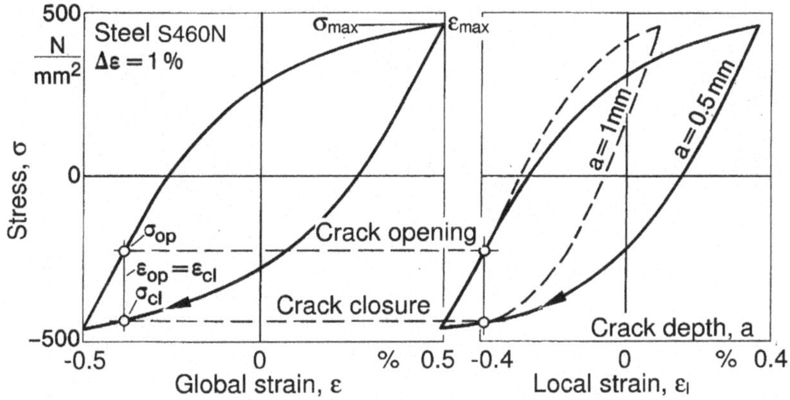

Fig. 4.34 Stress-strain hysteresis loops under large-scale yielding conditions for the experimental determination of crack opening and crack closure at a surface crack of depth a; (Vormwald and Seeger 1991)

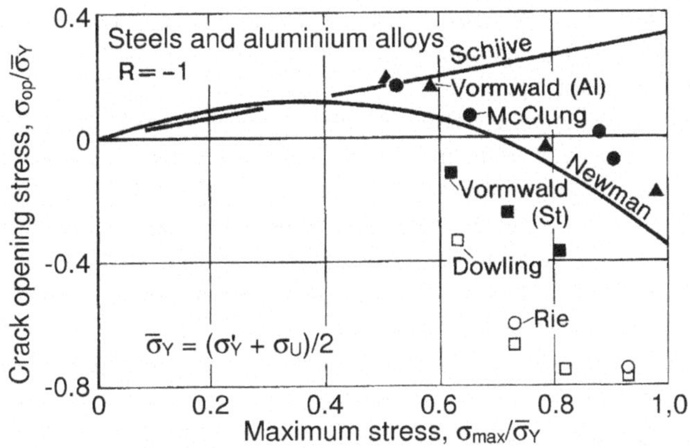

Fig. 4.35 Stabilised crack opening stress dependent on maximum stress, both related to substitute yield limit $\bar{\sigma}_Y$, steels and aluminium alloys; experimental results (point symbols) and two approximation formulae (curves); (Vormwald 1989, Vormwald and Seeger 1991, Vormwald et al. 1992)

$$\frac{\sigma_{op}}{\sigma_{max}} = B_0\left[1 - \left(\frac{\sigma_{max}}{\sigma_Y}\right)^2\right] + B_1 R \tag{4.63}$$

where the parameters B_0 and B_1 should be identified by fitting the experimental data.

A comparison of calculated (according to Newman) and measured crack opening stresses is shown in Fig. 4.35. The general, experimentally confirmed

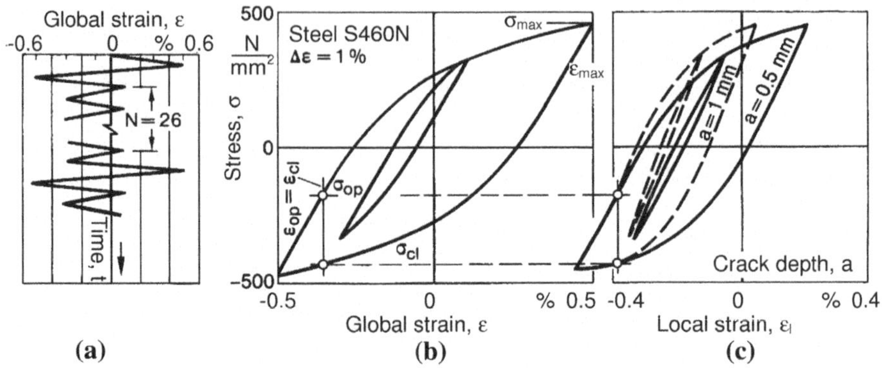

Fig. 4.36 Stress–strain hysteresis loops under variable amplitude large-scale yielding conditions (**a**, **b**); open short crack after large cycle (**c**); (Vormwald and Seeger 1991)

trend of decreasing crack opening stresses with increasing stress amplitudes is well mirrored by the approximation formula. The wide scatter of the experimental data is due to the complicated measuring technique with its associated uncertainties and to the definition of crack opening itself. Not in all cases is the complete loss of contact defined as crack opening.

Crack opening and crack closure are observed at nearly identical global strains. This phenomenon is even more important than the accuracy of an approximation equation for crack opening stresses. The value of the global crack opening and closure strain, $\varepsilon_{op} = \varepsilon_{cl}$, is highly sensitive to the load sequence. Overloads cause its decrease and therefore an increase in the effective ranges. This is illustrated by Fig. 4.36. During $N = 26$ load cycles with strain amplitude $\varepsilon_a = 0.2$ %, the crack is completely open. In a constant amplitude test with the same strain amplitude ε_a and stress ratio R_σ, the crack was closed during a considerable part of the cycle. The consequences of this effect on the fatigue life can be assessed based on a simple rule (Miner 1945). From the constant amplitude lives of $N_f = 7.7 \times 10^3$ cycles for $\varepsilon_a = 0.5$ % and $N_f = 2 \times 10^5$ cycles for $\varepsilon_a = 0.2$ % a two-level test life of N_{Block} can be calculated. The Miner-type damage sum for the sequence is $D = (1/7700 + 26/200000) \approx 2.6 \times 10^{-4}$. The fatigue failure criterion is met when the damage sum amounts to $D = 1.0$. According to this hypothesis, the sequence of one high amplitude cycle and 26 low amplitude cycles must be applied about $1/(2.6 \times 10^{-4}) \approx 3850$ times. The number of sequence repetitions ($N_r = 3850$) multiplied with the number of cycles in one sequence ($N = 27$) amounts to the predicted number of cycles to failure, $N_f \approx 104000$ cycles. However, the Miner rule overestimates the life by a factor of nearly four. The experimentally determined life amounts to only 26500 cycles.

In case the overload occurs less frequently, the low-amplitude cycles will restore their own constant amplitude crack opening and closure strain level. Corresponding measurement and simulation results are shown in Fig. 4.37, the simulation results from a more generally applicable algorithm are explained in the

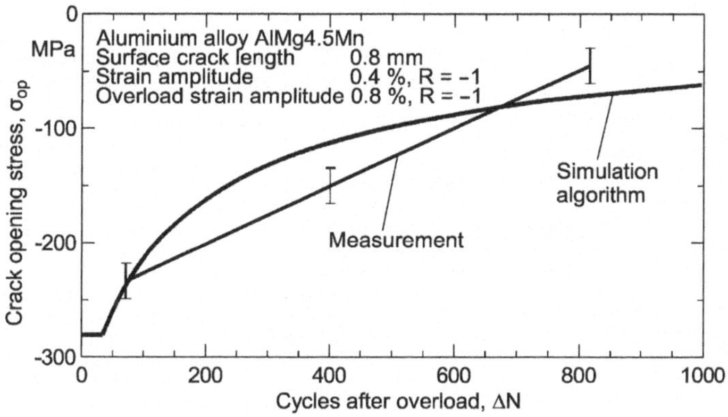

Fig. 4.37 Crack opening stress dependent on number of cycles after overload strain amplitude in aluminium alloy AlMg4.5Mn; comparison of simulation and measurement results; initial surface crack length $2a_0 = 0.8$ mm; (Vormwald and Seeger 1991)

following. This algorithm for estimating the sequence-dependent crack opening and closure strain is defined as follows (Vormwald and Seeger 1991):

- The crack opening and closure strain levels are completely described by one single global strain value, $\varepsilon_{op} = \varepsilon_{cl}$.
- This crack opening strain is calculated on a cycle-by-cycle basis. A cycle is defined by a closed stress over global strain hysteresis loop.
- No change in the crack opening strain is assumed in cases where the upper reversal point strain of the loop is lower than the previously established crack opening strain.
- A change in the crack opening strain always depends on the constant amplitude crack opening strain $\varepsilon_{op,c}$ which would be produced by a constant amplitude load sequence with the magnitude of the cycle under consideration. The strain $\varepsilon_{op,c}$ is determined by calculating the constant amplitude crack opening stress according to Eq. (4.15) and the corresponding strain on the ascending hysteresis loop branch. For completely compressive cycles, the constant amplitude crack opening strain $\varepsilon_{op,c}$ is set equal to the strain value at the upper reversal point.
- In case a new absolute maximum is reached within the course of a load sequence, the next closed cycle adjusts ε_{op} to the constant amplitude crack opening strain of this cycle.
- If the constant amplitude crack opening strain of a cycle is higher than the crack opening strain established by previous cycles, the effective stress and strain ranges are computed with the previous crack opening strain. After completion of the actual cycle, a new crack opening strain has developed, $\varepsilon_{op,i+1}$. An exponential decay of the difference between the (natural) crack opening strain of a cycle, $\varepsilon_{op,c}$, and the actual (previous) crack opening strain $\varepsilon_{op,i}$ is modelled. The Miner-type damage D is the characteristic decay variable instead of the newly produced crack length, see Eq. (4.64). This detail may be subject to revision.

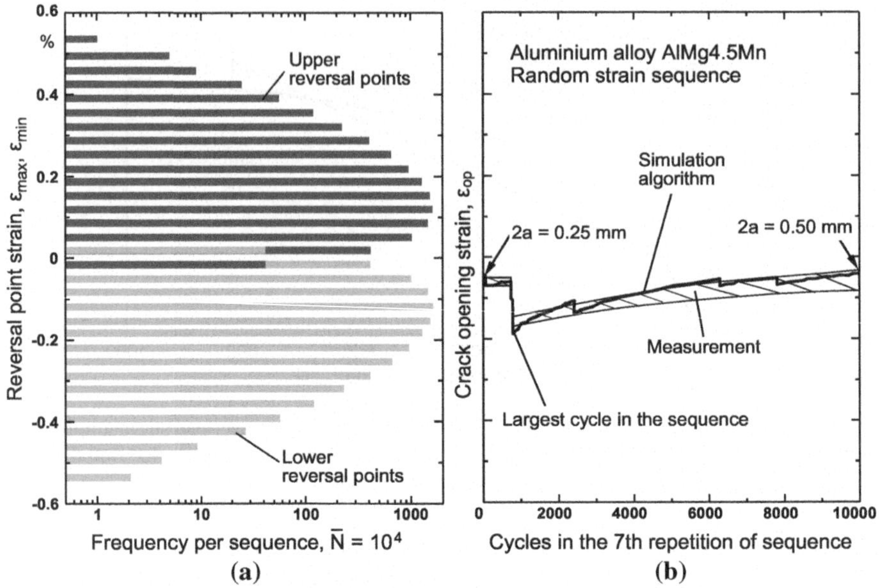

Fig. 4.38 Frequency distribution of upper and lower strain reversal points in partial pseudo-random sequence (**a**) and crack opening strain ε_{op} in 7th repetition of partial sequence (**b**), (Vormwald and Seeger 1991)

- If the constant amplitude crack opening strain of a cycle is lower than the crack opening strain that previous cycles have left, two cases have to be distinguished A large stress amplitude ($\sigma_a \geq 0.4\sigma_Y$) resets the crack opening strain to its constant amplitude value. Small stress amplitudes ($\sigma_a < 0.4\sigma_Y$) establish the stress opening strain according to Eq. (4.64).

The crack opening strain mentioned above under the last but one point is defined by the following relationship expressing a damage-correlated decay of the crack opening strains:

$$\varepsilon_{op,i+1} = \varepsilon_{op,c} - (\varepsilon_{op,c} - \varepsilon_{op,i}) \exp(-15D) \tag{4.64}$$

In Fig. 4.38, the measured and simulated crack opening strain behaviour is compared for a random load sequence. The graph on the left hand side shows the spectrum with a Gaussian distribution and a spectrum content of 10^4 cycles. The graph on the right hand side presents measured and calculated crack opening strains. The correspondence between the two is very good.

Life Analysis

The ΔJ-integral approximation according to Eq. (4.62) was combined with the closure definition shown in Fig. 4.34. For a crack growth analysis, the effective

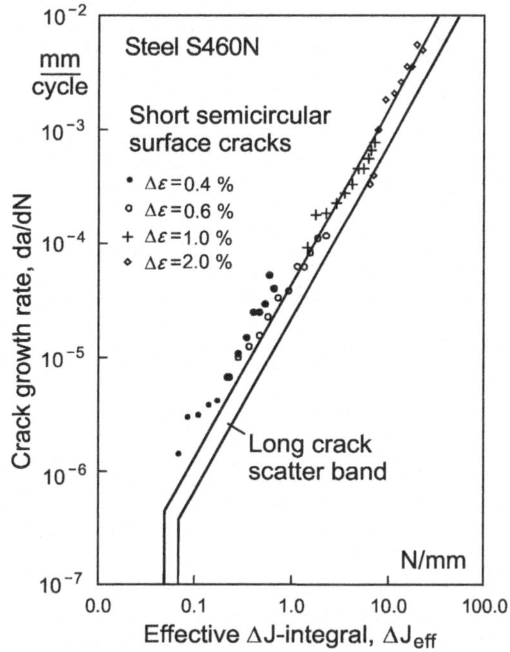

Fig. 4.39 Crack growth rate of long and short cracks, corresponding to small-scale and large-scale yielding conditions, dependent on effective ΔJ-integral; (Vormwald and Seeger 1991)

ranges were taken from the descending hysteresis loop branch in order to meet the path independence criterion. The experimentally gained crack growth rates are plotted over ΔJ_{eff} in Fig. 4.39. Obviously, the fatigue crack growth rates are uniquely described over four orders of magnitude under both small-scale and large-scale yielding conditions.

The corresponding power law crack propagation equation has the following form (Zheng and Liu 1986):

$$\frac{\mathrm{d}a}{\mathrm{d}N} = C_J(\Delta J_{\text{eff}})^{m'} \tag{4.65}$$

where C_J and m' are material parameters. In the limit of Paris law conditions, the relation $m' = m/2$ is valid.

In case of variable amplitude loading, the load sequence dependent simulation of crack opening strains and associated effective ranges considerably improves the prediction of fatigue lives. An example is shown in Fig. 4.40. In variable amplitude situations in connection with large-scale yielding, the cycle with the largest range in the stress-strain diagram moves the position of an enveloping hysteresis loop towards the stress-ratio $R = -1$. All smaller cycles produce loops which are nested in this enveloping loop. Large cycles are therefore both tension and compression overloads. Under small-scale yielding conditions, the fatigue crack growth plastic deformation remains on the crack flanks. Under large-scale yielding conditions, this plastic deformation is squeesed away during the compression half-

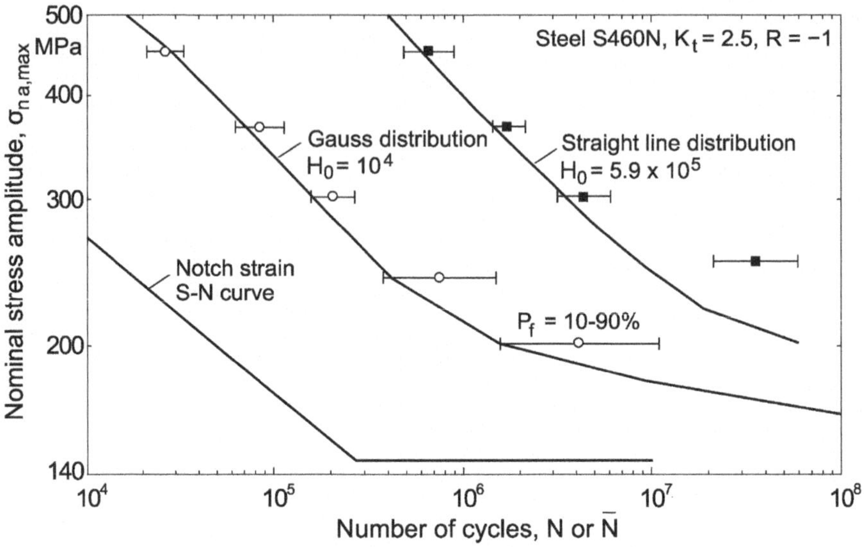

Fig. 4.40 Fatigue life data of an unnotched specimen simulating a mildly notched specimen ($K_t = 2.5$) of steel under variable amplitude random loading (number of cycles H_0 in partial sequence); comparison of calculation results (curves) and experimental results (point symbols with scatter range); simulated constant-amplitude notch strain life curve with initial crack depth $a_0 = 15$ μm; crack growth life up to $a_f = 0.25$ mm; (Vormwald 1989)

cycle of an overload. Compressive plastic shortening on crack flanks is a wide-ranging effect. Humps from variable amplitude load sequences are deleted. The consequence is an acceleration of the crack growth rates. This is very different from what is observed in small-scale yielding where the retardation effects dominate.

So far, only crack closure and crack growth in unnotched specimens has been discussed. This procedure is representative for the initiation of a technical crack at the root of mildly notched specimens or structures. In these members, further cycles may be applied before failure. The ΔJ_{eff}-based approach has been applied to calculate the crack growth process from the very beginning up to final fracture of a component (Savaidis et al. 2010). Special attention is given to the simulation of the plasticity-induced opening and closure behaviour of cracks when growing in the non-uniformly stressed notch root regions. The study is based on previous analytical and experimental work (McClung 1991[(3)]; Savaidis et al. 1995; Dankert et al. 1999; Brüning et al. 2006; Saal 1971).

One key element of the modelling approach is the treatment of closure of cracks in the inhomogeneous notch stress field. The opening stress for a crack with its tip at the position $x = a$ (distance from the notch root) is calculated using Eq. (4.15). The applied stresses, however, are the stresses at $x = a$ in the uncracked state. An elastic-plastic analysis is required to make these quantities available. An approximation by the Neuber rule (Neuber 1961) is used neglecting the multiaxial

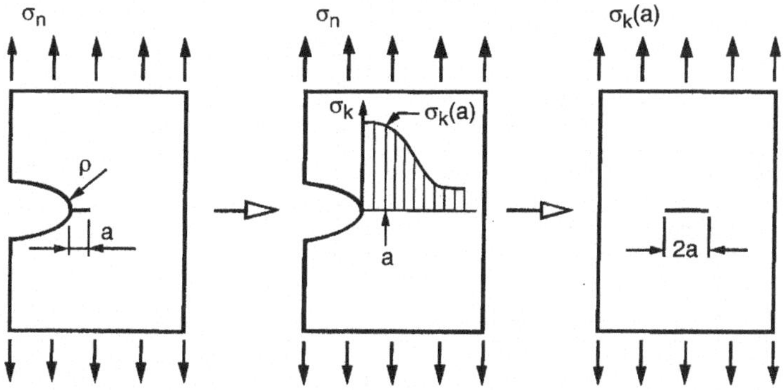

Fig. 4.41 Calculation of crack opening stress for short crack corresponding to large-scale yielding conditions (**a**), based on elastic-plastic notch stress distribution (**b**) converted to an homogeneous stress field (**c**); (McClung 1987, 1991[(3)], Savaidis et al. 1995)

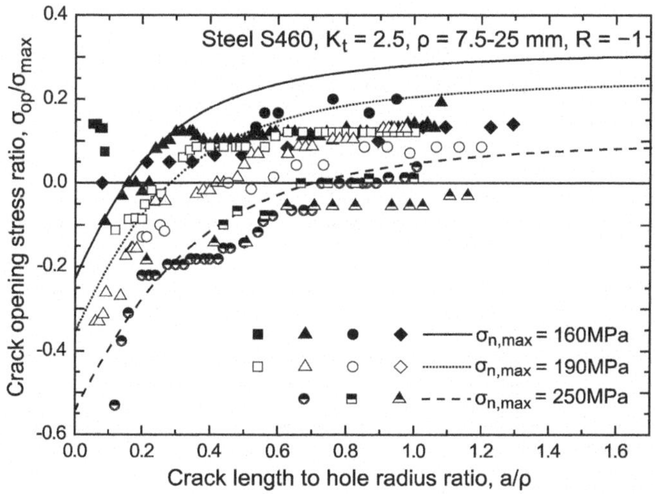

Fig. 4.42 Crack opening nominal stress ratio dependent on crack length to hole radius ratio; mildly notched specimen ($K_t = 2.5$), load ratio $R = -1$, steel S460; comparison of calculation results (curves) and experimental results (point symbols); (Savaidis et al. 1995)

local situation for convenience. The procedure is illustrated by Fig. 4.41. Results for cracks growing from the root of a centre hole specimen are shown in Fig. 4.42. The approximation accuracy of the analysis is acceptable against the background of experimental scatter.

Finally, Fig. 4.43 gives an overview of crack growth lives for various stress concentration factors K_t and nominal stress ratios R. This example is intended to show the capabilities of simulation of fatigue crack growth in the large-scale yielding regime for notched structures.

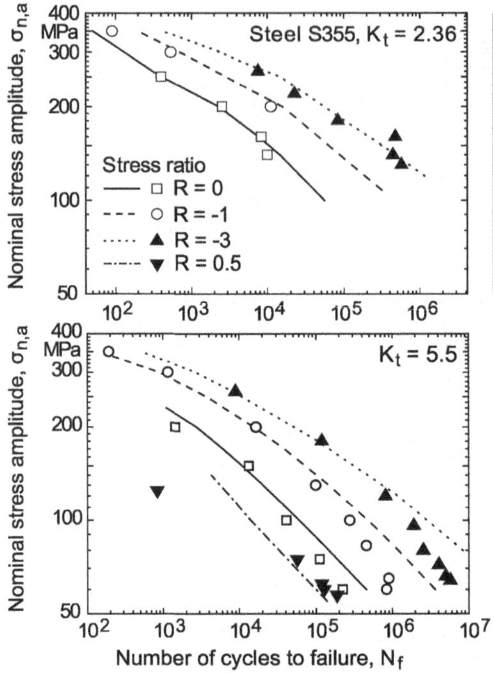

Fig. 4.43 Crack propagation life data of notched specimens ($K_t = 2.36$, 3.6, 5.5) of steel S355 under constant amplitude loading ($R = -3, -1, 0, 0.5$); comparison of calculation results (curves) and experimental results (point symbols); (Savaidis et al. 2010)

Due to the dominating acceleration effects which are taken into account conservatively, variable amplitude lives may be underestimated. Currently, there is no approach available which would allow the crack flanks to locally store the plastic deformation history in a similarly simple way as does the strip-yield model. For variable amplitude loading encompassing large-scale yielding conditions, a rigorous finite element based simulation (just as for small-scale yielding conditions) is not a realistic option.

4.3.4 Multiaxial and Mixed Mode Loading

Crack Path Considerations

Under small-scale yielding conditions, it is a widely confirmed observation that fatigue cracks grow along a plane or path where a pure mode I situation adjusts itself. The advanced software in the field heavily relies on this experience, e.g. FRANC2D (Swenson and Ingraffea 1988; Bittencourt et al. 1996), FRANC3D (Martha 1989, Carter et al. 2000), ZENCRACK (Zentech Int 2012; Timbrell and

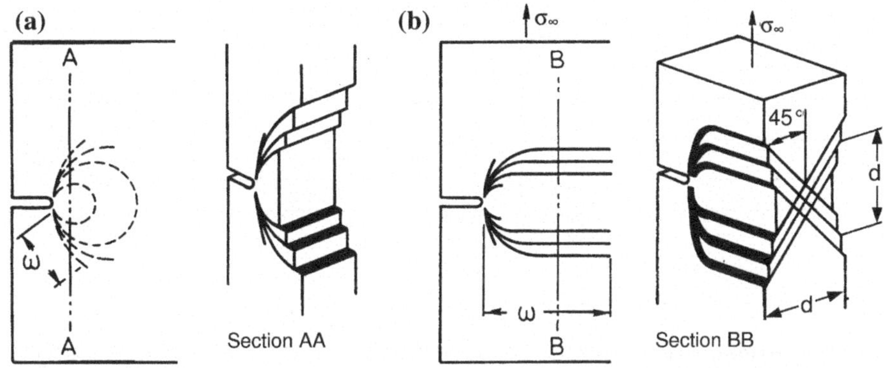

Fig. 4.44 Basic plastic deformation mechanisms at crack tip, theoretically analysed and experimentally confirmed (Hahn and Rosenfield 1965); hinge-type plastic deformation pattern under plane strain conditions of thick plates (**a**) and 45° through-thickness pattern under plane stress conditions of thin plates (**b**); (Schwalbe 1980)

Cook 1997), ADAPCRACK3D (Fulland 2003; Schöllmann et al. 2003) and BEASY (Mellings et al. 2002). For general prediction purposes, mixed mode criteria are implemented. Details of the approaches may be found in a special monograph (Richard and Sander 2009).

There does not exist much experience with fatigue cracks growing under large-scale yielding and non-pure mode I conditions. At first glance the mode I preference will also be maintained in large-scale yielding. However, there are observations opposing this statement. First, the well-known effect of shear lip formation is mentioned (Zuidema and Krabbe 1997). Even in simple laboratory test specimens the crack front escapes from the mode I plane towards the end of the test when cyclic plastic deformations increase considerably. Shear lips emerge from the surfaces. The crack front is under combined mode I and mode III loading; rigorous evaluation would also reveal mode II contributions.

Plastic deformation in metals occurs along planes of maximum shear stress. The basic mechanism of plastic deformation at the crack tip in a plate specimen of thickness d under remote tensile stress σ_∞ (mode I loading of the crack front) has been analysed and compared with experimental findings (Hahn and Rosenfield 1965). Under the plane strain conditions of thick plate specimens, the maximum shear stress planes generate a hinge-type plastic deformation pattern which extends to the surface of the specimen, Fig. 4.44a. This pattern is also initially occurring in thin-plate specimens. Under the following load increase, the hinge type pattern soon changes into the 45° through the thickness pattern which correlates with the maximum shear stress planes under plane stress conditions, Fig. 4.44b. Under large-scale yielding conditions, the fatigue crack growth planes seem to find their orientation in planes of maximum shear stress which may explain the onset and extent of shear lips.

Another example of the influence of cyclic plastic deformation on the crack geometry has been published (Zerres et al. 2011). A thin-walled tube specimen

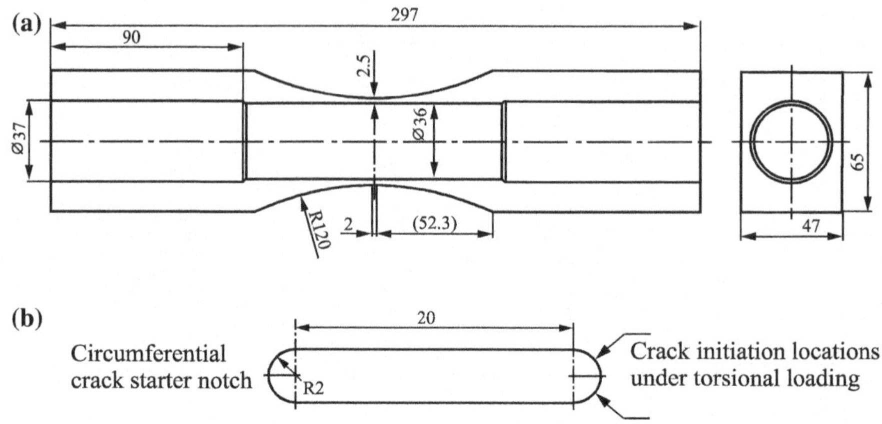

Fig. 4.45 Thin-walled tube specimen (**a**) with circumferential crack starter notch (**b**); crack initiation locations at notch edge under torsional loading; (Zerres et al. 2011)

with hour glass shape, Fig. 4.45, was subjected to alternating torsional loads, $R = -1$. Fatigue cracks growing from a machined sharp crack starter notch (circumferentially arranged) were observed. Under small-scale yielding conditions, a helical curling of the cracks with a 45° slope against the specimen axis occurs. However, the cracks under large-scale yielding conditions turned into planes perpendicular to the axis. An explanation may be found in plastic deformation arguments. Due to the hour glass shape of the specimen, the wall thickness increases and the cyclic plastic deformation decreases with increasing distance from the midplane. But fatigue cracks primarily grow where the cyclic plastic deformation is high.

The two introductive examples demonstrate that experience gained for multiaxiality and mixed mode influences cannot readily be transported from the small-scale to the large-scale cyclic yielding regime.

Biaxiality Effect on Crack Closure

In the area of multiaxial fatigue, it is common practice to distinguish between proportional and non-proportional loading cases. The distinction can be related to the external loading. However, the distinction can also refer to the local field variables, e.g. stresses or strains. Multiaxiality itself has an aspect of external and internal loading conditions. A uniaxial external load may cause locally multiaxial stress and strain states. In the case of external loading, the term 'combined loading' instead of 'multiaxial loading' is preferred. Pure torsion, for example, is not combined loading, however it causes a multiaxial stress state in tubes or cylindrical bars.

With only one load component acting, the loading state is always proportional. Superimposed on the torsional moment, the tube may be loaded by a tension force (or bending moment) which may then be applied proportionally or non-proportionally. Even under non-proportional combined tension and torsion loads, proportional or uniaxial stress states may appear at fatigue critical locations. In the hour glass shaped thin-walled tube specimen, Fig. 4.45, the fatigue cracks initiate at the uniaxially stressed edge of the crack starter notch. An early fatigue crack growth is observed under a nearly pure mode I (and therefore proportional) loading state, independently of what the external load combination is. However, the actual load sequences interfere very soon. The crack may be subjected to an increasing degree of mixed mode loading and it may tend to avoid it by curving. In this state, cyclic plastic deformation may intervene with demands on the growth direction not compatible with the mixed mode requirements of minimising the mode II crack growth component.

The general problem of large-scale cyclic yielding fatigue crack growth under non-proportional combined loading with local non-proportional mixed mode loading states is far from being solved either theoretically or empirically. It has to be noted that the general case is also still unsolved for small-scale yielding conditions. The research community has tackled some simpler special cases with increasing success.

It has been a natural choice to start investigating multiaxiality effects with a plane, centre-cracked structure ($a/W \approx 1/8$) under far-field biaxial loading. The ratio of the far-field or applied stresses in the direction of the crack-ligament line, σ_{x0}, and perpendicular to it, σ_{y0}, is termed the biaxiality ratio $\Lambda = \sigma_{x0}/\sigma_{y0}$. Relevant crack opening and crack closure investigations based on the finite element method in combination with the nodal release technique have been performed (McClung 1989). Biaxiality ratios $\Lambda = -1, -1/2, 0, +1/2$ and $+1$ are considered. Later on these investigations were revisited reporting the general trends found for $R = 0$ and $R = -1$ loading, Fig. 4.46 (McClung et al. 1999). The analysis results were consistent with experimental data (McClung and Sehitoglu 1988; Brown and Miller 1985; Hoshide et al. 1981). The restults can be summarised as follows:

- At lower applied stresses ($\sigma_{max}/\sigma_Y < 0.4$) the biaxiality ratio has only a minor effect on the crack opening stresses.
- At larger applied stresses, there is a trend to lower crack opening levels for increasingly negative biaxiality ratios, and to higher crack opening levels for increasingly positive ones.
- The effect of biaxiality is more pronounced for negative biaxiality ratios than for positive ones.

It is proposed to approximate the crack opening stresses by applying Eq. (4.15) with the following modification (McClung et al. 1999): Positive Λ-values are treated as being uniaxial, no alterations are required. For negative Λ-values, a semi-empirical, preliminary approximation formula is suggested. This equation calculates a closure-equivalent uniaxial stress, σ_{cl-eq}, which generates the same crack opening response as the original biaxial loading stress $\sigma_{y0}^{(2)}$ (with exponent (2) for 'biaxial'):

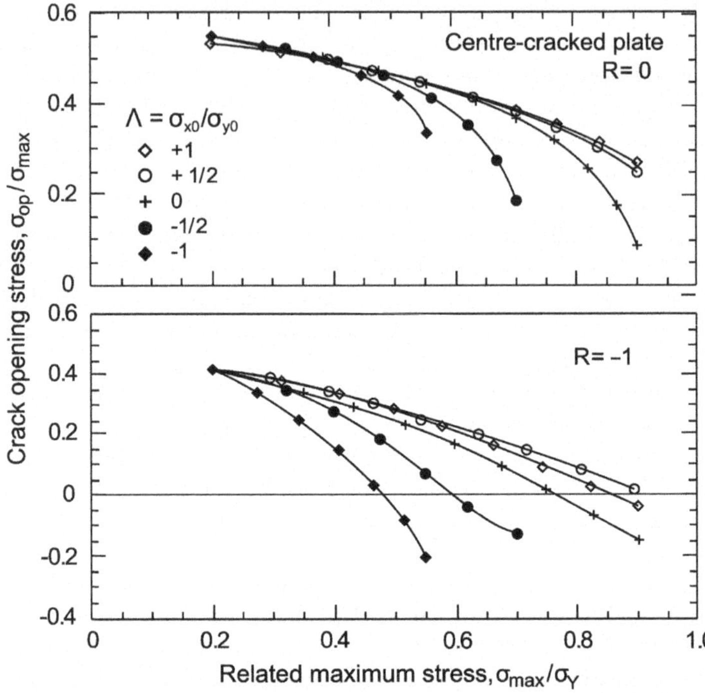

Fig. 4.46 Crack opening stress dependent on maximum stress for various biaxiality ratios; calculation results based on the finite element method; (McClung et al. 1999)

$$\frac{\sigma_{\text{cl-eq}}}{\sigma_{\text{Y}}} = \frac{\left(\sigma_{\text{y0}}^{(2)}/\sigma_{\text{Y}}\right)(1 - \sigma_{\text{co}}/\sigma_{\text{Y}}) + \sigma_{\text{co}}/\sigma_{\text{Y}}\left(\sigma_{\text{y0}}^{(2)}/\sigma_{\text{eq}} - 1\right)}{\sigma_{\text{y0}}^{(2)}/\sigma_{\text{eq}} - \sigma_{\text{co}}/\sigma_{\text{Y}}} \tag{4.66}$$

where σ_{eq} is the von Mises equivalent stress (in which the biaxiality ratio Λ is considered) and σ_{co} is the 'cut-off' stress, below which the influence of biaxiality vanishes. Rough estimates of $\sigma_{\text{co}}/\sigma_{\text{Y}} \approx 0.25$ for $R = -1$ and $\sigma_{\text{co}}/\sigma_{\text{Y}} \approx 0.5$ for $R = 0$ are specified. It is emphasised that no numerical results are available for other crack geometries.

The most important application-relevant crack geometry is the surface crack. In recognition of McClung's numerical results (McClung 1989), Savaidis decided to adopt them as a basis for estimations of crack closure of surface cracks (Savaidis 1995; Savaidis and Seeger 1996, 1997). A proportional plane stress situation with a naturally initiated crack perpendicular to the first principal stress direction was assumed. Their rather complex empirical modification of the Newman crack opening equation reads as follows:

$$\frac{\sigma_{1,\text{op}}^{(2)}}{\sigma_{1,\max}} = [z_1 + (z_2 - z_1)(R + 1)]\frac{\sigma_{1,\text{op}}^{(1)}}{\sigma_{1,\max}} \tag{4.67}$$

Table 4.1 Coefficients referring to Savaidis' crack opening approximation; (Savaidis and Seeger 1996, 1997)

i	a_i	b_i	c_i	d_i	i	a_i	b_i	c_i	d_i
0	0.535	0.535	0.535	0.535	2	−1.062	−2.585	−0.224	−0.816
1	−0.105	0.308	0.107	0.366	3	−0.331	−0.760	−0.421	−0.916

where the factor in brackets refers to negative Λ-values whereas positive Λ-values are treated as uniaxial. The exponents (1) and (2) designate uniaxial and biaxial states. The subscript 1 of the σ terms indicates that the first principal stress is exclusively considered. The auxiliary functions z_1 and z_2 are defined by the following expressions which refer to a further auxiliary function $g(\Lambda, R)$:

$$z_1 = 1 + \frac{g(\Lambda, -1) - g(0, R)}{\sigma^{(1)}_{1,\mathrm{op}}/\sigma_{1,\max}}, \quad z_2 = 1 + \frac{g(\Lambda, 0) - g(0, R)}{\sigma^{(1)}_{1,\mathrm{op}}/\sigma_{1,\max}} \tag{4.68}$$

$$\begin{aligned} g(\Lambda, R) = {} & \sum_{i=0}^{3} b_i \left(\frac{\sigma_{1,\max}}{\sigma_Y}\right)^i + \sum_{i=0}^{3} (a_i - b_i)\left(\frac{\sigma_{1,\max}}{\sigma_Y}\right)^i (\Lambda + 1) + \\ & + \left[\sum_{i=0}^{3} (d_i - b_i)\left(\frac{\sigma_{1,\max}}{\sigma_Y}\right)^i + \sum_{i=0}^{3} (c_i - a_i - d_i + b_i)\left(\frac{\sigma_{1,\max}}{\sigma_Y}\right)^i (\Lambda + 1)\right](R + 1) \end{aligned} \tag{4.69}$$

The coefficients a_i, b_i, c_i, d_i are presented in Table 4.1.

The Eqs. (4.67) to (4.69) are intended to reproduce McClung's numerical results (McClung 1989). Besides considering the proportional biaxial loading case above, Savaidis had to introduce a deviating, shear dominated failure mode. Crack growth in the shear modes II and III is experimentally found in cases of dominating torsional loading. These cases must be treated separately. It is not clear whether the shear-dominated failure mode has a mechanical reason or whether material anisotropies produced by the manufacturing process are the origin.

Biaxiality Effect on the ΔJ-Integral

So far, biaxiality effects on the crack opening level have been discussed. Cyclic plastic deformation is one origin of this influence. Additionally, the question has to be answered in which way the crack driving force—be it the ΔJ-integral or any other suitable parameter—is also affected by the biaxial stress state. Here, the discussion is restricted to the ΔJ-integral.

Once more, the J-integral is separated into an elastic and a plastic component, $J = J_e + J_p$. The elastic component J_e is not influenced by the biaxiality condition. The plastic component J_p can be written in the following form (He and Hutchinson 1983):

$$J_{\rm p} = \sigma_{\rm eq}\varepsilon_{\rm eq}a h_1(n, \Lambda) \tag{4.70}$$

The von Mises equivalent stresses and equivalent (plastic) strains are defined in tensor notation as follows:

$$\sigma_{\rm eq} = \sqrt{\frac{3}{2}s_{\rm ij}s_{\rm ij}}, \quad \varepsilon_{\rm eq} = \sqrt{\frac{2}{3}e_{\rm ij}e_{\rm ij}} \tag{4.71}$$

where the symbols $s_{\rm ij}$ and $e_{\rm ij}$ denote the components of the deviatoric stress tensor and plastic strain tensor, respectively. A power law relates the stresses to the total strains in the uniaxial loading case, compare Eq. (4.50):

$$\frac{\varepsilon}{\varepsilon_0} = \alpha\left(\frac{\sigma}{\sigma_0}\right)^n \tag{4.72}$$

The basic multiaxial approximation scheme is obtained in the following form (He and Hutchinson 1983; Dowling 1987):

$$J_{\rm p} = \alpha\sigma_0\varepsilon_0 a\, h_0^{(1)}\left(\frac{\sigma_{\rm eq}}{\sigma_0}\right)^{n+1} \tag{4.73}$$

The multiaxial function $h_0^{(2)}$ is derived from its uniaxial form $h_0^{(1)}$. For a wide centre-cracked plate under plane stress conditions, the substitution reads:

$$h_0^{(2)} = h_0^{(1)}\left(\sqrt{1 - \Lambda + \Lambda^2}\right)^{n-1} \tag{4.74}$$

A more detailed presentation may again be found in the literature (McClung et al. 1999). This includes further stress states, influence of finite boundaries and comparison of the approximation scheme with numerical results.

Using this information and Eq. (4.62), a surface crack estimate $\Delta J_{\rm eff}^{(2)}$ was constructed by Savaidis arriving at the following expression for proportional biaxial loading conditions (Savaidis 1995; Savaidis and Seeger 1996, 1997):

$$\Delta J_{\rm eff}^{(2)} = \left[1.24\frac{\Delta\sigma_{1,\rm eff}^2}{E} + \Delta\sigma_{1,\rm eff}^2\frac{2.046}{\sqrt{n'}}\left(\frac{1}{2K'}\right)^{1/n'}\left(\Delta\sigma_{1,\rm eff}\sqrt{1 - \Lambda + \Lambda^2}\right)^{1/n'-1}\right]a \tag{4.75}$$

The Vormwald-Seeger algorithm for the calculation of the sequence dependent crack opening strains is adapted for determining the effective parameter ranges in Eq. (4.75) (using the descending half-cycles). Calculated and experimentally determined fatigue lives are shown in Fig. 4.47. The thin-walled tube specimen, Fig. 4.45, however, without a crack starter notch, has been fatigue-loaded under proportional strain-controlled tensile and torsional loading. The strain amplitude ratio was $\gamma_{\rm a}/\varepsilon_{\rm a} = 1.732$. The numerical procedure gives realistic fatigue lives for short surface cracks under proportional variable amplitude loading.

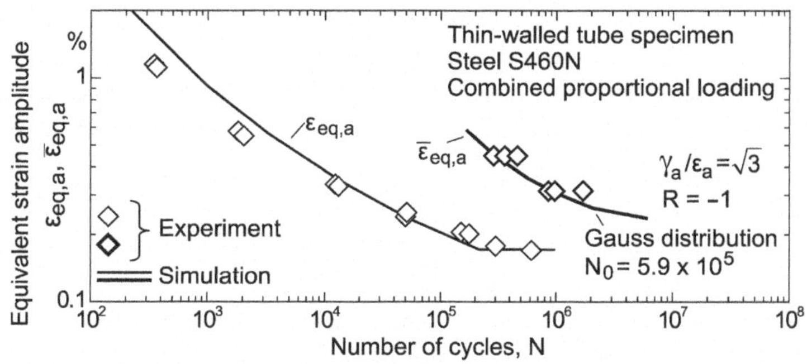

Fig. 4.47 Crack propagation life of thin-walled tube specimens of steel S460N under combined proportional constant amplitude and variable amplitude loading amplitudes ($\gamma_a/\varepsilon_a = 1.732$); equivalent plastic strain amplitude, $\varepsilon_{eq,a}$ or $\bar{\varepsilon}_{eq,a}$, dependent on number of cycles between initial crack length $a_0 = 22$ μm and final crack length $a_f = 0.25$ mm; comparison of calculation results (curves) and experimental results (point symbols); (Savaidis 1995)

Several further crack driving force parameters for short cracks (which are inevitably under large-scale cyclic yielding conditions) have been proposed. Socie suggested an equivalent strain intensity factor range $\Delta K_{\varepsilon,eq}$. It is defined based on the maximum shear strain range $\Delta\gamma_{max}$ and the strain range normal to the associated shear plane $\Delta\varepsilon_{\perp}$ (Socie et al. 1987; Socie and Furman 1996):

$$\Delta K_{\varepsilon,eq} = \sqrt{(Y_{II} G \Delta\gamma_{max})^2 + (Y_I E \Delta\varepsilon_{\perp})^2}\sqrt{\pi a} \tag{4.76}$$

where Y_I and Y_{II} are the geometry factors of linear-elastic fracture mechanics. The model is applicable to proportional loading. Crack closure is not addressed.

Reddy and Fatemi linked the Fatemi–Socie parameter, which is applicable on non-proportional loading cases too, with the strain intensity factor (Reddy and Fatemi 1992; Fatemi and Socie 1988):

$$\Delta K_{\varepsilon,eq} = YG\Delta\gamma\left(1 + k\frac{\sigma_{\perp\,max}}{\sigma_Y}\right)\sqrt{\pi a} \tag{4.77}$$

For the biaxial loading case, a 'universal' ΔJ-integral has been proposed which is claimed to have the material dependency eliminated (Tchankov et al. 2008):

$$\Delta J = \frac{1}{0.03E^{0.85}}\frac{\Delta\varepsilon_e}{\Delta\varepsilon_{tot}}\Delta J_e + \frac{1}{1200(\sigma'_Y)^{1.1}}\frac{\Delta\varepsilon_p}{\Delta\varepsilon_{tot}}\Delta J_p \tag{4.78}$$

with the modulus of elasticity E in GPa and the cyclic yield stresses σ'_Y in MPa. The various numbers in the equation have been identified by a parametric study which included four materials tested at elevated temperatures. Besides normalisation by the material-dependent factors, the total strain range $\Delta\varepsilon_{tot}$ has been split up, as normal into an elastic part $\Delta\varepsilon_e$ and a plastic part $\Delta\varepsilon_p$. The ratio of partitioned to total strain ranges has been introduced to take account of the contribution of each component of the 'universal' ΔJ-integral.

Hoshide and Socie assume proportionality between the ΔJ-integral and a strain energy density parameter according to the following equation (Hoshide and Socie 1988):

$$\Delta J = 2\pi a M_J \left(\frac{\Delta\sigma^2}{2E} + f(n') \frac{\Delta\sigma_{eq}\Delta\varepsilon_{p,eq}}{1+n'} \right) \tag{4.79}$$

Finally, this list of driving force parameters is concluded with the McDowell crack growth equation (McDowell 1994; McDowell and Bennett 1996):

$$\frac{da}{dN} = \left[C_1 (\Delta\tau_{II}\Delta\gamma_{e,max})^{m_1} + C_2 (\Delta\tau_{II}\Delta\gamma_{p,max})^{m_2} \right] \left(\frac{2a}{k'd} \right)^m \tag{4.80}$$

where $\Delta\gamma_{e,max}$ and $\Delta\gamma_{p,max}$ are the maximum elastic and plastic shear strain ranges, $\Delta\tau_{II}$ is the shear stress range in that plane and d is the grain diameter. The factor k' provides the maximum size of a microstructurally short crack. The parameters C_1, C_2, m_1, m_2 and m are related to the ranges of stresses and strains as well as to the cyclic material properties under tensile and shear loading.

The Eqs. (4.75) to (4.80) are listed with the intention that the reader might recognise the common basis and purpose of modelling the fatigue crack growth behaviour under large-scale cyclic yielding conditions. The crack driving force parameters are applicable to cracks in structures where the boundaries are sufficiently far away from the crack tip or crack front. Under these conditions, the driving force parameters are composed of two factors: a simple function of the crack length and a factor including the hardening behaviour of the material together with the far-field loading. The far-field loading is expressed in terms of the stresses and strains in the uncracked material state. The crack length independent factor is called damage parameter in conventional fatigue assessment approaches according to the local strain approach. The application-relevant crack closure effects are included in Eq. (4.75), but they are not explicitely addressed in Eqs. (4.76) to (4.80).

Non-proportional Loading Effect on the ΔJ-Integral

Multiaxial damage parameters have been shown to correlate calculated and experimentally determined fatigue lives also under non-proportional loading conditions. However, critical plane approaches are required. For example, the Fatemi–Socie parameter has to be evaluated for many planes in the material element and the plane which shows the maximum parameter value is assumed to be critical in the sense of delivering the shortest lives. In this plane, the fatigue crack growth rate is the highest. The crack driving force can be estimated by multiplying the multiaxial damage parameter with the actual crack length (in the case of ΔJ) or its square root (in the case of ΔK). In case of proportional loading,

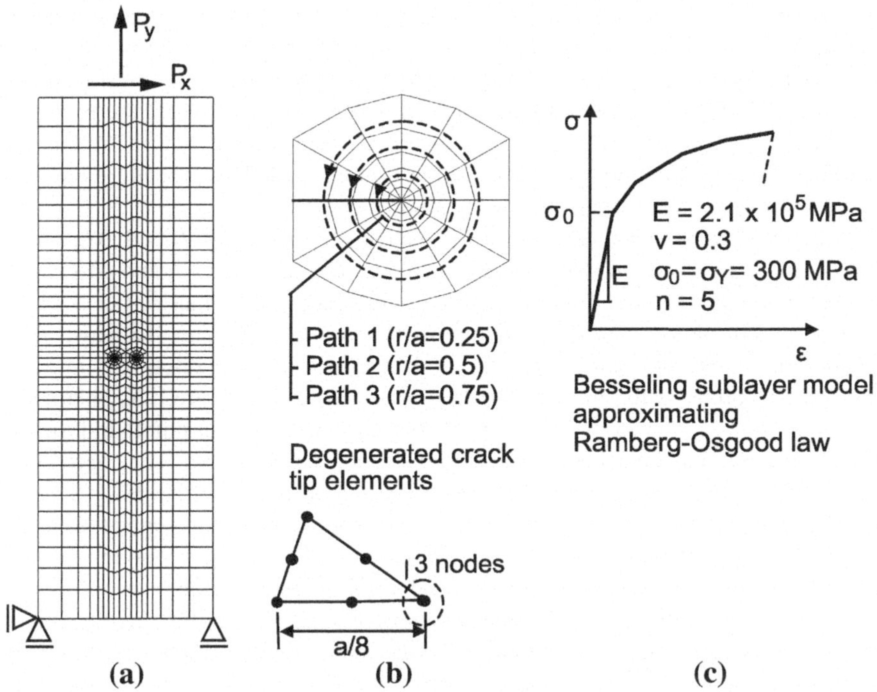

Fig. 4.48 Finite element model of centre-cracked specimen under combined tensile and shear loading (**a**); integration paths and degenerated crack tip elements (**b**); Besseling sublayer model approximating the Ramberg–Osgood material law (**c**); (Hertel et al. 2004)

such a crack driving force definition is backed by theoretical arguments. The crack driving force can be interpreted as the ΔJ-integral, for example. For non-proportional loading this theoretical background is widely lost. Despite this drawback, the usefulness of the critical plane multiaxial damage parameters is beyond doubt.

A numerical study on the performance of ΔJ-integral evaluating non-proportional cyclic loading conditions has been presented (Hertel et al. 2004). The calculations have been carried out for a centre cracked plate specimen subjected to combined tensile and shear loading, Fig. 4.48. A finite element model has been created and the crack tip has been modelled with degenerated quadrilateral elements in order to achieve realistic crack opening displacements. Initially, the crack tip nodes share the same location but they are not connected,. All calculations have been performed with the finite element program ANSYS. The Besseling sublayer model approximating the Ramberg–Osgood material law is applied resembling kinematic hardening. The crack is assumed as stationary; no nodal release algorithm is applied. Crack closure is not considered; under compressive loading, overlapping crack flanks are allowed. The ΔJ-integral has been calculated by numerical integration of Eq. (4.42) for three different paths encircling the crack tip. Three circular paths with different radii are applied in order to prove the path-independence of the results. In case of non-proportional loading the load reversal points, in terms of ΔJ, are detected by the following algorithm. If the change in ΔJ after the integration of Eq. (4.42) turns out to be negative, it is presumed that the last equilibrium state was a load reversal point. In consequence, the reference variables are redefined with stress, strain and displacement values of the last equilibrium state. Thereupon, the integration is continued. This procedure provides the monotonic increase of ΔJ for every load step.

In addition to the total ΔJ, the mode related components ΔJ_{I} and ΔJ_{II} are of special interest in a mixed mode I and II analysis. The integrals ΔJ_{I} and ΔJ_{II} can be derived by decomposing the stress, strain and displacement values of the crack tip field into a symmetric (mode I) and antisymmetric (mode II) part (Ishigawa et al. 1979). The symmetric and antimetric field variables are separately fed into Eq. (4.42). The values of the mode related integrals ΔJ_{I} and ΔJ_{II} are thus obtained.

In case of proportional loading, Fig. 4.49a, the load reversal points in terms of ΔJ, ΔJ_{I} and ΔJ_{II} have been determined by the proposed algorithm. They agree with the reversal points at the minima and maxima of the external load. Due to increasing plastic deformation under superimposed loading, the separated values of ΔJ_{I} and ΔJ_{II} are larger than the corresponding values for pure mode I or pure mode II loading, but they add up to the value of ΔJ as expected.

In case of non-proportional loading with phase shift, Fig. 4.49b, the ΔJ-integral remains practically path-independent. External loading with 90° phase shift results in predicted load reversal points in terms of ΔJ_{I} and ΔJ_{II} displaced from the maxima or minima of the external loads σ and τ. The load reversal points in terms of ΔJ are occuring between the maximum of stress and the maximum of strain. The result for ΔJ is similar to the result of ΔJ_{I} because the selected ratio of the external loads leads to a mode I dominated crack tip field. ΔJ_{I} and ΔJ_{II} are different from the correspondending values for pure mode I and pure mode II loading and cannot simply be added to obtain ΔJ.

In the case of non-proportional loading with different frequencies, Fig. 4.49c, the ΔJ-integral remains practically path-independent. The load reversal points in terms of ΔJ_{I} and ΔJ_{II} show the same frequencies as the external loads but are displaced relative to the maxima and minima of the external loads expressed by σ and τ. The load case is again mode I dominated and the results for ΔJ are similar to those for ΔJ_{I}.

Using the described algorithm, it is possible to calculate the practically path-independent crack tip parameter ΔJ based on finite element calculations even for non-proportional cyclic loading. The calculated ΔJ values should maintain their sound physical basis to describe fatigue crack growth even under non-proportional loading conditions. A superposition of solutions taken from pure mode I and pure mode II loading does not lead to reasonable results for a mixed mode load case, even under proportional loading conditions. Calculating ΔJ_{I} and ΔJ_{II} separately by decomposing the crack tip field in a symmetric and antimetric part is possible. Under proportional loading conditions, the separately calculated parts add up again

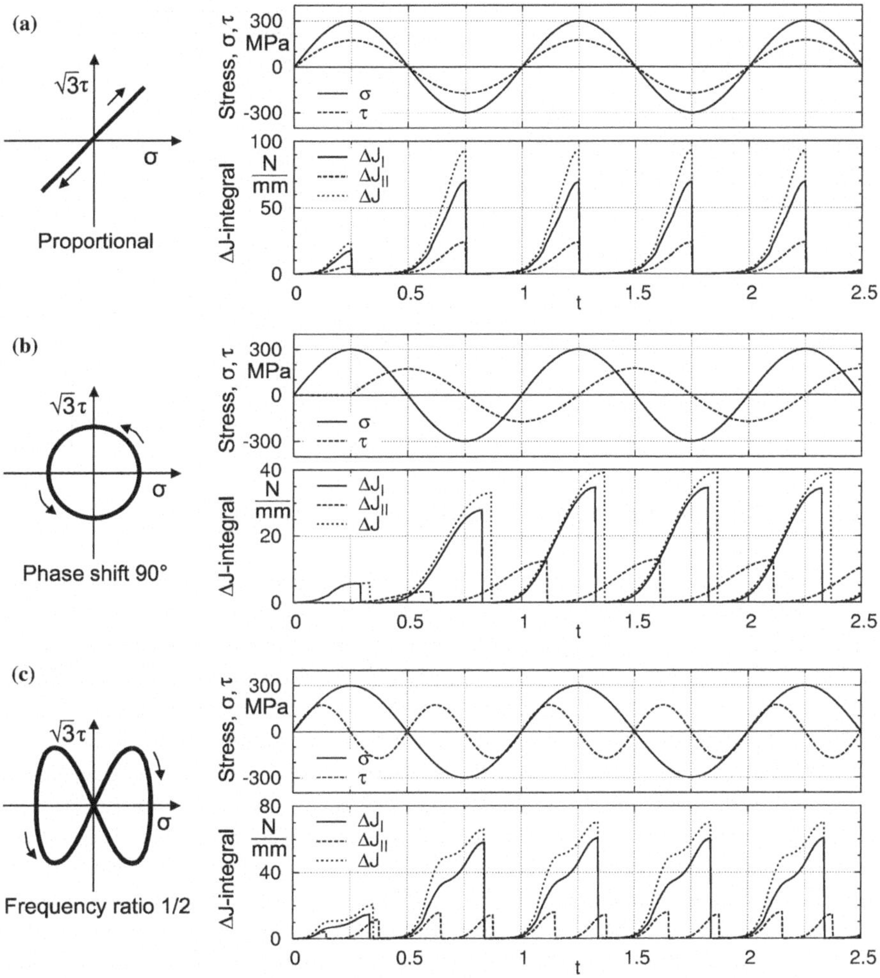

Fig. 4.49 Calculated integrals ΔJ, ΔJ_{I}, ΔJ_{II} for centre crack length $2a = 2.0$ mm in plate specimen under combined tensile and shear loading ($\sigma_a/\tau_a = 1.732$); proportional loading (**a**), phase shift 90° (**b**) and frequency ratio 1/2 (**c**); (Hertel et al. 2004)

to ΔJ. Suitable approximative solutions for ΔJ, ΔJ_{I} and ΔJ_{II} should be found, because elastic-plastic finite element calculations for cyclic loading are complicated and time-consuming. Such approximative solutions would have to take the interaction between mode I and mode II loading into account. Further investigations on the definition of a load cycle and on an appropriate mixed mode crack growth criterion including crack closure applicable to non-proportional loading should be performed.

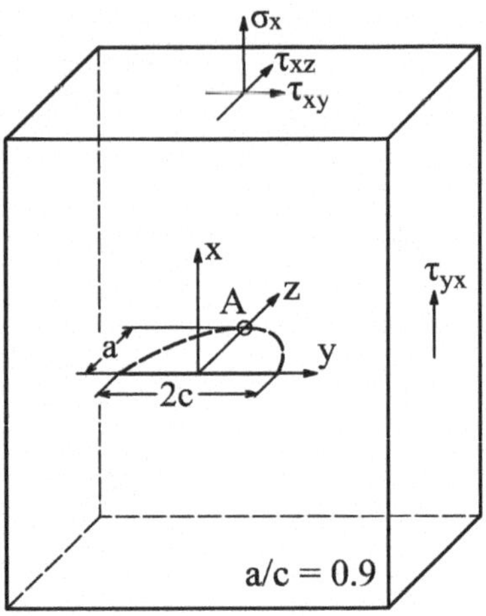

Fig. 4.50 Plate specimen subjected to coordinate-related tensile and shear loads; semicircular surface crack with analysis point A on crack front subjected to loading modes I, II and III; (Radaj, unpublished)

Crack Propagation Life Under Non-proportional Loading

The problem of fatigue crack propagation under non-proportional loading starting with a short, nearly semicircular surface crack, Fig. 4.50, under conditions of large-scale yielding, crack closure, and variable amplitude loading has been tackled by Döring and Hertel (Döring 2006; Döring et al. 2006; Vormwald and Döring 2009; Hertel and Vormwald 2011). Their model follows the critical plane approach. The fatigue crack growth simulations described below are performed for a variety of potential growth planes. The plane with the shortest life is decisive. Only point A is considered in the following presentations and the aspect ratio is assumed to be constant. The ΔJ-integral is used as the theoretical basis. However, the rigorous algorithms for its calculation as sketched previously have not been applied (Hertel et al. 2004). Instead, approximations—or better intelligent guesses—were used. A pseudo-nonproportional mixed-mode hypothesis was introduced. Each mode is assumed to contribute individually to the crack growth rate according to the following equation:

$$\frac{\mathrm{d}a}{\mathrm{d}N} = C_J \left(\Delta J^m_{\mathrm{I,eff}} + \Delta J^m_{\mathrm{II,eff}} + \Delta J^m_{\mathrm{III,eff}}\right) \tag{4.81}$$

The hypothesis behind the equation awaits validation. Next, the mode-related ΔJ-integrals were linked to the coordinate-related portions of the strain energy density W by the following expressions:

$$\begin{aligned} \Delta J_{\mathrm{I,eff}} &= 2\pi 0.7^2 \Delta W_{x,\mathrm{eff}} a \\ \Delta J_{\mathrm{II,eff}} &= \pi 0.62^2 \eta \Delta W_{xz} a/(1+\nu) \\ \Delta J_{\mathrm{III,eff}} &= \pi 0.58^2 \eta \Delta W_{xy} a \end{aligned} \tag{4.82}$$

The interrelation between the loading modes and the coordinates is shown in Fig. 4.50. The quantity $\Delta W_{\mathrm{x,eff}}$ is the elastic-plastic strain energy density portion when the crack remains fully open. It is obtained by integrating the stresses normal to the crack surface over the corresponding strains. The integration is performed along the descending hysteresis branch. The points of crack opening and closure are calculated on the basis of an extended Newman formula, Eq. (4.15). The quantity ΔW_{xz} is the strain energy density portion connected with the shear stresses τ_{xz} and shear strains γ_{xz} in the crack plane. Similarly, the quantity ΔW_{xy} is the strain energy density portion connected with the shear stresses τ_{xy} and shear strain γ_{xy}. The strain energy density portions finally read:

$$\begin{aligned} \Delta W_{x,\mathrm{eff}} &= \int_{\varepsilon_{xx,\max}}^{\varepsilon_{xx,\mathrm{cl}}} (\sigma_{xx} - \sigma_{xx,\max})\mathrm{d}\varepsilon_{xx} \\ \Delta W_{xz} &= \int_{\gamma_{xz,\max}}^{\gamma_{xz,\min}} (\tau_{xz} - \tau_{xz,\max})\mathrm{d}\gamma_{xz} \\ \Delta W_{xy} &= \int_{\gamma_{xy,\max}}^{\gamma_{xy,\min}} (\tau_{xy} - \tau_{xy,\max})\mathrm{d}\gamma_{xy} \end{aligned} \tag{4.83}$$

The factor η in Eq. (4.82) ranges between 0 and 1. It incorporates the indentation and friction of the crack surfaces, i.e. microstructural effects on the one hand and the influence of a compressive stress in the crack plane on shear crack growth on the other hand. It takes into account crack tip shielding by crack closure. The factor is set equal to the ratio of the crack face sliding displacements of a mode II crack with and without friction stresses on the crack surfaces, the former due to crack closure. The displacements are approximated by the Dugdale model extended to mode II loading. Then, the factor η results as follows:

$$\eta = \ln\left(\cos\left(\frac{\pi}{2}\frac{\Delta\tau/2 - \tau_{\mathrm{fr}}}{\tau_{\mathrm{Y}} - \tau_{\mathrm{fr}}}\right)^{-1}\right) \Big/ \ln\left(\cos\left(\frac{\pi}{2}\frac{\Delta\tau/2}{\tau_{\mathrm{Y}}}\right)^{-1}\right) \tag{4.84}$$

where $\Delta\tau$ represents the shear stress range $\Delta\tau_{xy}$ or $\Delta\tau_{xz}$. The shear yield stress τ_{Y} and the friction shear stress τ_{fr} on the crack surface are defined as follows:

$$\tau_{\mathrm{Y}} = \sigma_{\mathrm{Y}}/\sqrt{3} \tag{4.85}$$

$$\tau_{\mathrm{fr}} = (\tau_{\mathrm{fr}0} - \mu\sigma_{\perp\max}) \tag{4.86}$$

where $\sigma_{\perp\max}$ is the maximum stress normal to the crack surface over one shear strain cycle, τ_{fr0} is the constant friction stress due to crack face indentation, and μ is the friction coefficient. The latter two parameters are fitted to experimental fatigue life data under cyclic shear stress or strain with superimposed static stress normal to the crack surface.

In the actual version of the model, load sequence effects are taken into account by tracing sequence-dependent crack opening strains. Basically, the same algorithm is applied as described in Sect. 4.3.2. The experimentally observed and the calculated fatigue lives of thin-walled tube specimens under combined tension and torsion loads are plotted in Fig. 4.51. In case of in-phase constant amplitude loading, the experimentally observed and the calculated fatigue lives show very good correspondence.

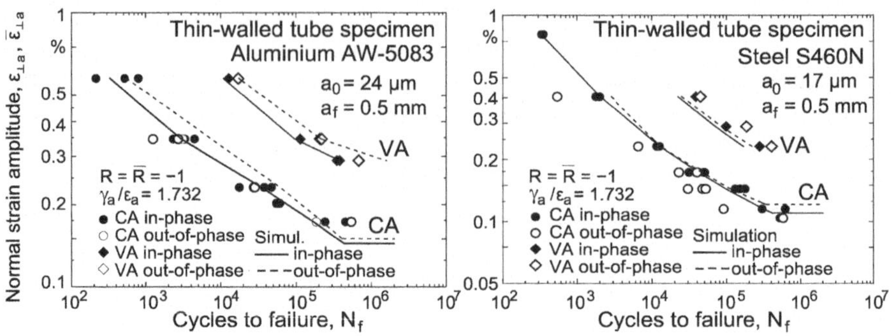

Fig. 4.51 Crack propagation life data of thin-walled tube specimens of aluminum alloy AW-5083 (**a**) and steel S460N (**b**) under in-phase and out-of-phase, constant amplitude (CA) and pseudo-random variable amplitude (VA) loading; comparison of calculation results (curves) and experimental data (point symbols); (Hertel and Vormwald 2011)

This holds true for two different materials, fine-grained steel S460N and aluminium alloy AW-5083. The experimentally observed fatigue crack growth mainly occurs in the planes of maximum normal strain, inclined between 20 and 30° against the specimen axis Fig. 4.52. The calculated plane of minimum crack growth life is inclined 25° against the specimen axis, which corresponds well with the experimental results.

Under 90° out-of-phase constant amplitude loading for the aluminium alloy specimen, the fatigue lives according the proposed short crack model are approximately twice as large as those under in-phase loading. The phase shift has no effect in case of the fine-grained steel. However, this is not observed experimentally. The experiments reveal that the phase shift has no effect in case of the aluminium alloy, whereas it results in shorter fatigue lives in case of the fine-grained steel. An investigation of the crack growth angles reveals the complexity of the out-of-phase loading case. The calculated crack growth angle for the aluminium alloy differs from the experimentally obtained value. However, the calculated and experimentally obtained crack angles of the fine-grain steel are in good correspondence, despite the fact that the calculated fatigue lives are too large.

Considering variable amplitude loading in experiments, the 90° phase shift results in longer lives compared to in-phase loading. This effect is more pronounced for the aluminium alloy than for the fine-grained steel. The fatigue lives, calculated using the short crack model, show the same tendency and are in good correspondence with the experimental findings. For the fine-grained steel with the larger plastic deformations the short crack model is conservative. Unfortunately, the experimentally observed crack growth angle in specimens under out-of-phase variable amplitude loading is only available for the aluminium alloy. The calculated crack plane inclination for this case is nearly identical, Fig. 4.52.

Short cracks will grow and finally constitute long cracks. As stated above, the general problem of fatigue crack growth under large-scale cyclic yielding conditions and non-proportional combined loading with locally non-proportional mixed mode states is far from being solved either theoretically or practically. Modelling

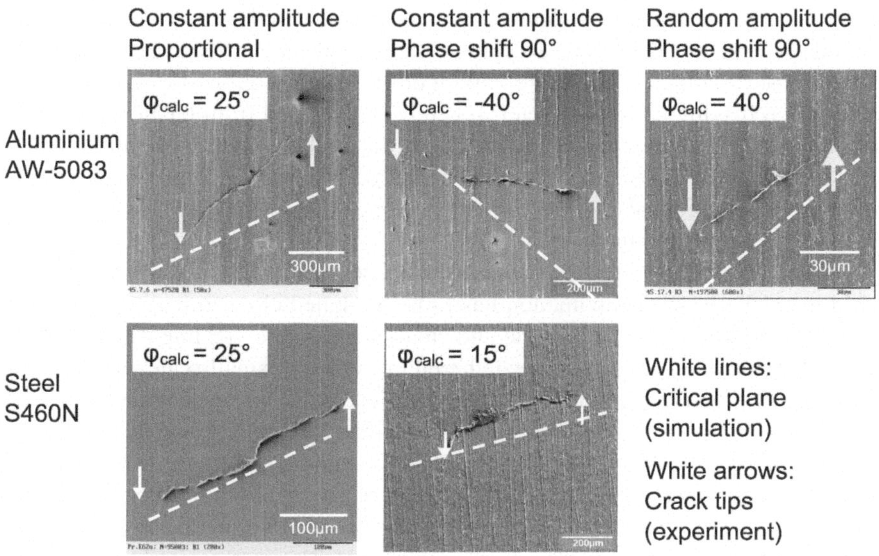

Fig. 4.52 Crack angles in thin-walled tube specimens of aluminium alloy AW-5083 and steel S460N under combined in-phase and out-of-phase, constant and pseudo-random variable amplitude loading; comparison of calculation results (φ_{calc}, white broken lines) and experimental results (crack tips marked by white arrows); (Hertel and Vormwald 2011)

efforts must keep in mind the interrelated areas of the required hypotheses, especially the crack driving force under mixed mode and non-proportional loading conditions, as well crack closure including curved or warped crack faces. Further phenomena have been detected in experiments, however, they have not yet been treated theoretically. Crack branching may occur, sometimes fraying a main crack into many separate cracks. Also, the opposite may happen: many separate cracks may join to form one main crack, like in the factory roof type failure of sharply notched round bars under cyclic torsional loading. The mechanics of fatigue crack growth provides a bulk of challenges for further theoretical and modelling work.

4.3.5 Conclusions

Under large-scale cyclic yielding conditions the stress intensity factor range has to be substituted by a crack driving force parameter of elastic-plastic fracture mechanics. The three most popular ones are the cyclic strain intensity factor, the cyclic crack tip opening displacement and the ΔJ-integral. Whichever parameter is chosen, the pros and cons have to be kept in mind. The cyclic strain intensity factor is easy to calculate, however, it lacks theoretical backing as a driving force. The cyclic crack tip opening displacement is a valid near-field parameter which should provide an acceptable consolidation of fatigue crack growth rate data.

Unfortunately, its rigorous determination from calculated near-tip plastic deformations is usually far beyond acceptable numerical effort. When using approximations instead, much of its theoretical vigour is lost. Therefore, the ΔJ–integral is preferred as driving force parameter. It can be determined from far-field quantities and, due to its path independence, describes the near–tip field in a valid manner. However, the path independence must be assured or, vice versa, its violation must be avoided. Theoretical and experimental work of many researchers has shown that this requirement can be fulfilled.

Whatever crack driving force parameter is chosen for the calculation of fatigue life, the crack closure phenomenon has to be modelled close to reality. The crack closure behaviour under large-scale and small-scale cyclic yielding differs considerably. Knowledge and experience gathered in respect of small-scale yielding conditions is invalid in respect of large-scale yielding conditions. In general, the ratio of effective to total ranges is higher for large-scale than for low-scale cyclic yielding. In the case of large-scale cyclic yielding, it is essential to distinguish between the opening and closure levels. The differences are not negligible. This has special consequences under variable amplitude loading. The overall trend shows considerable acceleration of fatigue crack growth compared with constant amplitude loading.

Stress multiaxiality gains influence on fatigue crack growth in the large-scale cyclic yielding regime, while it does not have a major influence under small-scale cyclic yielding conditions. This can already be observed in the biaxial loading case with an additional stress acting in the crack-ligament direction. The crack tip cyclic plastic deformations are affected and thereby the driving force and the crack closure level. For such cases, models and algorithms have been provided—in general restricted to proportional loading and to cracks growing in mode I. Realistic life estimates may be achieved when the models are used with some experience. The case of non-proportional loading is an issue of actual research. Only for short cracks which are too short for curving or kinking, practical algorithms for life calculations are available. They are closely related to the research field of multiaxial critical plane damage parameters and associated hypotheses. No validated algorithms are available for long cracks growing under these complex loading conditions.

List of Symbols

A_0, A_1, A_2, A_3	Constants in Newman's crack opening equation
a	Crack length
a_0	Initial crack length
a_{f}	Final crack length
a_{eff}	Effective crack length
a_{cd}	Closure development crack length
Δa	Crack growth increment
B	Specimen thickness
B_0, B_1	Constants in DuQuesnay's crack opening equation

C	Coefficient in Paris law
C_{eff}	Coefficient in ΔK_{eff} Paris law
C_J	Coefficient in ΔJ Paris law
C_{p}	Wheeler's retardation factor
D	Miner-type damage
d	Grain size, specimen thickness
e_{ij}	Components of deviatoric (plastic) strain tensor
E, E'	Modulus of elasticity, modified for plane strain
f	Crack opening function, function in ΔJ expression
$f_{\text{I,ij}}, f_{\text{II,ij}}, f_{\text{III,ij}}$	Angular functions of near tip stress fields
G	Shear modulus
g	Influence function in crack opening analysis
h_1, h_0	Functions for geometry and hardening influence on J_{p}
J, J_{e}, J_{p}	J-Integral, elastic and plastic component
ΔJ	ΔJ-integral
ΔJ_{eff}	ΔJ-Integral with effective parameter ranges
ΔJ_{e}	Elastic component of ΔJ-integral
ΔJ_{p}	Plastic component of ΔJ-integral
ΔJ_{I}, ΔJ_{II}, ΔJ_{III}	Mode related ΔJ-integrals
K, K'	Hardening coefficient, monotonic, cyclic
K_{t}	Stress concentration factor
K	Stress intensity factor
K_{I}, K_{II}, K_{III}	Stress intensity factor, modes I, II, III
ΔK	Stress intensity factor range
ΔK_{eff}	Effective stress intensity factor range
ΔK_{ε}	Strain intensity factor range
ΔK_{th}	Threshold stress intensity factor range
K_{p}	Peak stress intensity factor
K_{op}, K_{cl}	Stress intensity factor at crack opening and closure point
K_{max}, K_{min}	Stress intensity factor at upper and lower reversal point
K_{Ic}	Critical stress intensity factor for plane strain
K_{c}	Critical stress intensity factor
$K_{\text{i max}}$, $K_{\text{i min}}$	Maximum and minimum stress intensity factor of cycle i
$K^*_{\text{i max}}$	Fictitious maximum stress intensity factor of cycle i
$K^{(\text{W})}_{\text{i max}}$, $K^{(\text{W})}_{\text{i min}}$	Willenborg's stress intensity factor of cycle i
k	Factor in expression for equivalent ΔK_{ε}
k'	Factor on grain size for microstructural crack
$l_{0\text{i}}$	Bar length in strip-yield model
M_J	Coefficient in ΔJ expression
m, m'	Exponents in Paris law or factor in $\Delta\delta_{\text{t}}$ expression
N	Number of cycles
N_{f}	Number of cycles to failure
N_{i}	Number of cycles to failure with block i amplitude
n_{i}	Number of cycles in load block i

n, n'	Hardening exponent, monotonic and cyclic
P, P_0	Load, ligament yield load
p	Pressure, empirical exponent
Δp	Pressure range
Δp_{eff}	Effective pressure range
p_{max}	Pressure at upper reversal point
p_{min}	Pressure at lower reversal point
q	Empirical exponent
R	Stress ratio, load ratio, stress intensity factor ratio
$R^{(\mathrm{W})}$	Willenborg's stress intensity factor ratio
r	Radial distance
r_{Y}	Radial distance with linear-elastic stresses above yield stress
s	Path coordinate
s_{ij}	Components of deviatoric stress tensor
T_{i}	Components of traction vector
U	Crack opening ratio
u, u_x	Displacement in *x*-direction
u_y	Displacement in *y*-direction
u_{i}	Components of displacement vector
W	strain energy density
W_x, W_{xy}, W_{xz}	Strain energy density portions related to coordinate system
W	Specimen width
v	Displacement in *y*-direction
Y	Geometry factor
x, y, z	Coordinates
z_1, z_2	Auxiliary functions in crack opening stress equation
α	Coefficient in Ramberg–Osgood relationship
α	Constraint factor in tension
β	Constraint factor in compression
γ, γ_{xy}, γ_{xz}	Shear strains
γ_{a}	Shear strain amplitude
$\Delta\delta_{\mathrm{t}}$	Crack tip opening displacement range
ε, ε_{xx}, ε_{yy}	Normal strains
ε_{a}	Normal strain amplitude
ε_{l}	Local strain
$\varepsilon_{\mathrm{op}}$, $\varepsilon_{\mathrm{cl}}$	Strain at crack opening and crack closure point
$\varepsilon_{\perp}$	Normal strain in shear plane
ε_0	Reference strain in power-law relationship
$\varepsilon_{\mathrm{ref}}$	Reference strain
$\Delta\varepsilon_{\mathrm{e}}$	Elastic strain range
$\Delta\varepsilon_{\mathrm{p}}$	Plastic strain range
$\Delta\varepsilon_{\mathrm{eq}}$	Equivalent strain range
η	Crack surface factor indicating effective sliding
Λ	Biaxiality ratio of far-field stresses
μ	Crack surface friction coefficient

ν	Poisson's ratio
ρ	Notch radius
σ, σ_{xx}, σ_{yy}	Normal stresses
σ_{x0}, σ_{y0}	Far-field normal stresses
σ_{co}	Biaxiality cut-off stress
σ_1, $\sigma_{1,\max}$	First principal stress, its maximum value
$\Delta\sigma$	Stress range
$\Delta\sigma_{\mathrm{eff}}$	Effective stress range
$\Delta\sigma_{\mathrm{eq}}$	Von Mises equivalent stress range
σ_{ij}	Components of stress tensor
$\sigma_{\max}$	Stress at upper reversal point
$\sigma_{\min}$	Stress at lower reversal point
$\sigma_{\perp\max}$	Maximum normal stress on crack surface
σ_{op}, σ_{cl}	Stress at crack opening and crack closure point
σ_0	Reference stress in power-law relationship
σ_{res}	Residual stress
σ_{ref}	Reference stress
σ_{U}	Ultimate tensile strength
σ_{Y}, σ'_{Y}	Monotonic and cyclic yield stress
θ	Polar coordinate, polar angle
τ, τ_{xy}, τ_{xz}	Shear stresses
τ_{fr}	Friction shear stress on crack surface
τ_{fr0}	Friction shear stress due to indentation
τ_{Y}	Shear yield stresses
$\Delta\tau_{\mathrm{II}}$	Shear stress in maximum shear strain plane
φ	Plasticity correction on crack length
φ_{calc}	Calculated critical plane angle
ω, ω_{c}	Plastic zone size, tensile and compressive
$\omega_{\max}$	Maximum plastic zone size
ω_{cyc}	Cyclic plastic zone size
ω_{p}	Peak load plastic zone size

References

Ainsworth RA (1984) The assessment of defects in structures of strain hardening material. Engng Fract Mech 19:633–642

Aliaga D, Davy A, Schaff H (1985) A simple crack closure model for predicting fatigue crack growth under flight simulation loading. In: Durability and damage tolerance in aircraft design, pp 605–630. EMAS, Warley, UK

Alzos WX, Skat AC, Hillberry BM (1976) Effect of single overload/underload cycles on fatigue crack propagation. ASTM STP 595:41–60

Anderson TL (1995) Fracture Mechanics – Fundamentals and Applications. CRC Press, Boca Raton, Fla

Andersson H, Persson C, Hansson T, Melin S, Järvstråt N (2004) Constitutive dependence in finite-element modelling of crack closure during fatigue. Fatigue Fract Mater Struct 27:25–87
Anthes RJ (1997) Ein neuartiges Kurzrißfortschrittsmodell zur Anrißlebensdauervorhersage bei wiederholter Beanspruchung. Inst Stahlbau Werkstoffmech, TH Darmstadt, Rep 57
Anthes RJ, Rodriguez-Ocaña J, Seeger T (1996) Crack opening stresses under constant and variable amplitude loading determined by finite element analyses. In: Fatigue'96, (Eds: Lütjering G, Nowack H), pp 1075–1080. Elsevier Science, Oxford
Antunes F, Rodrigues D (2008) Numerical simulation of plasticity induced crack closure: identification and discussion of parameters. Engng Fract Mech 75:3101–3120
ASTM 1820 (2007) Standard test method for measurement of fracture toughness. ASTM
Atluri SN, Nishioka T, Nakagaki M (1984) Incremental path-independent integrals in inelastic and dynamic fracture. Engng Fract Mech 20:209–244
Barenblatt GI (1962) The mathematical theory of equilibrium cracks in brittle fracture. Advances Appl Mech 7:55–129
Baudin G, Robert M (1984) Crack growth life time prediction under aeronautical type loading. In: Proc 5th Europ Conf on Fracture, Lissabon, pp 779–792
Beretta S, Carboni M, Madia M (2009) Modelling of fatigue thresholds for small cracks in a mild steel by 'strip-yield' model. Engng Fract Mech 76:1548–1561
Bernard PJ, Lindley TC, Richards CE (1976) Mechanisms of overload retardation during fatigue crack propagation. ASTM STP 595:78–100
Bicego V (1989) Low cycle fatigue life predictions in terms of an EPFM small crack model. Engng Fract Mech 32:339–349
Bittencourt TN, Wawrzynek PA, Ingraffea AR, Sousa JL (1996) Quasi-automatic simulation of crack propagation for 2D LEFM problems. Engng Fract Mech 55:321–334
Blackburn WS (1972) Path independent integrals to predict onset of crack instability in an elastic-plastic material. Int J Fract Mech 8:343–346
Blom AF, Holm DK (1987) Load interaction effects on fatigue crack propagation in steels of varying strength. In: Role of fracture mechanics in modern technology, (Eds: Sih CC, Nisitani H, Ishihara T). Elsevier Science, Oxford
Blom AF, Hedlund A, Zhao W, Fathulla A, Weiss B, Stickler R (1986) Short fatigue crack growth behaviour in Al2024 and A7475. In: The behaviour of short fatigue cracks (Eds: Miller KJ, de los Rios ER), pp 37–66. MEP, London
Boettner RC, Laird C, McEvily AJ (1965) Crack nucleation and growth in high strain low cycle fatigue. Trans AIME 233:379–387
Brocks W, Scheider I (2003) Reliable J-values – Numerical aspects of the path-dependence of the J-integral in incremental plasticity. Materialpruefung 45:264–275
Broek D (1988) Elementary Engineering Fracture Mechanics. Martinus Nijhoff Publ, Den Haag
Brown MW, Miller KJ (1985) Mode I fatigue crack growth under biaxial stress at room and elevated temperature. ASTM STP 853:135–152
Brown MW, de los Rios ER, Miller KJ (1988) A critical comparison of proposed parameters for high strain fatigue crack growth. ASTM STP 924:233–259
Brüning J, Hertel O, Vormwald M, Savaidis G (2006) Fatigue crack growth at notches considering plasticity induced crack closure. In: Proc 16th Europ Conf Fract, (Ed: Gdoutos EE), pp 245–246. Springer Science+Business Media
BS 7910 (2005) Guide to methods for assessing the acceptability of flaws in metallic structures. British Standards Institution
Budiansky B, Hutchinson, JW (1978) Analysis of closure in fatigue crack growth. J Appl Mech 45:267–276
Buschermöhle H, Memhard D, Vormwald M (1996) Fatigue crack growth acceleration or retardation due to compressive overload excursions. In: Proc 6th Int Congr Fatigue, pp 583–588. Elsevier Science, Oxford
Carter BJ, Wawrzynek PA, Ingraffea AR (2000) Automated 3D crack growth simulation. Int J Num Meth Engng 47:229–253

Chaboche JL (1986) Time-independent constitutive theories for cyclic plasticity. Int J Plast 2:149–188

Chaboche JL (1989) Constitutive equations for cyclic plasticity and cyclic viscoplasticity. Int J Plast 5:247–302

Chermahini R, Palmberg B, Blom A (1993) Fatigue crack growth and closure behaviour of semicircular and semi-elliptical surface flaws. Int J Fatigue 15:259–263

Dankert M, Greuling S, Seeger T (1999) A unified elastic-plastic model for fatigue crack growth at notches including crack closure effects. ASTM STP 1343:411–426

de Koning AU (1981) A simple crack closure model for prediction of fatigue crack growth rates under variable amplitude loading. ASTM STP 743:63–85

de Koning AU, Dougherty, DJ (1989) Prediction of low and high crack growth rates under constant and variable amplitude loading. In: Fatigue crack growth under variable amplitude loading, pp 208–217, Elsevier Science, Amsterdam

de Koning AU, Liefting G (1988) Analysis of crack opening behavior by application of a discretized strip yield model. ASTM STP 982:437–458

Dill HD, Saff CR (1976) Spectrum crack growth prediction method based on crack surface displacement and contact analysis. Fatigue Crack Growth under Spectrum Loads, ASTM STP 595:306–319

Dill HD, Saff CR, Potter JM (1980) Effects of fighter attack spectrum and crack growth. Effect of Load Spectrum Variables on Fatigue Crack Initiation and Propagation, ASTM STP 714: 205–217

Donahue RJ, McClark H, Atanmo P, Kumble R, McEvily AJ (1972) Crack opening displacement and the rate of fatigue crack growth. Int J Fract 8:209–219

Döring R (2006) Zum Deformations- und Schädigungsverhalten metallischer Werkstoffe unter mehrachsig nichtproportionalen zyklischen Beanspruchungen. Inst Stahlbau Werkstoffmech, TU Darmstadt, Rep 78

Döring R, Hoffmeyer J, Seeger T, Vormwald, M (2003) A plasticity model for calculating stress-strain sequences under multiaxial nonproportional cyclic loading. Comput Mater Sci 28: 587–596

Döring R, Hoffmeyer J, Seeger T, Vormwald M (2006) Short fatigue crack growth under nonproportional multiaxial elastic-plastic strains. Int J Fatigue 28:972–982

Dowling NE (1977$^{(1)}$) Crack growth during low-cycle fatigue of smooth axial specimens. ASTM STP 637:97–121

Dowling NE (1977$^{(2)}$) Geometry effects and the *J*-integral approach to elastic-plastic fatigue crack growth. ASTM STP 601:19–32

Dowling NE (1987) *J*-integral estimates for cracks in infinite bodies. Engng Fract Mech 26: 333–348

Dowling NE, Begley JA (1976) Fatigue crack growth during gross plasticity and the *J*-integral. ASTM STP 590:82–103

Dowling NE, Iyyer NS (1987) Fatigue crack growth and closure at high cyclic strains. Mat Sci Engng 96: 99–107

Dugdale DS (1960) Yielding of steel sheets containing slits. J Mech Phys Solids 8:100–108

DuQuesnay DL, Topper TH, Yu MT, Pompetzki MA (1992) The effective stress range as a mean stress parameter. Int J Fatigue 14:45–50

Earthman JC (1991) Characterization of small crack growth in 12 % CrMoV steel under high temperature, low cycle fatigue conditions. Mater Sci Engng A132:89–95

Eidinoff HL, Bell PD, (1977) Application of the crack closure concept to aircraft fatigue crack propagation analysis. In: Proc 9th ICAF Symp on Aeronautical Fatigue, Darmstadt, pp 5.2/1–5.2/58

Elber W (1970) Fatigue crack closure under cyclic tension. Engng Fract Mech 2:37–45

Elber W (1971) The significance of fatigue crack closure. ASTM STP 486:230–242

El Haddad MH, Smith KH, Topper TH (1979$^{(1)}$) A strain based intensity factor solution for short fatigue cracks initiating from notches. ASTM STP 677:274–289

El Haddad MH, Smith KH, Topper TH (1979(2)) Fatigue crack propagation of short cracks. J Engng Mater Techn 101:42–46

El Haddad MH, Dowling NE, Topper TH, Smith KN (1980) *J*-integral applications for short fatigue cracks at notches. Int J Fract 16:15–30

El Haddad MH, Mukherjee B (1983) Elastic-plastic fracture mechanics analysis of fatigue crack growth. ASTM STP 803:II-689–II-707

El-Zeghayar M, Topper TH, Conle FA, Bonnen JJF (2010) Modeling crack closure and damage in variable amplitude fatigue using smooth specimen fatigue test data. Int J Fatigue 33:223–231

El-Zeghayar M, Topper TH, Soudki KA (2011) A model of crack opening stresses in variable amplitude loading using smooth specimen fatigue test data for three steels. Int J Fatigue 33:1337–1350

Fatemi A, Socie DF (1988) A critical plane approach to multiaxial fatigue damage including out-of-phase loading. Fatigue Fract Engng Mater Struct 11:149–165

Fleck N, Newman JC (1988) Analysis of crack closure under plane strain conditions. ASTM STP 982:319–341

Forman RG, Mettu SR (1992) Behavior of surface and corner cracks subjected to tensile and bending loads in Ti6Al4 V alloy. ASTM STP 1131:519–546

Forsyth PJ, Ryder DA (1960) Fatigue fracture. Aircraft Engng 32:96–99

Führing H (1977) Berechnung von elastisch-plastischen Beanspruchungsabläufen in Dugdale-Rißscheiben mit Rißuferkontakt auf der Grundlage nichtlinearer Schwingbruchmechanik, Inst Statik Stahlbau, TH Darmstadt, Rep 30

Führing H (1982) Modell zur nichtlinearen Rißfortschrittsvorhersage unter Berücksichtigung von Lastreihenfolgeeinflüssen (LOSEQ). Fraunhofer Inst Betriebsfest, Darmstadt, Rep FB-162

Führing H, Seeger T (1979) Dugdale crack closure analysis of fatigue cracks under constant amplitude loading. Engng Fract Mech 11:99–122

Führing H, Seeger, T (1984) Fatigue crack growth under variable amplitude loading. In: Subcritical crack growth due to fatigue, stress corrosion and creep (Ed: Larson LH), pp 109–133. Elsevier Applied Science, Oxford

Fulland M (2003) Risssimulation in dreidimensionalen Strukturen mit automatischer adaptiver Finite-Elemente-Netzgenerierung. VDI Fortschrittbericht, Ser 18, No 280, VDI-Verlag, Düsseldorf

Gallagher JP (1974) A generalized development of yield zone models. AFFDL-TM-74-28-FBR. Wright Patterson Air Force Laboratory

Gamache B, McEvily A (1993) On the development of fatigue crack closure. In: Proc 5th Int Conf on Fatigue and Fatigue Thresholds, Montreal, pp 577–582

Gray TD, Gallagher JP (1976) Predicting fatigue crack retardation following a single overload using a modified Wheeler model. ASTM STP 590:331–344

Grin EA, Shur DM, Mazepa AG (1987) Study of the kinetics of growth of a low-cycle fatigue crack with the use of the *J*-Integral. Translation. Problemy Prochnosti USSR 19:1325–1329

Gross D, Seelig T (52011) Bruchmechanik – mit einer Einführung in die Mikromechanik. Springer-Verlag, Berlin

Hahn GT, Rosenfield AR (1965) Local yielding and extension of a crack under plane stress. Acta Met 13:293–306

Haigh JR, Skelton RP (1978) A strain intensity approach to high temperature fatigue crack growth and failure. Mater Sci Engng 36:133-137

Hatanaka K, Fujimitsu T, Shiraishi S (1989) An analysis of surface crack growth in circumferentially grooves components under low-cycle fatigue. JSME Ser 1(32):245–255

He MY, Hutchinson JW (1983) Bounds for fully plastic problems for infinite bodies. ASTM STP 803:277–290

Heitmann HH (1983) Betriebsfestigkeit von Stahl: Vorhersage der technischen Anrisslebensdauer unter Berücksichtigung des Verhaltens von Mikrorissen. Ph D thesis, TH Aachen

Heitmann HH, Vehoff H, Neumann P (1985) Life prediction for random load fatigue based on the growth behaviour of microcracks. In: Advances in fracture research (Eds: Valluri SR et al.), pp 3599–3606. Pergamon Press, Oxford

Hertel O, Vormwald M (2011) Short-crack-growth-based fatigue assessment of notched components under multiaxial variable amplitude loading. Engng Frac Mech 78:1614–1627

Hertel O, Döring R, Vormwald M (2004) Cyclic *J*-integral under nonproportional loading. In: Proc 7th Int Conf Biaxial/Multiaxial Fatigue Fract, pp 513–518. DVM, Berlin

Herz E, Thumser R, Bergmann JW, Vormwald M (2006) Endurance limit of autofrettaged Diesel-engine injection tubes with defects. Engng Fract Mech 73:3–21

Herz E, Hertel O, Vormwald M (2011) Numerical simulation of plastic deformation induced fatigue crack opening and closure for autofrettaged intersecting holes. Engng Fract Mech 78:559–572

Horikawa K, Cho SM (1987) Initial fatigue crack growth behaviour in a notched component, report II, Trans JWRI 16:167–175

Hoshide T, Socie DF (1987) Mechanics of mixed mode small fatigue crack growth. Engng Fract Mech 26:841–850

Hoshide T, Socie DF (1988) Crack nucleation and growth modeling in biaxial fatigue. Engng Fract Mech 29:287–299

Hoshide T, Tanaka K, Yamada A (1981) Stress ratio effect of fatigue crack propagation in biaxial stress field. Fatigue Engng Mater Struct 4:355–366

Hoshide T, Miyahara M, Inoue T (1988) Elastic-plastic behaviour of short fatigue crack in smooth specimen. ASTM STP 924:312–322

Hou C-Y (2004) Three dimensional finite element analysis of fatigue crack closure behavior in surface flaws. Int J Fatigue 26:1225–1239

Huang JS, Pelloux RM (1980) Low cycle fatigue crack propagation in Hastelloy-X at 25 and 760 °C. Metall Trans A11:899–904

Ibrahim FK (1989) The effect of stress ratio, compressive peak stress and maximum stress level on fatigue behaviour of 2024-T3 aluminium alloy. Fatigue Fract Engng Mater Struct 12:9–18

Ibrahim FK, Thompson JC, Topper TH (1986) A study of the effect of mechanical variables on fatigue crack closure and propagation. Int J Fatigue 8:135–142

Irwin GR (1956) Onset of fast crack propagation in high strength steel and aluminium alloys. In: Proc Sagamore Res Conf, vol 2, pp 288–305

Irwin GR (1957) Analysis of stresses and strains near the end of a crack traversing a plate. J Appl Mech 24:361–364

Ishikawa H, Kitagawa H, Okamura H (1979) *J*-integral of a mixed-mode crack and its application. Proc 3rd Int Conf Mech Behaviour Mat ICM 3, vol 3, pp 447–455

Jablonski DA (1989) An experimental study of the validity of a delta *J* criterion for fatigue crack growth. Nonlinear Fracture Mechanics, ASTM STP 995:361–387

Jacoby GH, Nowack H, van Lipzig HT (1976) Experimental results and a hypothesis for fatigue crack propagation under variable amplitude loading. Fatigue Crack Growth Under Spectrum Loads, ASTM STP 595:172–186

Jiang Y (1993) Cyclic plasticity with an emphasis on ratchetting. Ph D thesis. Univ Illinois, Urbana-Champaign

Jiang Y (1995) A new constitutive model in the theory of plasticity – Part I: Theory. J Press Vess Technol 117(4):365–70

Jiang Y, Feng M (2004) Modeling of fatigue crack propagation. Engng Mater Technol 126:77–86

Jiang Y, Sehitoglu H (1996$^{(1)}$) Modeling of cyclic ratchetting plasticity, Part I: Development of constitutive equations. J Appl Mech 63:720–725

Jiang Y Sehitoglu, H (1996$^{(2)}$) Modeling of cyclic ratchetting plasticity, Part II: Comparison of model simulations with experiments. J Appl Mech 63:726–733

Jolles M (1985) Effects of load gradient on applicability of a fatigue crack growth rate-cyclic *J* relation. ASTM STP 868:381–391

Kim KS, Orange TW (1988) A review of path-independent integrals in elastic-plastic fracture mechanics. ASTM STP 945:713–729

Kishimoto K, Aoki S, Sakata M (1980) On the path independent integral *J*, Engng Fract Mech 13:841–850

Kocak M, Webster S, Janosh JJ, Ainsworth RA, Koers R (2005) Fitness for service procedure FITNET. GKSS, Geesthacht

Kumar R, Singh K (1995) Influence of stress ratio on fatigue crack growth in mild steel. Engng Fract Mech 503:377–384

Kumar R, German MD, Shih CF (1981) An engineering approach for elastic-plastic fracture analysis. EPRI-Rep NP-1931

Kuna M (^{2}2010) Numerische Beanspruchungsanalyse von Rissen. Vieweg Teubner, Wiesbaden

Laird C (1967) The influence of metallurgical structure on the mechanisms of fatigue crack propagation. ASTM STP 415:131–180

Lalor P, Sehitoglu, H, McClung RC (1986) Mechanics aspects of small crack growth from notches – the role of crack closure. In: The behaviour of short fatigue cracks (Eds: Miller KJ, de los Rios ER), pp 369–386. Wiley, London

Lamba HS (1975) The J-integral applied to cyclic loading, Engng Fract Mech 7:693–703

Lankford J, Davidson DL (1976) Fatigue crack tip plasticity associated with overloads and subsequent cycling. J Engng Mater Technol 98:17–29

Lazzarin P, Tovo R (1998) A notch stress intensity factor approach to the stress analysis of welds. Fatigue Fract Engng Mater Struct 21:1089–1103

Lei Y. (2008) Finite element crack closure analysis of a compact tension specimen. Int J Fatigue 30:21–31

Ma CC, Huang, JI, Tsai CH (1994) Weight functions and stress intensity factors for axial cracks in hollow cylinders, J Press Vessel Technol 116:423–430

Martha LF (1989) Topological and geometrical modeling approach to numerical discretization and arbitrary fracture simulation in three dimensions. Ph D thesis. Cornell Univ

Matsuoka S, Tanaka K, Kawahara M (1976) The retardation phenomenon of fatigue crack growth in HT80 steel. Engng Fract Mech 8:507–523

McClung RC (1987) Fatigue crack closure and crack growth outside the small scale yielding regime. Rep 139, University of Illinois, Urbana-Champaign

McClung RC (1989) Closure and growth of mode I cracks in biaxial fatigue. Fatigue Fract Engng Mater Struct 12:447–460

McClung RC (1991$^{(1)}$) Finite element modeling of crack closure during simulated fatigue threshold testing. Int J Fatigue 52:145–157

McClung RC (1991$^{(2)}$) The influence of applied stress, crack length and stress intensity factor on crack closure. Metall Trans A22:1559–1571

McClung RC (1991$^{(3)}$) A simple model for crack growth near stress concentrations. J Pressure Vessel Technol 113:542–548

McClung RC (1999) Finite element analysis of fatigue crack closure — A historical and critical review. In: Proc 7th Int Fatigue Crack Conf, vol 1, pp 495–502

McClung RC, Sehitoglu H (1988) Closure behavior of small cracks under high strain fatigue histories. ASTM STP 982:279–299

McClung RC, Sehitoglu H (1991) Characterization of fatigue crack growth in intermediate and large scale yielding. J Engng Mater Technol 113:15–22

McClung RC, Sehitoglu H (1996) On the finite element analysis of fatigue crack closure, part 1, basic modeling issues. Engng Fract Mech 33:237–52

McClung RC, Chell GG, Lee Y-D, Russel DA, Orient GE (1999) Development of a practical methodology for elastic-plastic and fully plastic fatigue crack growth. Rep NASA/CR-1999-209428, Marshall Space Flight Center

McDowell DL (1994) A shear stress based critical-plane criterion of multiaxial fatigue failure for design and life prediction. Fatigue Fract Engng Mater Struct 11:1475–1485

McDowell DL, Bennett VP (1996) Microcrack growth law for multiaxial fatigue. Fatigue Fract Engng Mater Struct 19:821–837

McDowell DL, Berard JY (1992) A ΔJ-based approach to biaxial fatigue. Fatigue Fract Engng Mater Struct 15:719–741

McEvily AJ (1969) Fatigue crack growth and strain intensity factor. In: Proc Air Force Conf Fatigue Fract Aircraft Struct Mater, pp 451–458. NASA Rep AFFDL-TR-70-144

McEvily AJ (1988) On crack closure in fatigue crack growth. ASTM STP 982:35–43
McEvily AJ, Beukelmann D, Tanaka K (1974) On large scale plasticity effects in fatigue crack propagation. In: Proc Symp Mech Behavior Mater, vol 1, pp 269–281. SMSJ, Kyoto
McEvily AJ, Endo M, Murakami Y (2003) On the square root on area relationship and the short fatigue crack threshold, Fatigue Fract Engng Mater Struct 26:269–278
McMillan JC, Pelloux RM (1967) Fatigue crack propagation under program and random loading. ASTM STP 415:505–535
Mellings S, Baynham J, Adey RA (2002) Predicting residual strength using fully automatic crack growth, Engng Anal Boundary Elements 26:479–488
Mills WJ, Hertzberg RW (1975) The effect of sheet thickness on fatigue crack retardation in 2024-T3 aluminium alloy. Engng Fract Mech 7:705–711
Miner MA (1945) Cumulative Damage in Fatigue. J Appl Mech 12:A159–A164
Miura N, Fujioka T, Kashima K, Kanno S, Hayashi M, Ishiwata M, Gotoh N (1994) Low cycle fatigue and ductile fracture for Japanese carbon steel piping under dynamic loading. Nucl Engng Design 153:57–69
Moreno V, Jordan E (1986) Prediction of material thermomechanical response with a unified viscoplastic constitutive model. Int J Plast 2:223–245
Mowbray DF (1979) Use of a compact-type strip specimen for fatigue crack growth rate testing in the high-rate regime. ASTM STP 668:736–752
Mróz Z (1967) On the description of anisotropic work-hardening. J Mech Phys Solids 15:163–175
Mróz Z (1969) An attempt to describe the behavior of metals under cyclic loads using a more general work-hardening model. Acta Mechanica 7:199–212
Mu Z, Vehoff H, Brede M, Neumann P (1996) Behavior of short cracks in alloy 800HT under biaxial fatigue. Mater Sci Technol 12:416–420
Musava JK, Radon JC (1980) An elastic-plastic crack growth analysis using the *J*-integral concept. In: Proc 3rd Coll on Fract, pp 129–141
NASGRO reference manual, version 4.02 (2002) NASA Johnson Space Center, http://www.swri.org/4org/d18/mateng/matint/nasgro
Neuber H (1961) Theory of stress concentration for shear-strained prismatical bodies with arbitrary nonlinear stress-strain law. J Appl Mech 28:544–550
Newman JC (1974) Finite element analysis of fatigue crack propagation – including the effect of closure. Ph D thesis. Virginia Polytech Inst State Univ, Blacksburg, Va
Newman JC (1976) Finite element analysis of fatigue crack closure. ASTM STP 590:281–301
Newman JC (1981) A crack closure model for predicting fatigue crack growth under aircraft spectrum loading. ASTM STP 748:53–84
Newman JC (1982) Prediction of fatigue crack growth under variable amplitude and spectrum loading using a closure model. ASTM STP 761:255–277
Newman JC (1983) A nonlinear fracture mechanics approach to the growth of small cracks. In: Behaviour of short cracks in airframe components. AGARD Conf Proc 328, pp 6.1–6.26
Newman JC (1984) A crack opening stress equation for fatigue crack growth. Int J Fract 24:R131–R135
Newman JC, Phillips EP, Everett RA ($1999^{(1)}$) Fatigue analysis under constant- and variable-amplitude loading using small-crack theory. NASA TM 209329, Washington
Newman JC, Wu XR, Swain MH, Zhao W, Phillips EP, Ding CF ($1999^{(2)}$) Small crack growth and fatigue life predictions for high-strength aluminium alloys, Part II: Crack closure and fatigue analysis. Fatigue Fract Engng Mater Struct 22:59–72
Nicholas T, Palazotto AN, Bednarz E (1988) An analytical investigation of plasticity induced closure involving short cracks. ASTM STP 982:361–379
Obrtlik K, Polak J (1985) Fatigue growth of surface cracks in the elastic-plastic region, Fatigue Fract Engng Mater Struct 8:23–31
Ohji K, Ogura K, Ohkubo Y (1974) On the closure of fatigue cracks under cyclic tensile loading. Int J Fract 10:123–124
Ohno N, Wang JD (1993) Kinematic hardening rules with critical state of dynamic recovery, part I, formulation and basic features of ratchetting behavior. Int J Plast 9:375–385

Padmadinata, UH (1990) Investigation of crack closure prediction models for fatigue in aluminium sheet under flight simulation loading. Ph D thesis. Techn Univ Delft

Pang CM, Song JH (1994) Crack growth and closure behavior of short fatigue cracks. Engng Fract Mech 47:327–343

Paris PC (1960) The growth of cracks due to variations in load. Ph D thesis, Lehigh Univ, Bethlehem, Pa

Paris PC, Erdogan F (1963) A critical analysis of crack propagation laws. J Basic Engng 85:528–539

Park SJ, Song JH (1999) Simulation of fatigue crack closure behaviour under variable-amplitude loading by a 2D finite element analysis based on the most appropriate mesh size concept. ASTM STP 1343: 337–348

Petit J, Zeghloul A (1986) On the effect of environment on short crack growth behaviour and threshold. In: The behaviour of short fatigue cracks (Eds: Miller KJ, de los Rios ER), pp 163–178. MEP, London

Petrak GS (1974) Strength level effects on fatigue crack growth and retardation. Engng Fract Mech 6:725–733

Pippan R, Zelger C, Gach E, Bichler C, Weinhandl H (2011) On the mechanism of fatigue crack propagation in ductile metallic materials. Fatigue Fract Engng Mater Struct 34:1–16

Radaj D, Vormwald M (2007) Ermüdungsfestigkeit – Grundlagen für Ingenieure. Springer-Verlag, Berlin

Rahman S, Wilkowski G, Mohan R (1997) Low-cycle fatigue crack growth considerations in pipe fracture analysis. Nucl Engng Design 168:105–118

Ramos MS, Pereira MV, Darwish FA, Motta SH, Carneiro MA (2003) Fatigue Fract Eng Mater Struct 26:115–121

Reddy SC, Fatemi A (1992) Small crack growth in multiaxial fatigue. ASTM STP 1122:276–298

Reger M, Remy L (1982) Influence of environment on microcrack propagation in high temperature low cycle fatigue. In: Proc 4th Europ Conf Fract, pp 531–538. EMAS, Warley, UK

Rice JR (1968) A path independent integral and the approximate analysis of strain concentration by notches and cracks, J Appl Mech 35:379–309

Richard HA, Sander M (2009) Ermüdungsrisse erkennen, sicher beurteilen, vermeiden. Vieweg Teubner Verlag, Wiesbaden

Rie KT, Schubert R (1987) Note on the crack closure phenomenon in low-cycle fatigue. In: Proc 2nd Int Conf Low Cycle Fatigue Elastic-plastic Behav Mater, pp 575–580

Rie KT, Wittke H (1996) New approach for estimation of ΔJ and for measurement of crack growth at elevated temperature. Fatigue Fract Engng Mater Struct 19:975–983

Ritchie RO, Yu W, Blom AF, Holm, DK (1987) An analysis of crack tip shielding in aluminium alloy 2124, a comparison of large, small, through-thickness and surface fatigue cracks. Fatigue Fract Engng Mater Struct 10:343–362

Saal H (1971) Einfluß von Formzahl und Spannungsverhältnis auf die Zeit- und Dauerfestigkeit und Rißfortschreitungen bei Flachstäben aus St52. Rep 17, Inst Stahlbau Werkstoffmech, TH Darmstadt

Sadananda K, Shahinian P (1979) A fracture mechanics approach to high temperature fatigue crack growth in Udimet 700. Engng Fract Mech 11:73–86

Sadananda K, Sarkar S, Kujawski D, Vasudevan A (2009) A two-parameter analysis of S–N fatigue life using $\Delta\sigma$ and σ_{max}. Int J Fatigue 31:1648–1659

Sanford RJ (2003) Principles of fracture mechanics. Pearson Education, Upper Saddle River, NJ

Savaidis G (1995) Berechnung der Bauteilanrißlebensdauer bei mehrachsigen proportionalen Beanspruchungen. Rep 54, Inst Stahlbau Werkstoffmech, TH Darmstadt

Savaidis G, Seeger T (1996) A short crack growth prediction model for multiaxial proportional elasto-plastic loading. In: Proc 6th Int Fatigue Congr, vol 2, pp 1001–1006. Elsevier, Oxford

Savaidis G, Seeger T (1997) Consideration of multiaxiality in fatigue life prediction using the closure concept. Fatigue Fract Engng Mater Struct 20:985–1004

Savaidis G, Dankert M, Seeger T (1995) An analytical procedure for predicting opening loads of cracks at notches. Fatigue Fract Engng Mater Struct 18:425–442

Savaidis G, Savaidis A, Zerres P, Vormwald M (2010) Mode I fatigue crack growth at notches considering crack closure. Int J Fatigue 32:1543–1558

Saxena A (1998) Nonlinear Fracture Mechanics for Engineers. CRC Press, Boca Raton, Flo

Schijve J (1973) Effect of load sequences on crack propagation under random and program loading. Engng Fract Mech 5:269–280

Schijve J (1976[(1)]) Observations on the prediction of fatigue crack propagation under variable-amplitude loading. ASTM STP 595:3–23

Schijve J (1976[(2)]) The effect of pre-strain on fatigue-crack growth and crack closure. Engng Fract Mech 8:575–581

Schijve J (1981) Some formulae for the opening stress level. Engng Fract Mech 14:461–465

Schijve J (1988) Fatigue crack closure – Observations and technical significance. ASTM STP 982:319–341

Schijve J (1992) Fundamental aspects of prediction on fatigue crack growth under variable-amplitude loading. In: Theoretical concepts and numerical analysis of fatigue, pp 111–130 EMAS, Warley, UK

Schijve J ([2]2009) Fatigue of structures and materials. Springer Science+Business Media

Schijve J, Vlutters AM, Ichsan SP, Provo-Kluit JC (1985) Crack growth in aluminium alloy sheet material under flight simulation loading. Int J Fatigue 7:127–136

Schöllmann M, Fulland M, Richard HA (2003) Development of a new software for adaptive crack growth simulations in 3D structures. Engng Fract Mech 70:249–268

Schwalbe KH (1980) Bruchmechanik metallischer Werkstoffe. Hanser-Verlag, München

Seeger T (1973) Ein Beitrag zur Berechnung von statisch und zyklisch belasteten Rißscheiben nach dem Dugdale–Barenblatt-Modell. Rep 21. Inst Statik Stahlbau, TH Darmstadt

Sehitoglu AGH, Gall K (1996) Recent advances in fatigue crack growth modeling. Int J Fract 80:165–192

Singh KD, Khor KH, Sinclair I (2008) Finite element and analytical modelling of crack closure due to repeated overloads. Acta Materialia 56:835–851

Singh KD, Parry MR, Sinclair I (2011) Variable amplitude fatigue crack growth behavior - a short overview. J Mech Sci Technol 25(3):663–673

Skelton RP (1982) Growth of short cracks during high strain fatigue and thermal cycling. ASTM STP 770, 337–381

Skinner J, Daniewicz S (2002) Simulation of plasticity induced fatigue crack closure in part-through cracked geometries using finite element analysis. Engng Fract Mech 69:1–11

Socie DF, Hua CT, Worthem DW (1987) Mixed mode small crack growth. Fatigue Fract Engng Mater Struct 10:1–16

Socie DF, Furmann S (1996) Fatigue damage simulation models for multiaxial loading. In: Proc 6th Int Fatigue Conf, pp 967–976

Solomon HD (1972) Low cycle fatigue crack propagation in 1018 steel. J Mater 7:299–306

Swenson DV, Ingraffea AR (1988) Modeling mixed-mode dynamic crack propagation using finite elements, theory and applications. Comput Mech 3:381–397

Tanaka K (1983) The cyclic *J*-integral as a criterion for fatigue crack growth. Int J Fracture 22:91–104

Tanaka K, Hoshide T, Sakai N (1984) Mechanics of fatigue crack propagation by crack-tip plastic blunting. Engng Fract Mech 19:805–825

Tchankov D, Sakane M, Itoh T, Hamada N (2008) Crack opening displacement approach to assess multiaxial low cycle fatigue. Int J Fatigue 30:417–425

Thumser R (2009) Simulation des Rissfortschritts in autofrettierten und nicht autofrettierten Bohrungsverschneidungen auf der Grundlage der linear-elastischen Bruchmechanik, Ph D thesis, Univ Weimar

Thumser R, Bergmann J, Herz E, Hertel O, Vormwald M (2008) Variable amplitude fatigue of autofrettaged diesel injection parts. Mater Wissensch Werkstofftech 39:719–725

Timbrell C, Cook G (1997) 3D FE fracture mechanics analysis for industrial applications. In: Seminar on inelastic finite element analysis, Inst Mech Eng, London

Tomkins B (1975) The development of fatigue crack propagation models for engineering applications at elevated temperatures. J Engng Mater Technol 97:289–297

Tomkins B (1980) Micromechanisms of fatigue crack growth at high stress. Metal Sci 14:408–417

Trebules VW, Roberts R, Hertzberg RW (1973) Effect of multiple overloads on fatigue crack propagation in 2024-T3 aluminium alloy. ASTM STP 536:115–146

Vardar O (1982) Fatigue crack propagation beyond general yield. J Engng Mater Technol 104:192–199

Verreman Y, Bailon JP, Masounave J (1986) Fatigue short crack propagation and plasticity induced crack closure at the toe of a fillet welded joint. In: The behaviour of short fatigue cracks (Eds: Miller KJ, de los Rios ER), pp 387–404. Wiley, London

Vormwald M (1989) Anrißlebensdauervorhersage auf der Basis der Schwingbruchmechanik für kurze Risse. Rep 47, Inst Stahlbau Werkstoffmech, TH Darmstadt

Vormwald M (2011) Ermüdungslebensdauer von Baustahl unter komplexen Beanspruchungsabläufen am Beispiel des Stahles S460. Mater Test 53:98–108

Vormwald M, Döring R (2009) Deformations and damage to metallic materials under multiaxial non-proportional loading. Comp Mater Sci 46:555–560

Vormwald M, Heuler P, Seeger T (1992) A fracture mechanics based model for cumulative damage assessment as part of fatigue life prediction. ASTM STP 1122:28–43

Vormwald M, Heuler P, Krä C (1994) Spectrum fatigue life assessment of notched specimens using a fracture mechanics based approach. ASTM STP 1231:219–231

Vormwald M, Seeger T (1991) The consequences of short crack closure on fatigue crack growth under variable amplitude loading. Fatigue Fract Engng Mater Struct 14:205–225

Wang GS (1993) The plasticity aspect of fatigue crack growth. Engng Fract Mech 46:909–930

Wang GS, Blom AF (1991) A strip model for fatigue crack growth predictions under general load conditions. Engng Fract Mech 40:507–533

Wang SZ, Yang Z, Kang MK (1983) Fatigue crack growth rate under full yielding condition for 15CDV6 steel. Engng Fract Mech 18:895–902

Wells AA (1961) Unstable crack propagation in metals, cleavage and fast fracture. In: Proc Crack Prop Symp Cranfield, pp 210–230

Wheeler OE (1972) Spectrum loading and crack growth. J Basic Engng 94:181–186

Willenborg J, Engle RM, Wood HA (1971) A crack growth retardation model using an effective stress concept. Techn Mem 71-1-FBR, Wright-Patterson Air Force Base, Dayton, Oh

Williams ML (1957) On the stress distribution at the base of a stationary crack. J Appl Mech 24:109–114

Wüthrich C (1982) The extension of the *J*-integral concept to fatigue cracks. Int J Fracture 20:R35–R37

Wüthrich C, Hoffelner W (1984) Fatigue crack growth at high strain amplitudes. In: Proc 4th Int Conf Mech Behav Mater, vol 2, pp 911–917. Pergamon Press, Oxford

Yamada Y, Lacy T, Newman J Jr (2007) Effects of crack closure on fatigue crack-growth predictions for 2024-T351 aluminum alloy panels under spectrum loading. Int J Fatigue 29:1503–1509

Yang S, Lu Y (1989) Surface crack growth in a plate under the remote high strain controlled cyclic loading. Fatigue Fract Engng Mater Struct 12:399–407

Yoon KB, Saxena A (1991) An interpretation of the ΔJ for cyclically unsaturated materials. Int J Fract 49:R3–R9

Zentech Int ZENCRACK (2012) Tool for 3D fracture mechanics simulation. http://www.zentech.co.uk/zencrack.htm

Zerbst U, Madia M, Hellmann D (2011) An analytical fracture mechanics model for estimation of *S–N* curves of metallic alloys containing large second phase particles. Engng Fract Mech. doi:10.1016/j.engfracmech.2011.12.001

Zerres P (2010) Numerische Simulation des Ermüdungsrissfortschrittes in metallischen Strukturen unter Berücksichtigung zyklischer Plastizitätseffekte. Rep 88, Inst Stahlbau Werkstoffmech, TU Darmstadt

Zerres P, Döring R, Vormwald M (2011) Simulation of fatigue crack growth under consideration of cyclic plasticity. Mater Wissensch Werkstofftech 42:1045–1118

Zerres P, Vormwald M (2012) Finite element based simulation of fatigue crack growth with a focus on elastic-plastic material behaviour. Comp Mater Sci 57:73–79 (in press, available online: doi: 10.1016/j.commatsci.2012.01.018)

Zhang J, Bowen P (1998) On the finite element simulation of three-dimensional semi-circular fatigue crack growth and closure. Engng Fract Mech 60:341–60

Zhao L, Tong J, Byrne J (2004) The evolution of the stress-strain fields near a fatigue crack tip and plasticity-induced crack closure revisited. Fatigue Fract Engng Mater Struct 27:19–29

Zheng M, Liu HW (1986) Fatigue crack growth under general-yielding cyclic loading. J Engng Mater Technol 108:201–205

Ziegler B (1959) A modification of Prager's hardening rule. Quart Appl Math 17:55–65

Ziegler B, Yamada Y, Newman JC (2008) Crack growth predictions using a strip-yield model for variable-amplitude and spectrum loading. Mater Wissensch Werkstofftech 39:711–718

Zuidema J, Krabbe JP (1997) The effect of regular and irregular shear lips on fatigue crack growth in Al 2024. Fatigue Fract Engng Mater Struct 20:1413–1422

Index

Index of Chapter 1

D. Radaj and M. Vormwald, *Advanced Methods of Fatigue Assessment*,
DOI: 10.1007/978-3-642-30740-9, © Springer-Verlag Berlin Heidelberg 2013

Index of Chapter 2

Index of Chapter 3

Index of Chapter 4

The manufacturer's authorised representative in the EU is Springer Nature Customer Service Centre GmbH, Europaplatz 3, 69115 Heidelberg, Germany. If you have any concerns regarding our products, please contact ProductSafety@springernature.com

Printed and bound by CPI Group (UK) Ltd, Croydon, CR0 4YY
25/09/2026
02222881-0013